*Springer Series in* **Materials Science** 5

Edited by Manuel Cardona

# Springer Series in *Materials Science*

Editors: U. Gonser · A. Mooradian · K. A. Müller · M. B. Panish · H. Sakaki

M. S. Dresselhaus · G. Dresselhaus
K. Sugihara · I. L. Spain · H. A. Goldberg

# Graphite Fibers and Filaments

With 226 Figures

Springer-Verlag Berlin Heidelberg New York
London Paris Tokyo

Professor Dr. Mildred S. Dresselhaus
Dr. Gene Dresselhaus
Dr. Ko Sugihara

Massachusetts Institute of Technology,
Cambridge, MA 02139, USA

Professor Dr. Ian L. Spain

Colorado State University,
Fort Collins, CO 80523, USA

Dr. Harris A. Goldberg

Celanese Hoechst Corp.,
Summit, NJ 07901, USA


*Guest Editor:* Professor Dr., Dres. h. c. Manuel Cardona

Max-Planck-Institut für Festkörperforschung, Heisenbergstrasse 1
D-7000 Stuttgart 80, Fed. Rep. of Germany


*Series Editors:*

Prof. Dr. *K. A. Müller*

IBM, Zürich Research Lab.
CH-8803 Rüschlikon, Switzerland


Prof. Dr. *U. Gonser*

Fachbereich 12/1
Werkstoffwissenschaften
Universität des Saarlandes
D-6600 Saarbrücken, FRG

Dr. *M. B. Panish*

AT&T Bell Laboratories,
600 Mountain Avenue,
Murray Hill, NJ 07974, USA


Prof. *H. Sakaki*

Dr. *A. Mooradian*

Leader of the Quantum Electronics Group, MIT,
Lincoln Laboratory, P.O. Box 73,
Lexington, MA 02173, USA

Institute of Industrial Science,
University of Tokyo,
7-22-1 Roppongi Minato-ku,
Tokyo 106, Japan

ISBN-13:978-3-642-83381-6    e-ISBN-13:978-3-642-83379-3
DOI: 10.1007/978-3-642-83379-3

Library of Congress Cataloging-in-Publication Data. Dresselhaus, M.S. Graphite fibers and filaments. (Springer series in materials science ; v. 5). Bibliography: p. Includes index. 1. Graphite fibers. I. Title. II. Series. TA455.G7D74 1988 620.1′93 88-4593

© Springer-Verlag Berlin Heidelberg 1988
Softcover reprint of the hardcover 1st edition   1988

2153/3150-543210

# Preface

This book was begun after three of the present authors gave a series of invited talks on the subject of the structure and properties of carbon filaments. This was at a conference on the subject of optical obscuration, for which submicrometer diameter filaments with high length-to-diameter ratios have potential applications. The audience response to these talks illustrated the need of just one scientific community for a broader knowledge of the structure and properties of these interesting materials.

Following the conference it was decided to expand the material presented in the conference proceedings. The aim was to include in a single volume a description of the physical properties of carbon fibers and filaments. The research papers on this topic are spread widely in the literature and are found in a broad assortment of physics, chemistry, materials science and engineering and polymer science journals and conference proceedings (some of which are obscure). Accordingly, our goal was to produce a book on the subject which would enable students and other researchers working in the field to gain an overview of the subject up to about 1987.

Most of the present commercial production consists of fibers manufactured from polyacrylonitrile (PAN). These fibers have high strength and relatively low extensional moduli, making them particularly suited for applications in structural composites. For specialized applications, a smaller production of fibers from mesophase pitch precursors is also important. These ex-polymer PAN and pitch fibers have extraordinary potential as high strength-to-weight components for aerospace applications.

In the last few years a newer form of non-continuous filament, prepared by a catalytic chemical vapor deposition (CCVD) process has been studied extensively. These filaments have a more ordered structure than their polymer-based counterparts, and are more suited for research applications. In addition, there may be increased commercial use of the CCVD fibers in applications where short lengths can be used. This is particularly important since their cost may be considerably lower than present commercial carbon fibers. Accordingly, applications of the CCVD filaments may go well beyond the aerospace industry.

This book contains information on both ex-polymer and CCVD fibers, thereby distinguishing it from previous reviews, which have not included the newer CCVD fibers. This volume specifically considers the preparation, microstructure and defects, electronic structure, lattice, thermal, mechanical, magnetic, electrical and high temperature properties of carbon fibers, together with modifications induced by intercalation and ion implantation, and finishes with a brief discussion of applications. Over 500 references are included, which are chosen as much as possible from the recent literature.

This book is aimed at an audience of practitioners using carbon fibers in their many applications, and wanting information about their properties. Another audience is the researchers in the field of carbon science generally, who need to make contact with other carbon-based materials. The book will be of particular interest to students entering the field of carbon science and carbon fiber technology, since much of the material in the book is tutorial. Some of the results are new, having been developed to answer specific questions which arose as the book was being written.

It is the hope of the authors that this book has brought together a large body of material in a coherent fashion, and that it will stimulate further advances in the fascinating field of carbon fibers by presenting both the fundamental concepts and the engineering applications.

The MIT authors wish to acknowledge AFOSR (# F49620-85-C-0147) for support of this work and Ms. Phyllis Cormier for the long hours she worked in preparing the manuscript. The MIT authors wish to thank Professor M. Endo for his advice and encouragement throughout the preparation of this manuscript and for his inspiration which has been largely responsible for the interest of many physicists in this field. Ian Spain wishes to acknowledge B.J. Wicks, R.A. Coyle, D.J. Johnson, J.W. Johnson, and A. Oberlin for donation of figures, Mr. Charles Bowers who helped prepare figures, and W.N. Reynolds who supplied unpublished data. Ian Spain also thanks the Centre d'Etudes Nucléaires Grenoble for hospitality, and AFOSR for a grant (# F49620-86-C-0083) for support in writing this review. Thanks are due to Dr. I.L. Kalnin for introducing Harris Goldberg to the field of carbon fibers and for many years of collaborative research. In addition, the help and support of numerous colleagues and collaborators at the R.L. Mitchell Technical Center are gratefully acknowledged by Harris Goldberg.

Cambridge, MA, January 1988                                    *The Authors*

# Contents

# 1. Introductory Material on Graphite Fibers and Filaments

The purpose of this book is to review the physical properties of carbon filaments and fibers. These materials have been developed in the last 25 years, and new types are still being discovered and tested. It is hoped that in this book we will demonstrate that scientifically interesting and technologically important physical research has been, and can be carried out on these novel materials.

Carbon fibers are of technological importance today because fiber/resin composites have strength-to-weight properties superior to those of any other materials. This is bringing about a revolution in aircraft design, so that modern civilian and military aircraft are increasingly using carbon fiber composites in all but their main structural components. Several designs which employ essentially 100% carbon fiber composites for the airframe [B.W. Anderson 1987] are in the stage of certification for civilian use. Of course, experimental aircraft of this type have been flying for some time, and one of them, the "Voyager", completed a non-stop around-the-world flight in 1986 without refueling.

While their use in airframes has pushed carbon fiber development, other applications are current, and envisioned. For instance, carbon fibers in bulk carbon composites are used in rocket engines, while automobile manufacturers are possibly about to follow the aviation industry in using fiber/resin composites to reduce the weight of vehicles and reduce the time and expense required to change styles and designs. The critical factor for this latter application is the cost. Accordingly, research is being carried out on new ways of producing strong and tough carbon fibers with costs as low as 10% of current values.

Research has been of the utmost importance in bringing about this technological revolution. Although many of the details of processing conditions (which are absolutely essential for superior mechanical properties) have not been released by manufacturers, it is fortunate that a large body of literature has appeared concerning the relationship between structure and physical properties. Even so, much research carried out by companies has not been published. We acknowledge the fact that some of the material on ex-polymer fibers appearing in this review may be out-of-date when viewed from the perspective of leading carbon fiber manufacturers. It is hoped, however, that this

1

review illuminates the major principles governing the physical properties of these and related materials.

There are many different types of carbon fibers and filaments. It has been known since the last century that cellulose-based threads can be carbonized in inert atmospheres to form carbon filaments. Edison used ex-cotton and ex-bamboo carbon threads as lamp filaments until tungsten was found to be superior. These ex-cotton carbon filaments do not possess high enough mechanical strength to make them technologically interesting. The factor which is of crucial importance to the fiber strength is the alignment of the carbon hexagons along the fiber axis, with structurally coherent fibrils extending over long distances. Ex-cotton filaments are essentially isotropic, so that the high in-plane bond strength of graphite is not the controlling factor in their strength, but rather the relatively weak inter-planar bond.

Carbon fibers with well-aligned carbon hexagons became commercially available in the early 1960s after extensive research in Great Britain, Japan, and the USA. These filaments are in most cases derived from organic precursors by extrusion into polymeric fibers, followed by heat treatment (stabilization) and subsequent carbonization (heat treatment above $\sim$1000°C) and then further heat treatment up to $\sim$3000°C in an inert atmosphere. We refer to these fibers as "ex-polymer fibers". Correspondingly, "ex-rayon" fibers are carbon fibers prepared from rayon precursors, etc. Ex-rayon fibers were produced in quantity first [Bacon et al. 1966]. It was soon demonstrated by Shindo [1961a, 1961b, 1964] that polyacrylonitrile (PAN) was a better starting material. This PAN precursor became of increased importance when English workers [Watt et al. 1966; Standage and Prescott 1966] showed that high strength fibers could be obtained by oxidizing the PAN precursor while under strain without the need for a hot-stretching stage after carbonization. More recently, a new type of fiber has appeared, based on pitch [Hawthorne et al. 1970; Otani et al. 1970]. These mesophase pitch-based carbon fibers attain their alignment through a liquid-crystal-like state (mesophase). The resulting spun fibers have high elastic moduli ($\sim$500–800 GPa) after carbonization and heat treatment. However, the major commercial use for carbon fibers is for relatively low modulus ($\sim$200 to 300 GPa), high strength ($>$3 GPa) applications where ex-PAN fibers have superior properties. An isotropic pitch-based carbon fiber can be spun at lower cost; these fibers are finding application in high bulk, low performance applications, such as reinforced cement.

An important milestone in the understanding of the potential of these commercial, ex-polymer fibers was the study of a highly organized form of carbon filament (called a carbon whisker) obtained from carbon arcs [R. Bacon 1960]. Extremely high breaking strengths were obtained, which have only been fractionally realized in commercial filaments. A comparison of the mechanical properties of steel and carbon fibers and whiskers is given in Table 1.1. This table provides an indication of the potential improvements which can still be made in commercial materials.

2

**Table 1.1.** Comparison of densities and strengths of typical steel and carbon fibers[a]

| Material | Density [kg/m$^3$] | Tensile strength [GPa] | Normalized strength[b] |
|---|---|---|---|
| Steel | 7870 | 1.5 | 1.0 |
| High modulus carbon fiber[c] | 1800 | 2.4-2.6 | 7.0-7.6 |
| High strength carbon fiber[c] | 1800 | 3.1-4.6 | 9.0-13.5 |
| Carbon whiskers[d] | 2240 | up to 20 | up to 47 |

[a] See [Riggs 1985].
[b] Normalized strength compares Tensile strength/Density relative to steel.
[c] See Table 2.2.
[d] [R. Bacon 1960].

Another kind of filament has been developed recently [Endo 1988b], produced by decomposing a hydrocarbon on a heated substrate in the presence of transition metal catalysts [catalytic chemical vapor deposition (CCVD) carbon filaments]. These filaments are not continuous like ex-polymer fibers, but their lengths can reach several hundred millimeters. It is possible that their production costs will be substantially lower than those of ex-polymer fibers, making them of interest for chopped or milled fiber applications, with aspect ratios (length/diameter) of over 100. CCVD filaments have been the subject of intense research in the last few years, and will be systematically reviewed here for the first time.

In addition to the pristine (unmodified) fibers and filaments, it is possible to modify their properties in two major ways. Firstly, they can be coated with carbon, metals, insulators, superconductors, etc. (Chap. 12). Secondly, they can be intercalated with many elements and molecules, thereby changing their properties. (Intercalation is the process of inserting chemical species between the layer planes – see Chap. 10). Potential applications of modified fibers and filaments include high strength electrical conductors and superconductors, magnetic materials with favorable shape factors for magnetic shielding, etc.

This review is perhaps somewhat unusual for a book on this subject as it spans the interdisciplinary area between physics, materials science, and, to a lesser extent, chemistry. The underlying physical phenomena will be stressed, and it will be seen that many physical models have been applied to these materials in unique ways. In some instances new theories have been applied.

In reviewing the literature we have been overwhelmed by the number of publications, and have been forced to make decisions on omitting material which might appear arbitrary to some readers. To some extent, this is a personal record, which inevitably stresses some of the areas that have been investigated by our own research groups. Wherever possible, we have used previous research reviews, and summarized them, rather than repeating material in detail. As a result, the properties of the newer CCVD filaments are emphasized at the expense of the older ex-polymer fibers. We apologize to those whose work has not been fairly emphasized as a result of this process.

The review is organized in a way that begins with synthesis (Chap. 2), structural aspects (Chap. 3), then passes to physical phenomena such as lattice properties (Chap. 4), thermal properties (Chap. 5), mechanical properties (Chap. 6), electronic structure (Chap. 7), electronic transport properties (Chap. 8), and high temperature properties (Chap. 9), thence to modified filaments by intercalation (Chap. 10) and by ion implantation (Chap. 11), and ends with a brief discussion of applications (Chap. 12). However, before proceeding to the main body of the review, a brief description of the overall structure of the different kinds of filaments will be given. This will enable some essential concepts and terminology to be introduced. In this book we will attempt to conform to the recently adopted standardized terminology [Donnet et al. 1982, 1983, 1985, 1986, 1987a, 1987b].

## 1.1   Introductory Discussion of Structure of Carbon Filaments

As mentioned above, there are several types of carbon filaments, and this preliminary discussion will consider their structure.

### 1.1.1   Ex-polymer Fibers

Ex-polymer fibers are manufactured by extruding a polymer through a nozzle into a continuous filament, after stabilization treatment at about 200°–350°C in air, then heat treatment of the filaments to temperatures ($T_{HT}$) on the order of 1000°C in order to carbonize the filaments by the removal of H, O, N, ... etc. Further heat treatment to temperatures typically between 1300° and 3000°C modifies their mechanical properties. Heat treatment above

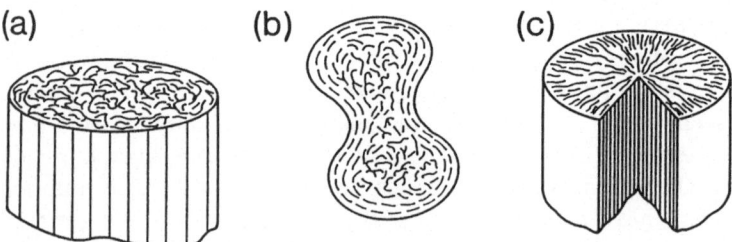

Fig. 1.1a-c. Illustration of observed ex-polymer fiber cross-sectional macroscopic morphology: (a) circular section, showing a disordered arrangement of ribbons, (b) dog-bone section observed in partially graphitic ex-PAN fibers, with a circumferential arrangement of ribbons in the sheath region, and random in the core, (c) ex-pitch (mesophase pitch), "PAC-man" section showing a radial arrangement of the ribbons

**Fig. 1.2.** Sketch of a typical graphene plane in a carbon fiber showing vacancy-cluster defects [Fourdeux et al. 1971]

5 nm

~2200°C is referred to as a "graphitization" step. These fibers are manufactured commercially, and are produced from several starting polymers. The most important today are ex-polyacrylonitrile (ex-PAN) and ex-pitch fibers, while ex-rayon fibers were of importance until very recently.

The most obvious feature of carbon fibers is their small diameters, which are typically about 7 $\mu$m. This compares with diameters of human hairs of ~70 $\mu$m. Carbon fibers can barely be seen by eye, and appear as featureless black filaments under optical magnification. Under closer examination afforded by scanning electron microscopes, the cross sections are often found to be circular (Fig. 1.1a). Sometimes, particularly in the case of ex-PAN fibers, the cross section is double-lobed or "dog-bone" shaped (Fig. 1.1b). Ex-pitch fibers sometimes have a roughly circular cross section, in which a segment has been removed, sometimes called a "PAC-man" section (Fig. 1.1c).

Under even closer scrutiny using x-ray diffraction and transmission-electron microscopy, it is found that the basic structural unit of carbon fibers is a planar network of connected benzene rings, illustrated in Fig. 1.2. This is similar to the planar structure occurring in graphite, in which the carbon atoms lie in hexagonal honeycomb arrays. Note that the planar structure of the carbon fiber is not perfect. We sometimes refer to these two-dimensional planar arrays of carbon atoms as "graphene" planes. Though the planar structures are imperfect, the planar arrays nevertheless lie roughly aligned along the direction of the fiber, and are responsible for the strength of these materials. Coulson (1952), for example, has pointed out that the C–C bond in graphite is the strongest bond occurring in nature.

5

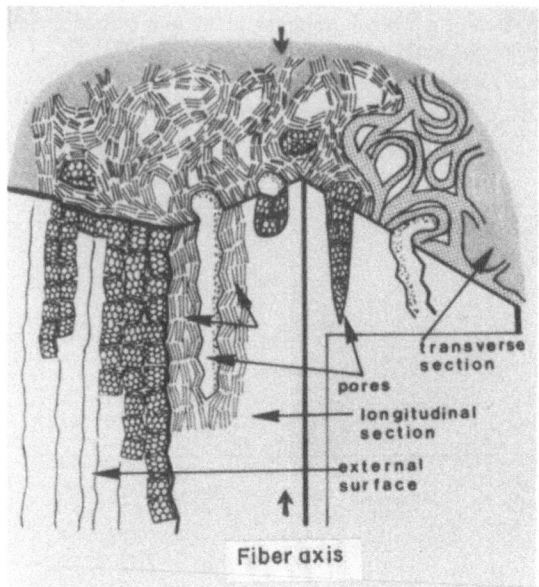

**Fig. 1.3.** Artist's conception of the structure of a high strength ex-polymer fiber [adapted from Guigon et al. 1984b]

In the carbon fiber, these graphene planes stack on top of each other similar to pages in a newspaper. They are ribbon-like, having typical widths of several tenths of micrometers. Figure 1.3 illustrates the typical arrangement of these planes in an ex-PAN fiber. If a cross section were being viewed along the fiber axis, the ribbons (stacks of graphene planes) would be seen to undulate, and separate, but most would be roughly aligned along the fiber axis (Fig. 1.4). For example, the mean angle made by these ribbons to the fiber axis would be about 20° in typical high strength fibers used in aircraft. Note in Fig. 1.3 the presence of voids, which are typically needle-shaped. These voids reduce the density below the value $(2.26\,\mathrm{g/cm^3})$ expected for an ideal planar hexagonal array of carbon atoms. Also indicated in Fig. 1.4 are the "crystallite" dimensions $L_a$ and $L_c$, representing the extent of relatively straight portions of the lattice planes along their length, and the stacking height of the graphene planes in the ribbons, respectively.

Increasing the heat-treatment temperature ($T_{HT}$) allows the graphene planes to grow. (The heat treatment temperature is sometimes denoted by HTT in the literature.) Figure 1.5 illustrates a high resolution electron micrograph of the surface layers of an ex-PAN fiber [D.J. Johnson 1980]. When carbonized at 1000°C, (Fig. 1.5a), the graphene planes are apparently less than 10 nm in length. They are roughly aligned along the fiber axis. At 1500°C (Fig. 1.5b), the graphene planes are more extensive, better aligned and in sharper contrast. Further improvements in the alignment occur on heat treatment at 2500°C (Fig. 1.5c). In this case it can be seen that the ribbons are much more extensive both along and perpendicular to the fiber axis.

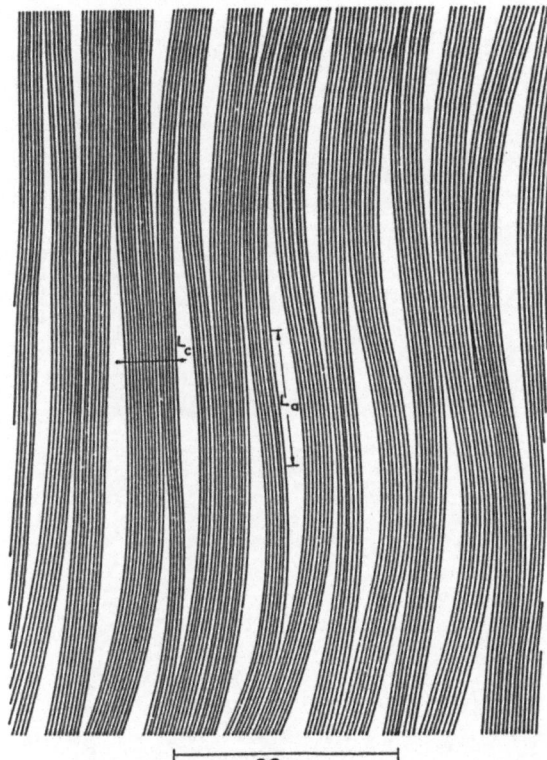

**Fig. 1.4.** Sketch of the cross section of an ex-PAN fiber along the axis direction [Fourdeux et al. 1971]. Here the in-plane and c-axis structural coherence lengths $L_a$ and $L_c$ are indicated

20 nm

Even after high temperature heat treatment some crystallites remain badly misaligned (see Chap. 3).

A cross section across the fiber axis, as shown also in Fig. 1.3, indicates a much higher state of disorder than in the longitudinal section (Fig. 1.4). The cross section shows that the planes are twisted over short distances compared to the ribbon widths, and there is a random orientation of planes over the diameter of the fiber. The arrangement of planes over this section varies with fiber type and processing conditions. For instance, some ex-mesophase pitch fibers are found in which the planes are roughly arranged in a radial fashion (see Fig. 1.1c), while a circumferentially preferred arrangement can be found in ex-PAN fibers, as indicated for the outer regions in Fig. 1.1b.

These twists in the planar arrays prevent them from stacking in a regular fashion, as in hexagonal graphite. The stacking of the planar arrays is typically random, although inter-layer correlations are developed by heating the fibers to high temperature $T_{HT}$. It is also necessary to consider time-at-temperature (or the residence time), since the graphitization processes are kinetic in nature. Most heat-treatment processes for fibers are carried out over a time period on the order of 10 minutes or more, and this will be assumed unless otherwise stated.

7

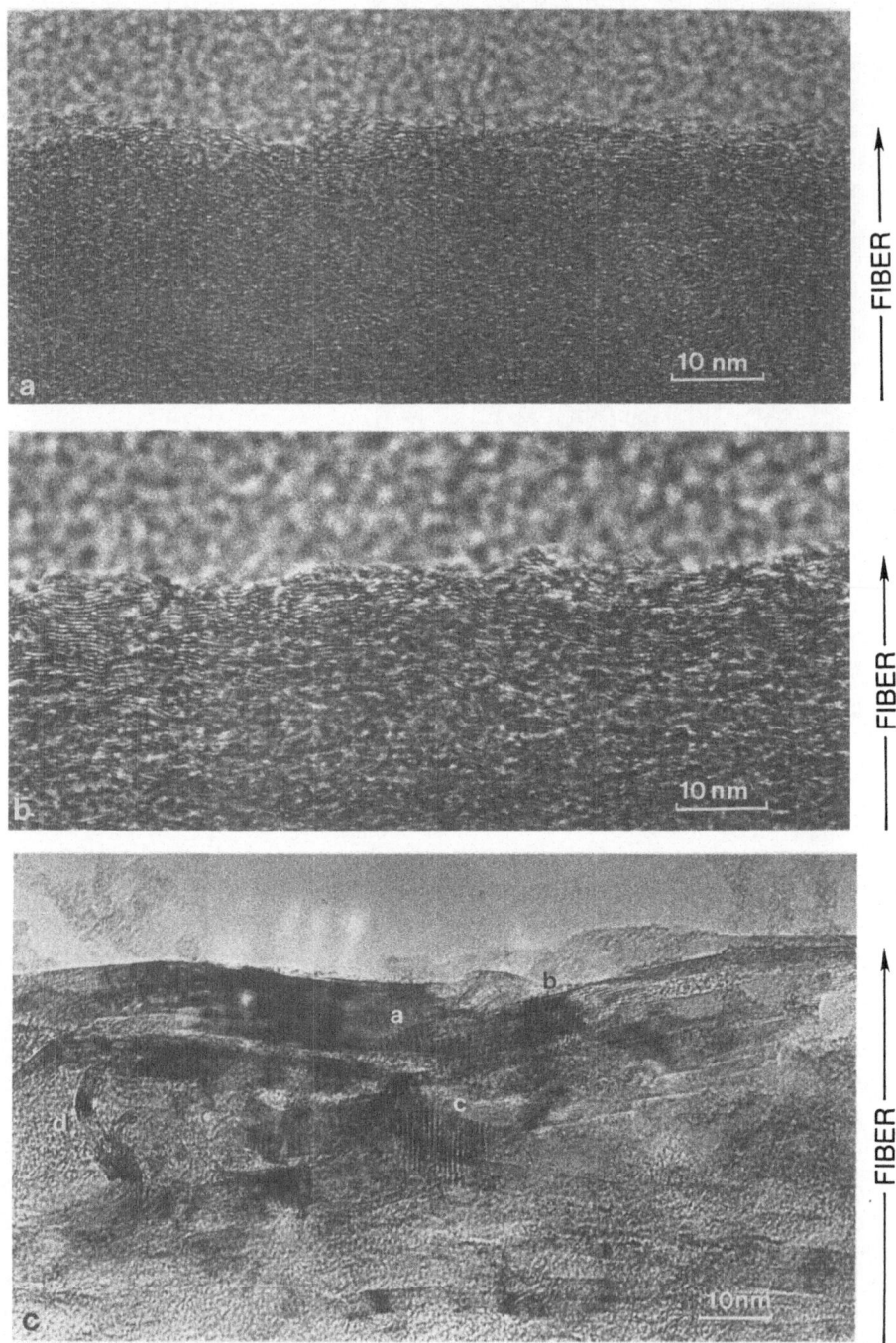

**Fig. 1.5** Figure caption see opposite page

When the carbon layers are disordered, as in book pages which lie flat with respect to their neighbors but are randomly rotated, the carbon is said to be turbostratic. Fibers with $T_{HT} = 1300°\,C$, for example, are turbostratic, and are called "carbon fibers". Ex-PAN fibers which are heat treated to this temperature can have relatively high strengths (Table 1.1), and are called "high strength" fibers.

Heat treatment to very high $T_{HT}$ (e.g., $\sim 3000°\,C$) allows inter-layer correlations to develop as demonstrated by the appearance of weak $(hkl)$ x-ray diffraction peaks when the interlayer spacing becomes less than 0.34 nm; in this case, the fiber is described as partially graphitic, which is usually abbreviated to "graphite fibers" in the popular literature. Fibers with higher $T_{HT}$ values generally have higher Young's moduli of elasticity. However, their breaking strengths are often observed to be lower than for high strength fibers (Table 1.1). Ex-PAN fibers of this type are called "high modulus fibers".

The particular graphene planar network illustrated in Fig. 1.2 contains defects in the form of multiple vacancies. These voids are probably formed as the polymer molecules fuse together to form larger graphite ribbons. Impurities, such as H and N would be attached to both the dangling bonds at these vacancies and the edges of the ribbons. The vacancies and impurities are progressively removed at higher heat treatment temperatures $T_{HT}$.

Another kind of disorder occurs in the graphene planes, and is not shown in Fig. 1.2. In this figure the hexagons are shown to be all of the same orientation. However, defects exist which produce changes in the bond directions, so that all orientations of the hexagons would be found over the length of a ribbon. This, coupled with stacking disorder, ensures that all orientations of the in-plane axes occur in the small regions of the fiber probed by x-ray or electron diffraction measurements.

### 1.1.2 Arc-Grown Carbon Whiskers

Carbon whiskers are produced in carbon arcs under argon gas pressure, and are of interest because their physical properties are close to "ideal" graphite values. The structure of arc-grown carbon whiskers [R. Bacon 1960] is much more regular than that in ex-polymer fibers or even in as-grown ex-CCVD filaments (described below). As shown in Fig. 1.6, the carbon planes are arranged in a circular fashion about the whisker axis. X-ray diffraction patterns show single-crystal spots, indicating a highly correlated layer-stacking sequence.

---

Fig. 1.5a-c. High resolution electron micrographs of the surface layers of an ex-PAN fiber, (a) $T_{HT} = 1000°\,C$, (b) 1500°C, (c) 2500°C. The arrows in the right margin delineate the fiber region. Letters $a$–$d$ in figure (c) refer to badly misoriented regions that are discussed in Chap. 3. (Kindly supplied by D.J. Johnson [Bennett and D.J. Johnson 1979; D.J. Johnson 1980])

FIBER
AXIS

Fig. 1.6. Sketch illustrating the scroll structure of a carbon whisker [R. Bacon 1960]

Based on this and other observations, it was proposed that the layers are scrolled, and that growth occurs by adding atoms to the scroll edges. In order for the planes to stack in this graphitic fashion it is necessary that the planes be flat, so that the scroll would be composed of straight sections separated by tilt boundaries (see Sect. 3.1). Along the fiber axis, the whiskers can have the properties of single crystal graphite.

The term "whisker" is usually restricted to filaments with highly crystalline structure, often single crystal with only a central screw dislocation. In this sense, the term is not completely appropriate to the above materials. In this review, "graphite whisker" will be used in conformity with existing practice, and we thus distinguish these more ideal filaments from the more disordered filaments.

### 1.1.3 Catalytic Chemical Vapor-Deposited Filaments

Catalytic chemical vapor-deposited filaments are prepared by the decomposition of gas-phase hydrocarbons onto a catalyzed surface, typically at $1100°C$. It is thought that these non-continuous filaments may have important commercial applications as a result of their potentially low cost and high mechanical performance. After heat treatment to above $\sim3000°C$ the structural reorganization of the CCVD filaments is sufficiently massive that external surfaces exhibit facetting (Fig. 1.7). This is again a consequence of the relatively high structural order that allows this reorganization to occur over laboratory time scales.

These filaments have certain similarities to whiskers in that the layers are arranged in a circular fashion about the core (Fig. 1.7a). However, as-deposited filaments ($T_{HT} \sim1100°C$ typically) exhibit poor inter-layer correlations, and the graphene planes are relatively straight only over short distances (Sect. 2.2). Heat treatment to $\sim3000°C$ results in almost complete graphitization, as the crystallite-size increases. This tendency to graphitize more readily

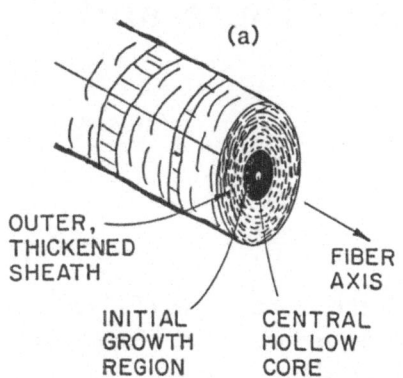

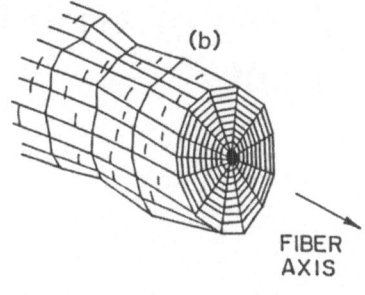

(a)

OUTER,
THICKENED
SHEATH

FIBER
AXIS

INITIAL
GROWTH
REGION

CENTRAL
HOLLOW
CORE

(b)

FIBER
AXIS

Fig. 1.7a.b. Sketch illustrating the structure of CCVD filaments: (a) as-deposited at 1100°C, (b) after heat treatment to 3000°C

than ex-polymer fibers is related to the much greater structural organization occurring in as-grown CCVD filaments than in ex-polymer fibers heat treated to the same temperature. The circular arrangement of the planes in this case is thought to be tree-ring, rather than scroll-type as a result of the growth process, which occurs initially from the ends. After heat treatment to sufficiently high $T_{HT}$, facetting is observed (Fig. 1.7b). The highly ordered structural arrangement of these filaments has permitted interesting physical investigations to be carried out with a unique sample geometry.

The above descriptions are only intended to guide the reader into the following sections, and a more complete description of structure will follow in Chap. 3.

# 2. Synthesis of Graphite Fibers and Filaments

This chapter considers the methods by which the various types of fibers are produced. It is noted that ex-polymer fibers are produced by methods which lie in the area of applied polymer chemistry. Several reviews of their manufacture have been written. Therefore, only a brief description of their preparation is given here, enabling the reader to relate structural features to growth methods. The methods of growing CCVD filaments is considered in greater detail. The growth of the CCVD filaments in lengths of hundreds of millimeters has not been reviewed previously. It is emphasized that commercial carbon fibers are manufactured from polymers, and that one type (ex-PAN) dominates the commercial market, though low cost isotropic pitch fibers are also beginning to find wide application in high bulk, low performance applications. The present review of fiber growth would therefore seem to be out-of-balance with respect to current commercial utilization. However, the most interesting physical experiments have been carried out on the newer CCVD fibers, thereby providing the rationale for the balance we have chosen.

## 2.1 Carbon Fibers from Polymeric Precursors

Although many organic polymers can be used to produce carbon and graphite fibers, only three have had any commercial significance: rayon, polyacrylonitrile (PAN), and pitch. The former was developed first, but has been replaced largely by PAN fibers. The more recently developed mesophase-pitch-based fibers have only recently been able to match the superior mechanical properties of those made from PAN. However, these fibers have high Young's moduli and relatively low breaking strengths compared to high strength PAN fibers (see Table 6.1 and the discussion in Chap. 6). For this reason the major market for fibers utilizes high strength, ex-PAN fibers, which are best suited to applications in composites (Chap. 12). An excellent review of fiber processes and properties has recently been given by Riggs [1985].

The processing of carbon fibers has a number of steps which are common to all fibers made from polymeric precursors. These are:

- spinning (extrusion of the polymer melt or solution) into fine fibers;
- conversion of the fibers into a chemical form which will not melt and can

12

withstand higher temperature heat treatments (called stabilization);
- carbonization at temperatures usually ~1000°C in order to form material which is predominantly made up of hexagonal networks of carbon.

Further heat treatment to temperatures of up to 3000°C is used to increase the degree of order in fibers when very high modulus is desired. This step is usually referred to as graphitization. The stages of carbonization and graphitization are similar for almost all organics. The major differences are the degree of orientation and crystallinity which can be achieved at a given temperature. Thus, as discussed in Chap. 1, some precursor materials are very difficult to order structurally even at 3000°C (such as rayon) while others achieve a fairly well ordered crystalline structure (such as materials made from mesophase pitch).

### 2.1.1  Ex-rayon Fibers

A summary of the processing of rayon-based fibers can be found in the book by Gill [1972]. The first low temperature treatment ($T_{HT}$ is typically ~300°C) converts the structure to a form which is stable to higher processing temperatures, and involves polymerization and formation of cross-links. In this process, 50%–60% of the fiber mass is lost to decomposition products, including $H_2O$, CO, and $CO_2$. The carbonization step is usually carried out at ~1500°C, resulting in further mass loss. As shown in Table 2.1, the yield is typically 20%–25% of the original polymer mass after the carbonization step. The mechanical properties of carbonized rayon are poor, however, as a result of poor alignment of the hexagonal networks. At this stage the fiber is essentially isotropic, and the mechanical properties can be improved only by stretching the fiber during graphitization, which is an expensive process. The combination of this factor, together with their relatively poor mechanical properties, and the high mass loss during processing are the reasons that ex-rayon carbon fibers have not proved competitive in the market place.

Table 2.1. Weight loss during processing for different polymer precursors [Riggs 1985]

| Precursor | Structure | Percentage weight retention[a] |
|---|---|---|
| Rayon | $C_6H_{10}O_5$ | 20-25 |
| PAN | $CH_2$-CH-CN | 45-50 |
| Mesophase pitch | Polynuclear aromatics | 75-85 |

[a]Based on weight of precursor.

## 2.1.2 Ex-PAN Fibers

Polyacrylonitrile (PAN) can be spun into well-oriented polymer fibers. The nominal chemical structure is shown in Fig. 2.1 and the major steps in the preparation of PAN carbon fibers in Fig. 2.2. It is emphasized that the chemistry of the process is extremely complex, so that Fig. 2.1 is a great simplification of the actual process. Also, the processes depicted in Fig. 2.2 can vary somewhat in actual fiber production.

The fibers are heated in an oxygen-containing atmosphere at temperatures between 200° and 300°C to stabilize them for the subsequent carboniza-

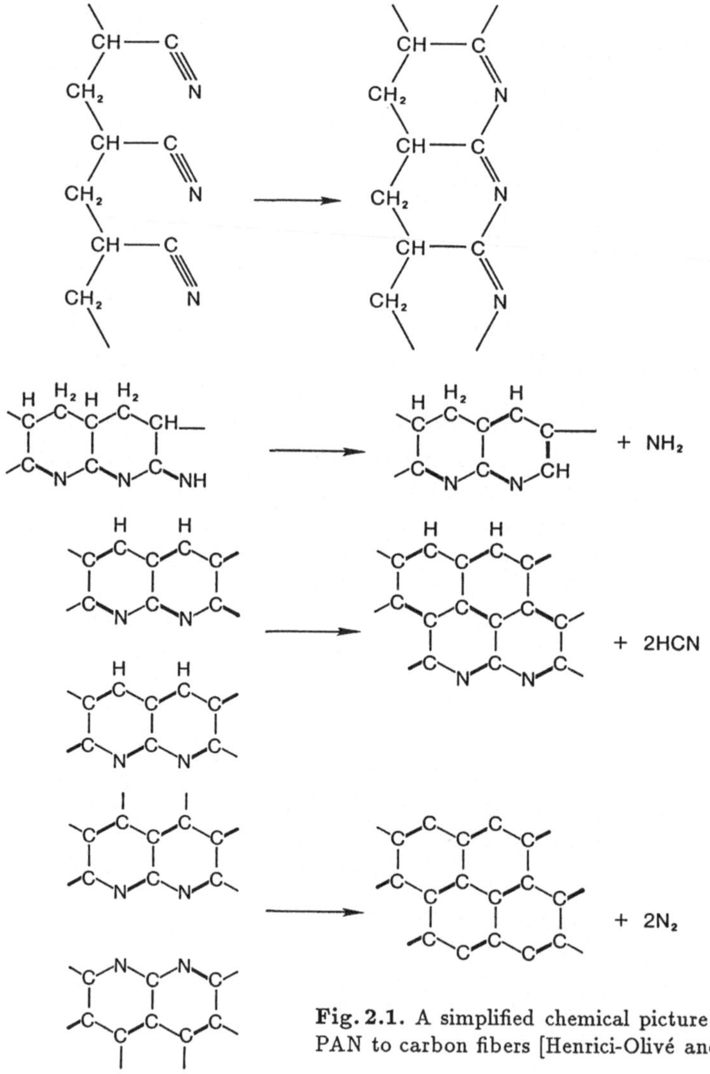

Fig. 2.1. A simplified chemical picture of the evolution of PAN to carbon fibers [Henrici-Olivé and Olivé 1983]

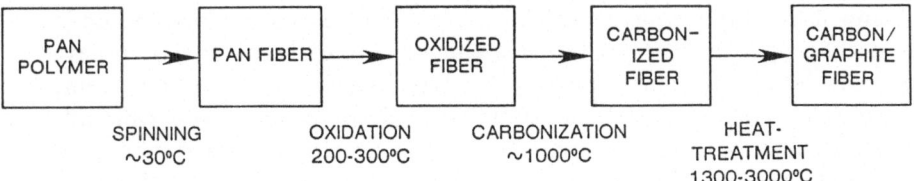

```
┌──────────┐    ┌──────────┐    ┌──────────┐    ┌──────────┐    ┌──────────┐
│   PAN    │ →  │PAN FIBER │ →  │ OXIDIZED │ →  │ CARBON-  │ →  │ CARBON/  │
│ POLYMER  │    │          │    │  FIBER   │    │  IZED    │    │ GRAPHITE │
│          │    │          │    │          │    │  FIBER   │    │  FIBER   │
└──────────┘    └──────────┘    └──────────┘    └──────────┘    └──────────┘
      SPINNING         OXIDATION        CARBONIZATION       HEAT-
       ~30°C           200-300°C          ~1000°C        TREATMENT
                                                        1300-3000°C
```

**Fig. 2.2.** Schematic of the processing steps in the preparation of PAN-based carbon fibers [Chwastiak et al. 1979]

tion step. The chemistry of the stabilization and carbonization of carbon fibers from PAN is very complex and has been reviewed by Henrici–Olivé and Olivé [1983]. Two of the important processes which occur during stabilization are: (a) the nitrile groups react to form closed ring structures, and (b) the oxygen helps to crosslink the chains. The reaction of the nitrile groups is extremely exothermic, and must be done at a controlled rate. The stabilization kinetics are one of the factors that limit the practical diameter of ex-PAN carbon fibers to about 10 $\mu$m. The orientation of the rings can be maintained if this process is done under tension (called hot stretching). Carbonization is carried out at temperatures between 1000° and 1500°C (typically ~1000°C in commercial processes and then heat treated to ~1300°C for high strength fibers). After carbonization, well-defined hexagonal networks develop. Considerable gas evolution takes place during this process, which may be partially responsible for cracks which lower the tensile strength of the material (see Chap. 6). The hexagonal carbon networks are typically aligned with respect to the fiber axis, making an average angle of ~20°. We refer to the angle between the planes and the fiber axis as a "misorientation angle" which is defined in Fig. 3.15. Further improvements in the orientation (and thus the modulus) can be achieved by higher heat-treatment temperatures (i.e., graphitization) as shown in Table 2.2.

**Table 2.2.** Some physical and mechanical properties[a] of polymer-based high strength and high modulus carbon fibers [Hughes 1987; Endo 1988a]

| Fiber type | Ex-PAN | | | Ex-pitch | | | | | |
|---|---|---|---|---|---|---|---|---|---|
| | High str. | High mod. | Ultra-high mod. | High str. | High mod. | Ultra-high mod. P-100  P-120 | | Moderate mod. HM50 HM60 HM80 | |
| $T_{HT}$ [°C] | 1300 | 1800 | 2500 | 1300 | 1800 | 2500 | | | |
| $d[\mu m]$ | 5.5-8 | 5.4-8 | 8.5 | 10 | 10 | 10 | | | |
| $\varrho_m[gm/cm^3]$ | 1.8 | 1.8 | 1.96 | 2.02 | 2.06 | 2.15 | | | |
| $\sigma_T$ [GPa] | 3.1-4.6 | 2.4-2.6 | 1.9 | 1.9 | 2.1 | 2.2 | 2.4 | 2.8  3.0  3.5 | |
| $E_Y$ [GPa] | 230-260 | 360-390 | 520 | 380 | 520 | 690 | 830 | 490  590  790 | |
| $\varepsilon_{max}$ | 1.3-1.8 | 0.6-0.7 | 0.4 | 0.5 | 0.4 | 0.3 | 0.3 | 0.6  0.5  0.4 | |
| $N(C)$ [%] | 92-95 | 99-99+ | 99+ | 99 | 99 | 99+ | | | |

[a]The properties are denoted as follows: $T_{HT}$ is the heat-treatment temperature, $d$ is the fiber diameter, $\varrho_m$ is the mass density, $\sigma_T$ is the tensile strength, $E_Y$ is the Young's (tensile) modulus, $\varepsilon_{max}$ is the elongation at break [%], and $N(C)$ is the percentage carbon content.

15

Many of the improvements which have been made over the last decade in mechanical properties without hot stretching in the carbonization step have arisen from details in the preparation procedures. These include improvements in the starting polymers, and the careful elimination of metal contaminants using filters. Another factor of importance is the control of surface chemistry, and this extends to the improvement of surfaces for compatibility with resins. It is inappropriate to address these issues in the present book and the reader is referred to the above references for further details.

### 2.1.3 Ex-pitch Fibers

Ex-pitch fibers are generally produced from petroleum asphaltene or coal tar, which consists partially of fused aromatic molecules (i.e., containing several carbon rings). Their recent importance arises from the observation that a highly oriented, optically anisotropic liquid crystal state (mesophase) occurs when some pitch materials are heated for a period of time near 350°C. This mesophase pitch material can then be spun into fibers with a highly preferred orientation. Ex-pitch fibers have potentially lower cost than ex-PAN fibers for three reasons: the starting material is reasonably inexpensive; they are melt processed (PAN is typically spun from solution); and they have lower weight loss during processing (see Table 2.1). However, the process of preparing the mesophase pitch-based fibers is expensive, so that only isotropic-pitch-based fibers have low cost. Some mechanical and structural parameters of the isotropic pitch fibers are given by D.J. Johnson et al. [1975]. These isotropic pitch fibers are used for relatively low performance, low cost applications and will not be discussed further in this volume. The development and properties of fibers made from mesophase pitch (see Table 2.2) have been reviewed by Singer [1978] and R. Bacon [1980].

In the synthesis of mesophase pitch carbon fibers, the mesophase pitch is prepared by first hydrogenating a coal tar precursor material and then heat treating the precursor above the softening point (308°C) to evaporate highly volatile components. The resulting mesophase pitch material is then spun to form carbon fibers by extrusion through a capillary tube as shown in Fig. 2.3 [Hamada et al. 1987b]. The parameters controlling the extrusion are the temperature of the pitch (which determines the viscosity), the flow rate of the pitch and the diameter of the extruding capillary. If the extrusion occurs without stirring, a radial cross-sectional structure is formed [Bright and Singer 1979]; but with stirring, other cross-sectional structures can be achieved, such as random, and onion-type (Fig. 2.4), depending on the nozzle shape, stirring conditions and viscosity of the pitch [Hamada et al. 1987a, 1987b]. After extrusion, the fibers are thermoset by oxidation in dry air while slow heating to temperatures in the range 200°–300°C, and then cooling down rapidly in an argon atmosphere. The fibers are subsequently heat treated in the temperature range 2500°–2700°C to carbonize and graphitize the fibers.

16

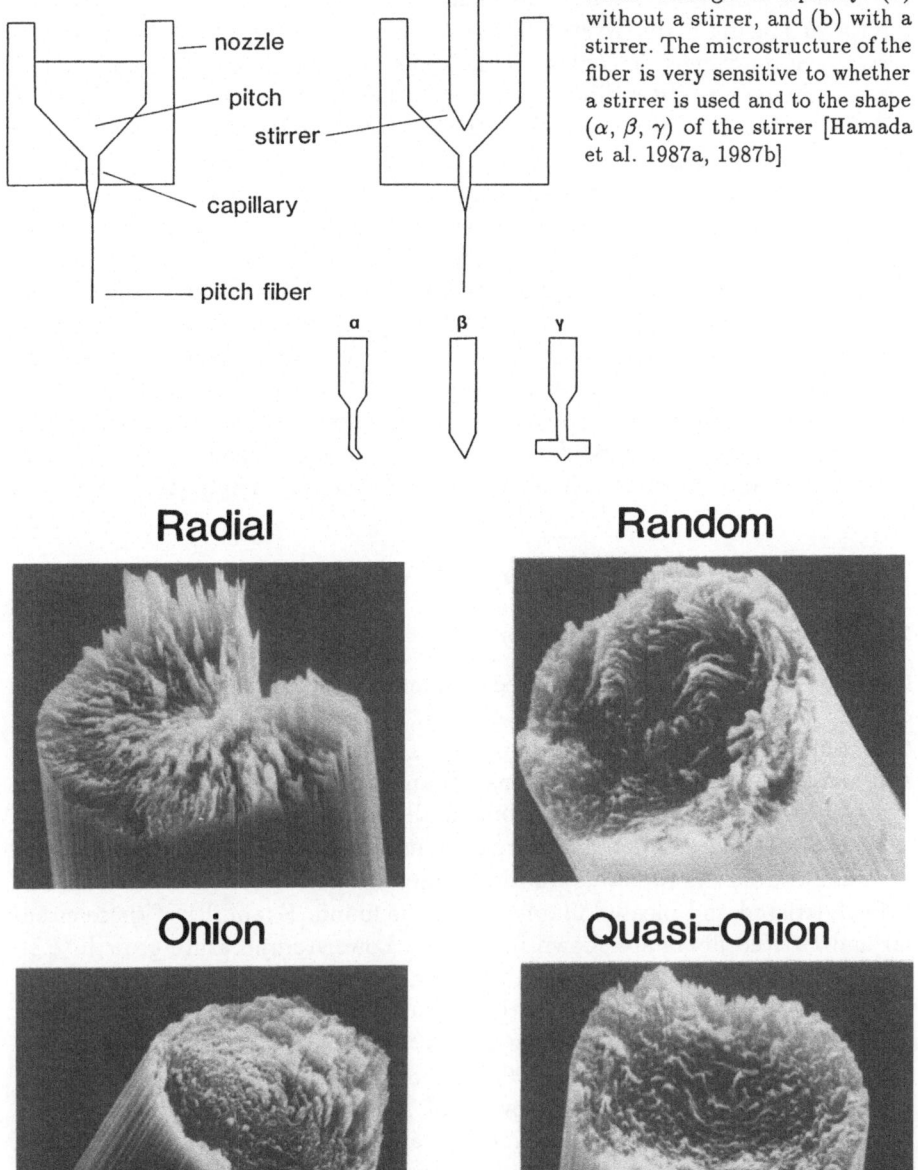

(a)

nozzle

pitch

stirrer

capillary

pitch fiber

(b)

**Fig. 2.3a,b.** Extrusion of pitch fibers through a capillary: (a) without a stirrer, and (b) with a stirrer. The microstructure of the fiber is very sensitive to whether a stirrer is used and to the shape ($\alpha$, $\beta$, $\gamma$) of the stirrer [Hamada et al. 1987a, 1987b]

α    β    γ

## Radial

## Random

## Onion

## Quasi–Onion

**Fig. 2.4.** Typical transverse microstructures of ex-pitch fibers showing radial, random, onion-skin and quasi-onion-skin structures [Hamada et al. 1987a, 1987b]

17

If the fibers are spun while stirring, the degree of structural ordering of the resulting fibers is reduced, as characterized by such parameters as the interlayer spacing $\tilde{c}$, the crystallite size $L_c$, the resistivity, and the magnetoresistance. Exploiting this control of the cross-sectional microstructure, a new type of mesophase pitch fiber has recently been developed by the Kashima Oil Co. of Japan (Carbonic HM50, HM60 and HM80) with highly desirable strength and elongation characteristics (see Table 2.2) [Endo 1988a]. Though based on the same type of precursor material as the Thornel P-55 and P-100 fibers, superior mechanical properties are achieved through a different cross-sectional microtexture, namely a radial arrangement of wavy graphitic ribbons (see Fig. 2.4a), in contrast to the more planar arrangement of the graphitic ribbons in the Thornel P-100 ex-pitch fibers. The high strength pitch fibers characteristically have a turbostratic arrangement of the ribbons and exhibit a large amount of distortion of the basic hexagonal structural units within the ribbons. The correlation between the structure and the mechanical properties (Sect. 6.2.2) has been established through microstructure (Sect. 3.7.2) and magnetoresistance (Sect. 8.4) measurements [Endo 1988a].

## 2.2  Carbon Filaments by CCVD

### 2.2.1  General Considerations

Catalytic chemical vapor deposited filaments grow from the decomposition of hydrocarbons at high temperature (e.g., $\sim 300° - 2500°C$) in the presence of a metallic catalyst (e.g., Fe, Ni, Co). They may grow on many types of substrates (e.g., carbon, silicon, quartz) and from many hydrocarbons (e.g., acetylene, benzene, natural gas), but in all cases, growth is favored in a hydrogen atmosphere. Filament diameters may range from about 100 nm to several hundred micrometers. In many cases the cross section is circular, but helical, twisted and pleated filaments may be found. Examples of these circular and twisted fibers are shown in Fig. 2.5. Lower-temperature growth (e.g., $<1000°C$) often produces filaments with a "wormy" or "vermicular" shape, whereas higher-temperature growth produces relatively straight filaments.

Various CCVD filaments have been studied for nearly 100 years [Schützenberger and Schützenberger 1890; Pelabon 1905] and there have been many studies since then. A detailed review of growth characteristics and mechanisms has been given by R.T.K. Baker and Harris [1978]. In this section, a brief summary of the Baker and Harris review will be given with more extensive discussions of recent work, including the exciting development of relatively long fibers and very thin filaments from benzene or natural gas. The thin filaments in particular may be of commercial interest. Many detailed physical measurements have been made on such filaments, and they have been successfully intercalated (Sect. 10.1).

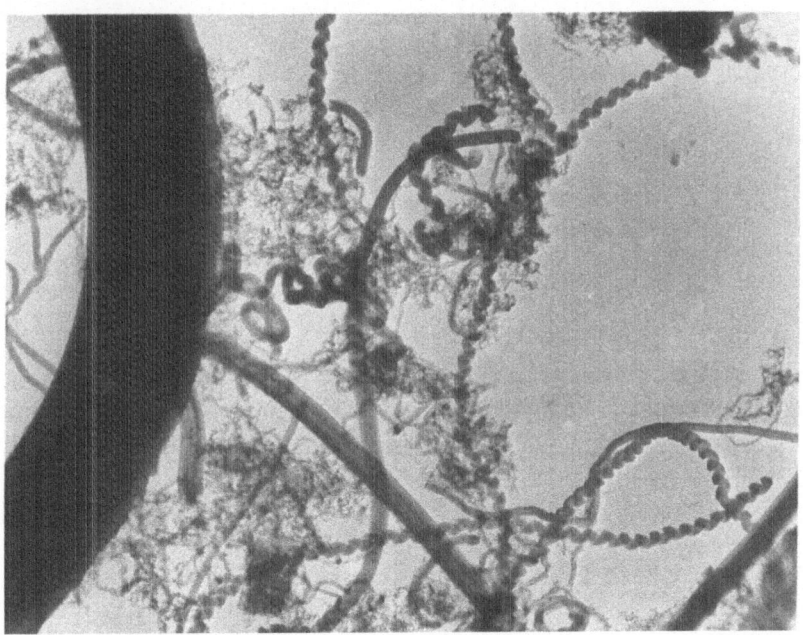

**Fig. 2.5.** Electron micrographs of CCVD carbon filaments of different sizes, with relatively straight, vermicular and twisted filaments

## 2.2.2 Detailed Considerations

The major results on the growth and structure of CCVD filaments are as follows:

- The growth of filaments occurs via a catalyzed dehydrogenation reaction of a hydrocarbon in several steps. This includes the activation of the catalyst, the initial growth and elongation of a filament in which the catalyst particle is often carried at the tip of the filament, and a subsequent thickening process in which carbon is directly deposited on the side of the filament [Tesner et al. 1970; S.D. Robertson 1970, 1972]. It was shown by R.T.K. Baker and coworkers [1972, 1973; R.T.K. Baker and Harris 1973] that only a part of the catalyst particle at the tip of the filament was exposed to the hydrocarbon, see Fig. 2.6, and that a coating of carbon could cover the tip completely, arresting growth. Further growth could occur if this surface layer was removed by oxidation.
- Several different kinds of CCVD filaments can be grown. Vermicular filaments (i.e., with an irregular, wormy, form) are favored at lower temperatures (e.g., $\leq 900°$C, but depending on the hydrocarbon), while straighter filaments are grown at higher temperature. Baird et al. [1971, 1974] studied the structure using high resolution electron microscopy,

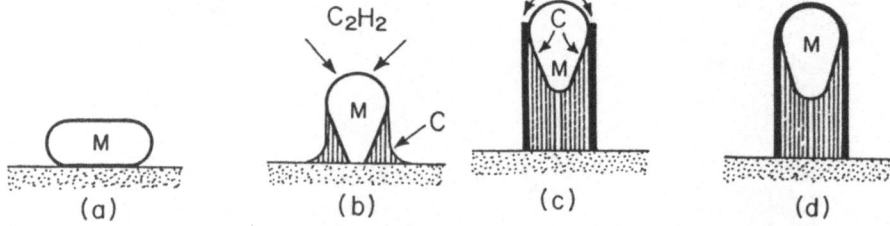

**Fig. 2.6.** Schematic representation proposed for carbon filament growth from the pyrolysis of acetylene ($C_2H_2$) on a metal particle (M) where C denotes carbon. [From Baker et al. 1972]

finding that vermicular filaments tended to have an amorphous structure, and straight filaments a more graphitic one. There are differences in the structure of filaments grown from different precursors such as butadiene, acetylene, methane and propane.

- The initial diameter of filaments growing with catalyst particles at the tips is closely related to the diameter of the catalyst particles. Vermicular filaments often have a low density, easily oxidized core, enclosed by a sheath of material which is resistant to oxidation [R.T.K. Baker et al. 1972, 1973; R.T.K. Baker and Harris 1973]. In contrast, graphitic filaments have hollow cores and a minimum diameter as discussed below [Oberlin et al. 1976a; Tibbetts 1984].

- The growth rate of filaments can be described by an Arrhenius equation. Baker [1987] suggested that the enthalpy obtained from the growth data was similar to that obtained for the diffusion of carbon in the catalyst metal, and proposed that diffusion was the rate-limiting step in the growth of filaments. The growth rate is also inversely proportional to the diameter of the filament in its initial stage [Endo and Komaki 1983; Tibbetts 1987].

### 2.2.3 Growth Mechanisms

To account for the growth of carbon filaments on catalyst surfaces, various mechanisms have been proposed for different systems. A mechanism for the growth of carbon filaments from pyrolysis of acetylene on isolated nickel particles was postulated by R.T.K. Baker et al. [1972], see Fig. 2.6. According to this mechanism, acetylene decomposes on the front-exposed surfaces of the metal particle to release hydrogen as gas and carbon, which dissolves in the particle. The dissolved carbon diffuses through the particle and is precipitated at the trailing end from the body of the filament. This mechanism depends on the temperature gradient across the catalyst particle due to the exothermic decomposition of hydrocarbons. Since the solubility of carbon in a metal is temperature dependent, precipitation of excess carbon will occur at the colder zone behind the particle, thus allowing the filament to grow (Fig. 2.6c). Such a process will continue until the leading tip of the catalyst particle is poisoned (Fig. 2.6d) after which the filament will cease to grow in length. Support for the diffusion model comes from experiments on the kinetics of growth of filaments from acetylene at ~1000°C which yield an activation energy of 140 kJ/mole [R.T.K. Baker et al. 1973]. The concentration of carbon in the iron catalyst particle is probably close to that at the eutectic (4.3% C) and the activation energy for diffusion is 133 kJ/mole [Tibbetts 1983]. Similarly, the enthalpy for the growth of filaments with $\alpha$-Fe, $\gamma$-Fe, Ni, Co, Fe/Ni and Cu catalysts is similar to the enthalpy of diffusion [R.T.K. Baker et al. 1987].

To explain the tubular structure of the carbon filament, it was proposed by Baird et al. [1974] that the filament is formed entirely by a catalytic process involving the surface diffusion of carbon around the metal particle rather than bulk diffusion of carbon through the catalytic particle. A similar model was proposed by Oberlin et al. [1976a, 1976b], and is illustrated in Fig. 2.7. Such transport would be easier if the catalyst particle were liquid. It is interesting to note, for instance, that benzene-derived fibers are typically produced between 1100° and 1300°C [Koyama 1972; Koyama et al. 1972], which is below the melting temperature of Fe (1534°C) but is often above that for the 4.3 % C eutectic ($T_E$ =1147°C). It has also been observed that carbon filaments grown from acetylene on carbon substrates at 850°C with nickel catalysts are non-graphitic ($T_E$ = 1300°C) but those grown on silicon are graphitic ($T_E$ for Ni/Si ~960°C) [Bowers et al. 1988]. Since the decomposition of acetylene is exothermic, a 100°C temperature rise due to the chemical reaction is possible and the catalytic particle may be in a molten state. It is emphasized, however, that lower temperature growth must be associated with a solid catalyst particle. It is to be noted that not all dehydrogenation reactions are exothermic (e.g., methane), so that a temperature rise of the catalyst particle cannot be essential to growth, but may modify it.

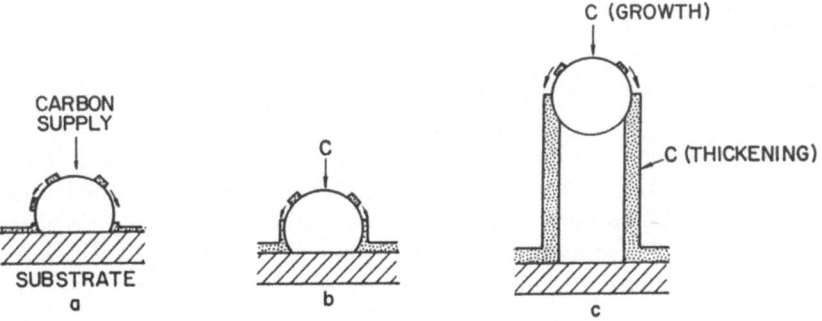

**Fig. 2.7.** Schematic illustration of the growth mechanism proposed for carbon fibers from benzene through a catalytic effect. [From Oberlin et al. 1976a]

A series of papers [de Bokx et al. 1985; Kock et al. 1985; Boellard et al. 1985] has been published recently by workers at Utrecht. Using the technique of magnetization and temperature-programmed hydrogenation, they concluded that high carbon content carbides are a prerequisite for growth. There are many conflicting reports concerning the composition of the catalyst particle, which are reviewed by these workers. In the case of a Fe catalyst for example, a hexagonal form with composition $Fe_{2.2}C$, or $Fe_2C$ was postulated, rather than $Fe_3C$ identified after cool-down by Oberlin et al. [1976b]. The induction period associated with growth was then related to the formation of the carbide. Such a carbide would be compatible with the rapid diffusion of carbon required to explain the observed growth rates, and would be driven by a concentration gradient.

Audier et al. [1980, 1981] carried out a study to ascertain the structural relationships between the catalyst particle and the growing filament. In the case of the Fe catalyst, it was found that the alloy was oriented with its (100) axis along the filament axis, but the (110) axis was found for the case of fcc alloys such as Fe/Co or Fe/Ni. The preference of the carbon to grow on certain faces was proposed as a reason that filaments are hollow.

The above discussion illustrates the lack of information available on the state of the catalyst particle. It is therefore important to regard the state of models for the growth process as tentative. For example, it is not possible to specify the diffusion enthalpy in the catalyst particle until its state (i.e., liquid or solid), composition (i.e., metal or carbide), or structure (if solid) is known. Diffusion parameters can also depend sensitively on the dimension of small particles, being modified by surface effects. It is possible that the similarity of kinetic and growth enthalpies is only fortuitous.

A model for the formation of cylindrical layers was put forward by Tibbetts [1984]. According to this model for the minimum diameter of the central filament core, the graphitic planes are elastically strained in forming a tree-ring structure. This gives rise to an extra elastic term in the free energy equation. If $\delta G$ is the change in Gibbs free energy when a length $dl$ of filament with inner radius $r_i$ and outer radius $r_o$ is precipitated, then

$$\delta G = 2\pi(r_o + r_i)\sigma\, dl + \frac{1}{12}\pi E_Y \tilde{c}^2 \ln(\frac{r_o}{r_i})dl - \frac{\Delta\mu_0 dv}{\Omega} \quad , \qquad (2.1)$$

where $\sigma$ is the energy required to form a unit area of basal surface, $E_Y$ is Young's modulus, $\Delta\mu_0$ is the chemical energy change when a carbon atom precipitates from the catalyst particle, $\tilde{c}$ is the interlayer spacing, $dv$ is the volume change and $\Omega$ the atomic volume of graphite. The change in chemical energy during the reaction $\Delta\mu$ may be written as

$$\Delta\mu = \Delta\mu_0 - \frac{2\sigma\Omega}{(r_o - r_i)} - \frac{E_Y \tilde{c}^2 \Omega}{12(r_o^2 - r_i^2)}\ln(\frac{r_o}{r_i}). \qquad (2.2)$$

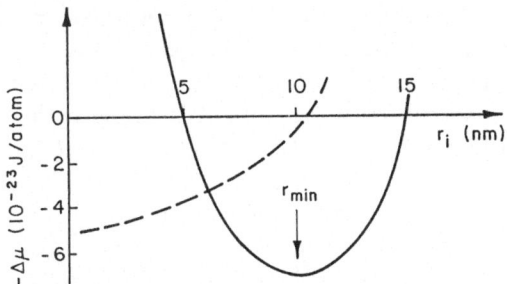

Fig. 2.8. Plot of (2.2) with $r_o = 20$ nm, $E_Y = 10^{10}$ GPa, $\sigma = 7.7$ mJ/m$^2$, $\tilde{c} = 0.335$ nm. In the evaluation $\Delta\mu_0$ is arbitrarily taken as $40 \times 10^{-23}$ J/atom. The dashed line is a hypothetical metastable phase of carbon without a strain energy term [Tibbets 1984]

Assuming reasonable values for $E_Y, \tilde{c}, \sigma, \Omega, \Delta\mu_0$, Tibbetts showed that the free energy change would preclude formation of carbon filaments with $r_i < 5$ nm and he obtained good agreement between experimental values for $r_i$ and $r_o$ and those predicted by his model.

Experiments suggest that $r_o$ is determined by the catalyst particle size (Sect. 3.7.3) in the early stages of growth. Assuming typical values for the parameters, Tibbetts's three-dimensional diagram simplifies to that depicted in Fig. 2.8 for $-\Delta\mu$ vs $r_i$. In this diagram the sign of $\Delta\mu$ is the negative of that in (2.2) in accord with normal chemical thermodynamic notation, where a negative change of free energy is needed for a reaction to proceed. The upward curvature of the dashed line at $r_i > 0$ corresponds to the strain energy term. A broad minimum for $-\Delta\mu$ occurs at $r_i \approx 10$ nm corresponding to the most probable value for $r_i$. It is to be noted, however, that a metastable phase of lower density could form in the core region, with a radius-independent free energy term (no strain energy) as indicated in Fig. 2.8. This may explain the origin of non-graphitic inner cores which oxidize easily (see Fig. 3.37).

Yet another model for the hollow tube has been proposed by Boellard et al. [1985], who also studied the structure of the filament near the tip. Cleavage planes near the tips of the filaments suggested that the basal planes were at an angle to the filament axis. They observed that the Ni catalyst particle was aligned with its (11$\bar{2}$) axis parallel to the filament axis, in contrast to

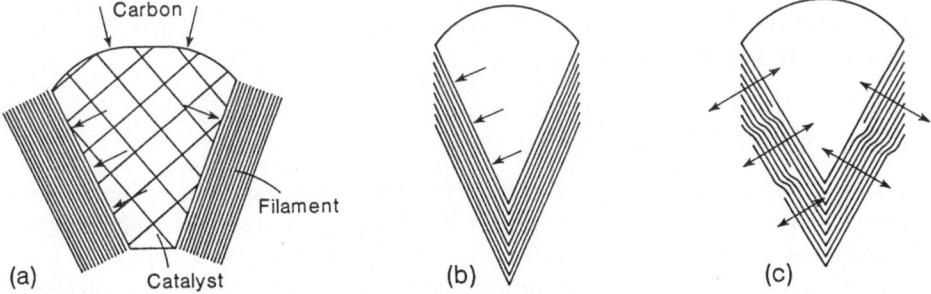

Fig. 2.9a-c. Schematic illustration of the growth mechanism proposed by Boellard et al. [1985]; (a) cross-section, (b) side-view, and (c) illustration of dislocation formation due to non-uniform growth

the general rule above. The electron diffraction results were interpreted in terms of a model in which the carbon was excreted from the catalyst so as to form planes parallel to the catalyst surface as shown in Figs. 2.9a and b. The graphitic cones were then postulated to slip over each other. If the growth rate were non-uniform, then edge dislocations (Sect. 3.1.4) would be formed, as in Fig. 2.9c.

The above description of the growth mechanism applies to many of the cases studied, but is not universal. A related mechanism was proposed for carbon filament growth from the Fe-Pt/$C_2H_2$ system [R.T.K. Baker and Waite 1975] and from the pyrolysis of natural gas in stainless steel tubes [Tibbetts 1983]. In this case growth occurs through the rapid movement of carbon through the catalyst by a diffusion process to form a filament (Fig. 2.10). In this process the carbon filaments grow from their bases rather than from their tips. This conclusion is based on the observation of the growth of irregular filaments with adhering carbonaceous particles [Tibbetts 1983].

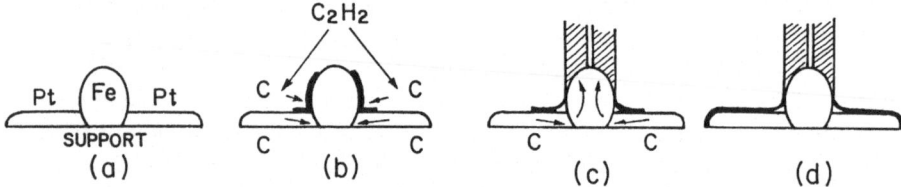

**Fig. 2.10.** Schematic representation proposed for carbon filament growth from the Pt-Fe/$C_2H_2$ system. [From R.T.K. Baker and Waite 1975]

A survey of the literature on carbon filament formation shows that there are many contradictory results, as discussed by R.T.K. Baker and Harris [1978]. These authors pointed out that factors such as the purity of the gas can be very important in determining whether or not filaments grow. Also caution should be exercised in interpreting results obtained after the reaction vessel has been cooled and atmospheric gases allowed to interact with the filaments. Finally, the diversity of phenomena observed in the growth of carbon fibers should lead to caution in attempting to generalize results obtained under fairly specific conditions.

### 2.2.4 Growth Conditions

Many of the earlier studies of carbon filaments were made in order to understand the growth conditions for CCVD filaments so that filament growth could be avoided in fuel and nuclear industry applications. It was later shown that relatively long (e.g., ~250 mm) filaments could be prepared by thermal decomposition of benzene between ~1150° and 1300°C [Koyama and Onuma 1963; Koyama 1972; Koyama et al. 1972]. The reproducible synthesis of long fibers [Koyama and Endo 1982; Endo et al. 1976] by controlled catalytic pyrolysis of benzene led to many structure/properties studies as well as to developments of technological interest. This section discusses growth conditions for these long filaments with diameters of $\sim 10\,\mu$m. Since many of the recent studies of the synthesis of CCVD filaments have been based on the decomposition of benzene, this precursor hydrocarbon will be emphasized in the present discussion. An apparatus for growing CCVD filaments is sketched in Fig. 2.11.

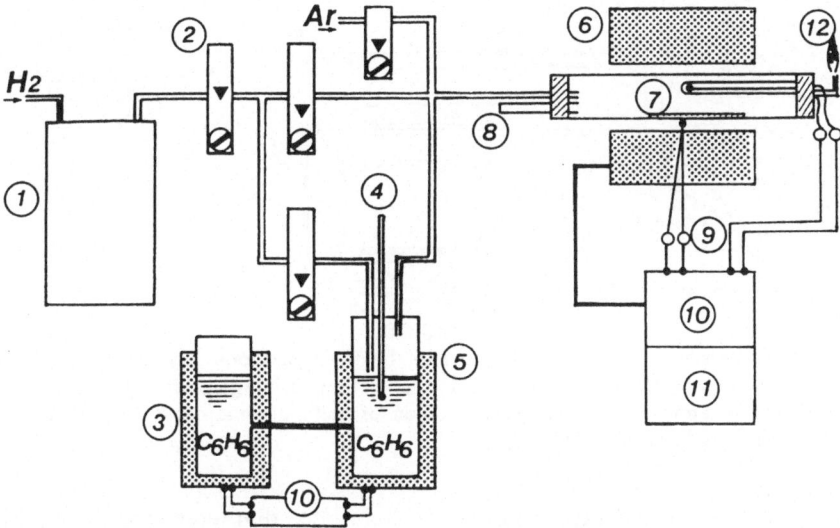

Fig. 2.11. Schematic diagram of apparatus for preparing carbon fibers by thermal decomposition of hydrocarbons [Endo 1975]: (1) hydrogen purification unit, (2) flow meter, (3) benzene reservoir, (4) thermometer, (5) thermostatic bath, (6) electrical furnace, (7) substrate, (8) peephole, (9) thermocouples, (10) temperature controller, (11) temperature recorder and (12) gas exhaust

The growth rate of filaments is very sensitive to the precise reaction conditions such as the partial pressure of vaporized benzene, the purity and flow rate of the hydrogen gas, the residence time for thermal decomposition and the temperature of the furnace (see Fig. 2.11). The partial benzene pressure in the benzene and hydrogen mixture is conveniently controlled by the temperature of the liquid benzene (hydrocarbon). Typical fiber growth conditions are $\sim 300$–$500$ cc/min for the gas flow rate, and $\sim 47°$–$57°$C for the benzene temperature in the apparatus. Under these growth conditions, large densities of fibers with diameters ranging from 3 to 140 $\mu$m and lengths up to $\sim 25$ cm are obtained on the substrate. The yield of carbon fibers is enhanced by roughening the substrate surface prior to the decomposition reaction.

Catalytic particles can be introduced onto a substrate by evaporating Fe, Ni or other metal catalysts. When heated in a hydrogen atmosphere, the

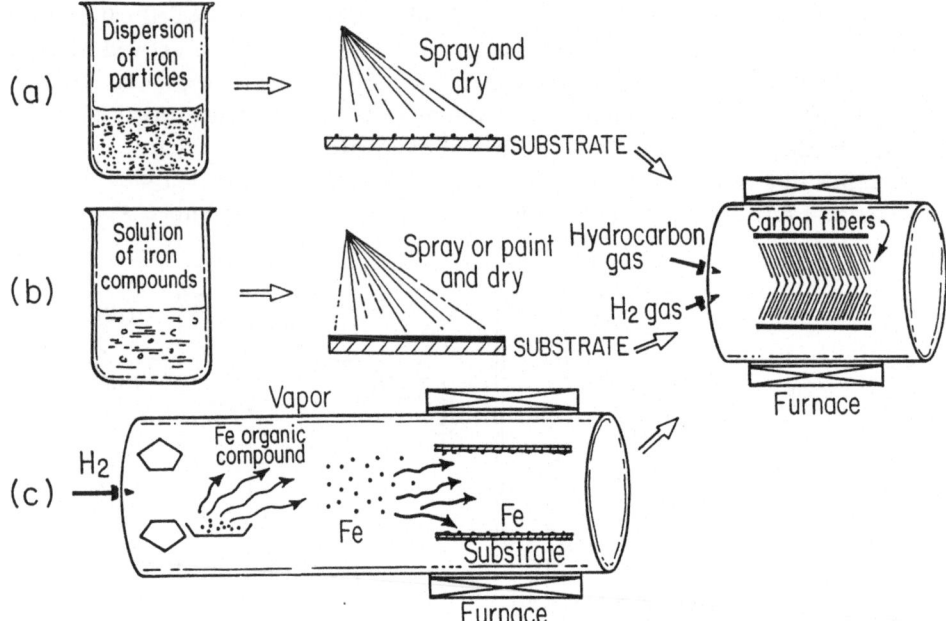

**Fig. 2.12.** Dispersion of iron catalyst particles by three methods: (a) spray and dry a suspension of iron particles, (b) thermal decomposition of inorganic iron compound on heated substrate and (c) thermal decomposition of metallorganic compound on heated substrate [Endo and Shikata 1985]

metal film is converted into small spherules whose diameter can be roughly controlled by the preparation parameters. Three other approaches for introducing the catalytic particles [Endo and Shikata 1985; Endo and Ueno 1984] are indicated in Fig. 2.12: (a) Superfine catalyst particles of Fe ($\leq$10 nm diameter) are evaporated from a suspension (e.g., containing alcohol) which is sprayed and dried over the substrate; (b) solution of an inorganic compound (such as iron nitrate) is sprayed onto a hot substrate ($\sim$1000°C) and the thermal decomposition of the compound leaves behind a residue of superfine particles on the substrate; (c) an organometallic solution such as ferrocene is sprayed with benzene or another hydrocarbon and the ferrocene is decomposed on the hot substrate, leaving behind ultrafine iron particles on the substrate. In one variation of method (c) the seeding and fiber formation process can be integrated in the spray so that fiber formation can occur over a volume rather than a surface, thus increasing the speed and efficiency of the growth process. By dispersing the iron particles without aggregation, the number density of the nucleated fibers can be increased significantly. The importance of using ultrafine catalytic particles is demonstrated in Fig. 2.13, where it is shown that the relative fiber yield decreases as the catalytic particle diameter increases. Figure 2.13 shows that use of a broader distribution

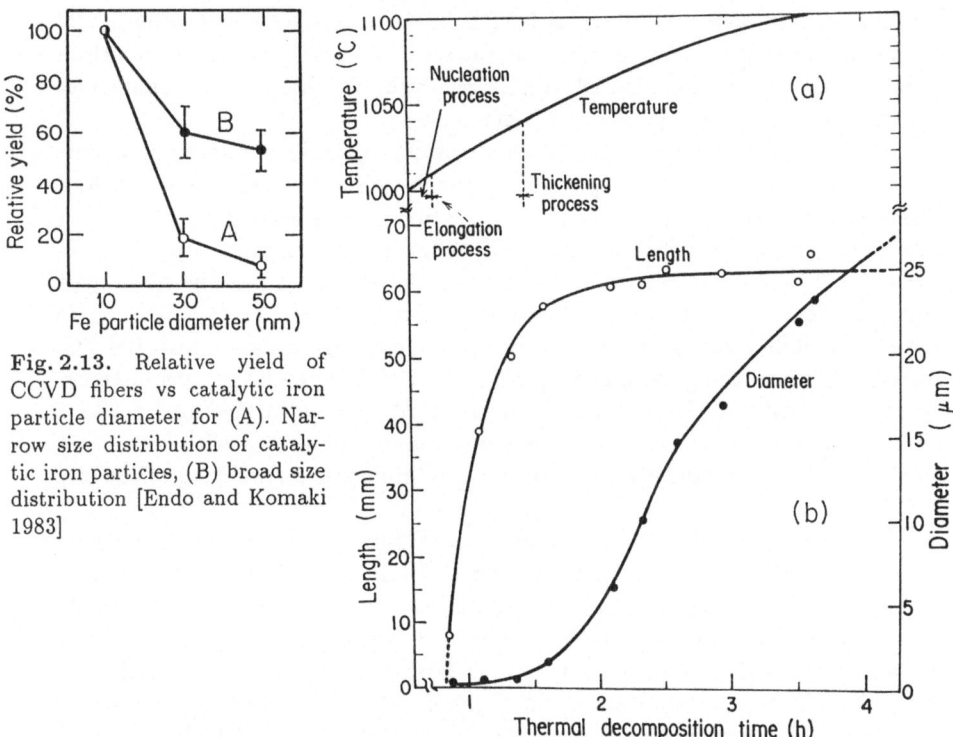

**Fig. 2.13.** Relative yield of CCVD fibers vs catalytic iron particle diameter for (A). Narrow size distribution of catalytic iron particles, (B) broad size distribution [Endo and Komaki 1983]

**Fig. 2.14.** Growth curves of carbon fibers from benzene by a CCVD process: (a) substrate temperature vs growth time showing the temperature regimes for the nucleation, elongation, and thickening processes, (b) fiber length and thickness vs growth time corresponding to the temperature-time schedule shown in (a). [From Endo and Koyama 1977]

of particle sizes enhances the yield, presumably because of the higher concentration of the smaller diameter particles in the broader distribution [Endo and Ueno 1984; Endo and Shikata 1985].

In the fiber growth process, the fiber nucleation and elongation is initiated at ∼1000°C (see Fig. 2.14), while the fibers are subsequently thickened at the somewhat higher decomposition temperatures of the hydrocarbons. The increases in fiber length and diameter associated with the lengthening and thickening processes are shown in Fig. 2.14 corresponding to the particular temperature schedule shown in the figure.

Tibbetts [1985] studied the statistical distribution of CCVD filament lengths with time for growth at 1045°C from natural gas. They found that all length histograms could be fit to an expression

$$n(l) = N \exp(-2.18 \, l) \quad , \tag{2.3}$$

where $N$ is the number of filaments growing at zero time, $n(l)$ is the number growing within each millimeter of length, and $l$ is the length in millimeters of the filaments. This distribution was shown to be consistent with growth in which filaments lengthen linearly with time, and for which the probability of growth cessation is independent of time.

High resolution electron-microscopy results strongly support the occurrence of a two-stage growth process of the benzene-derived fibers (BDFs). The first stage is responsible for the formation of the inner core containing long, straight and parallel carbon layers cylindrically oriented around a hollow tube. The second phase corresponds to the thickening of the fiber by a pyrolytic deposit on the core. The growth curve (see Fig. 2.14) exhibits two distinct regions with a transition between these regions at about 1040°C, corresponding to the growth in length ($T \leq 1040°$C) and the growth in thickness ($T \geq 1040°$C). The growth rate in length was estimated to be about 1 mm/min, and the thickness of fibers was found to increase with increasing partial pressure of benzene and substrate temperature [Endo and Koyama 1977].

For the mass production of low cost CCVD fibers, a volume seeding (floating catalyst) technique has been developed [Endo and Ueno 1984; Endo and Shikata 1985] as shown in Fig. 2.15. Ultrafine metal catalyst particles (5–25 nm diameter) are introduced into the reactor either directly (Fig. 2.15a) or indirectly through a carrier fluid such as ferrocene (Fig. 2.15b), and the ensuing fiber growth takes place in the three-dimensional space of the reaction chamber at the same basic temperature (1100°C) as is used with the substrate method shown in Fig 2.12. Dry nitrogen gas is used for purging the furnace

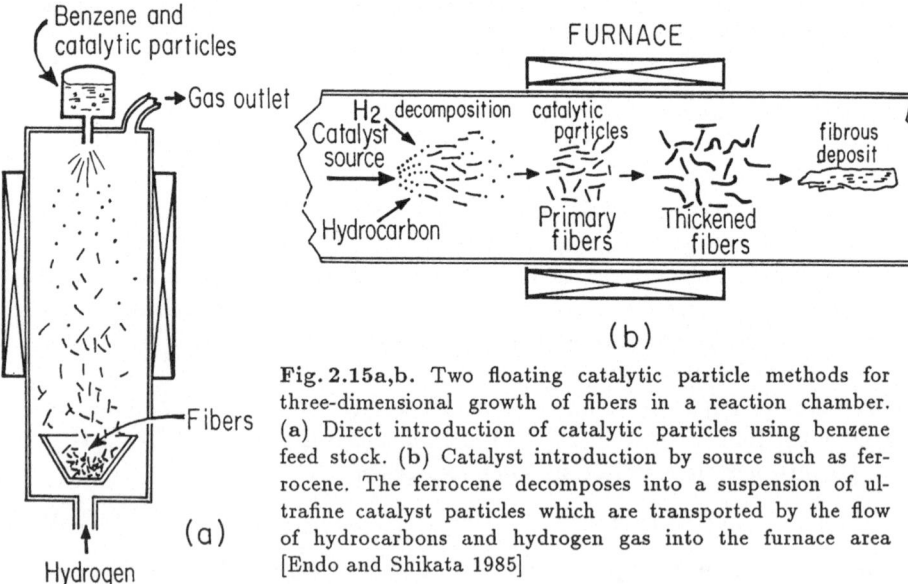

Fig. 2.15a,b. Two floating catalytic particle methods for three-dimensional growth of fibers in a reaction chamber. (a) Direct introduction of catalytic particles using benzene feed stock. (b) Catalyst introduction by source such as ferrocene. The ferrocene decomposes into a suspension of ultrafine catalyst particles which are transported by the flow of hydrocarbons and hydrogen gas into the furnace area [Endo and Shikata 1985]

until it reaches 1000°C. After the furnace reaches 1100°C, the reactants are introduced into the furnace either by transferring the grafoil boat containing the ferrocene (catalyst source) or by flow of the interacting ferrocene vapor, as shown in Fig. 2.15b. The growth mechanism is believed to be similar for the substrate and floating catalyst methods [Endo and Shikata 1985].

The catalyst/hydrocarbon feed ratio determines the fiber diameter and aspect ratio of the CCVD filaments. Fiber formation is enhanced as the catalyst particle size decreases (see Fig. 2.13). Since fiber growth ceases when the particle surface becomes covered either with carbon layers, oxygen or other impurities, high chemical purity is essential for the growth of long filaments. By the substrate method, a growth density of several hundred fibers/mm$^2$ can be realized.

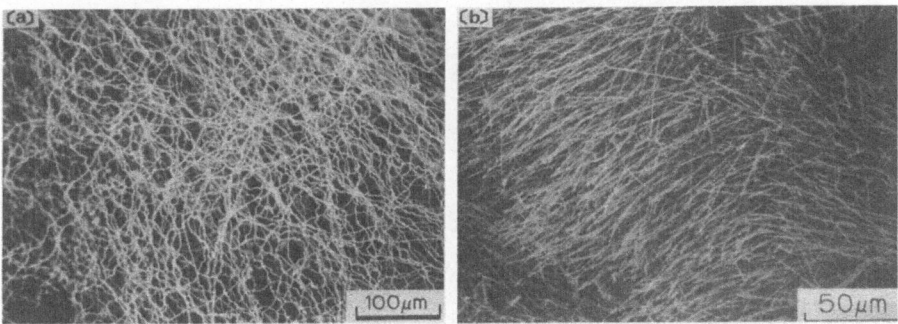

Fig. 2.16a,b. SEM pictures showing typical aggregations of benzene-derived CCVD carbon fibers prepared by the floating seed method; (a) meandering and (b) straight fibers [Endo et al. 1987a]

When the volume seeding method of Fig. 2.15 is operated as a continuous CCVD process, ultrathin short filaments are produced [Endo et al. 1987a]. Because of the very large increase in fiber yield, this process may be of industrial interest as a potential technique for making carbon fibers at low cost. For the continuous CCVD process the fibers collect in a sponge-like deposit (see Fig. 2.16), consisting of vermicular carbon fibers with diameters between 0.1 and 1.5 $\mu$m and typical lengths of about 1 mm. At the tip of the fibers small catalytic particles are observed with diameters of about 5 nm, encapsulated in a carbon deposit of vapor-grown carbon. Along the length of the fibers is a characteristic central hollow tube. The floating time of the particles coming from the pyrolysis of ferrocene is ~1–2 minutes, which is also an estimate of the fiber growth time. The reaction yield for the floating particle method is high (~80% of the benzene supply is converted into fibers) [Endo and Shikata 1985].

Very little difference is observed in the morphology of the CCVD fibers prepared by the substrate and floating catalyst methods. In both cases the

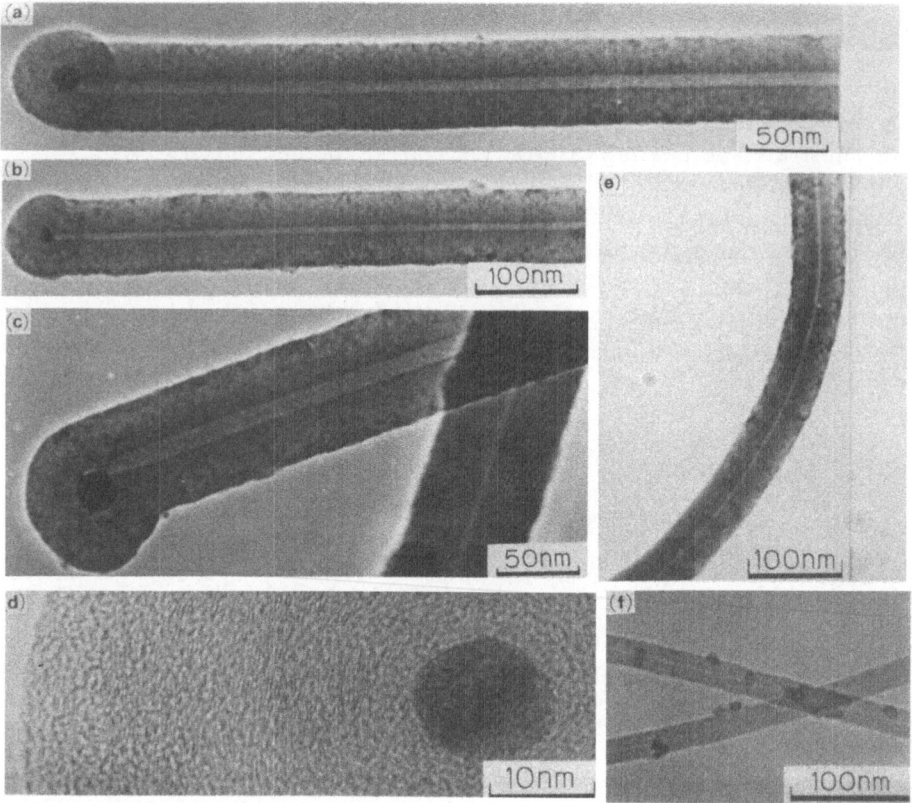

Fig. 2.17a-f. TEM pictures of the carbon fibers. (a-c) The tip of the fibers and hollow tube, (d) catalytic particle, (e) curved fiber and (f) contaminated fibers with catalytic seeds [Endo et al. 1987a]

fibers prepared in a given batch have uniform diameters and little variation in fiber diameter between different fibers (Fig. 2.16). Also for fibers grown by the floating seed method, the fiber diameter and length can be controlled by the growth conditions (Fig. 2.14). High resolution TEM studies (Fig. 2.17) show that the diameter of the hollow tube of the ultrathin fibers is slightly less than the diameter of the catalytic particle. The hollow tube is remarkably straight and of constant diameter along the fiber length (Fig. 2.17a-c). After the thickening process, the fibers may be either straight or curved (see Fig. 2.17e) if the pyrolytic deposition is inhomogeneous [Endo et al. 1987a]. The TEM micrographs show dense regions at the fiber tips, identified by these authors as a cementite ($Fe_3C$) particle probably formed during cooling of the furnace when the growth of fibers had ceased. The contamination of fibers with catalytic seeds resulting from too high a catalyst concentration interferes with normal fiber growth as shown in Fig. 2.17f [Endo et al. 1987a].

## 2.3 Carbon-Coated Filaments

CCVD filaments have superior structural characteristics relative to polymer-based filaments (Sect. 2.1). However, their main disadvantages are that the CCVD filaments can only be prepared in discrete lengths, whereas polymer-based filaments are continuous. The possibility of combining the advantageous properties of both types has been explored by Matsumura et al. [1985], who deposited carbon onto polymer-based filaments by the pyrolysis of benzene, cyanoacetylene and by the plasma-assisted CVD method [Shioya et al. 1987]; this major advance will be discussed further in Chap. 12.

Commercial polymer-based fibers of initial diameter $\sim 7\,\mu$m were thickened to $\sim 30\,\mu$m at 1200°C. Preparations were carried out as a function of monomer pressure. Using the electrical conductivity $\sigma$ as a tool to estimate structural perfection (see Sect. 8.3), a pressure of $\sim 100$ Pa ($\sim 1$ mm Hg) was found to maximize the conductivity for both benzene and cyanoacetylene $[\sigma_{max}(300\,\mathrm{K}) \sim 1.6 \times 10^5 (\Omega\mathrm{m})^{-1}]$, corresponding to a relatively high degree of structural perfection. Heat treatment to 3000°C (30 min residence time) produced a conductivity of $1.1 \times 10^6 (\Omega\mathrm{m})^{-1}$, almost comparable to that of single crystal graphite $[\sigma_{graphite}(300\,\mathrm{K}) = 2.5 \times 10^6 (\Omega\mathrm{m})^{-1}]$, see Sect. 8.3. Furthermore, Raman spectra, x-ray diffraction, magnetoresistance, and conductivity measurements of these fibers after intercalation were consistent with a high degree of crystalline perfection.

It is possible that these coated filaments may have applications, if the cost of coating filaments is not prohibitive, and if continuous filaments can be coated to diameters such that the mass is increased several-fold. In addition, relatively high structural perfection is achieved without (expensive) heat treatment to greater than 2000°C (Chap. 12).

## 2.4 Filaments Prepared from Carbon Arcs

Graphite filaments may be grown from carbon arcs under a high pressure of inert gas [R. Bacon 1958, 1960]. Specifically, a positively charged upper electrode strikes an arc against a copper block. Carbon vapor from the electrode deposits on the block, building up a deposit underneath it, of diameter slightly larger than the electrode. The maintainence of the arc necessitates the use of an electrode feed mechanism. The deposit grows like a stalagmite at the rate of $\sim 10$ mm/min. Growth only occurs at high temperature and over a pressure range of a few bars, the optimum being about 92 bars. This is close to the triple point ($p \sim 100$ bar and $T \sim 4500$ K) of carbon [Bundy et al. 1973]. Larger needles were grown at higher pressures, possibly from the liquid phase [R. Bacon 1960].

Graphite filaments, called whiskers by R. Bacon [1960], grow inside the deposit, and can be removed by breaking the deposit open, thereby revealing a forest of filaments protruding up to 30 mm from the exposed surface. Their diameters generally range from 1 to 5 $\mu$m. Electrical resistivities at room temperature, and their temperature dependence (reviewed in Sect. 8.3) indicate structural perfection close to that of single crystal graphite, while the sides show a high reflectivity, also characteristic of basal planes of single crystal graphite, and appearing like Christmas tinsel. Structural analysis of the graphite whiskers confirmed that the layer planes were in the form of a scroll (see Fig. 1.6), although prismatic facetting may be present. Sears [1959] has proposed a growth model. Very high breaking strengths (up to 20 GPa) were reported. This very high value has been important in guiding subsequent work aimed at producing very high strength ex-polymer fibers, discussed in Sect. 6.2.

## 2.5 Synthesis by Ion Bombardment

When metals are bombarded by energetic ions (e.g., 100 eV Ar$^+$ ions) the etching process may not be homogeneous, forming sputter cones, particularly if the surface is seeded with impurities. Similar bombardment of carbon surfaces leads to unique *growth* features, which are filament-tipped cones (Fig. 2.18) [Cuomo and Harper 1977]. The diameters of the filaments are typically ~200–2000 Å, and lengths are in the range ~1–10 $\mu$m after bombardment to doses of ~$10^{23}$–$10^{24}$ ions/m$^2$ (at 500–1000 eV). Such growth rates are two orders of magnitude higher than the sputter-etch rate.

The main phenomena which characterize such filamentous growth can be summarized as follows [Floro et al. 1983; Solberg 1986; Solberg et al. 1987; Van Vechten et al. 1987]:

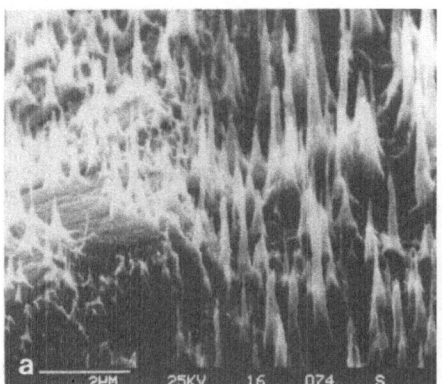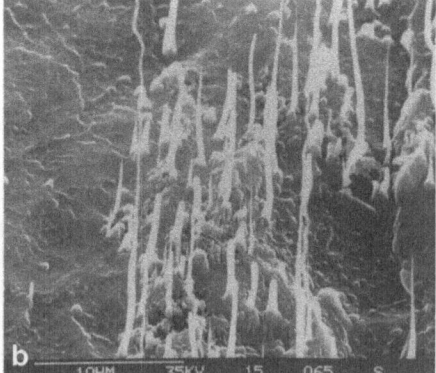

**Fig. 2.18a,b.** Carbon filaments grown by bombarding a pyrolytic carbon surface with 700 eV Ar$^+$ ions, to a dose of $4.5 \times 10^{23}$m$^{-2}$ for the carbon surface at (a) 275°C and (b) 100°C [Solberg et al. 1987]

1. Filaments may be grown on several types of hexagonal carbon surfaces, but not on diamond, nor glassy carbon.
2. Filament growth is not dependent on impurity seeding. Indeed, a metallic impurity flux $\geq 1\%$ of the primary bombardment flux is sufficient to suppress growth completely.
3. The direction of filamentous growth is parallel to the ion beam. Accordingly, if the beam is tilted with respect to the normal to the surface, the filament growth direction changes accordingly, even though the electric field from the plasma to the surface remains parallel to the normal.
4. After an initiation period, corresponding to a dose of $\sim 0.5 \times 10^{23}$ ions/m$^2$, an initial growth spurt is observed [(0.5–2) $\times 10^{23}$ ions/m$^2$] followed by slower growth in which the filament both elongates and thickens.
5. If the substrate temperature is controlled during bombardment, no growth is observed below $\sim 270$ K. Growth with increasing density is found in the temperature range $270 \leq T \leq 600$ K, with a concentration plateau ($\sim 10^{12}$ filaments/m$^2$) above that temperature. A transition in shape occurs at $T_c \sim 500$ K, with more cylindrical structures below $T_c$ (Fig. 2.18a) and more conical structures above $T_c$, some structures having a small filament at the tip (Fig. 2.18b).
6. The surface filament density increases, and the volume per filament decreases with increasing ion energy.
7. No effect of ion current density on growth could be detected for the limited ion current density range 2–10A/m$^2$.

The growth of these filaments clearly requires an initial nucleation and a flux of atoms to the growing cones and filaments. Van Vechten et al. [1987] have conjectured that the carbon atoms become highly mobile on the bombarded surface, and that nucleation sites can become "supersaturated" [Van Vechten 1985]. This requires a critical ion density to supply carbon atoms at a sufficient rate so that an initial growth feature is formed, and that this feature continues to grow. A screw dislocation was proposed as a typical nucleation site, persisting through the cone of the filament so that further growth is preferred at the top, rather than the sides of the filament. This model requires mobile carbon atoms to diffuse from the bombarded surface to the tip along the length of the cone and filament, and this length can be considerable (up to $\sim 20\,\mu$m).

This model appears to be in conflict with item 7 above, while the temperature dependence of the surface filament density, listed as item 5, needs explanation. It would be anticipated that the density of screw dislocations ($\sim 10^{10}$–$10^{12}$/m$^2$) for pyrolytic graphite [Hennig 1952, 1965] would, for example, be independent of temperature. Quite clearly, the temperature dependence of the surface density of filaments is at variance also with a model based on enhanced thermal diffusion.

A possible adaptation of the supersaturation model [Van Vechten 1985] has recently been proposed [Van Vechten et al. 1987; Solberg 1986]. Firstly it is possible that impurities poison nucleation sites at lower temperatures. A possible impurity candidate for this is hydrogen, which cannot be removed from chambers typically used for ion processes. It has been shown that the hydrogenation of carbon on graphite surfaces is enhanced under electron bombardment [Brice and Ashby 1984] and that the reaction rate increases above 230°C. This is about the same temperature at which a substantial rise in filament density was observed, and is possibly due to the increased availability of hydrogen-free sites above this temperature.

Secondly, a calculation of the arrival rate of sputtered carbon atoms at the conical bases suggested that this was the mechanism by which growth occurred after the establishment of surface features. Accordingly, growth occurs in two phases – a nucleation stage in which carbon atoms arrive at the preferred site, supersaturate it, and form a conical growth feature, followed by a second stage in which growth of the cone and of the filament occurs with sputtered carbon serving as the source. This second stage still requires carbon atoms to be mobile on the filament surfaces, but diffusion lengths are greatly reduced over those required if the bombarded surface were the sole source of carbon atoms. A key experiment would be to nucleate conical features at 300°C, for example, then cool below −30°C and ion bombard to ascertain whether or not growth continues via the sputter mechanism.

The above models are clearly at a preliminary stage, and much needs to be done to establish their veracity, even on a qualitative basis.

# 3. Structure

The main types of carbon filaments have been introduced in the previous chapter, and a brief description has been given of their principal structural features. This chapter considers the main types of measurements that have been carried out to elucidate their structure. In all cases, unless otherwise stated, it is assumed that the basic structural unit is the hexagonal plane of carbon atoms in the honeycomb arrangement. In highly perfect filaments, these planes are stacked in a regular $ABAB$ (graphitic) sequence, while in less perfect filaments, the stacking is random (turbostratic). For disordered filaments, the planes are bent, or twisted, and defects such as vacancies, interstitials, dislocations, grain boundaries, voids, impurities, etc., are present. The purpose of structural information is, as far as possible, to characterize the basic structural unit and the defects which are present. This chapter will therefore attempt to show in each case what structural information can be achieved by each of the characterization techniques.

This chapter is organized as follows: Firstly, the crystal structure of graphite and related defects are introduced, followed by consideration of the types of defects that are expected in filaments. Then the different types of structural measurements that have been carried out on filaments are presented.

The structure of fibers has been reviewed by several authors: ex–rayon fibers by R. Bacon [1973], ex-PAN by Reynolds [1973] and D.J. Johnson [1987a, 1987b], and ex-pitch fibers by R. Bacon [1980]. Kelly [1981] has given a detailed overview of the structure of graphite, and of the types of defects. Again, the purpose of the present section is to outline the important developments rather than give an exhaustive survey of the references.

## 3.1 Graphite and Its Defect Structures

### 3.1.1 Structure of Ideal Graphite

The crystal structure of single crystal graphite is hexagonal, in which planes of honeycomb carbon hexagons are stacked in an $ABAB\ldots$ sequence (Fig. 3.1) with $P6_3/mmc$ symmetry. Accepted values of the lattice constants [Kelly 1981] are $a_0 = 0.2462$ nm and $c_0 = 0.6707$ nm at room temperature, so that the in-plane bond length is 0.1421 nm and the interplanar separation, $\tilde{c}_0$, is 0.3354 nm.

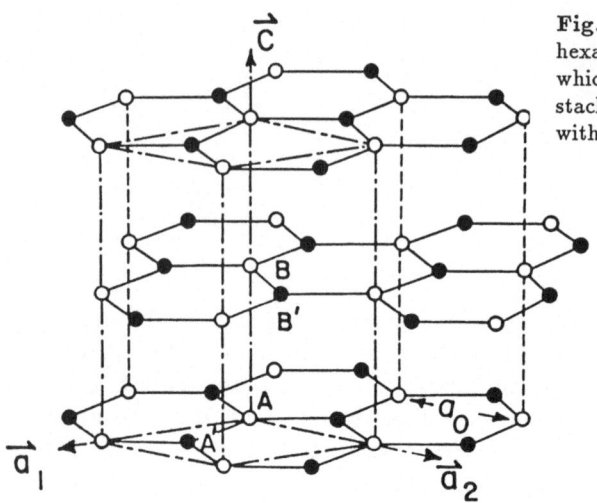

Fig. 3.1. The crystal structure of hexagonal single crystal graphite, in which planes of carbon hexagons are stacked in an $ABAB...$ sequence with $P6_3/mmc$ symmetry

An alternative stacking sequence ($ABCABC...$) can occur for carbon, resulting in a rhombohedral unit cell (space group R$\bar{3}$m). This rhombohedral form of graphite always occurs together with the hexagonal one, and can be induced by grinding single crystal graphite into a fine powder [Lipson and Stokes 1942]. Basal dislocations in graphite split into two partials with this $ABC$ stacking sequence. The importance of this phase is its relationship to disordered structures, considered in Sect. 3.1.7.

### 3.1.2 Basic Scattering Concepts

Diffraction provides the dominant technique for structural characterization, both for the ideal structure and for the various types of defects of importance to the structural characterization of carbon fibers. Considering diffraction from a monochromatic x-ray or electron beam of wavelength $\lambda$, and wave vector $k_0 = 2\pi \hat{k}_0/\lambda$, then diffraction occurs when the scattered wave vector $k_s = 2\pi \hat{k}_s/\lambda$ obeys the relationship

$$\Delta k \equiv (k_s - k_0) = G \quad , \tag{3.1}$$

where $G$ is any reciprocal lattice vector of the crystal, $\hat{k}_0$ and $\hat{k}_s$ are unit vectors denoting the wave-vector directions and $\Delta k$ is the change in wave vector in the scattering event. This is discussed in many standard texts (e.g., Kittel [1985]) and is very useful for understanding diffraction from carbon fibers, particularly using the Ewald sphere construction. The reciprocal lattice points for a hexagonal system are often labeled with 4 indices ($hkil$) but the relationship $h + k + i = 0$ allows only three to be used ($hkl$), and this practice will be adhered to in the present work, so that, for example, the ($20\bar{2}3$) reflection is labeled (203).

The coherent scattering (diffraction) from a crystal is given by

$$I \propto f(G) \sum_{j=1}^{n} \exp(\mathrm{i}G \cdot r_j) \quad , \tag{3.2}$$

where $I$ is the intensity of diffracted x-rays, $f(G)$ is the atomic scattering factor, or form factor (the Fourier transform of the electronic charge distribution of the carbon atoms) and the sum over the $n$ atoms in the unit cell yields the structure factor, where $r_j$ denotes the vector coordinates of the atoms in the unit cell, so that the structure factor corresponds to the Fourier transform of the real space structure.

If the crystal is infinite in extent, with atoms in perfectly periodic positions, then the Fourier transform consists of a set of points. These points are broadened by the finite number of atoms in the crystal, and by atomic position defects such as thermal vibrations, vacancies, stacking faults, etc. The role of x-ray and electron diffraction characterization is therefore an attempt to describe these defects from measurements of diffraction intensities, linewidths, and line shapes.

It is particularly noted that not all reciprocal lattice vectors in (3.2) result in diffraction intensity. A convenient way of representing the allowed diffraction conditions and line intensities has been introduced by Ruland [1968] and is represented in Fig. 3.2. All reciprocal lattice points have been rotated about the $(00l)$ axis and appear in the $(h0l)$ plane, with the size of the spot representing a cylindrical average of the diffracted intensity. The dashed lines indicate the volume of reciprocal space available to Cu $K_\alpha$ radiation, and the dot-dashed line for Mo $K_\alpha$.

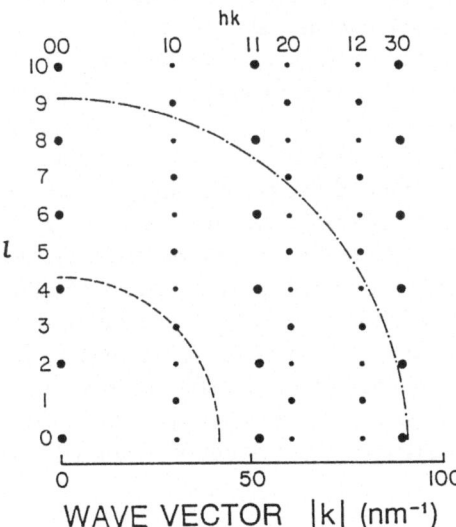

**Fig. 3.2.** Interference peaks in the reciprocal lattice of graphite, projected onto a plane. Sizes of the spots represent intensities (not taking into account the atomic scattering factor). The dashed line is the projection of the Ewald sphere centered on (000) at $|k| = 2\pi/\lambda$ for Cu $K_\alpha$ radiation and the dot-dashed line is for Mo $K_\alpha$ radiation [Ruland 1968]

### 3.1.3 Point Defects

Lattice defects have been mentioned in earlier sections. Only a brief review of the main types of defects will be given here, since Kelly [1981] has already summarized this area for bulk graphite. The interstitial, vacancy, and interstitial-vacancy pair (Frenkel defect) are the basic types of point defects. Reviews have been given by Thrower and Mayer [1978], D.J. Bacon and Nicholson [1975, 1977], and Nicholson and D.J. Bacon [1975]. The formation and migration energies $E_{fj}$ and $E_{mj}$ for vacancies ($j = v$) and interstitials ($j = i$) in graphite have been tabulated by Thrower and Mayer [1978], see Table 3.1. The formation energy is sufficiently large so that equilibrium concentrations are extremely small at normal temperatures. Thus, high metastable concentrations of vacancies and interstitials can be produced by fast-neutron or high-energy electron bombardment, or from processing conditions, such as in the case of fibers. Figure 1.2 illustrates a typical pre-graphitic network in which single, di- and multiple vacancies (vacancy loops) occur.

**Table 3.1.** Energies of formation ($E_{fj}$) and migration ($E_{mj}$) for vacancies and interstitials in graphite [from Thrower and Mayer 1978]

|  | Vacancy [eV] | Interstitial [eV] |
|---|---|---|
| $E_{fj}$ | $7.0 \pm 0.5$ | $7.0 \pm 1.5$ |
| $E_{mj,a}$ (basal direction) | $3.1 \pm 0.2$ | $< 0.1$ |
| $E_{mj,c}$ (c-direction) | $> 5.5$ | $> 5$ |

The energy of formation of a vacancy $E_{fv}$ in graphite has been calculated by Coulson et al. [1963], Coulson and Poole [1968] and D.J. Bacon and Nicholson [1975, 1977]. The calculation is very difficult, since it contains terms involving (a) the breaking of carbon bonds, (b) the elastic energy created by the relaxation of surrounding atoms towards the vacancy, (c) $\sigma$- and $\pi$-electron energy changes due to the modification of wave functions near the vacancy, (d) an estimate of the energy required to rebond the atom to the edge of the crystal and (e) a defect-induced change in the interlayer interaction energy. This latter energy differs for a vacancy on a type $A$ or $B$ site (see Fig. 3.1), but the difference is not significant in terms of the total energy of formation. The basic problem with the calculation of $E_{fv}$ is that the various terms described above differ in sign, and some are larger than $E_{fv}$ itself, so that uncertainties in the calculation are significant. Coulson and Poole [1968] also estimated that divacancy formation from the combination of two vacancies would result in an energy release of $\sim 5$ eV. This indicates that formation of vacancy loops would be strongly favored.

38

Kelly [1981] also reviewed theoretical calculations for interstitials and found the situation far less satisfactory than for vacancies. Interstitials can diffuse readily at room temperature along the basal directions (see Table 3.1) so that interstitial loops are formed in irradiated graphite (Sect. 3.1.4). However, interstitials are not likely to be produced from preparation steps in filaments, while interstitials formed from irradiation of fibers would probably diffuse to vacancy or ribbon-edge sinks, rather than form loops. However, interstitial clusters, or loops, would form readily as atomic clusters join together in the carbonization process.

### 3.1.4 Dislocations

The concept of dislocations in carbon fibers is perhaps best approached by considering the ideal case of graphite, and then applying these concepts to the more defective material. A review of the description and observation of dislocations in graphite has been given by Amelinckx et al. [1965], Heerschap et al. [1964], and Heerschap and Delavignette [1967], while more recent references are included in Kelly's book [1981], classifying such dislocations in terms of the directions of the dislocations $s$ and their Burgers vectors $b$ [Fujita and Izui 1961]:

1. $s$ and $b$ in the basal plane. This is the most common dislocation found in graphite, and is glissile. This ease of motion is responsible for the anomalously low value of $C_{44}$ found at low acoustic frequencies (e.g., $\leq 100\,\mathrm{MHz}$) (see Sect. 4.3.1 for discussion). The dislocations are easy to observe in transmission electron microscopy (TEM) studies since they split into partials separated by ~0.05–0.1 nm, between which is a region of stacking faults ($ABC$, rhombohedral stacking) (Fig. 3.3). The total dislocation has a Burgers vector $b = (a/3)$ $\langle \overline{1}\,\overline{1}\,2\,0 \rangle$ splitting into partials

$$\tfrac{1}{3}a\langle 1\,0\,\overline{1}\,0 \rangle + \tfrac{1}{3}a\langle 1\,\overline{1}\,0\,0 \rangle = \tfrac{1}{3}a\langle 2\,\overline{1}\,\overline{1}\,0 \rangle \quad . \tag{3.3}$$

The formation of the partials arises because the energy required to deform bonds for a single, total dislocation would be prohibitive. The energy can be lowered by spreading the deformation over a ribbon. The partials are said to repel each other. The finite width of the ribbon then arises because the decrease in in-plane bonding energy is balanced by an increased out-of-plane bonding energy (stacking fault energy).

These dislocations are observed in both edge ($s \perp b$) and screw ($s \parallel b$) configurations. The ribbons can split, forming junctions of these ribbons so that a network of intersecting dislocations is set up. Amelinckx et al. [1965] have also discussed other configurations. The graphitization process described in Sect. 3.1.6 can be thought of as the gradual removal of stacking faults, and therefore of these basal dislocations, by annealing.

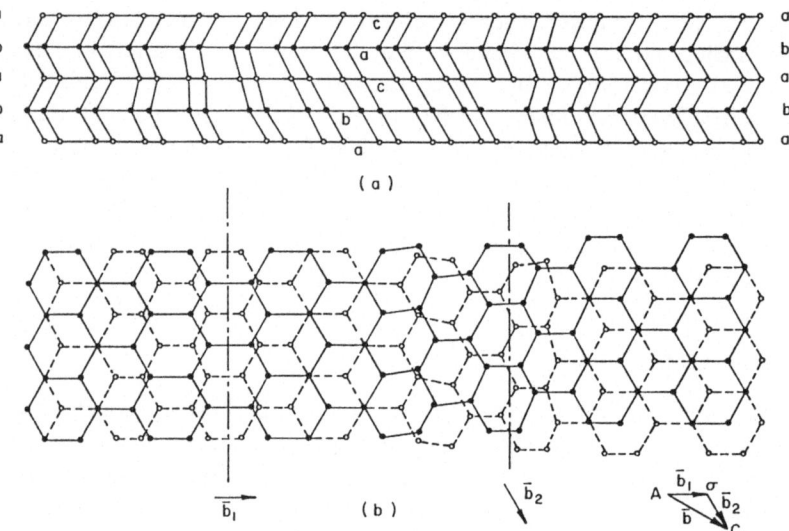

**Fig. 3.3a,b.** Schematic of (a) an extended basal dislocation in graphite, showing (b) the stacking sequence between partials. The partials are separated by ~0.1 μm, and this is much reduced in the figure [Amelinckx et al. 1965]

**2.** *s* in the basal plane, *b* parallel to the *c*-axis. This edge dislocation is associated with the termination of a plane, or the formation of extended interstitial or vacancy loops. The radius of curvature for the vacancy loops is estimated to be >10 nm [Tsuzuku 1961]. Examples of such dislocations can be seen in Figs. 3.4a and 3.4b.

**3.** *s* in the *c*-direction. Both screw ($b \parallel c$) (see Fig. 3.4c) and edge ($b \perp c$) dislocations have been observed with low densities ($\sim 10^7$–$10^9 \mathrm{m}^{-2}$) in single crystal graphite [Thomas and Roscoe 1968]. However, neither are expected to be of importance in filaments, except for dislocations that pass through only one layer [Kelly et al. 1966]; these are discussed below in terms of boundaries.

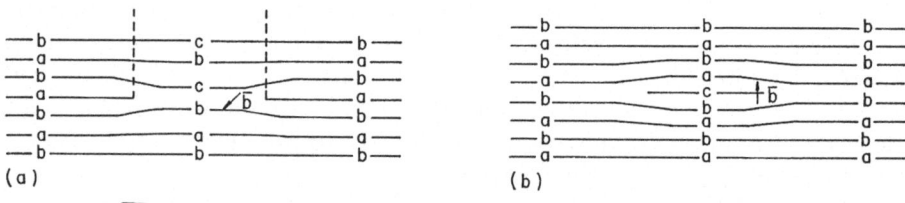

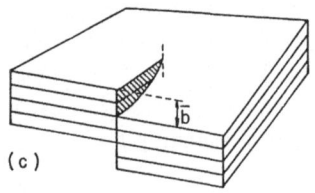

**Fig. 3.4a-c.** Schematic of dislocations with Burgers vectors in the *c*-direction. (a) Edge dislocation in the form of a vacancy loop; (b) edge dislocation in the form of an interstitial loop; (c) screw dislocation with $b = c$ [Amelinckx et al. 1965; Heerschap and Delavignette 1967]

### 3.1.5  Boundaries

Boundaries are discontinuities across which the crystal axis changes. The simplest boundary can be envisioned by holding a pile of papers and creasing them so that half of the pile makes an angle of $2\theta$, say, with respect to the other half. Certain angles are favored in any material, allowing the atomic stacking to be continued correctly on either side of the boundary. This type of discontinuity is referred to as a twin boundary.

Twin boundaries (twins) occur in graphite, and were described in terms of partial dislocations lying in the boundary by Kennedy [1960] and Freise and Kelly [1963]. These latter authors showed that twins could be derived from the parent crystal by a rotation of 20°48′ about a $\langle 1\bar{1}00 \rangle$ direction (Fig. 3.5). The boundary plane is $(11\bar{2}0)$ and each hexagonal plane contains a partial dislocation with Burgers vectors $(a/3)[10\bar{1}0]$ and $(a/3)[01\bar{1}0]$ alternating between the planes. This boundary could be formed in filaments with tree-ring morphology as $T_{\mathrm{HT}}$ increases, enabling the strain energy associated with the radius of curvature to be lowered. The lowest energy configuration would then occur for an 18-sided polygon, with only a slight distortion of the tilt angle.

Another kind of boundary can be formed in which the $c$-axes are not tilted but the $a$-axes on each side of the boundary are tilted with respect to each other [Roscoe and Thomas 1966]. Such boundaries which lie in the basal plane can be constructed from non-basal dislocations, and are illustrated by the examples in Fig. 3.6. Such boundaries, extending over only one plane, are probably extremely important in filaments, allowing planar regions to grow and join, even though their $a$-axes are misaligned. Thus, over a ribbon,

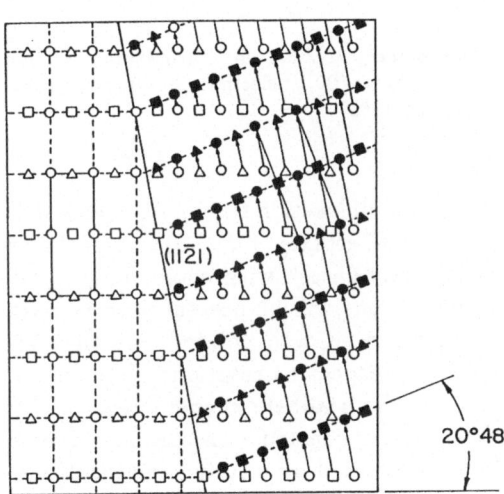

Fig. 3.5. Schematic diagram of a twin boundary in graphite [Freise and Kelly 1963]

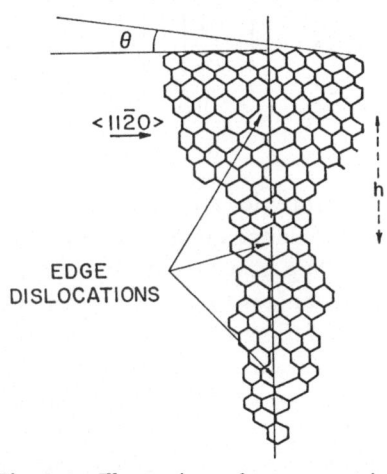

Fig. 3.6. Illustration of a symmetrical tilt boundary where $h$ is the distance between non-basal dislocations [Roscoe and Thomas 1966]

41

all orientations of *a*-axes would occur with respect to rotations about the perpendicular to the ribbon. This is referred to as *a*-axis mosaicity.

### 3.1.6 Turbostratic Graphite

Turbostratic graphite refers to a random stacking of graphene layers, so that in the ideal case a two-dimensional structure with non-interacting layer planes is formed. This leads to a problem of nomenclature, since the lattice cell constant, $c_0 = 0.6708$ nm appropriate to graphite, would be inappropriate to turbostratic carbon. The symbol $\bar{c}$ will therefore be used to denote the interlayer separation, to avoid confusion. It is noted that $\bar{c}$ does not represent a lattice constant, since turbostratic carbon is not stacked in a regular sequence. Thus $\bar{c}$ represents an average value for the interlayer separation, which varies from 0.3354 nm for the ideal graphite structure to more than 0.344 nm for low $T_{HT}$ carbon fibers.

If individual carbon layers were arranged in a completely random manner with respect to rotation about the axis perpendicular to the planes, while remaining parallel to each other (turbostratic carbon), then reciprocal lattice points for this two-dimensional structure would be rods, with indices $(h, k)$ (Fig. 3.7), except for the $(00l)$ reflections, which would remain discrete spots.

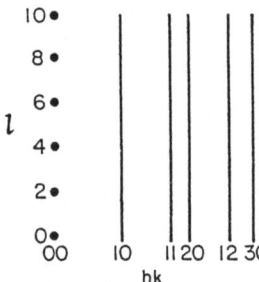

Fig. 3.7. Interference maxima for turbostratic carbon showing sharp spots for $(00l)$ reflections and lines for $(hkl)$ reflections [Ruland 1968]

Consider now the case of x-ray or electron diffraction from a fiber in which the incident beam is at right angles to the fiber axis (Fig. 3.8). As outlined in Chap. 1, a carbon fiber typically has "crystallites" with *c*-axes roughly at right angles to the fiber axis and all orientations will be present with respect to rotation about the fiber axis. In addition, it was pointed out that all orientations of the *a*-axes are present with respect to rotations about these *c*-axes as a result of boundaries that lie in the graphene planes (Sect. 3.1.5). If the fiber is graphitic, then the Ewald sphere construction can be made as follows. The *a*-axis mosaicity would be represented by rotating the reciprocal lattice points about the $(00l)$ axis, forming hoops in the case of $(hkl)$ points for $k \neq 0$, while the *c*-axes lying in the plane perpendicular to the fiber axis would be represented by rotation of the $(hkl)$ hoops and $(00l)$ spots about the

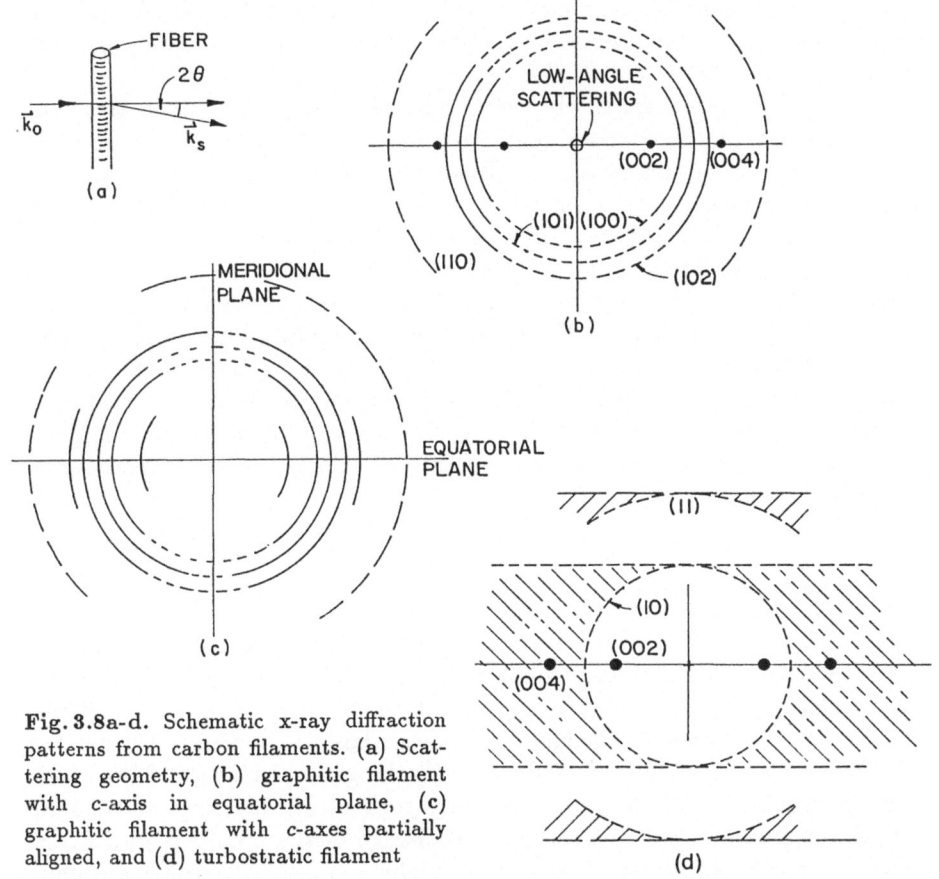

**Fig. 3.8a-d.** Schematic x-ray diffraction patterns from carbon filaments. (a) Scattering geometry, (b) graphitic filament with c-axis in equatorial plane, (c) graphitic filament with c-axes partially aligned, and (d) turbostratic filament

$(hk0)$ axis. The intersection of the $(00l)$ spots with the Ewald sphere would be points, and of the hoops would be rings. However, these rings would have maximum intensity in the direction of the $(00l)$ spots on the photographic plate (the equatorial plane), falling to zero in the perpendicular direction (the meridional plane). The x-ray pattern would consist of $(00l)$ spots and $(hkl)$ rings of non-uniform intensity (Fig. 3.8b). As a result of the disorder, the $(hk0)$ line shape is asymmetric as discussed below.

If the c-axes lie with a distribution of angles about the normal plane (Sect. 3.4.3), then the $(00l)$ diffraction spots would spread into segments of arcs, with their centers on the $(00l)$ plane. The intensity variation along the $(hkl)$ rings would change, with a tendency towards a uniform distribution (Fig. 3.8c). In the case that the c-axes are completely misaligned, both the $(00l)$ and $(hkl)$ diffractions will be smeared into uniform rings. This is a typical polycrystalline x-ray diffraction pattern, which could be obtained from well-aligned graphitic filaments by chopping or grinding them into small pieces.

Such an experiment is often carried out to obtain the lattice parameters. Finite "crystallite" dimensions would also result in the broadening of spots and lines (Sect. 3.2.1).

If the fiber were completely turbostratic (see Fig. 3.8d), then the $(hk)$ rods in reciprocal space would generate tubes when rotated about the $(00l)$ axis, to allow for $a$-axis mosaicity. When rotated about the $(hk0)$ axis, to accommodate $c$-axis isotropy in the normal plane, the intersections of these tubes with the Ewald sphere would be smeared out in the directions parallel to those of the $(00l)$ spots, with a minimum diffraction angle for each $(hk)$ reflection (Fig. 3.8d). Note that the $(hk0)$ line shape for this case would be asymmetric. However, the $(00l)$ reflections would again produce spots, which would be broadened by the finite "crystallite" size and the misorientation of the $c$-axes from the fiber normals. The $c$-axis spacing would be increased from the single crystal value of 0.3354 nm to 0.344 nm. (Higher values for $\bar{c}$ are sometimes encountered, due to the presence of substantial concentrations of impurities, such as H, O, N.)

### 3.1.7 Partial Graphitization

Many carbons have disorder between that of single crystal graphite and turbostratic carbon. Our understanding of the scattering from partially graphitized and turbostratic structures dates back to earlier work on bulk carbons, and has been reviewed by Ruland [1968]. The basic experimental facts that need to be incorporated into models are (i) the gradual change of the $c$-axis spacing from $\bar{c} = 0.344$ nm, characteristic of turbostratic layering, to 0.335 nm for perfect graphite, and (ii) the gradual change in the broadening of Bragg peaks. Typical interference peaks for partially graphitized carbon are illustrated in Fig. 3.9. Characteristically, the $(00l)$ and $(hk0)$ lines are sharpest,

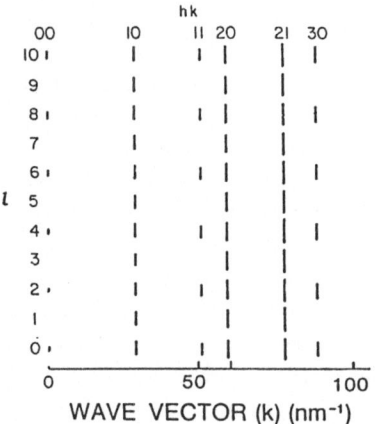

Fig. 3.9. Interference peaks for partially graphitized carbons showing line broadening for the various peaks [Ruland 1968]

with linewidths increasing as their distance from the origin increases. General $(hkl)$ lines are broader, with those for which $(h-k) \neq 3n$ ($n$ = integer) being broader than those for which $(h-k) = 3n$.

Several attempts have been made to quantify the degree of turbostratic disorder. The simplest approach is to use an empirical graphitization index $g_p$, which depends linearly on the interplanar separation $\bar{c}$, which is generally measured by the $(002)$ x-ray reflection $d_{002}$ [Maire and Mering 1958]:

$$ g_p = \frac{0.344 - \bar{c}}{0.344 - 0.3354} \quad . \tag{3.4} $$

Other workers have attempted to relate $d_{002}$ for bulk carbons to the disorder using models for the interatomic interactions (see Hutcheon [1970] for a review). Ruland [1968] has critically reviewed these models.

Ruland [1965] has developed a model to treat the degree of disorder in partially graphitic carbons and a similar treatment has been outlined by Gay and Gasparoux [1965]. This model is based on the idea of an autocorrelation function describing the positions of unit cells within the crystal structure. Ruland considered the disorder that was induced by displacing carbon hexagon planes parallel to the layers and he was able to describe the experimental diffraction lines in terms of 4 parameters: $\bar{c}$, the mean interlayer spacing; $\sigma_{\parallel}^2$ and $\sigma_{\perp}^2$, the variances of the probability distribution function for layers parallel and perpendicular to the planes; and $\alpha$, the probability of a rhombohedral stacking fault.

A useful method for determining the root mean square local lattice disorder parameter $\sigma$ was suggested by Ergun [1970]. The integral breadth $B_h$ of a line in the direction $\widehat{k}_h$ is defined as

$$ B_h = \int_0^\infty I(k_h) \, \frac{dk_h}{I_{\max}} \quad , \tag{3.5} $$

where $I(k_h)$ is the intensity profile with maximum intensity $I_{\max}$ and $\widehat{k}_h$ is the direction in $k$-space along which the scan is taken. If more than one $(00l)$ line is present, then

$$ B_h = \frac{K}{L_c} + \frac{2\pi^2 \sigma^2}{l^2} \quad . \tag{3.6} $$

The first term on the right hand side of this equation is related to the "crystallite" dimension, $L_c$, where $K$ is the Scherrer constant discussed in Sect. 3.4.4, and $\sigma$ provides a measure of the fluctuations $(\Delta \bar{c}/\bar{c})$ in the interplanar separation $\bar{c}$. As an example of the application of (3.5) and (3.6), D.J. Johnson [1980] studied the $c$-spacing of an ex-PAN fiber and found a value of $\sigma = 9 \times 10^{-3}$ for the variance.

It is stressed that the Ruland model treats the order-disorder process as a continuous one in which graphitic ordering is achieved by a gradual increase in the correlation between planes, through their mutual displacement. Thus the probability of a rhombohedral stacking fault, $\alpha$, increases smoothly from zero for graphitic stacking to 0.5 for turbostratic stacking, so that the order-disorder process may be regarded as a continuous phase transition. The causes of the displacements of the layers with respect to each other are probably interstitials, vacancy clusters, dislocations, and layer plane bending (without regular twinning) so that the graphitization process can be regarded as one in which these structural defects are gradually eliminated.

Although these models have been developed for bulk carbons, it is clear that they can also be applied to carbon fibers. Experimental difficulties arise because other defects also produce broadening effects in the Bragg peaks which must be separated before the degree of graphitization can be inferred.

## 3.2 Structure of Fibers and Filaments

### 3.2.1 Defects in Filaments with Partially Graphitic Structure

The description of defects given above pertains to crystalline material. Several of these concepts can still be used for filaments which are partially graphitized, in particular:

1. Vacancies and vacancy clusters are expected, in concentrations well above equilibrium values. The vacancies tend to anneal out at high heat treatment temperatures.

2. Interstitials and small interstitial clusters are not expected to occur in significant concentrations, since their migration energy is sufficiently low to enable them to diffuse to vacancies and "crystallite" edges.

3. Extra planes of relatively small dimension (e.g., 10–100 nm radius) occur and are similar to interstitial clusters. The boundary may be described as a dislocation loop ($s \perp c$, and $b \parallel c$) but this description is of limited use, since the extended planes on either side of the "interstitial loop" are not in registry.

4. Hexagonal networks will contain boundaries running along the networks, which may be described in terms of non-basal dislocations extending over the planes ($s \perp c$, and $b \parallel c$), see Fig. 3.6.

5. The concept of the tilt or twin boundary separating regions at an angle of 20°48′ about the $\langle 1\bar{1}00 \rangle$ direction is clearly of application only to highly graphitic filaments such as those vapor grown from benzene or methane, with $T_{HT} \sim 3000°C$. In the case of turbostratic materials, the lack of interplanar registry allows plane bending to occur continuously without the need for tilt boundaries to lower the energy (see discussion below on disclinations).

6. The concept that basal dislocations ($s, b \perp c$) can be split into partials and produce regions of stacking fault energy is applicable to highly

graphitic filaments. Freise and Kelly [1963] pointed out that the resulting regions of rhombohedral stacking faults, occurring at random, would not give rise to the extra diffraction lines observed for this structure, but would only broaden the hexagonal lines. Filaments with a more pronounced turbostratic structure correspond to an interlayer disorder which cannot be described only by a higher density of dislocations, but the relative displacements in this case can be described as a combination of displacements from the two symmetry types, at least as a first approximation, as outlined in Sect. 3.1.7.

7. In addition to the above defects, impurities play an important role in fibers. For instance, impurity concentrations are at least several parts per thousand in commercial fibers and often several percent, while hydrogen is believed to be present in CCVD filaments, but the hydrogen concentrations are not known. The impurities probably reside at the edges of fibrils and vacancy clusters, although some impurities (e.g., boron) may occupy lattice sites.

8. Turbostratic disorder is progressively removed by higher temperature heat treatments.

### 3.2.2 Highly Disordered Fibers and Filaments

Highly disordered fibers and filaments can be described in terms of a number of concepts pertaining to disordered materials. Two concepts of importance are the disclination, and bond disorder.

Massive rearrangements of graphene planes which are difficult to describe with dislocations can be described using disclination structures. (For a review of disclinations, see Nabarro [1967].) Zimmer and White [1982] have discussed the formation of disclinations in the mesophase, and suggested that they are prominent features in ex-mesophase pitch fibers, as evidenced by fracture behavior [White and Buechler 1985]. Figure 3.10 illustrates the main types

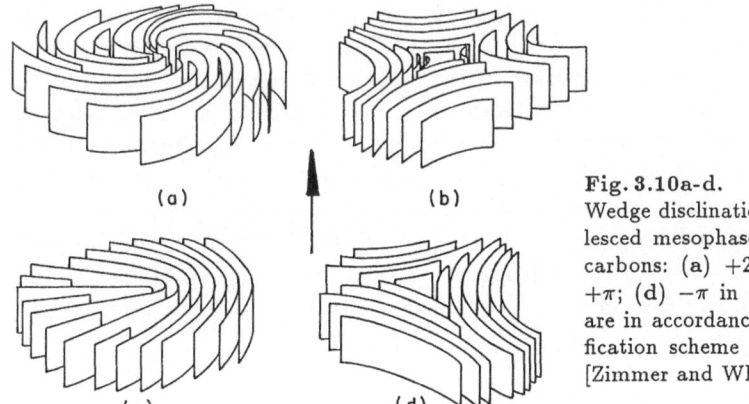

(a)  (b)

(c)  (d)

Fig. 3.10a-d.
Wedge disclinations found in coalesced mesophase and anisotropic carbons: (a) $+2\pi$; (b) $-2\pi$; (c) $+\pi$; (d) $-\pi$ in which the angles are in accordance with the classification scheme of Friedel [1922]. [Zimmer and White 1982]

47

of wedge disclinations, subdividing them according to the accepted scheme [Friedel 1922]. The fiber axis would lie along the direction of the arrow. The graphene planes in the fiber would then twist and rejoin other planar regions as shown. It is clear that the stacking of the planes cannot be graphitic. Heat treatment, leading to graphitization, would involve the formation of grain boundaries with well-defined angular relationships, as discussed above. However, planar separations and rejoining may still occur in this way. Although the concept has been developed for carbons prepared via a mesophase, such disclinations may also be important in other ex-polymer fibers.

Carbon exists in a number of polytypes in which the bonding differs from the trigonal ($120°$) $sp^2$ form, found in single crystal graphite. The most common form is tetrahedral $sp^3$ bonding found in diamond. The $sp^1$ bond occurs in linear hydrocarbons. All of these types of bonding can be found in disordered carbons in varying amounts.

The most complete studies of bonding types have been made for hydrogen-containing carbon films, which are sometimes called "diamond-like", but would be more appropriately referred to as "disordered, hydrogenated carbon" (for a recent review see J. Robertson [1987]).

The most sensitive technique to monitor the $sp^2$ bonding in a system with predominantly $sp^3$ bonding is Raman spectroscopy because the Raman cross section for the vibration associated with $sp^2$ bonding (see Sect. 4.5) is two orders of magnitude greater than that for $sp^3$ bonding. Many other techniques have also been used to characterize the bonding in "diamond-like" films. While Raman spectroscopy is most sensitive in the low $sp^2$ bonding limit, other techniques are useful in the low $sp^3$ bonding limit.

A study of the bonding types in carbon films has been made by Dischler et al. [1983] using infrared spectroscopy and by Fink et al. [1984] using electron energy loss spectroscopy (EELS). Both groups concluded that about 60%–70% of the bonds were of $sp^3$, 30%–40% of $sp^2$, and a few percent of $sp^1$ character in their materials, which were prepared in a glow discharge. The formation of $sp^3$ bonds is attributed to the presence of atomic hydrogen which is introduced into the reaction chamber by the glow discharge or some other plasma mechanism (dc plasma, rf plasma or microwave plasma, etc.). The proportion of $sp^2$ bonds increased with heat treatment, since the graphite phase has the lowest free energy. Raman studies [Dillon et al. 1984] allowed the increase in $sp^2$ bonding to be monitored by heat treatment, but the Raman frequency associated with the $sp^3$ bonds is masked by that of the disorder profile (Sect. 4.5), as discussed above.

It is anticipated that certain kinds of carbon fibers and filaments may contain $sp^3$ and $sp^1$ bonds in addition to $sp^2$, especially those prepared at low temperature and in the presence of a plasma discharge. Specific reference to EELS data on ion-grown filaments is made in Sect. 3.7.4.

## 3.3 Density

The mass density of carbon fibers $\varrho_m$ is lower than that of single crystal graphite for two reasons: larger interlayer spacing $\tilde{c}$ because of departures from ideal 3D layer stacking and the presence of small pores in the fibers. In practice, the two effects can be separated since x-ray diffraction is sensitive to $\tilde{c}$, and measurement of the geometrical density is sensitive to both effects.

The geometrical density of carbon filaments can be measured by dividing their mass by the geometrical volume. The former can be measured easily with a sensitive balance for which a buoyancy correction is made, while the latter can be estimated in several ways, including the displacement volume, or the immersion weighing method. Both of these methods are subject to the problem of whether or not the fluid penetrates the pores of the filament. Helium at atmospheric pressure, or mercury at 10 000 psi ($\sim$70 MPa) used in commercial pycnometers, does not penetrate the pores in polymer-derived carbons, since the pores are generally less than $\sim$3 nm in diameter, even for fibers with low heat treatment temperatures. Accordingly, the major changes observed in the measured density as a function of $T_{HT}$ are a consequence of the change in pore volume, with a subsidiary effect due to the change of the fibril material. The density of the fibrillar material can be estimated from x-ray diffraction measurements (Sect. 3.4) so that the geometrical density compared to the x-ray density gives an estimation of the relative pore volume.

Figure 3.11 illustrates density data for several types of fibers as a function of their Young's modulus, and compares these densities to that of single crystal graphite ($\varrho_m = 2.26$ gm/cm³). It can be seen that there is a trend to higher density with increasing modulus, and also with fiber type from ex-rayon, through ex-PAN and ex-pitch, to CCVD fibers. X-ray densities for fibers are approximately 2.20 gm/cm³, so that geometrical densities correspond to relative pore volumes of $\sim$40% for low-modulus, to $\sim$15% for high-modulus filaments. Further information on pore sizes is given by low-angle x-ray scattering, discussed in Sect. 3.5.

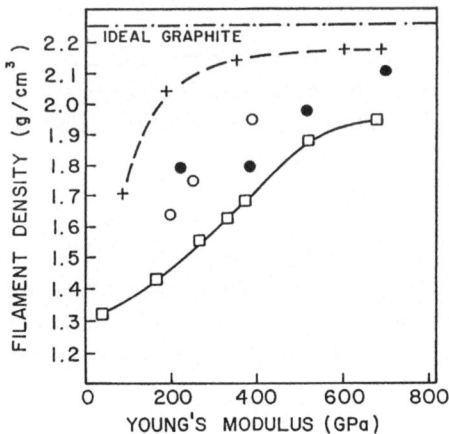

**Fig. 3.11.** The mass density of various fibers as a function of their Young's modulus, compared to single crystal graphite: ($\square$) ex-rayon [R. Bacon 1973]; ($\circ$) ex-PAN [D.J. Johnson 1987b]; ($\bullet$) ex-pitch [Tanabe et al. 1987]; ($+$) CCVD [Endo et al. 1976]

## 3.4 X-Ray Diffraction

X-ray diffraction, the most important of the scattering techniques, has been used as a dominant tool to study the defect structure of all types of carbon fibers and filaments. For example, the linewidth of the $d_{00l}$ reflections give the c-axis crystallite size $L_c$. The a-spacing and in-plane crystallite size is provided by study of the (100) diffraction line and its half width at half-maximum intensity, while the observation of the (101) and (112) profiles provides information on the interplanar correlation, since turbostratic graphite does not show well-defined (101) and (112) diffraction lines but rather shows "Bragg rods". The information obtained from x-ray diffraction studies can be classified as:

1. the density on a micrometer scale (see Sect. 3.3),
2. the extent to which neighboring carbon layers are stacked in a correlated $ABAB...$ sequence (degree of graphitization), or are turbostratically stacked in a random sequence,
3. the spatial extent of the crystalline structure, commonly denoted by the in-plane crystallite dimensions $L_{a\parallel}$ and $L_{a\perp}$, parallel and perpendicular to the fiber axis, and the crystallite dimensions along the c-axis $L_c$,
4. the preferred orientation of the above crystalline, or turbostratically stacked regions with respect to the fiber axis.

From a structural standpoint, the occurrence of turbostratic stacking of planar structural units is remarkable and unique to graphite.

A comprehensive review of x-ray diffraction from bulk carbons has been given by Ruland [1968], while R. Bacon [1973] has considered x-ray diffraction information obtained on ex-rayon fibers, and Reynolds [1973] and D.J. Johnson [1987b] have reviewed ex-PAN fibers. Work on these fibers will be summarized briefly and more recent work on CCVD filaments will be reviewed for the first time. Most of the basic concepts are common to both x-ray and electron diffraction.

### 3.4.1 Interlayer Spacing

The first characterization measurement carried out with x-ray diffraction is often the determination of the lattice constants, a and c, using Bragg's law. It is particularly useful to obtain $\bar{c}$ to give information about the degree of graphitization (see Sect. 3.1.7)

$$n\lambda = 2d_{hkl} \sin \theta_{hkl} \quad , \tag{3.7}$$

where $2\theta_{hkl}$ is the scattering angle and $d_{hkl}$ is the spacing between $(hkl)$ lattice planes. In general, to determine the lattice constants, the carbon fibers are pulverized with standard Si powder (for calibration purposes). Measurement of $d_{002}$ using the Bragg law gives the interlayer spacing between graphene layers $\bar{c}$.

From this measurement, the density of the material on a microscopic level can be deduced, although a small correction is necessary for defects such as vacancies, etc. The most noticeable feature of these measurements is the increase in the interlayer spacing $\tilde{c}$ from its single crystal value (3.354 Å), to values such as 3.44 Å associated with turbostratic disorder. Higher values for $\tilde{c}$ are sometimes encountered, due to the presence of substantial concentrations of impurities (Sect. 11.3).

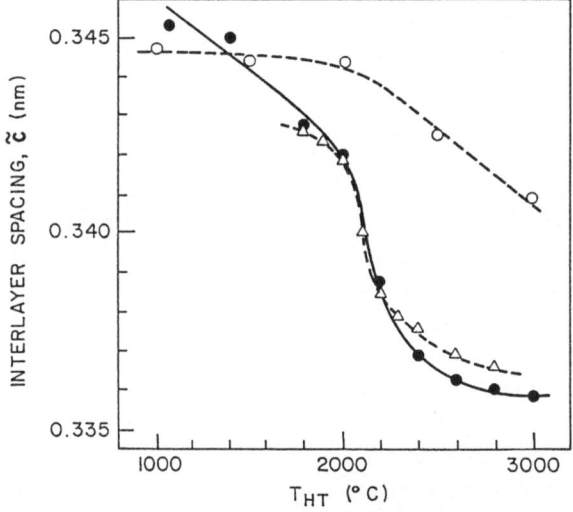

**Fig. 3.12.** Plot showing decrease of the interlayer spacing $\tilde{c}$ with increasing $T_{HT}$ for various filaments of different crystalline perfection: (•) CCVD fibers; (o) ex-PAN fibers; (△) anthracene char [Endo and Koyama 1977]

Figure 3.12 illustrates measurements of the average interplanar separation $\tilde{c}$ for ex-PAN carbon fibers, benzene-derived CCVD filaments, and bulk carbons (anthracene char). The major points to be noted are:

- Benzene-derived filaments graphitize more readily than ex-PAN fibers, and somewhat more readily than ex-pitch fibers heat-treated to the same temperature.
- At low $T_{HT}$ the interplanar spacing of both filaments and fibers lies above the value of 0.344 nm, characteristic of turbostratic carbon. This is due to impurities.

### 3.4.2 X-Ray Studies of Turbostratic and Partially Graphitized Carbon

Early studies showed that rayon-based filaments are non-graphitizable [Ruland 1969a]; that is, they do not reach a completely graphitized stage even af-

**Table 3.2.** Structural and physical parameters for ex-PAN fibers from x-ray diffraction measurements

| $T_{HT}$ [°C] | Density [kg/m$^3$] | Young's modulus [GPa] | $\phi$ [°] | $L_c$ [nm] | $L_{a\parallel}$ [nm] | $L_{a\perp}$ [nm] | $d_{002}$ [nm] |
|---|---|---|---|---|---|---|---|
| 1000[a] | 1650 | 210 | 44 | 1.2 | 3.2 | 1.9 | – |
| 1500[a] | 1750 | 250 | 41 | 1.7 | 3.9 | 2.7 | – |
| 2500[a] | 1950 | 390 | 20 | 5.3 | 9.8 | 8.0 | – |
| 1300[b] | – | 210 | 21 | 2.8 | – | – | 0.354 |
| 2000[b] | – | 345 | 11.5 | 5.8 | – | – | 0.342 |
| 2500[b] | – | 480 | 7.5 | 11.3 | – | – | 0.342 |
| 3500[b] | – | 690 | 5.8 | 23 | – | – | 0.329 |

[a] [D.J. Johnson 1980]
[b] [Goldberg 1985]

ter heat treatment above 3000°C. Similarities between rayon-based filaments and PAN-based fibers were also reported [Ruland 1969a; Perret and Ruland 1970]. Using the graphitization characterization parameter $g_p$ from (3.4), it is found that fibers graphitize more readily in the following ordering sequence; ex-rayon, ex-PAN, ex-mesophase pitch, to ex-benzene derived CCVD fibers. This ordering sequence is substantiated by other measurements, discussed below.

A set of x-ray diffractograms from ex-PAN and ex-pitch fibers is illustrated in Fig. 3.13, and the resulting structural parameters obtained from these diffractograms are given in Table 3.2 [Goldberg 1985]. These diffractograms were obtained using the geometrical arrangement described in Fig. 3.13, with a flat-plate camera to record the data.

Note that it should be possible in principle to resolve the (110) and (112) reflections more easily than the (100) and (102) reflections. This is clearly the case in the flat-plate diffraction photographs of the PAN fibers with modulus ~700 GPa shown in Fig. 3.13d. In these photographs one can also see the gradual appearance of the (104) spot as the modulus is increased. Similar results are obtained for pitch fibers (Fig. 3.13e-g) as a function of modulus, although the degree of 3D order in the ex-pitch fibers is somewhat higher than in ex-PAN fibers. Careful densitometer traces of these photographs show no evidence for 3D order in ex-PAN fibers with modulus 350 GPa and below. These flat-plate photographs also clearly show the relationship of the crystallite orientation to the modulus via examination of the (00$l$) reflection arcs as already discussed. An example of an asymmetric ($hk$0) line shape is found in the photograph in Fig. 3.13a for the completely turbostratic 210 GPa ex-PAN fiber.

A variety of structural studies of ex-pitch Carbonic HM50, HM60, and HM80 fibers have been carried out, from which a consistent picture of the structure-property relations has been deduced [Endo 1988a]. The lesser de-

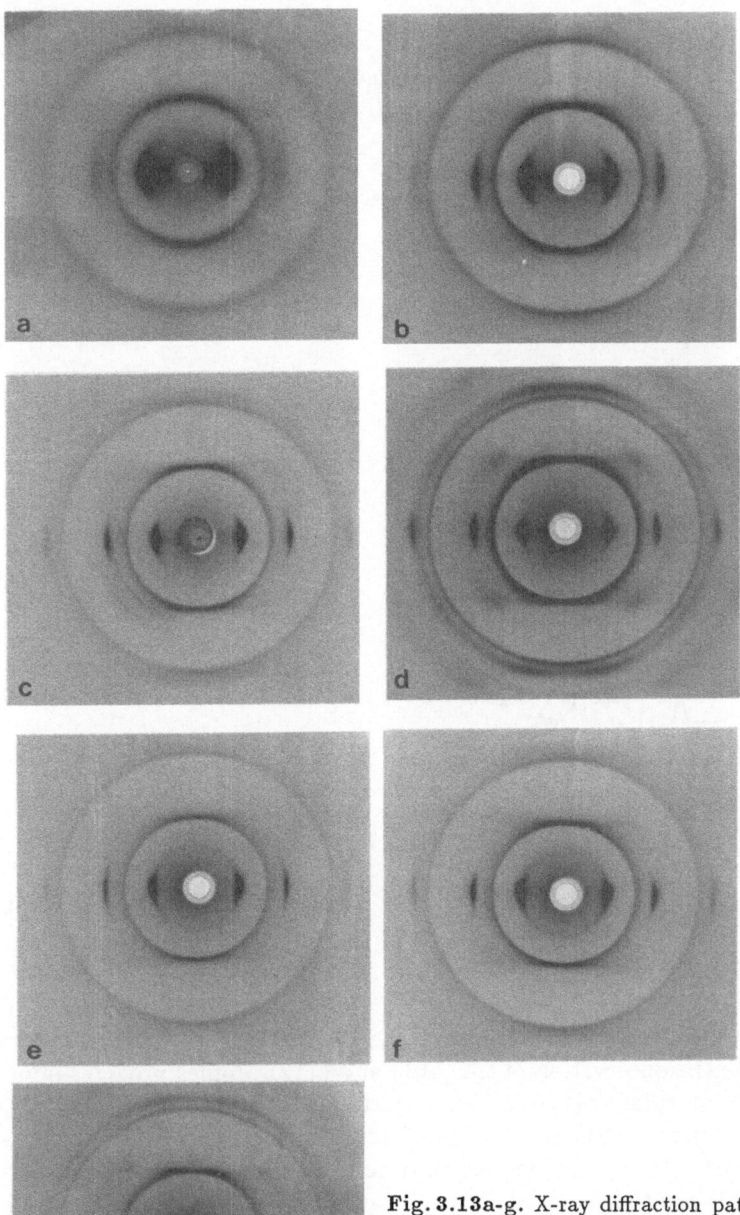

**Fig. 3.13a-g.** X-ray diffraction patterns of ex-PAN fibers for various Young's moduli: (a) 210 GPa, (b) 350 GPa, (c) 480 GPa, (d) 690 GPa, and ex-pitch carbon fibers of various moduli: (e) 350 GPa, (f) 480 GPa, (g) 690 GPa. Patterns were recorded with Mo $K_\alpha$ radiation with the incident beam perpendicular to the fiber axis [Goldberg 1985]

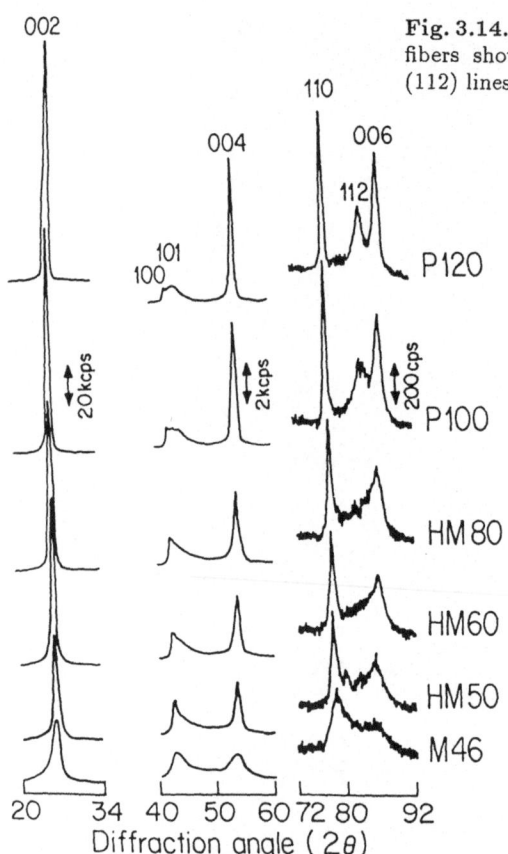

**Fig. 3.14.** X-ray diffractograms for various ex-pitch fibers showing the $(00l)$, $(100)$, $(110)$, $(101)$, and $(112)$ lines [Endo 1988a]

gree of structural order of the Carbonic fibers relative to the Thornel ex-pitch P-100 and P-120 fibers is confirmed by the larger interplanar $d_{002}$ spacings and greater linewidths (smaller $L_c$ values) as measured by the x-ray $(00l)$ diffractograms (Fig. 3.14). The $a$-spacing and in-plane crystallite size is provided by study of the $(100)$ diffraction line and its half width at half-maximum intensity, while the observation of the $(101)$ and $(112)$ profiles provides information on the interplanar correlation; in particular, turbostratic carbons do not show well-defined $(101)$ and $(112)$ diffraction lines. X-ray diffractograms taken with fibers of different degrees of ordering (from the most highly ordered Thornel P-120 to the least ordered Torayca M46) are shown in Fig. 3.14, where a broadening of the $(00l)$, $(100)$, and $(110)$ lines is seen with increasing disorder. Note that the $(101)$ and $(112)$ lines are seen only for the P-120 and P-100 ex-pitch fibers that exhibit 3D interplanar correlations. Consistent with these results are the selected area diffraction TEM patterns, showing more distortion of the honeycomb planar networks in the Carbonic fibers, leading to larger elongations of these fibers prior to fracture, see Sect. 6.2.2.

**Table 3.3.** Correlation of $d_{002}$ with $L_c$ for various high modulus carbon fibers[a]

| Fiber name | $d_{002}$ [nm] | $L_c$ [nm] |
|---|---|---|
| Thornel P-100 | 0.3392 | 24 |
| Thornel P-120 | 0.3378 | 28 |
| Carbonic HM50 | 0.3423 | 13 |
| Carbonic HM60 | 0.3416 | 15 |
| Carbonic HM80 | 0.3399 | 18 |
| Torayca M46 | 0.3434 | 6.2 |
| Celion G-50 | 0.342 | 5.8 |
| Celion GY-70 | 0.342 | 11 |

[a] The Thornel and Carbonic fibers are pitch-based while the Torayca and Celion fibers are PAN-based. The data for the Celion fibers are from Goldberg [1985] while those for the other fibers are from Endo [1988a].

The degree of graphitization can be correlated with a number of other parameters. For example, Touzain and Bagga [1984] have correlated $g_p$ in (3.4) with a number of properties such as $L_c$, the resistivity, Young's modulus, and the tensile strength, while Endo [1988a] has correlated $d_{002}$ directly with $L_c$, see Table 3.3.

### 3.4.3 Preferred Orientation of the c-Axes

Many carbon filaments are characterized by complete rotational disorder of the c-axes in the plane perpendicular to the filament axis, but partial disorder in the axial direction. In the case of complete axial disorder, the $(00l)$ interferences would be spread into spheres, and diffraction rings would result for all reflections, as in the case of polycrystalline samples.

However, if the c-axes are preferentially aligned, so that the probability function describing the preferential alignment $P(\phi)$ goes to zero at an angle $\phi < 90°$, then the $(00l)$ reflections for graphitic or turbostratic carbon are spread into arcs, from which the probability function $P(\phi)$ can be deduced. Here $\phi$ refers to the angle between the layer planes and the fiber axis, see Fig. 3.15, and is referred to as the misorientation angle in the text. The evaluation of $P(\phi)$ is usually done using the (002) reflection, because its intensity is higher than other $(00l)$ reflections. Therefore we use the notation $I(\phi)$ for the x-ray intensity profile to determine the probability function $P(\phi)$. [Note that $(hkl)$ reflections ($h$ or $k > 0$) cannot be used since the interference maxima are spread into rods.] In practice, this is done either using a photographic plate exposure, with the exposure density then read with a densitometer (Fig. 3.16), or using an x-ray diffractometer [Ruland 1967]. The

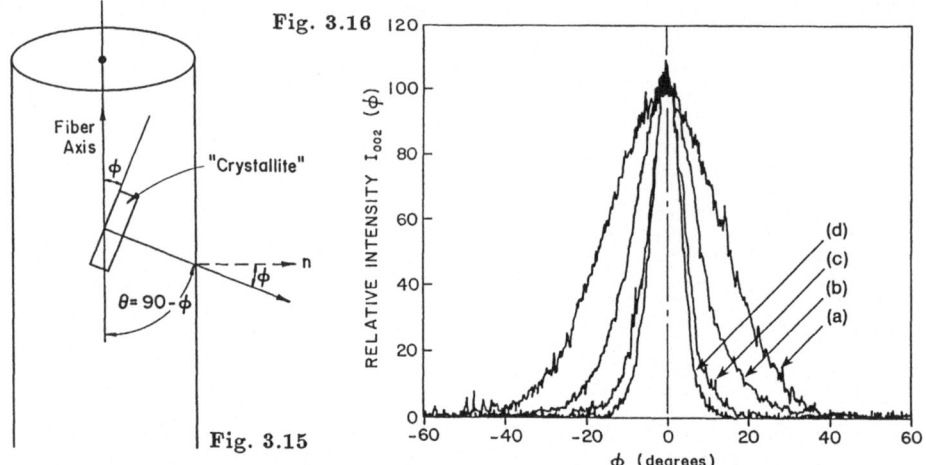

Fig. 3.16

Fig. 3.15

**Fig. 3.15.** A sketch showing the definition of the crystallite orientation angle $\theta$ and the corresponding "misorientation" angle $\phi$ for crystallites in carbon fibers. Note that $\theta = \pi/2 - \phi$

**Fig. 3.16.** Typical $I_{002}(\phi)$ x-ray line-shape data for heat-treated ex-PAN fibers of varying crystalline perfection: (a) Celion 3000, $E_Y = 210$ GPa; (b) Celanese GY-50, $E_Y = 350$ GPa; (c) Celanese GY-70, $E_Y = 490$ GPa; (d) Celanese GY-100, $E_Y = 690$ GPa, where $E_Y$ denotes the Young's modulus [Stamatoff et al. 1983]

results yield an intensity function $I_{002}(\phi)$ which has been measured by several workers for polymer-derived fibers. These measurements are of importance for fibers because their mechanical properties are directly related to the development of preferred orientations with increasing $T_{HT}$ (see R. Bacon [1973], Reynolds [1973], and Sect. 6.2).

The intensity distribution function $I_{002}(\phi)$, corrected for other broadening effects, can be fitted reasonably well with a $\cos^m\phi$ function for bulk carbons [G.E. Bacon 1956a, 1956b]. However, this distribution is characterized by the near equality of the full width at half maximum (FWHM) intensity and the full width at half integral intensity (FWHII). This does not appear to be the case either for rayon fibers, for which Ruland [1967] used an expression of the (Poisson kernel) form

$$I_{002}(\phi) = \frac{(1 - q^2)}{1 + q^2 - 2q\cos\phi} \quad , \tag{3.8}$$

where $q$ is an adjustable parameter, or for ex-PAN fibers [Stamatoff et al. 1983; Goldberg 1987] for which a Gaussian gives a better fit [FWHM = 1.74 FWHII] (see Table 3.2). The FWHM is often used to express the mean misorientation angle $\langle\phi\rangle$ according to FWHM = $2\langle\phi\rangle$.

Two processing factors influence the preferred orientation strongly: the processing temperature ($T_{HT}$) and stretching. Ruland [1967] showed that the

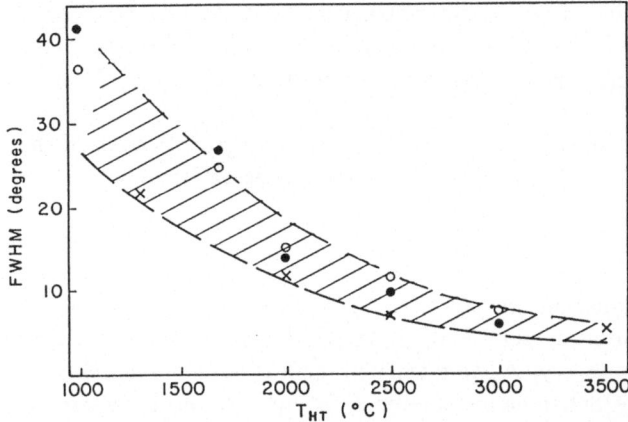

**Fig. 3.17.** Variation of the FWHM of the x-ray (002) reflection with $T_{HT}$ for ($\times$) ex-PAN fibers [Stamatoff et al. 1983], and for ex-pitch fibers [($\circ$) radial structure, ($\bullet$) random structure] [Bright 1979]. The ex-pitch data are from single stage and double stage thermosetting fibers

preferred orientation in rayon-based fibers is almost completely destroyed by processing at 1000°C, but that higher $T_{HT}$ results in increased preferred orientation. Also, the preferred orientation could be increased substantially with hot stretching during processing. These fibers attain good alignment after carbonization, and the alignment improves after further heat treatment. Figure 3.17 includes data for FWHM of ex-PAN [Goldberg 1987] and ex-pitch [Bright 1979] fibers. The decrease in FWHM at higher $T_{HT}$ is noted, leading to higher elastic moduli as discussed in Sect. 6.2. The value of FWHM depends on the processing conditions. Values given by D.J. Johnson [1980], which are listed in Table 3.2, are considerably higher, for the same $T_{HT}$ than those of Goldberg [1987] and his collaborators. This is probably due to different processing conditions, as well as differences in analysis methods.

### 3.4.4 Crystalline Structures of Finite Size

Line broadening of a different kind is also produced by the presence of crystallites of finite size, which are characterized by two parameters, $L_a$ and $L_c$, corresponding to crystallite dimensions in and perpendicular to the graphene planes, respectively. Here $L_a$ is further subdivided into $L_{a\perp}$ (perpendicular to the fiber axis) and $L_{a\parallel}$ (parallel to the fiber axis). Using a Fourier transform technique, Ruland (1968) demonstrated that all Bragg peaks are broadened in three dimensions in the same way. Furthermore, the broadening in each direction of reciprocal space is proportional to the Fourier transform of the crystallite shape function, which bears an inverse proportion to the dimensions of the sample. Therefore, crystallite dimensions can in principle be determined from the analysis of the line shape of one diffraction peak. In practice, other factors affecting line broadening (e.g., orientation, layer-plane disorder) also have to be taken into consideration, as is for example discussed by Houska and Warren [1954].

The crystallite dimension $L_c$ is measured from the integral breadth of the $(00l)$ lines denoted by $B_{(00l)}(2\theta)$ through use of the Scherrer formula

$$L_c = \frac{K\lambda}{B_{00l}(2\theta)\cos\theta} \quad , \tag{3.9}$$

where $K$ is the Scherrer constant (see Sect. 3.1.7), $\lambda$ is the x-ray wavelength, $\theta$ is the Bragg angle and the x-ray scan is taken in the direction normal to that for which broadening due to crystallite misorientation is present, which is in the equatorial plane. The Scherrer parameter $K$ is equal to unity when the integral width $B_h$ given by (3.5) is used in (3.9). The Scherrer formula gives a "weighted" average $L_h$ for the crystallite size. If the distribution function for crystallites of size $L$ is $n(L)$, then

$$L_h = \frac{K}{B_h} = \frac{\int_0^\infty n(L)L^2 dL}{\int_0^\infty n(L)L\,dL} \quad . \tag{3.10}$$

The crystallite dimension $L_a$ is normally estimated from the x-ray line width of the (10) or (11) bands associated with the turbostratic material, and is related to the half width $B_{1/2}(2\theta)$ by the following relation [Warren and Bodenstein 1966]:

$$L_a = \frac{1.77}{B_{1/2}(2\theta)\cos\theta} \quad , \tag{3.11}$$

where $\theta$ is the Bragg angle. However, Ruland and Tompa [1968] suggest other values for the Scherrer parameter $K$ closer to 2.0. With regard to (3.11), the meridional width of the (100) reflection, $B_{100}$, is used to measure the layer plane length parallel to the fiber axis, $L_{a\|}$, while the equatorial width is used to measure the width perpendicular to the fiber axis, $L_{a\perp}$. Table 3.2 gives values of $L_{a\|}$ and $L_{a\perp}$, and $L_c$ for ex-PAN fibers.

Such formulae must be used with caution, since other lattice defects can also cause line broadening. In particular, a turbostratic arrangement of layer planes can affect the broadening of $(00l)$ lines (Sect. 3.4.2) as seen from (3.6). Thermal vibrations, lattice vacancies and interstitials, impurity atoms, and dislocations, etc., can also contribute to line broadening as well as instrumental broadening. The formulae given above do however serve as a lower limit.

Values for $L_a$ and $L_c$ deduced from x-ray diffraction measurements for polymer-based fibers vs $T_{HT}$ are sketched in Fig. 3.18. A monotonic increase in crystallite size is noted with increased $T_{HT}$. However, the values for $L_a$ and $L_c$ that are obtained are much smaller than what one expects from the strength of the filaments. This will be discussed in greater detail in Sect. 3.5,

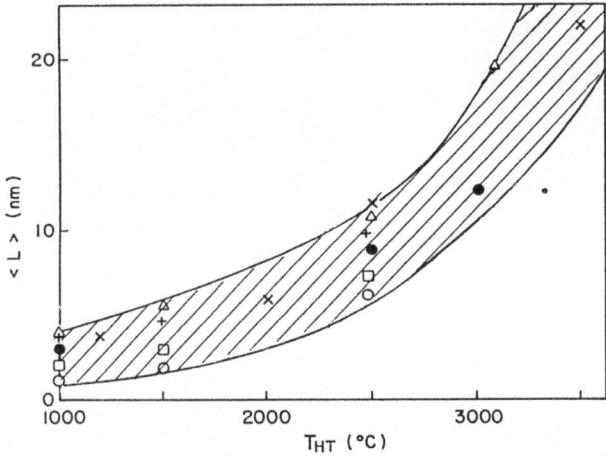

**Fig. 3.18.** Increase of average crystallite size $\langle L \rangle$ with increasing $T_{HT}$ for various fibers: (o) $L_c$ [D.J. Johnson 1980]; (+) $L_{a\parallel}$ [D.J. Johnson 1980]; ($\square$) $L_{a\perp}$ [D.J. Johnson 1980]; ($\triangle$) $L_a$ [Shindo 1961b]; ($\times$) $L_c$ [Stamatoff et al. 1983]

where it is shown that the ribbons extend over hundreds of nanometers, and that the x-ray technique is measuring the average length of relatively straight portions of these ribbons.

## 3.5   Small Angle Scattering

Small angle x-ray scattering enables one to probe electron density variations on a length scale up to several thousand angstroms. This means that the porosity and other sources of heterogeneity in carbon and graphite fibers can be studied using this technique.

The origin of this scattering is the electron density discontinuities at the pore boundaries. Consider, for example, the case of oriented, elongated pores in a high $T_{HT}$ ex-PAN fiber. If the incident x-ray beam is perpendicular to the fiber axis, then opposite, parallel walls of pores will scatter the photons effectively in the equatorial plane, but not in the meridional plane. Those photons with wavelength equal to half the distance between pore walls will scatter strongly. Accordingly, the detailed shape of the small angle pattern can provide important information on the microstructure of the fibers. The pore dimensions are typically 0.5–3 nm across and several tens of nanometers long [Fourdeux et al. 1971]. The actual size, shape, orientation, and number of pores will of course depend upon the fiber heat treatment and precursor. Figure 3.19 shows the small angle pattern for a low modulus and a high modulus ex-PAN fiber. As the polymer carbonizes, the two lobes in the small angle diffraction pattern become enlarged, consistent with the creation of pores due to outgassing. Note the large increase in the orientation of the

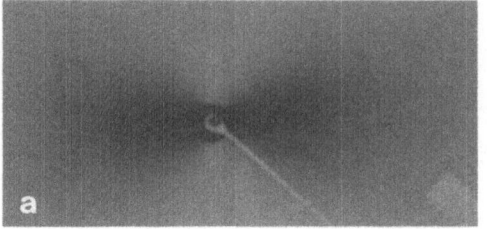

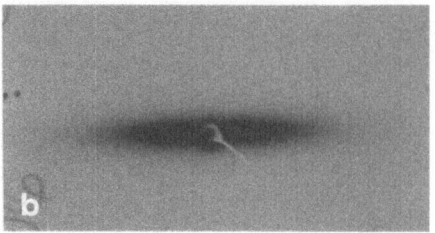

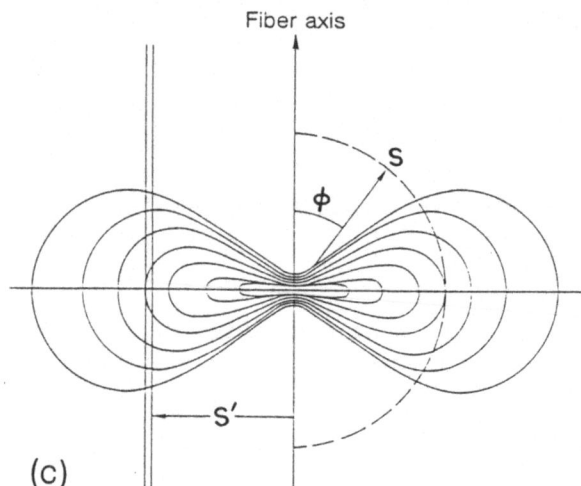

Fiber axis

(c)

**Fig. 3.19a-c.** The small angle pattern for (a) low modulus and (b) high modulus ex-PAN fibers. (c) Typical constant intensity contours showing pinhole patterns on the right hand side with a path of constant $S$ used to obtain $I(\phi)$. The slit pattern on the left hand side is used to obtain pore dimensions [Goldberg 1985; Tillgner and Ruland 1987]

scattering in the high modulus fibers. This means that the pore structure becomes oriented, as well as the crystallites that make up the fiber. The scattering from low modulus fibers is more diffuse, but still provides evidence for some degree of orientational alignment.

Early work on fibers made from rayon precursors [Fourdeux et al. 1969] and from PAN [Perret and Ruland 1970; Fourdeux et al. 1971] clearly demonstrated that one can quantitatively understand the density of carbon fibers in terms of the elongated pores which cause small angle x-ray scattering. In addition, it was shown that the orientation of the pores correlated well with the orientation of the basal planes in the fiber. These results, along with wide angle x-ray diffraction and electron microscopy, formed the basis for the Ruland ribbon model of carbon and graphite fibers (see Figs. 1.3 and 1.4). According to the ribbon model, the fibers consist of twisted, undulating ribbons of 2D hexagonal carbon networks. Regions which have uniform $c$-axis spacing are the crystallites seen in wide angle x-ray diffraction. The spaces which invariably exist between the ribbons (and microfibrillar clusters of ribbons) are the pores which are most easily observed with small angle x-ray scattering.

In a two-phase media (carbon plus pores) with sharp, smooth boundaries, the intensity of the small angle scattering $I(S)$ should asymptotically approach

$$I(S) = c_A/S^4 \quad , \tag{3.12}$$

where $c_A$ is a constant proportional to the total surface area of the pores. It is conventional to use $S \doteq \Delta k/2\pi$ for small angle scattering experiments where $\Delta k$ is given by (3.1), and this convention is followed here. This description, known as Porod's law [Porod 1951, 1952], neglects any fluctuations or variations in the electron density within one of the phases. D.J. Johnson and Tyson [1969, 1970] suggested that deviations from Porod's law in ex-PAN fibers were due to the presence of an amorphous carbon phase between the turbostratic graphite regions. Perret and Ruland [1970] argued, however, that one would expect such amorphous carbon regions to contribute a $1/S^3$ component to $I(S)$. They therefore plotted the $S^3 I(S)$ versus $S^2$ showing that the large $S$ behavior did not asymptotically approach a constant as expected from Johnson and Tyson's model, but was approximately linear in $S$ (Fig. 3.20). Perret and Ruland [1970] further argued that this behavior was due to fluctuations in the spacing between the hexagonal networks. In the general case, fluctuations in electron density contribute a term in $I(S)$ proportional to $1/S^D$, where $D$ is the dimensionality of the fluctuations. Thus if the fluctuations were due to unoriented amorphous carbon, we would have $D = 3$, while fluctuations in the spacing between basal planes occur in only one dimension (i.e., $D = 1$).

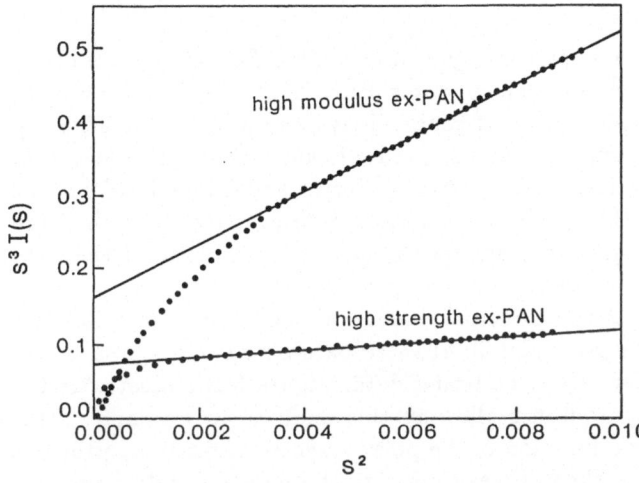

Fig. 3.20. Plot of $S^3 I(S)$ vs $S^2$ for two types of carbon fibers [Perret and Ruland 1970]

Perret and Ruland [1970] subtracted out the small angle scattering coming from these fluctuations, and then applied a Fourier transform to the remaining data to obtain the pore distribution function $g(r)$ in real space (Fig. 3.21). This function was then fitted to an exponential distribution of pores given by

$$g(r) = [\exp(-r/l_p)]/l_p^2 \quad , \tag{3.13}$$

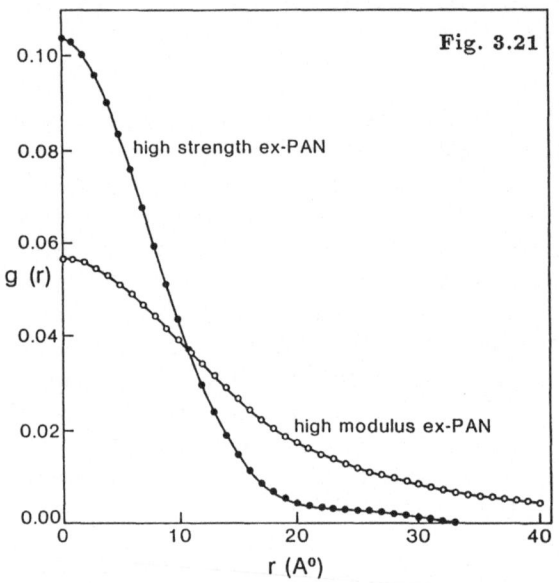

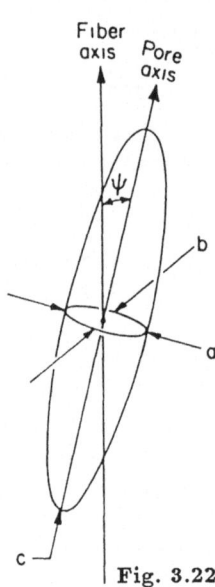

**Fig. 3.21.** Plot of the normalized pore distribution function $g(r)$ vs $r$ for two types of fibers. See (3.13) and discussion in Sect. 3.5. [Perret and Ruland 1970]

**Fig. 3.22.** A sketch illustrating a simple model for a pore of length $c$, dimensions $a, b$ in the section, at an angle $\psi$ to the fiber axis

where the length $l_p$ characterizes the typical pore size. This analysis leads to a pore size of about 15 Å for high modulus PAN fibers, see Fig. 3.21, with an overall porosity of about 10%. Of course, it is clear from the small angle photographs (Fig. 3.19) that the pore structure is anisotropic. The length of the pores in these materials is about 30 nm [Perret and Ruland 1970]. The numerous assumptions that go in to an absolute determination of the pore structure from small angle x-ray scattering make the accuracy of the pore dimensions somewhat uncertain.

Figure 3.22 illustrates the principal dimensions and the orientation of a pore within the fiber. In principle, all of these parameters $a, b, c$, and $\psi$ can be deduced from the low-angle scattering data. Figure 3.19c illustrates the constant x-ray intensity contours allowing this analysis to be done. If the intensity is measured as a function of the polar angle $\phi$ for constant values of $S$ using a pinhole source, then curves similar to those on the right hand side of Fig. 3.19c are obtained, and can be analyzed by an equation:

$$S^2 B_{\text{obs}}^2 = S^2 B_\psi^2 + 1/c^2 \quad , \tag{3.14}$$

where $B_{\text{obs}}$ is the measured width of the distribution and $B_\psi$ is the width of the orientation distribution of the pores. A plot of $S^2 B_{\text{obs}}^2$ versus $S^2$ gives $B_\psi$ from the slope, and $1/c^2$ from the intercept. The pore sizes $a$ and $b$ are measured from intensity measurements similar to those above, but using a slit camera (Kratky camera) with the slit oriented along the fiber axis. The intensity measured along the profile at $S = S'$ (see Fig. 3.19c on the left hand side) can be analyzed using the equation

$$I(S') = \exp\left[-(a^2 + b^2)\pi^2 S'^2/6\right] \quad .$$ (3.15)

Usually $a > b$, so that the contributions from the two pore widths can be separated [Perret and Ruland 1969, 1970].

These techniques have been applied recently to several carbon fibers derived from rayon, PAN, and pitch [Tillgner and Ruland 1987]. It was concluded that ex-mesophase-pitch fibers showed a lower porosity and density fluctuation than the other types. This is consistent with the mass density curve (Fig. 3.11) and indicates that these fibers have less structural imperfection than the other types. However, the average length of the micropores $(c)$ was similar for all of the fiber types for a given value of $T_{HT}$.

The intensity contours could be characterized with three dimensions $a$, $b$, $c$, for ex-PAN and ex-rayon fibers, with aspect ratios $c/a$ between 1.5 and 3.5, and $c/b$ between 4 and 24. There were significant differences between the results for these two materials. It was conjectured that the presence of larger pores (mesopores) in ex-pitch fibers did not permit the micropores to be analyzed by this model to give $a$ and $b$.

Another interesting result of their analysis was that the orientation function of the micropores gave a wider distribution than that obtained using wide angle x-ray diffraction measurements, discussed in Sect. 3.4.3 (i.e., $\psi > \langle\phi\rangle$). It would be expected that the same result should be obtained from wide and small angle scattering experiments since the ribbons are enclosing the pores, as illustrated in Figs. 1.3 and 1.4. The explanation of this is that (3.14) weights more heavily the contributions from the smaller pores, which are typically aligned at higher angles to the fiber axis.

In recent years it has become increasingly clear that disordered materials often exhibit properties suggesting non-integer fractal dimensionality. In such cases the material is called a fractal. The value of the non-integer dimension is called the fractal dimension. There can be many different fractal dimensions for a single system, depending upon the property of the system under study [Stanley 1984]. The Hausdorff dimension $D$ describes the way the total mass $M$ of a system depends on its size $L$, i.e.,

$$M = L^D$$ (3.16)

and is the only definition of the fractal dimension we will use here. The scattering intensity due to fluctuations in electron density of a system characterized by a Hausdorff dimension $D$ is proportional to $S^{-D}$. Thus Perret and Ruland [1970] attempted to fit the deviation's from Porod's law to the special cases of $D = 1$ (c-axis spacing fluctuations) and $D = 3$ (uniform density amorphous carbon).

Recently, Tang et al. [1986] re-examined the small angle scattering of three types of carbon fibers, and interpreted the results in terms of fractal concepts. They pointed out that there are two ways in which fractal behavior may influence the small angle scattering in carbon and graphite fibers. As discussed above, the carbon "crystallites" may be fractal, thus giving rise to a component of $I(S)$ proportional to $S^{-D}$. In addition, the surface of the elongated pores is not likely to be smooth. A smooth surface has a fractal dimension of 2, while a rough surface is characterized by a fractal dimension greater than 2. Bale and Schmidt (1984) showed that in the general case of porous material where the pore surface is characterized by the fractal dimension $D$, Porod's law can be generalized to

$$I(S) = c_A/S^{6-D} \quad . \tag{3.17}$$

Thus Tang et al. [1986] were able to interpret the deviations from Porod's law in terms of the fractal nature of both the surface of the pores (at small $S$, 0.16–0.4 nm$^{-1}$) and the fractal nature of the carbon aggregates comprising the fiber (at large $S$, 1.5–3 nm$^{-1}$).

The extent to which the carbon crystallites are fractal depends on the degree of order in the fiber. P-50 fibers (a 350 GPa ex-pitch fiber) have

$$I(S) = c_A/S^3 \tag{3.18}$$

for $S$ between 1.5 and 3.0 nm$^{-1}$. This implies that the carbon crystallites which comprise this fiber are not fractal. On the other hand, at small $S$, Tang et al. [1986] found that the surface of the pores is not smooth but is characterized by a fractal dimension of 2.6.

In a high strength, low modulus carbon fiber made from PAN (AS4 Magnamite fiber made by Hercules Corp.), the scattering intensity for $S$ between 1.5 and 3 nm$^{-1}$ indicates that the carbon crystallites have a fractal dimension of 2.3. The fact that this is significantly lower than 3 is consistent with the highly disordered nature of this type of fiber. It is clear, however, that most of the carbon in the fiber cannot be contributing to this small angle scattering. This is because a fractal material has a density which decreases with increasing size. A material which is fractal over a change in length scale by a factor of two will have its density decrease to 62% of its initial value. If all of the carbon in a carbon fiber were fractal over the lengths probed by the small angle scattering with momentum transfer between 1.5 and 3 nm$^{-1}$ as suggested by Tang et al. [1986], then the density of the fiber would have to be less than 1.4 g/cm$^3$. This is clearly not the case. Thus, if fractal concepts have any applicability to carbon fibers, they must apply to only a portion of the carbon making up the fibers. The fractal picture proposed by Tang et al. [1986] must therefore be a heterogeneous model similar to that proposed by D.J. Johnson and Tyson [1970].

One of the least understood aspects of carbon fibers is the nature of the disorder and its role in determining physical properties. Small angle x-ray scattering is one of our most direct probes of at least some types of disorder. Further work is clearly needed to clarify the nature of the heterogeneities and disorder.

## 3.6 Optical Microscopy

Optical microscopy has been of limited usefulness for studying the microstructure of typical filaments, since the wavelengths of optical photons (300–700 nm) are not small enough to resolve structural features such as pores with sufficient detail. The most important optical microscopy work has used the polarized-light technique to probe the arrangement of the ribbons across transverse sections of ex-polymer fibers. This work is complementary to the more detailed electron microscope studies, discussed in the next section. The principle of the method [Woodrow et al. 1965] is that light reflected from the prismatic edges of graphene planes has a maximum intensity when the $E$ vector for the light is parallel to the a-axes. Interesting reflection patterns from ex-PAN fiber sections were obtained by Watt and W. Johnson [1970], Butler and Diefendorf [1969], Knibbs [1971], Diefendorf and Tokarsky [1975], and Wicks and Coyle [1976], on the basis of which these authors proposed a core/sheath structure for ex-PAN fibers (illustrated in Fig. 1.1b), which will be discussed further in Sect. 3.7. Figure 3.23 shows a set of polarized light microscopy photographs illustrating this feature [Wicks and Coyle 1976]. Bright and Singer [1979] reported polarized optical microscopy results for ex-pitch fibers, using these results to select fibers with predominantly radial or random cross-sectional arrangements of the ribbons.

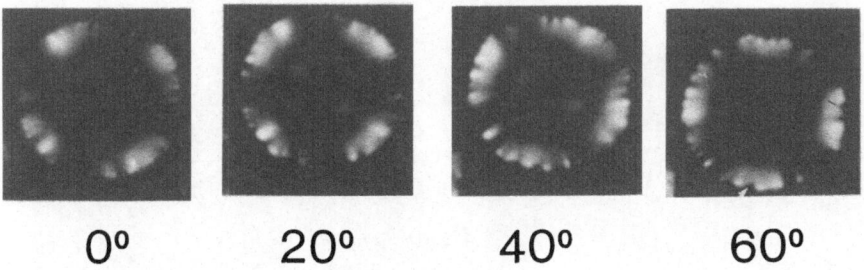

$0^\circ$        $20^\circ$        $40^\circ$        $60^\circ$

**Fig. 3.23.** Polarized light microscope photographs of the cross section of an ex-PAN fiber showing the skin/core structure. The polarization angles $0^\circ$, $20^\circ$, $40^\circ$ and $60^\circ$ refer to the angle between the optical $E$ field and an arbitrary axis in the section [Wicks and Coyle 1976]

## 3.7 Electron Microscopy

Since the fiber cross sections are typically 10 $\mu$m in diameter, the observation of many of the interesting structural features requires the use of electron microscopy techniques. For gross features, scanning electron microscopy (SEM) provides very general structural information and is used routinely in conjunction with many studies as a general characterization technique. Microscopic information close to the atomic level is provided by the higher resolution transmission electron microscopy (TEM) studies, and this technique has been extensively applied to study carbon fibers. Because of the highly planar structure of graphite at an atomic level and the stiffness of the graphite layer planes, graphite surfaces have provided a prototype material for scanning tunneling microscopy (STM) studies. It is likely that STM studies of carbon fibers will soon be carried out, providing important new information about the defects in carbon fibers on an atomic level.

### 3.7.1 SEM Characterization

Of particular interest have been the SEM micrographs of the microstructure of the fiber cross sections, showing the extent and texture of the lamellar organization as well as the general organization of the graphitic layers, whether they are in a circumferential or radial preferred orientation. Other related microstructural features that are typically investigated by SEM include: information on microcracks, flaws and voids connected with fiber processing; characterization of the porous structures associated with the exfoliation process; information on the fracture cross sections of fibers arising from various

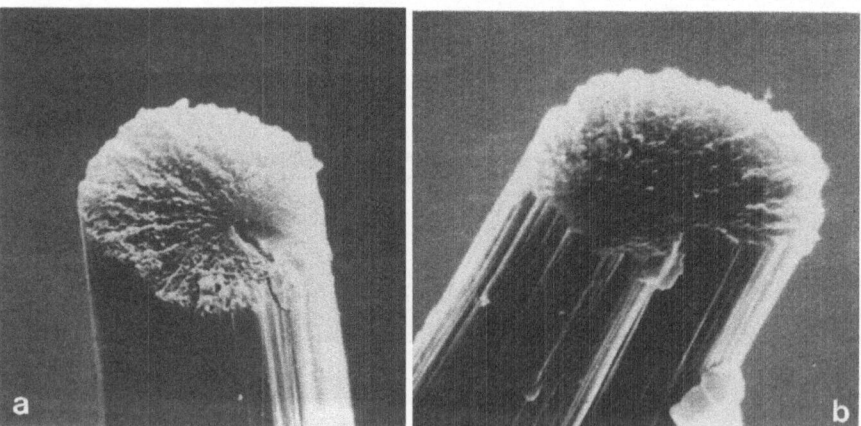

Fig. 3.24a,b. SEM photograph of fracture surfaces of ex-Acrylan fibers showing (a) a circular, low modulus fiber with an internal flaw, (b) bean-shaped, high modulus fiber [D.J. Johnson and Tyson 1969]

failure modes, and measurement of the fiber cross-sectional area and shape studies. Only a few representative examples are given here, since the technique is used too widely to discuss all of the interesting examples.

Figure 3.24 illustrates SEM photographs of fracture surfaces of two ex-Acrylan (PAN) fibers [J.W. Johnson 1969; D.J. Johnson and Tyson 1969; J.W. Johnson and Thorne 1969]. The SEM photograph of Fig. 3.24a shows an internal flaw in a high strength fiber with roughly circular cross section, and Fig. 3.24b shows a surface flaw in a higher modulus fiber with a bean shape. The relationship of such flaws to fracture is discussed extensively in Sect. 6.2. Sharp and Burney [1971] also have compared defects in ex-PAN fibers using SEM and TEM techniques, to be discussed in Sect. 6.2.1. Many similar cross sections obtained on ex-PAN fibers show that the cross section is typically circular for low $T_{HT}$(e.g., 1300°C), high-strength filaments; however, the dog-bone shape is common for higher $T_{HT}$ (e.g., 2200°C), high modulus filaments. One should remember that the shape of ex-PAN carbon fibers is determined by the shape of the polymer precursor, and not the processing temperature. The arrangement of the ribbons is usually random. However, the SEM photograph in Fig. 3.25 shows a circular filament in which the central region (core) has an apparently radial structure, while the outer regions (sheath) have a circumferential arrangement. Further remarks regarding the core-sheath arrangement are made in the next section, where high-resolution TEM has resolved the question of heterogeneity.

The microstructure of the fiber cross section for ex-pitch has many possibilities depending on the extrusion parameters of the pitch material [Hamada

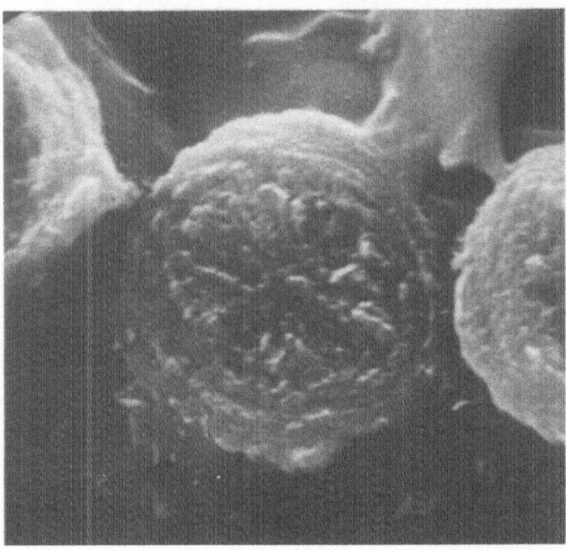

Fig. 3.25. SEM photographs of the section of an ex-PAN fiber of diameter about 7 $\mu$m suggesting a core-sheath structure [Wicks and Coyle 1976]

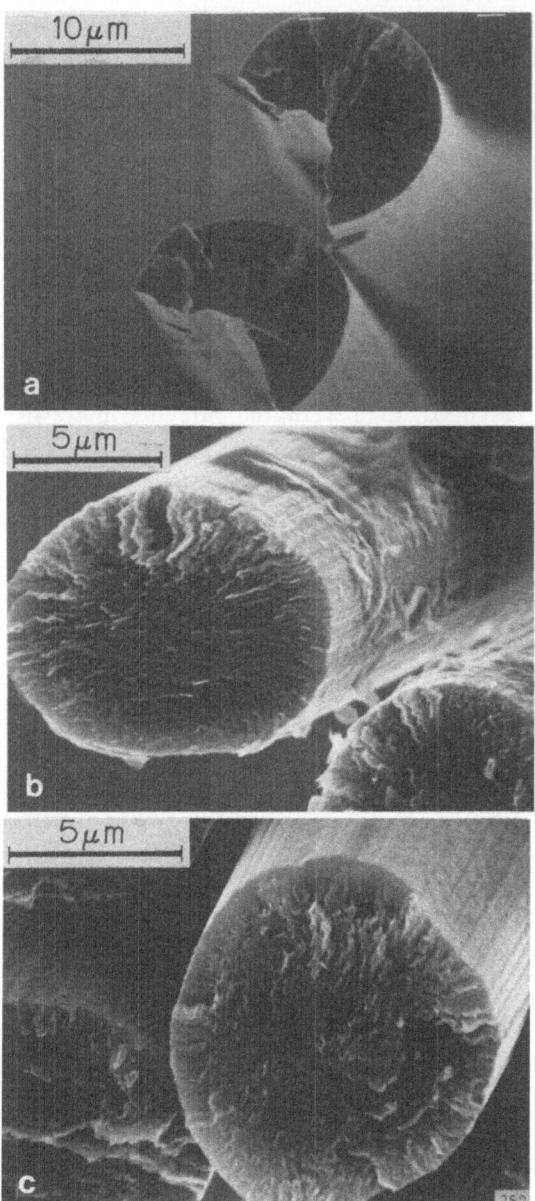

**Fig. 3.26a-c.** SEM photographs showing illustrative cross sections for ex-pitch fibers: (a) Thornel P-100, $\sigma_T \sim 2$ GPa; (b) Thornel P-120 XS, $\sigma_T \sim 2$ GPa; (c) Carbonic HM60, $\sigma_T \sim 3$ GPa [Endo 1987]

et al. 1987a,b]. The microstructure also depends on whether or not the pitch is stirred above the capillary tube where extrusion occurs (see Fig. 2.3), the shape of the stirring tip, the viscosity (temperature) of the pitch, and the stirring conditions. Some of the microstructures that could be achieved with the same pitch precursor include: radial (no stirrer); random (nozzle $\beta$); and onion type (low viscosity/high temperature). Some representative types of microstructures common for ex-pitch fibers are shown in Fig. 3.26, namely the (a) radial (PAC-man), (b) "PAN-AM" onion type, and (c) radial-ribbon type. The "PAC-man" fiber would be circular before heat treatment. The fiber, however, changes its shape upon relieving strain and reducing mass due to the loss of volatile components [Edie et al. 1986]. These authors also have prepared ex-pitch fibers spun with 3 and 8 lobes, finding improved mechanical properties. Further remarks about the arrangement of the layers in ex-pitch fibers will be made in Sect. 3.7.2.

Figure 3.27 shows a set of SEM photographs of benzene-derived carbon filaments heat-treated to temperatures up to 3500°C [Yoshida et al. 1985; Endo and Shikata 1985]. These SEM photographs clearly show a change from circular to polygonal cross sections as $T_{HT}$ is increased. Also, the tree-ring arrangement of the graphene layers can be seen at low $T_{HT}$, changing to relatively straight regions in which the strain energy is reduced at high $T_{HT}$. These SEM photographs were taken with the unusual operating conditions of 2–5 kV primary beam energy to reduce the electron penetration depth in the carbon. This low beam energy enhances the clarity of the image.

The importance of these SEM studies cannot be overemphasized and further examples are given elsewhere in the text. However, some caution should be exercised in interpreting the structure observed in fracture surfaces in terms of the inherent structure of the filament, since more detailed studies sometimes indicate that a simple interpretation of the SEM sections is in error (see D.J. Johnson [1987b]).

### 3.7.2  High Resolution Transmission Electron Microscopy

A number of TEM techniques have been used to characterize carbon fibers, either as-grown or when subjected to various graphitization steps. These techniques include bright and dark field imaging, electron diffraction and selected area electron diffraction (SAD), lattice fringe imaging, phase contrast methods, and tilting experiments, among others. Dark field imaging of (10) and (11) beams has been shown [Oberlin 1979] to provide a powerful tool for differentiating between different microstructural models for PAN- and pitch-based fibers. Image simulation by the multislice technique to interpret phase contrast images has also been emphasized [Millward and Thomas 1979]. From these studies as a function of precursor material and heat treatment temperature ($T_{HT}$), many conclusions have been reached about the structure of

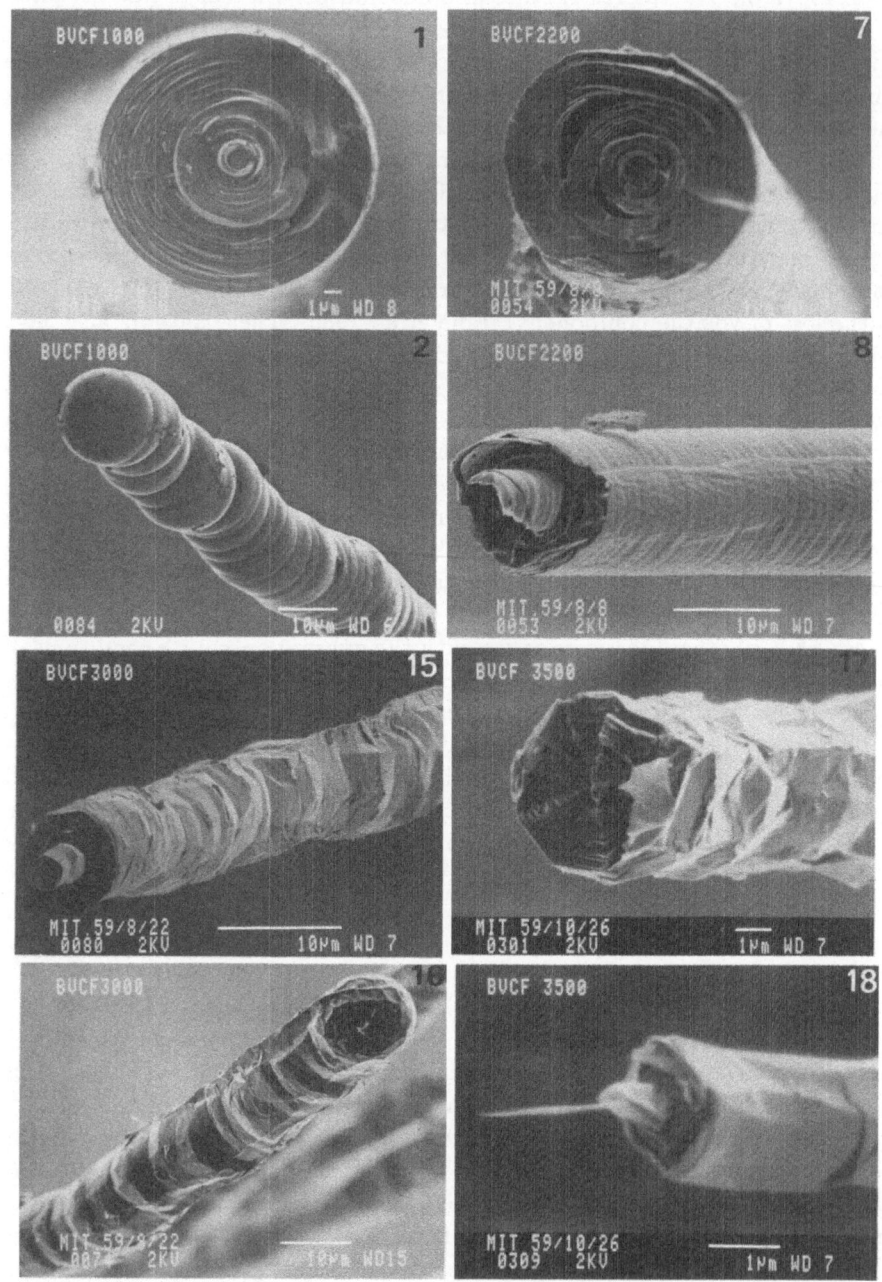

**Fig. 3.27a-d.** SEM photographs showing the morphology change on the surface and the cross section of benzene-derived filaments at different heat treatment temperatures: (a) as-deposited (1000°C); (b) 2200°C; (c) 3000°C; (d) 3500°C [Yoshida et al. 1985]

70

graphite fibers and the structural changes introduced by the various graphitization steps when the fibers are subjected to different heat treatment temperatures and conditions. Three structural parameters have been used to characterize the degree of disorder in a carbon fiber: $L_a$, $L_c$ and $r_t$, where $L_a$ and $L_c$ are respectively the in-plane and $c$-axis crystalline coherence lengths of the fibers, and $r_t$ is the transverse radius of curvature for zigzags in the lamellar ribbon structures [Guigon et al. 1984a, 1984b].

On the basis of detailed TEM observations of graphitizable carbons and of carbon fibers as a function of $T_{HT}$, the following steps in the graphitization process have been identified [Guigon et al. 1984a, 1984b], as indicated in Fig. 3.28. For $T_{HT} \leq 800°C$, the basic structural units, consisting of a short length ($\approx 10$ Å) of two parallel layers, start to pile up and form distorted columnar structures, as impurity atoms are released, mostly in gaseous form. For $800° \leq T_{HT} \leq 1500°C$, the columnar structures increase in length with a lesser degree of misorientation of the basic structural units (see Fig. 3.28). In this step, individual misoriented basic structural units become aligned and $L_c$ increases gradually. For $1500° \leq T_{HT} \leq 1900°C$, the columnar structure disappears as wavy ribbons or wrinkled layers are formed by hooking the adjacent columns together, and in this range of $T_{HT}$, both $L_a$ and $L_c$ increase rapidly. By $T_{HT} \sim 1700°C$, the zigzag structure of the wavy ribbons begins to disappear and a turbostratic layer arrangement starts to appear, with $L_a$ values higher than 200 Å at $T_{HT} \sim 2000°C$. In the range $1900 \leq T_{HT} \leq 2100°C$, the layers start to unwrinkle. By $2100°C$, most of the in-plane structural defects have been eliminated so that a rapid increase in $L_a$ can occur due to the disappearance of tilt and twist boundaries. When the dewrinkling is

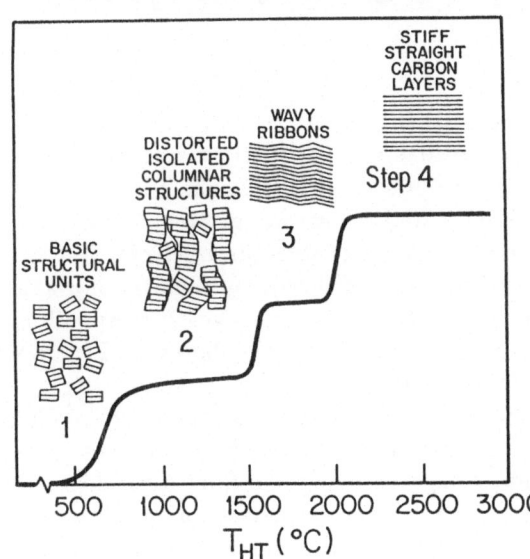

Fig. 3.28. Various steps of the graphitization process as a function of heat treatment temperature $T_{HT}$ [Goma and Oberlin 1980]

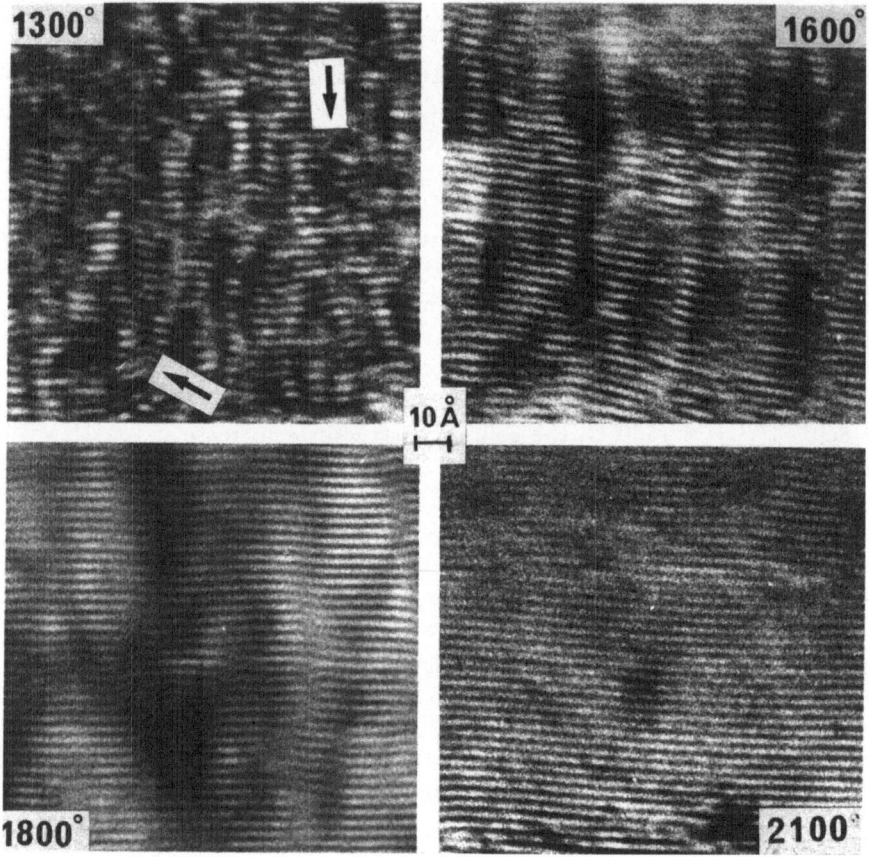

**Fig. 3.29a.** Caption see opposite page

complete, the interlayer spacing reaches $\sim 3.42$ Å. At this $T_{HT}$ value, there is an onset of 3D interplanar site correlations and many electronic properties exhibit sudden changes. Above $T_{HT} \simeq 2100°C$, stiff, straight carbon layers are observed in the lattice fringe images and 3D crystal growth commences. Graphitization thus corresponds to the removal of structural defects, first between layers, then a straightening of the layers followed by removal of $a$-axis misorientation. Lattice fringe patterns and dark field images corresponding to the various graphitization steps discussed above are shown in Fig. 3.29 [Rouzaud et al. 1983; Goma and Oberlin 1980].

In the graphitization of carbon fibers, the lamellae assume a ribbon-like structure and are folded along the fiber axis. As the graphitization process proceeds, the voids, which are elongated along the fiber axis, decrease in size and concentration. With regard to the graphitization process, the high tensile strength fibers (e.g., ex-PAN) behave similarly to the low $T_{HT}$

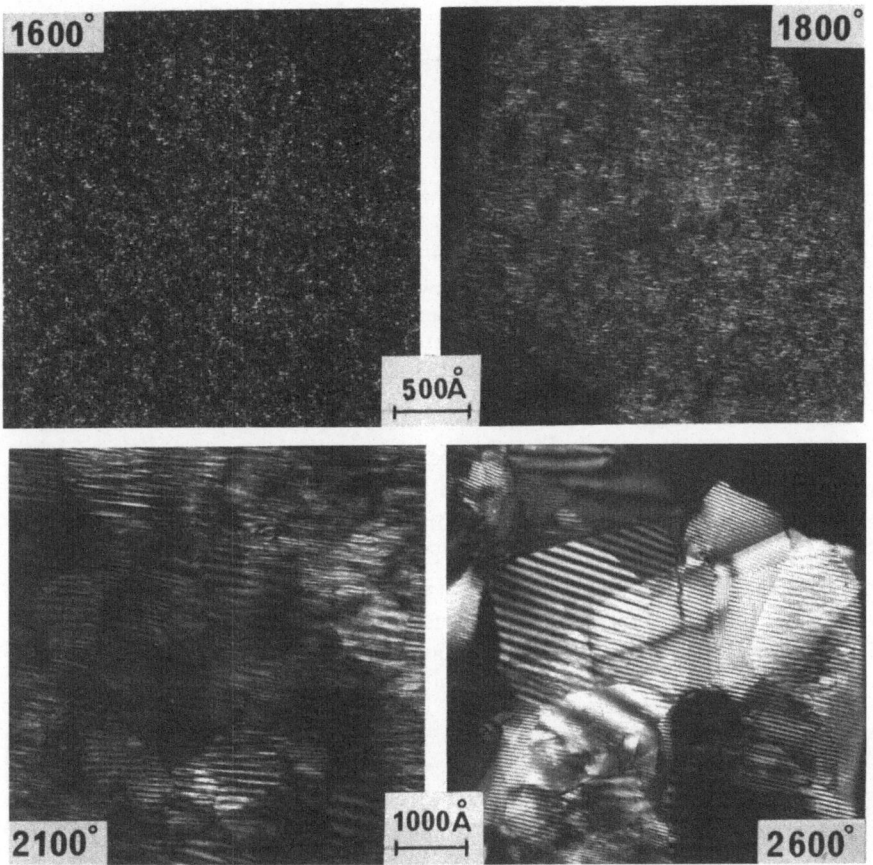

**Fig. 3.29a,b.** Progress in the carbonization process with increasing heat treatment temperatures as viewed by high resolution TEM: (a) (002) lattice fringes; (b) (11) dark-field micrographs. In (a) the 1300°C lattice fringe pattern shows distorted columns of layers with misoriented single basic structural unit (*arrows*), 1600°C shows long distorted layers, wrinkled in a zigzag; 1800°C shows less distorted layers; 2100°C shows stiff and straight layers. In (b) 1600°C and 1800°C show turbostratic rotational Moiré fringes and crystallites, while 2100°C and 2600°C show the growth of crystallites [Rouzaud et al. 1983; Goma and Oberlin 1980]

non-graphitizable carbons, while the high modulus fibers (e.g., ex-pitch) behave more like high temperature graphitizable carbons. For the high tensile strength fibers, the lamellae form distorted graphitic layers that are arranged in a wrinkled zigzag pattern.

The longitudinal and transverse radii of curvature can be estimated from the (002) dark field images (see Fig. 3.30). If the basic structural units of the fibers make folds in a direction parallel to the fiber axes and an electron beam

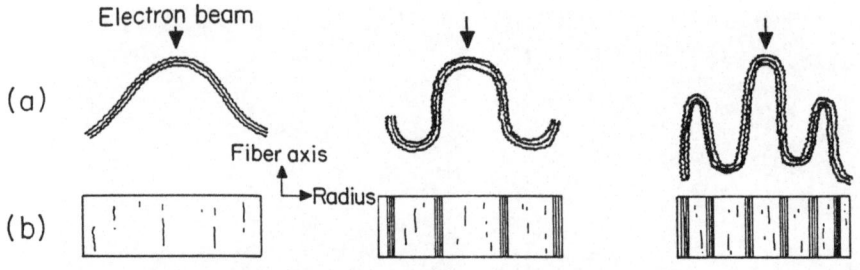

Fig. 3.30. Folded structure model showing how folds are detected by the TEM. A dark band is formed when the graphene planes line up along the electron beam. From these minima and maxima in the signal, the transverse and longitudinal radii of curvature are estimated [Guigon et al. 1984a, 1984b; Endo 1988a]

is directed perpendicular to the fiber axis, crystallites are observed as a bright band when the basal planes are parallel to the incident beam. By measuring the distance between the bright bands running parallel to the TEM image, the radius of the fold can be estimated (see Fig. 3.30). The thickness of the bright band gives the thickness of the parallel stackings of the carbon layers along the fiber radius. Such an analysis has been used to show that the basic structural unit for the P-100 and P-120 fibers is planar, while the Carbonic HM60 fibers

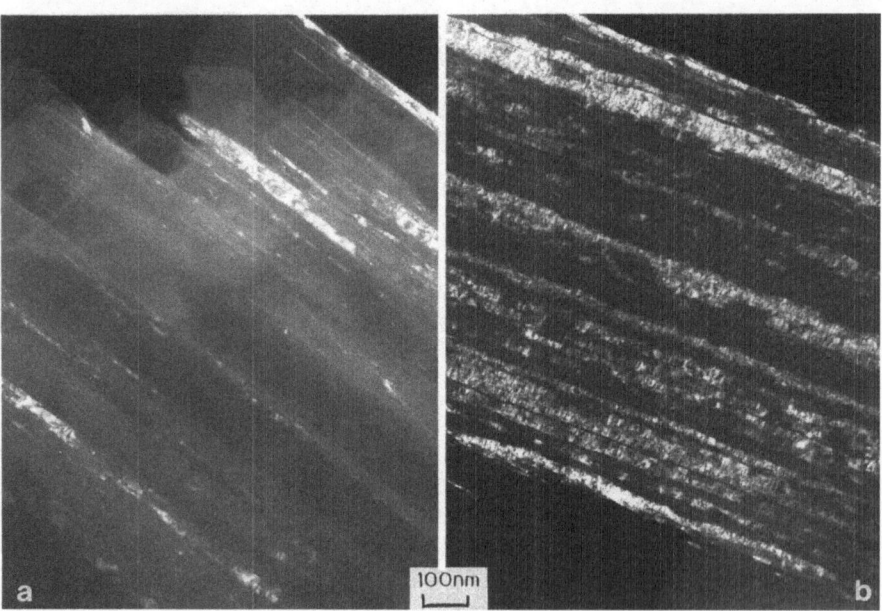

Fig. 3.31a,b. Dark-field images from (002) spots of two ex-pitch fibers; (a) P-120, (b) Carbonic HM60 [Endo 1988a]

show well-developed folds [Endo 1988a]. Bright and dark field TEM images confirm these basic conclusions, as illustrated in Fig. 3.31. The relationship of these folds to the mechanical properties is discussed in Sect. 6.1.2.

For the high modulus fibers, the distorted carbon planes are dewrinkled upon heat treatment and the layer stacks become folded parallel to the fiber axis. From the TEM pictures, the parameters $L_a$, $L_c$ and $r_t$ can be directly determined. A correlation of the structure with the mechanical properties shows that the tensile strength depends linearly on the average of the ratio of parameters $\bar{r}_t/\bar{L}_a$, while the bulk modulus decreases with increasing $(1/\bar{L}_a)$, and the resistivity increases as $(1/\bar{L}_a)$. A model for the microtexture of a high modulus fiber developed by Guigon et al. (1984a, 1984b) is presented in Fig. 3.32, showing the general short range orientation of the wavy ribbons. A similar figure for a high strength fiber appears as Fig. 1.3.

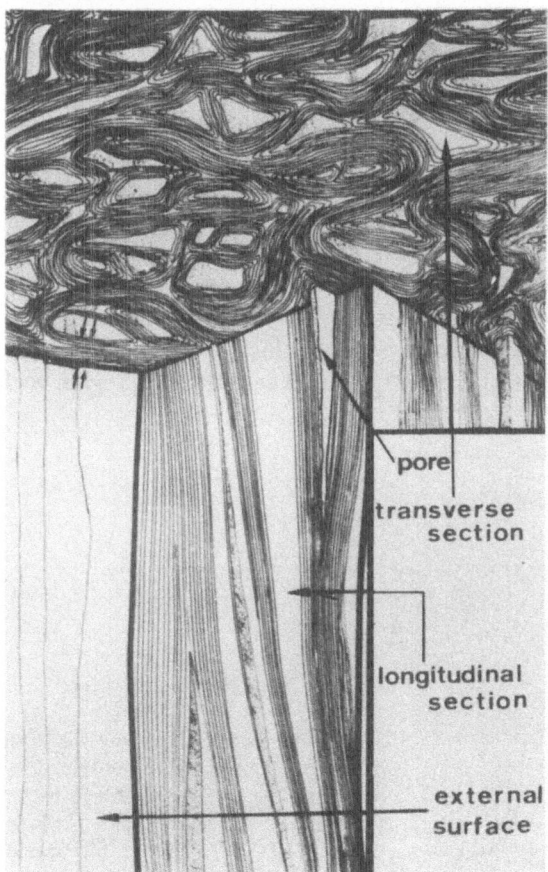

pore

transverse
section

longitudinal
section

external
surface

Fig. 3.32. An artist's conception of structural features of a high-modulus ex-PAN fiber. This can be compared to a similar sketch of a high strength ex-PAN fiber (Fig. 1.3) [Guigon et al. 1984b]

Electron diffraction is a valuable tool for studying the structure of carbon fibers. It gives information which is similar to that obtained from x-ray diffraction, but can probe regions of small volume. Firstly, the penetration of the electron beam is limited by energy losses, so that samples are usually less than 0.1 $\mu$m thick. Secondly, the beam is focused onto a small spot, which can be of diameter less than 0.1 $\mu$m. "Selected-area diffraction" (SAD) then probes the structural heterogeneity on a scale of the order of 0.1 $\mu$m.

The small scale of the diffracting volume can produce an interesting effect in the case of turbostratic carbon fibers. This can be understood by considering the Ewald sphere construction, described in Sect. 3.1.6. The $a$-axis averaging over the cross section of the fiber will not occur, so that intersection of the Ewald sphere with the cylinders which are formed by $a$-axis rotation of the turbostratic rods (see Fig. 3.8d) will only be circles for the particular case of $c$-axes oriented parallel to the incident beam. Non-circular diffraction arcs for the (10) reflections are a diagnostic of turbostratic layer packing.

There have been many studies of selected-area electron diffraction. Several patterns will be considered together with other diffraction information on fibers (in this section), intercalated fibers (in Sect. 10.2) and ion implanted fibers (in Sect. 11.3).

Ex-PAN fibers exhibit a skin (between 1000 Å and 1500 Å thick) that is somewhat more graphitic than the core region, see Fig. 3.33, as first suggested by Butler and Diefendorf [1969]. The subject of skin-core heterogeneity has been controversial, and has been reviewed by Bennett and D.J. Johnson [1979], and D.J. Johnson [1980, 1987b]. Early work using optical microscope techniques (Sect. 3.6), suggested that the relatively thick sheath comprises crystallites which are circumferentially aligned. The core was assumed by Watt and W. Johnson [1970] to arise from regions which were not fully oxidized in the pre-carbonization steps (Sect. 2.1). In this case the core region was

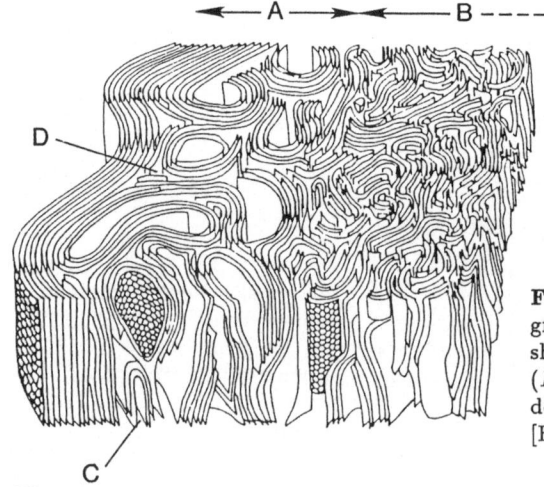

**Fig. 3.33.** An artist's conception of the graphene layers of a high-modulus fiber, showing: ($A$) a narrow surface region; ($B$) an interior region; ($C$) a "hairpin" defect; and ($D$) a wedge disclination [Bennett et al. 1983]

thought to be oriented radially. However, further studies of Knibbs [1971], using optical microscopy techniques (Sect. 3.6), showed a skin-core heterogeneity, even for fully stabilized fibers. SEM photographs in Fig. 3.25 by Wicks and Coyle [1976] show a near-surface sheath that appears to be $> 1\,\mu$m in thickness.

D.J. Johnson [1980] used dark-field images of longitudinal sections to show that the layers were randomly oriented throughout the section, and suggested that the origin of the optical birefringence effects arises from strain rather than structural differences. The strain differences across the fiber section are suggested to originate from differential thermal expansion. Bennett and D.J. Johnson [1979] carried out high-resolution TEM studies and showed that there was a relatively narrow skin region in fully stabilized fibers (100–150 nm) in which the structure differed from the core region.

Even in the skin region for high $T_{\mathrm{HT}}$ fibers, the crystallites of the layer planes are interlinked in a highly complex manner, and they enclose an intricate void system [Bennett and D.J. Johnson 1979]. The crystallites also exhibit a high axial preference. Typical values of $L_c$ and $L_a$ for PAN fibers heat treated at 2500°C are: $L_c \sim 50$ Å; $L_{a\parallel} \gg 100$ Å, (for both skin and core), $L_{a\perp} \sim 85$ Å in the skin and $\sim 50$ Å in the core [Bennett and D.J. Johnson 1979]. The folds are more extensive in the skin, having layers which are folded back by 180° in a hairpin fashion. A simple model which incorporates these features is illustrated in Fig. 3.34 [Johnson 1987b]. It is noted that this model does not include skin-core heterogeneity arising from a change from random to circumferential alignment, except in a very narrow region (100 nm) near the surface. It is the relative freedom of the surface layers to rearrange themselves, unimpeded by the constraints of neighboring crystallites, that permits the skin-core differences to occur.

**Fig. 3.34.** A sketch of the 3D arrangement of the layer planes in ex-PAN fibers near the surface, showing the skin-core heterogeneity proposed by Bennett and D.J. Johnson [1979]. The figure is from D.J. Johnson [1987b]

The formation of large misoriented crystallites in the skin region is believed to be responsible for fiber failure under applied stress [Reynolds and Sharp 1974]. Figure 1.5c illustrates the lattice fringe image for the surface layers, which includes a badly misoriented "crystallite". It is again important to stress that the skin-core heterogeneity is dependent on a number of processing factors, such as $T_{HT}$, the precursor polymer, the spinning process, and the degree of oxidation of the polymer, etc. Some of the differences between various studies arise from such processing conditions, while others from difficulties in the interpretation of observed phenomena.

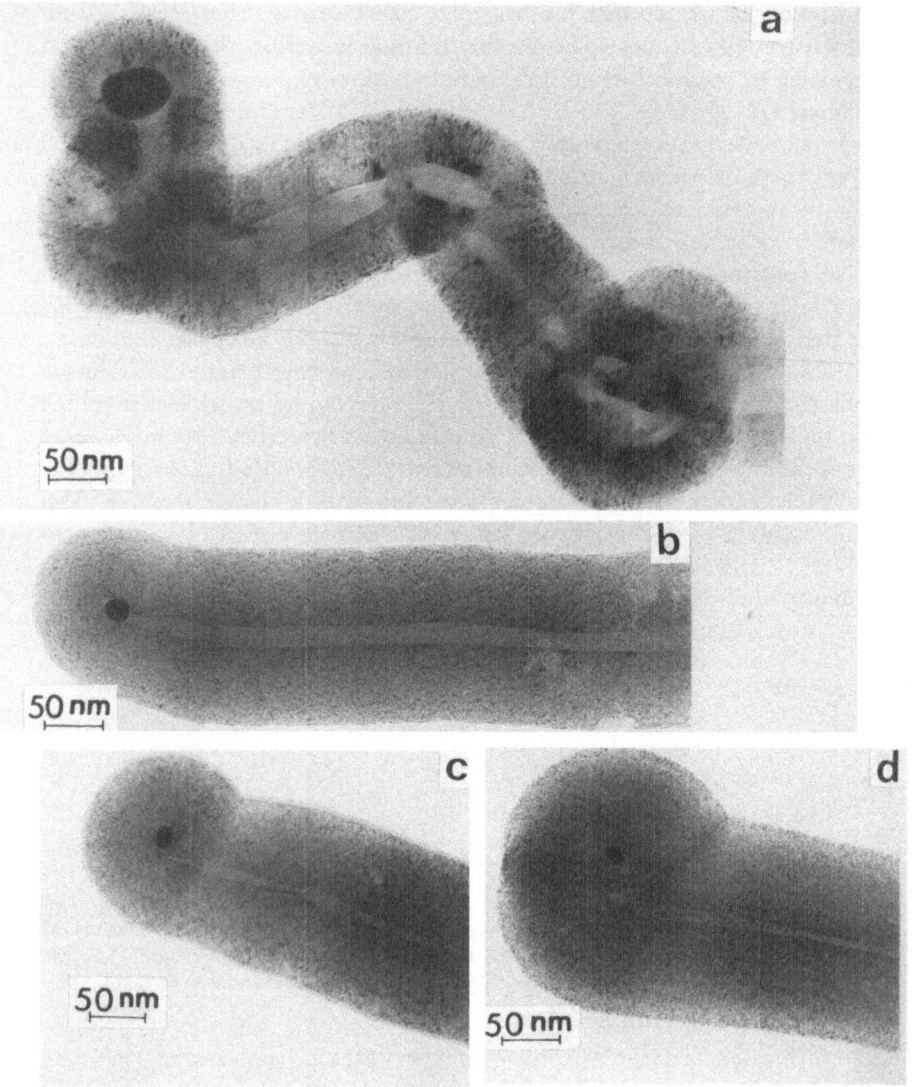

**Fig. 3.35.** Bright field TEM images of carbon fibers derived from benzene showing hollow tube structures and catalyst particles at fiber tips. [From Oberlin et al. 1976a]

### 3.7.3 Microstructure of the CCVD Filaments

In the past, there have been differing reports on whether or not the core of CCVD filaments is hollow, or of relatively low density amorphous carbon. However, more recent results on CCVD filaments have provided strong support for a hollow core in the cases studied [Oberlin et al. 1976a], with carbon layers forming a tree-ring structure around a central core region as shown in Figs. 3.35 and 3.36 [Baird et al. 1974; Koyama et al. 1974; Endo 1975; Endo et al. 1976, 1977a, 1982b; Oberlin et al. 1976a, 1976b; Tibbetts 1983]. In Fig. 3.35, we see the ends of several fibers containing a small catalytic Fe particle (from $\sim 3$ nm to $\sim 30$ nm diameter) leading to a hollow core of slightly smaller diameter than the catalytic particle. The catalytic particles are of surprisingly uniform dimensions and exhibit restricted field electron diffraction patterns characteristic of cementite when embedded in the as-grown fibers and cooled to room temperature. Sometimes a carbon fiber develops around

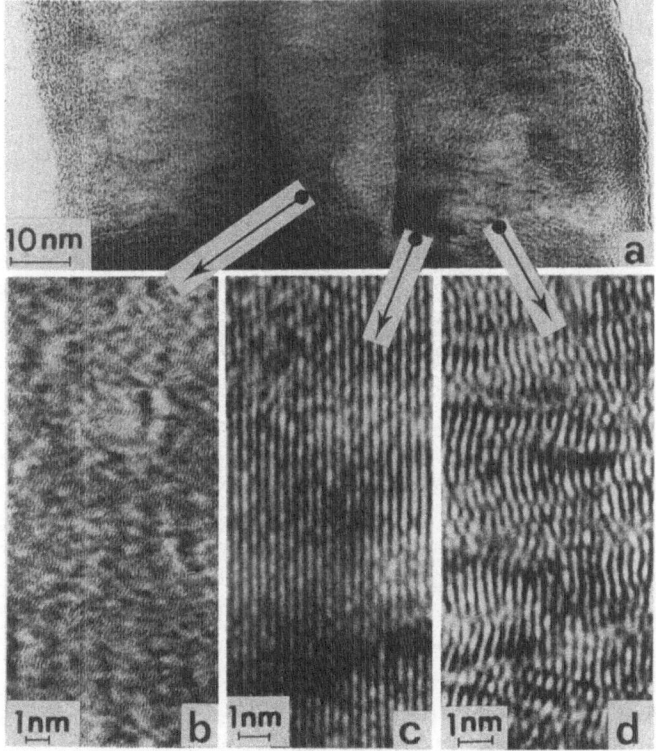

Fig. 3.36a-d. Lattice fringe images of carbon fibers derived from benzene: (a) view along the fiber axis, (b) fringes from a disordered region, (c) (002) fringes near the hollow tube, and (d) fringes in the external part of the fiber. [From Oberlin et al. 1976a]

a pair of catalytic particles. High resolution TEM lattice fringe images (see Fig. 3.36) show the inner fiber core to have a disordered structure, while the microstructure of the outer fiber core shows very well ordered, long parallel planes (Fig. 3.36c) and the microstructure in the as-grown sheath region (Fig. 3.36d) is somewhat more disordered than that shown in Fig. 3.36c. The high resolution TEM studies of the filaments [Oberlin et al. 1976a] show the spacing of the layers in the central tree ring structure to be significantly greater ($\sim$3.48 Å) than the normal interlayer carbon spacing (3.35 Å), indicating a two-dimensional organization of these core layer structures, consistent with the large curvatures (small radius of curvature) of the layers near the fiber cores. After heat treatment at elevated temperatures ($T_{HT} > 2500°C$), the sheath region becomes well graphitized with an interplanar distance of $\bar{c} \sim 3.36$ Å, similar to that of single crystal graphite, while the core region remains turbostratic due to the high radius of curvature, which prevents adjacent layers from establishing long range 3D interplanar registration of the graphene planes. R.T.K. Baker et al. [1972, 1973] and R.T.K. Baker and Harris [1973] have shown that the material near the core oxidizes more rapidly than that in the sheath, consistent with higher perfection in the sheath.

TEM studies on vapor-grown fibers show that stiff, nearly perfect graphitic layers are formed very early in the graphitization process. In fact, well-formed graphite layers are already present in the as-grown fibers prepared at $\sim$1100°C. The selected area diffraction pattern (Fig. 3.37b) of the tip of the as-grown fiber (Fig. 3.37a) shows broad, though well-defined (002) diffraction spots, typical of turbostratic graphite. The lattice fringe image pattern taken from the as-grown fiber (see Fig. 3.37c) shows rather straight continuous layers close to the hollow fiber core, with more disordered, zigzag lamellae observed away from the hollow core region. As the fibers are heat treated to 2800°C, the lamellae acquire a polyhedral cross section as shown in the bright field image of Fig. 3.37d, indicative of a major increase in $L_a$ and $L_c$, as indicated by the sharp (00$l$) spots observed in the selected area electron diffraction pattern of the fiber tip shown in Fig. 3.37e, where $l = 2, 4, 6,...$ . The formation of (101) arcs in the diffraction pattern is indicative of the onset of $AB$ interlayer correlations as 3D order is established (Sect. 3.4.2). The long in-plane coherence distance for the heat-treated fibers is seen in Fig. 3.37f, where the (002) lattice fringes are shown as stiff, regular flat layers [Oberlin et al. 1976a; Endo et al. 1977a].

These are but a few examples of the important role played by electron microscopy in the characterization of carbon fibers. These characterization techniques are also used extensively in the study of intercalated carbon fibers (see Sect. 10.2) and of ion-implanted fibers (see Sect. 11.3).

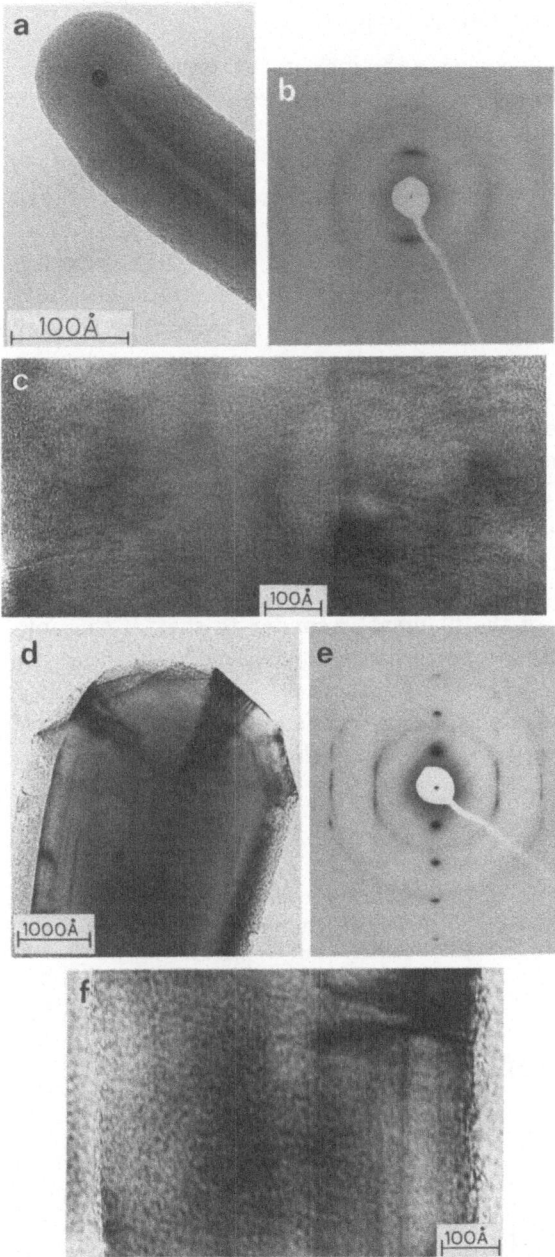

Fig. 3.37a-f. Graphitization of vapor-grown carbon fibers. (a-c) For the as-grown fibers: (a) bright field image, (b) selected area diffraction (SAD) pattern of (a) (turbostratic structure), (c) (002) lattice fringes showing distorted layers. (d-f) For the graphitized fibers: (d) tip of the fiber in bright-field showing its polyhedral shape, (e) SAD pattern of (d) (polycrystalline graphite), (f) (002) lattice fringes showing stiff and flat layers [Oberlin et al. 1976b; Endo et al. 1977a]

## 3.7.4 Electron Energy Loss Spectroscopy

Electron energy loss spectroscopy concerns the inelastic scattering of an electron beam by the solid, whereas the TEM studies discussed above concern elastic scattering. Electron energy loss spectroscopy (EELS) is carried out by directing an energetic (typically $\sim$100 keV) beam of electrons onto a thin region of material (typically $<$100 nm) and measuring the intensity of the transmitted beam as a function of energy loss ($\Delta E$). The bond type can be inferred from such measurements, probing less than 0.1 $\mu$m in lateral extent. The technique has recently been applied by Batson [1987] to carbon filaments prepared by ion-beam techniques (Sect. 2.5). In the case of carbon, two energy regions are of interest:

- $\Delta E \sim$280–300 eV. In this region the energy loss is associated with the excitation of a $1s$ core electron to the Fermi level of graphite, or to the conduction band of diamond. It can be seen that the spectra for these two cases are different (Fig. 3.38a) enabling the bond type to be distinguished. Also in the plot is the curve for amorphous, hydrogenated carbon [Fink et al. 1983], which has a mixture of predominantly $sp^2$ (graphitic) and $sp^3$ (diamond-like) bonds, and a clearly distinguishable signature. The spectrum of the filament is similar to that of this material. (For a review of the properties of amorphous carbon films, see J. Robertson [1986].)

- $\Delta E \sim$ 0–40 eV. In this region the energy loss is due to the excitation of plasmons, identifiable as arising from the $\pi$ electrons and the $\pi$ and $\sigma$ states of the $L$ shell (Fig. 3.38b).

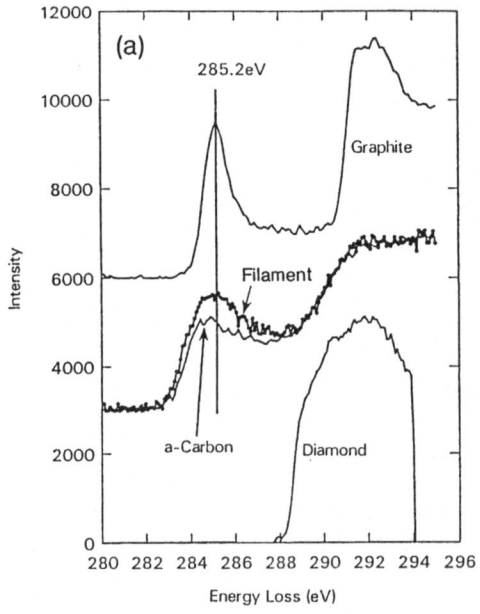

**Fig. 3.38a,b.** EELS spectra of graphite, diamond, amorphous (hydrogenated) carbon and an ion-grown carbon filament [Batson 1987]. (a) 280–296 eV for several carbon materials (b) 0–40 eV for the ion-grown filament only

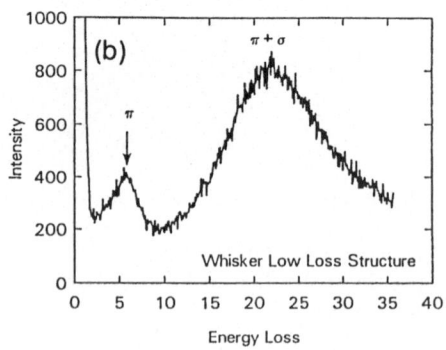

The spectra in both the low and high energy regions for ion beam grown filaments near the tip are consistent with an admixture of $sp^2$ and $sp^3$ bonds. Note that the peak is shifted to lower energy in Fig. 3.38 than that at 285.2 eV for graphite, but not as much as for amorphous carbon. This is consistent with an amorphous, hydrogenated carbon heated to $\sim$650°C by Fink et al. [1983] in which a bond ratio of $sp^2 : sp^3 = 2 : 1$ was inferred. The low energy electron energy loss spectrum for the same, annealed, amorphous hydrogenated carbon was also similar to that obtained for the filament (Fig. 3.38b). This type of structure is also consistent with high resolution electron microscope studies made on these filaments by the same author [Batson 1987].

## 3.8 Other Spectroscopies

Normally several spectroscopic techniques are used simultaneously to characterize carbon filaments. As an example, Electron Spectroscopy for Chemical Analysis (ESCA) or X-ray Photoelectron Spectroscopy (XPS) is used routinely to probe for chemical impurities, as are Secondary-Ion Mass Spectroscopy (SIMS), and Auger Electron Spectroscopy (AES). The usefulness of ESCA is discussed in Sect. 6.2 with respect to the importance of impurities in fracture processes [J.B. Jones et al. 1980]. Various spectroscopic tools are used to study the effects of surface treatments, which allow the fibers to bond more readily to matrix materials [Waltersson 1982]. X-ray Photoelectron Spectroscopy (XPS) has been used to study the chemical effects of surface treatments, also. This technique probes the near-surface region to a depth of about 3 nm with typical electron energies, and several studies on carbon fibers have been reported. References can be found in Ishitani [1981], Proctor and Sherwood [1983] and Denison et al. [1985, 1987].

Carbon-13 nuclear magnetic resonance (NMR) has been used to assess the orientation of graphene planes in ex-pitch fibers [Resing et al. 1985]. This can be done since the chemical shift depends on the direction of the magnetic field with respect to the c-axis ($\theta$). To carry out these measurements, a pitch fiber was intercalated with $AsF_5$, and $^{13}C$ NMR measurements were carried out as a function of $\theta$. The data could be fitted with a model in which 85 % of the graphene planes were closely oriented to the fiber axis, while 15 % were randomly oriented. These randomly oriented planes are of importance in controlling fracture behavior, as is discussed in Sect. 6.2.

## 3.9 Other Characterization Techniques

There are a number of other experimental techniques which can be used to characterize the defect structures of graphite, but not being structural measurements, they depend on structural techniques (e.g., x-ray diffraction) for their calibration. Examples of such characterization techniques are: Raman spectroscopy (Chap. 4), electrical resistivity and magnetoresistance (Chap. 8), diamagnetic susceptibility and electron spin resonance (Chap. 8) and photoconductivity (Chap. 4). These techniques may be called "secondary structural techniques" and are discussed in more detail in the chapters cited.

# 4. Lattice Properties

The lattice dynamics of a material are intimately connected with the elastic coefficients, which control the mechanical properties. The lattice dynamics in turn determine the phonon dispersion relations, which control the specific heat and thermal expansion, and strongly affect the transport properties through carrier scattering processes. Since the lattice dynamics for carbon fibers have not been considered explicitly, the results for crystalline graphite have conventionally been used with some modifications to take into account the small crystallite size, the high concentration of defects and the general disorder which are all prevalent in carbon fibers [Lespade et al. 1982]. To explain certain aspects of transport phenomena in carbon fibers (Chap. 8), the dispersion relations for crystalline graphite in the long wavelength approximation are used to treat electron-phonon scattering.

The general outline of this chapter will be to introduce the elastic properties and lattice dynamics of single crystal graphite in Sects. 4.1 and 4.2, then to discuss models for the elastic properties of fibers (Sect. 4.3), followed by characterization techniques for carbon fibers that are sensitive to lattice modes, such as Raman spectroscopy (Sects. 4.4 and 4.5) and photoconductivity (Sect. 4.6).

## 4.1 Elastic Parameters of Single Crystal Graphite

The elastic response of a medium depends on the elastic coefficients, which are defined by the stress-strain relationship in the linear regime. In a crystalline material the general linear tensor relationship between stress ($\sigma_{\mathrm{T}ij}$) and strain ($\varepsilon_{ij}$) is given by

$$\sigma_{\mathrm{T}ij} = \sum_{k,l=1}^{3} C_{ijkl}\varepsilon_{kl} \quad , \tag{4.1}$$

which is a generalized statement of Hooke's law. In (4.1) repetition of indices indicates summation over these indices. The $C_{ijkl}$ coefficients are the elastic stiffness coefficients of the material, and the elastic compliance coefficients $S_{ijkl}$ define the inverse of $C_{ijkl}$, i.e.,

$$\varepsilon_{ij} = \sum_{k,l=1}^{3} S_{ijkl}\sigma_{\mathrm{T}kl} \quad . \tag{4.2}$$

The 81 (or $3^4$) components of these tensors are usually greatly reduced by the crystalline symmetry. The following contracted notation is conventionally used:

$$C_{rs} \equiv C_{ijkl} \begin{cases} r = i, & \text{for } i = j \\ s = k, & \text{for } k = l \\ r = m + 3, & \text{for } i \neq j \neq m \\ s = p + 3, & \text{for } k \neq l \neq p \end{cases} \quad (4.3)$$

where $r, s = 1, 2, \ldots, 6$. In this notation there are 5 independent elastic constants for hexagonal symmetry; they are $C_{11}(C_{1111})$, $C_{12}(C_{1122})$, $C_{13}(C_{1133})$, $C_{33}(C_{3333})$, and $C_{44}(C_{3232})$ in which both the contracted and the full notation for the elastic constants are listed as $C_{rs}(C_{ijkl})$. The 5 independent elastic compliances for hexagonal symmetry are correspondingly $S_{11}(S_{1111})$, $S_{12}(S_{1122})$, $S_{13}(S_{1133})$, $S_{33}(S_{3333})$, and $S_{44}(4S_{3232})$ in which both the contracted and the full notation for the elastic compliances are given by Nye [1960]. From (4.1) and (4.3) we can write down the stress-strain relationship for hexagonal graphite as follows:

$$\sigma_{\mathrm{T}} = \begin{bmatrix} \sigma_{\mathrm{T}11} \\ \sigma_{\mathrm{T}22} \\ \sigma_{\mathrm{T}33} \\ \sigma_{\mathrm{T}32} \\ \sigma_{\mathrm{T}13} \\ \sigma_{\mathrm{T}12} \end{bmatrix} = \begin{bmatrix} C_{11} & C_{12} & C_{13} & 0 & 0 & 0 \\ C_{12} & C_{11} & C_{13} & 0 & 0 & 0 \\ C_{13} & C_{13} & C_{33} & 0 & 0 & 0 \\ 0 & 0 & 0 & C_{44} & 0 & 0 \\ 0 & 0 & 0 & 0 & C_{44} & 0 \\ 0 & 0 & 0 & 0 & 0 & \frac{1}{2}(C_{11} - C_{12}) \end{bmatrix} \begin{bmatrix} \varepsilon_{11} \\ \varepsilon_{22} \\ \varepsilon_{33} \\ 2\varepsilon_{32} \\ 2\varepsilon_{13} \\ 2\varepsilon_{12} \end{bmatrix} . \quad (4.4)$$

The strain is related to the stress by the inverse relation involving the compliance tensor $S_{rs}$ given by (4.2). Values for the elastic constants $C_{rs}$ and for the compliance coefficients $S_{rs}$ have been reviewed by Kelly [1981], and

Table 4.1. Elastic constants[a] for single crystal graphite [from Kelly 1981]

| | Elastic moduli [GPa] | | Elastic compliance $[10^{-12} \mathrm{Pa}^{-1}]$ |
|---|---|---|---|
| $C_{11}$ | 1060±20 | $S_{11}$ | 0.98±0.03 |
| $C_{12}$ | 180±20 | $S_{12}$ | −0.16±0.06 |
| $C_{13}$ | 15±5 | $S_{13}$ | −0.33±0.08[b] |
| $C_{33}$ | 36.5±1 | $S_{33}$ | 27.5±1.0 |
| $C_{44}$ | 4.5±0.5 | $S_{44}$ | 240±30 |

[a] Relationships between the elastic stiffness coefficients $C_{ij}$ and the elastic compliances $S_{ij}$ are

$$C_{11} = 0.5[(S_{11} - S_{12})^{-1} + X S_{33}] \qquad C_{13} = -X S_{13}$$
$$C_{12} = 0.5[-(S_{11} - S_{12})^{-1} + X S_{33}] \qquad C_{44} = S_{44}^{-1}$$

where $X = C_{33}(C_{11} + C_{12}) - 2C_{13}^2 = [S_{33}(S_{11} + S_{12}) - 2S_{13}^2]^{-1}$.
Numerically $X = (4.48 \pm 0.16) \times 10^4$ GPa$^2$.
[b] Thermal expansion data [Kelly 1981] are fitted best with $S_{13} = -1.8 \times 10^{-12} \mathrm{Pa}^{-1}$.

are given in Table 4.1 together with relations between them. The high values of $C_{11}$ and $C_{12}$ reflect the strong intra-layer $sp^2$ bonds, which are stronger for carbon than any other material [Coulson 1952]. The low values for $C_{33}$ and $C_{44}$ reflect the weak interlayer bonds. The uncertainty in $C_{13}$ arises from experimental uncertainties, while the uncertainty in $C_{44}$ arises from sample-dependent phenomena. It should be pointed out that the elastic constants can be determined directly from the phonon dispersion relations $\omega(q)$ discussed in Sect. 4.2. When $C_{13}$ and $C_{44}$ are determined from $\omega(q)$, no sample-dependent effects are expected. If, however, the elastic constants are measured by ultrasonic methods, the results for certain elastic constants are found to be sensitive to the defect structure. In general, the elastic constants exhibit only a very weak temperature dependence.

## 4.2  Lattice Dynamics of Single Crystal Graphite

A number of models for the lattice dynamics of graphite have been proposed. The work until 1982 has been reviewed [Dresselhaus and Dresselhaus 1979, 1982], and therefore is only briefly mentioned here. An early theory of the lattice vibrations of graphite by Komatsu and Nagamiya [1951] attempted to explain the observed $T^2$-dependent specific heat at low temperature. Since the theory of Komatsu and Nagamiya [1951] was based on the semi-continuum model, it is not valid for high frequencies or large wave numbers. Thus, Yoshimori and Kitano [1956] extended the theory using the Born-von Karman force constant model. More detailed calculations based on the Born-von Karman approach were later carried out by Nicklow et al. [1972], Maeda et al. [1979] and Al-Jishi and Dresselhaus [1982a]. An ab initio calculation of the elastic stiffness constants of graphite has been made recently by Jansen and Freeman [1987]. The most recent models provide a good fit to the experimentally observed phonon frequencies and elastic constants for crystalline graphite.

In order to fit the experimental elastic constant measurements for graphite, it was necessary to extend the Maeda model to include up to the fourth-neighbor in-plane and out-of-plane interactions [Al-Jishi and Dresselhaus 1982a]. Five types of force constants were introduced and evaluated by comparison with experiment, namely, $\phi_r^{(n)}$, $\phi_{ti}^{(n)}$, and $\phi_{to}^{(n)}$ representing, respectively, the radial, in-plane tangential, and out-of-plane tangential force constants between the $n^{\text{th}}$ nearest in-plane neighbors, while $\hat{\phi}_r^{(n)}$ and $\hat{\phi}_t^{(n)}$ denote the radial and tangential force constants between the $n^{\text{th}}$ out-of-plane nearest neighbors. The convergence of the neighbor expansion is slow. To obtain a faster convergence, the shell model, which includes screening effects due to the electronic charge, should be employed.

Calculated results for the phonon dispersion curves along certain high symmetry axes and for the corresponding phonon density of states relation

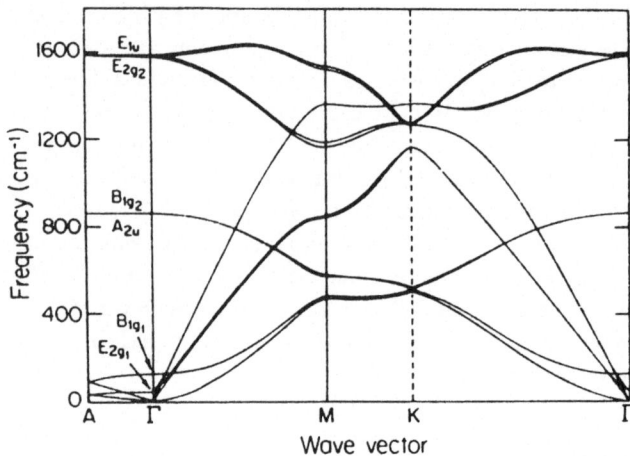

**Fig. 4.1.** Phonon dispersion curves for graphite calculated along certain high symmetry axes [Al–Jishi and Dresselhaus 1982a]

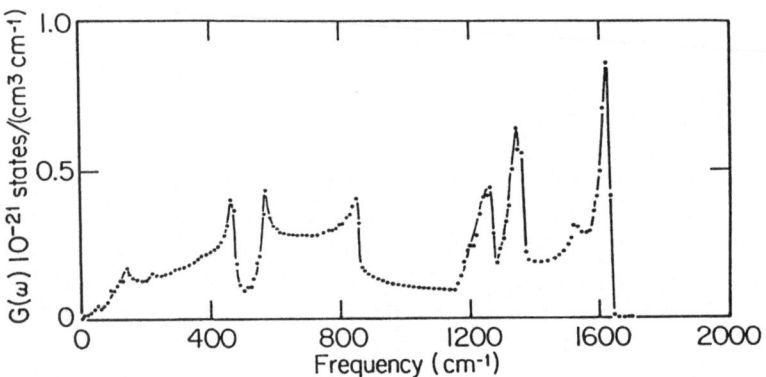

**Fig. 4.2.** Phonon density of states corresponding to the phonon dispersion curves for graphite in Fig. 4.1 [Al–Jishi and Dresselhaus 1982a]

**Table 4.2.** Values for phonon frequencies obtained fitting phonon data for graphite to a Born-von Karman model.[a]

| Phonon mode | Observed [cm$^{-1}$] | Calculated [cm$^{-1}$] |
|---|---|---|
| $\omega(E_{2g_2})$ | 1582 | 1582 |
| $\omega(E_{1u})$ | 1587 | 1587 |
| $\omega(A_{2u})$ | 868 | 867 |
| $\omega(E_{2g_1})$ | 42 | 42 |
| $\omega(B_{1g_1})$ | 127 | 127 |
| $\omega_{TA}(M\text{point})$ | ∼465 | 465 |
| $\omega_{TO}(M\text{point})$ | ∼480 | 475 |

[a] References for the experimental and calculated values listed in the table are given in Al-Jishi and Dresselhaus [1982a].

are shown in Figs. 4.1 and 4.2. Table 4.2 gives the observed lattice mode frequencies and a comparison of the observed values to the model calculations [Al–Jishi and Dresselhaus 1982a]. The calculations yield excellent agreement with infrared, Raman and neutron scattering measurements of lattice modes and with the measured elastic constants.

### 4.2.1 Lattice Vibrations in the Long Wavelength Approximation

We briefly describe the theory of the lattice vibrations in the long wavelength approximation. This topic has been reviewed by Kelly [1981]. The equations of motion for lattice vibrations are given by

$$
\frac{\partial^2 u_x}{\partial t^2} = v_l^2 \frac{\partial^2 u_x}{\partial x^2} + v_t^2 \frac{\partial^2 u_x}{\partial y^2} + v_l^2 \left(\frac{1+\sigma}{2}\right) \frac{\partial^2 u_y}{\partial x \partial y} + \zeta \left(\frac{\partial^2 u_x}{\partial z^2} + \frac{\partial^2 u_z}{\partial x \partial z}\right),
$$
$$
\frac{\partial^2 u_z}{\partial t^2} = \zeta \left(\frac{\partial^2 u_z}{\partial x^2} + \frac{\partial^2 u_z}{\partial y^2}\right) + v_z^2 \frac{\partial^2 u_z}{\partial z^2} + \zeta \left(\frac{\partial^2 u_x}{\partial x \partial z} + \frac{\partial^2 u_y}{\partial y \partial z}\right) ,
$$

(4.5)

in which the $z$ axis is parallel to the graphite $c$-axis, and $x$ and $y$ refer to the basal plane. Here $u = (u_x, u_y, u_z)$ is a displacement vector, $\zeta$ is a reduced elastic coefficient given by $\zeta = C_{44}/\varrho_m$ where $\varrho_m$ is the mass density, and $v_l$ and $v_t$ denote the longitudinal and transverse sound velocities associated with the in-plane vibrations, while $v_z$ corresponds to the sound velocity for the out-of-plane vibrations. The equation for $\partial^2 u_y/\partial t^2$ is obtained by interchanging $x \leftrightarrow y$ in the equation for $\partial^2 u_x/\partial t^2$. The magnitudes of other quantities in (4.5) are given by [Seldin and Nezbeda 1970]

$$
\begin{aligned}
v_l &= (C_{11}/\varrho_m)^{1/2} & &= 2.10 \times 10^4 \text{ m/s}, \\
v_t &= [(C_{11} - C_{12})/2\varrho_m]^{1/2} & &= 1.23 \times 10^4 \text{ m/s}, \\
v_z &= (C_{33}/\varrho_m)^{1/2} & &= 3.92 \times 10^3 \text{ m/s},
\end{aligned}
$$

(4.6)

where the mass density $\varrho_m = 2.26$ g/cm$^3$, and the Poisson ratio $\nu = C_{12}/C_{11}$. Here $C_{44}$ is related to the shearing force and is very sensitive to crystal perfection, especially to the density of stacking faults, as discussed in Sect. 3.2.

In the expression for $u_z$ in (4.5) the term related to the bonding between the honeycomb planar networks is neglected, since this term is negligible in the long wavelength limit [Yoshimori and Kitano 1956]. The calculations [Komatsu 1955; Yoshimori and Kitano 1956] show that the vibrational frequencies for the three acoustic modes are given by

$$
\begin{aligned}
\omega_1^2 &= v_l^2 q_a^2 + \tfrac{4\zeta}{\tilde{c}_0^2} \sin^2(\tilde{c}_0 q_z/2), \\
\omega_2^2 &= v_t^2 q_a^2 + \tfrac{4\zeta}{\tilde{c}_0^2} \sin^2(\tilde{c}_0 q_z/2), \\
\omega_3^2 &= \delta^2 q_a^4 + \tfrac{4v_z^2}{\tilde{c}_0^2} \sin^2(\tilde{c}_0 q_z/2) + \zeta q_a^2,
\end{aligned}
$$

(4.7)

where $q_a^2 = q_x^2 + q_y^2$ and $\bar{c}_0$ is the interlayer spacing (3.35Å for well-graphitized samples) and $\delta$ is a constant related to the bending of the honeycomb planar network, which has a magnitude of $\delta \sim 6.11 \times 10^{-7} \mathrm{m}^2/\mathrm{s}$ [Kelly 1967].

In the long wavelength approximation (4.7) becomes

$$\omega_1^2 \simeq v_l^2 q_a^2 + \zeta q_z^2 \,, \quad \omega_2^2 \simeq v_t^2 q_a^2 + \zeta q_z^2 \,, \quad \omega_3^2 \simeq v_z^2 q_z^2 + \zeta q_a^2 \,. \tag{4.8}$$

Usually, the terms proportional to $\zeta$ are unimportant and can be neglected. However, the term $\zeta q_a^2$ in the expression for $\omega_3$ is important and is responsible for the low temperature anomaly in the thermoelectric power of highly oriented graphite [Sugihara et al. 1986]. The quantity $\zeta$ is also important in determining the sound velocity $v_R$ of the Rayleigh wave in thin film carbons [Sugihara and Dresselhaus 1986, 1987], and $v_R$ is approximately given by $\sqrt{\zeta}$ (Sect. 8.4.2). The in-plane vibration is polarized along the basal plane, while the out-of-plane mode oscillates along the $c$-axis. Both modes are responsible for carrier scattering processes [McClure and Smith 1963; Sugihara and Sato 1963; Ono and Sugihara 1966].

## 4.3 Models for the Elastic and Lattice Properties of Fibers

As discussed in Chap. 3, a fiber typically consists of a collection of crystallites which are disordered both internally and with respect to each other. There are two basic problems in calculating the elastic behavior of these disordered systems:

1. Finding a set of elastic parameters which describe the elastic behavior of the crystallites. This requires that the effect of defects on $C_{ij}$ and on $S_{ij}$ be considered.
2. Defining a suitable averaging procedure for combining the $C_{ij}$ or $S_{ij}$ into an effective value descriptive of the whole fiber.

These problems are considered separately in Sects. 4.3.1 and 4.3.2.

### 4.3.1 Effect of Defects on the Elastic Parameters of Crystallites

Firstly, it is probably a good approximation to assume that the intra-layer compliances are less affected by disorder than the inter-layer coefficients. $C_{11}$ and $C_{12}$ are expected to diminish as point defects are introduced, but should not be sensitive to the inter-layer disorder. In contrast, $C_{33}$ and $C_{44}$ are expected to depend sensitively on disorder.

Good quality single crystals examined by ultrasonic techniques (frequency 10 MHz) give $C_{44}$ values as low as 0.2 GPa, but $C_{44}$ can increase to 4.5 GPa on neutron [Seldin and Nezbeda 1970] or electron [Ayasse et al. 1979] irradiation.

However, the value deduced from inelastic neutron diffraction experiments is 4.5 GPa [Nicklow et al. 1972]. C. Baker and A. Kelly [1964] proposed that glissile, basal dislocations were softening the elastic constant $C_{44}$ at low frequency. Ayasse et al. [1979] derived the resonance frequency of the dislocation and fitted the variation of $C_{44}$ with electron irradiation dose. This softening of the shear modulus below the resonance frequency ($\sim$0.1–10 GHz typically for graphite) also occurs in metals, but the effect in metals is much smaller (typically a few percent) [Granato and Lücke 1956] compared to a factor of 25 in graphite.

Another influence on $C_{33}$ and $C_{44}$ would be the decrease of interlayer bonding due to turbostratic disorder. This is exemplified by the increased inter-layer separation in disordered carbons. This effect was directly measured using inelastic neutron diffraction experiments, and a 15% decrease of $C_{33}$ was found for an ex-PAN fiber with $\bar{c} = 0.341$ nm. This corresponds to a partially graphitized fiber with $T_{HT} \sim 2000°$C (see Table 2.2). No corresponding results were reported for $C_{44}$. However, van der Hoeven and Keesom [1963] analyzed the low temperature specific heat of pregraphitic carbons with a high density of stacking faults, and concluded that $C_{44}$ was reduced to 0.7 GPa. This is indicative of the reduction that may occur in $C_{44}$ for pregraphitic filaments.

In the case of graphite fibers, one expects the following considerations to apply:

- $C_{33}$: The value of $C_{33}$ for graphite fibers (those with well-developed interplanar correlations, such as CCVD and ex-pitch fibers with $T_{HT} >$ 2500°C) is expected to be close to the value for single crystal graphite. As turbostratic disorder sets in, the value for $C_{33}$ is expected to decrease. As noted above, the decrease in $C_{33}$ relative to crystalline graphite could be 15% for a fiber with $\bar{c} = 0.34$ nm [Collins and Haywood 1960], and lower still for those with $\bar{c}$ approaching the fully turbostratic value $\bar{c} = 0.344$. In general, it is expected that $C_{33}$ can be directly correlated with $\bar{c}$.

- $C_{44}$: A correlation is again expected between $C_{44}$ and $\bar{c}$, so that increased disorder reduces $C_{44}$. The reduction in $C_{44}$ is expected to be somewhat less than that for $C_{33}$ since $C_{44}$ is less strongly dependent on the interlayer separation than is $C_{33}$ (see B.T. Kelly [1981]). It is possible that the low-frequency (ultrasonic) value of $C_{44}$ will also be reduced for well-graphitized fibers, as the dislocation density is reduced sufficiently to unpin the dislocations.

- $C_{13}$: The value of $C_{13}$ for graphite fibers is expected to be sensitive to defect structure, but the trend is difficult to ascertain. Fortunately, the Young's and torsional moduli of fibers are relatively insensitive to $C_{13}$.

- $C_{11}$ and $C_{12}$: The values of $C_{11}$ and $C_{12}$ are expected to decrease slightly (e.g., less than 10%) as the density of point defects increases with lower

$T_{HT}$ values. However, $C_{11}$ and $C_{12}$ are essentially insensitive to turbostratic disorder. It is possible that a correlation exists with $\bar{c}$, but only because point-defects and turbostratic disorder tend to increase together as $T_{HT}$ decreases.

### 4.3.2 Models for Young's and Torsional Moduli of Fibers

There are many types of disorder in fibers, so that it is difficult to construct a universal model to explain all the elastic data on fibers. The parameters measured experimentally are Young's modulus (or extensional modulus), $E_Y$, and the torsional modulus, $G$. Experimental data are reviewed in Sect. 6.1.2. The Young's modulus $E_Y$ has been closely related to the misalignment of the crystallites with respect to the fiber axis (see Sect. 3.4.3) and Fig. 3.15 illustrates a crystallite whose c-axis is misaligned at an angle $\phi$ with the azimuthal plane, or $\theta$ with the fiber axis ($\theta = 90° - \phi$). Most workers have used $\theta$, but $\phi$ is used here, since the formulae are easier to interpret with the $\phi$ notation. Also, $\phi$ is the angle which is measured using x-ray or electron microscopy studies, giving the probability function $I(\phi)$. See Fig. 3.15.

If one attempts to determine the elastic behavior of fibers by using the elastic properties of graphite and measured orientation dependence for the graphite crystallites, two extreme cases can be envisioned, corresponding to the uniform stress (Reuss) and uniform strain (Voigt) averages. Expressions for these cases were developed by Price [1965] and Goggin and Reynolds [1967]. These two limits represent upper and lower bounds for the elastic energy of the system. In a highly anisotropic medium such as graphite, the results for the two extreme cases are usually very different. Although clearly not completely correct, the uniform stress model is usually closer to experimental results than the uniform strain model, and is thus most often used. This is understandable in terms of the structure of the fibers, consisting of graphene planes, extending over relatively long distances along the fiber axis.

In order to describe the detailed elastic behavior using the uniform stress (or strain) model, all one needs to do is derive the relationship between the stress and strain for a crystallite oriented at an angle $\phi$ with respect to the fiber axis (see Fig. 6.4). In the uniform stress model, the stress will be known for every crystallite since it is the stress that is applied to the fiber. After determining the strain for a particular crystallite oriented at an angle $\phi$, one just averages over the misorientation distribution to determine the elastic behavior of the fiber.

For simplicity, we will define the primed coordinates pertaining to the whole fiber as follows:
- the 2′ axis is parallel to the fiber axis;
- the 3′ axis is perpendicular to the fiber surface;
- the 1′ axis is in the basal plane of the crystallite and is identical to the 1 axis.

The unprimed coordinates are defined in terms of the crystalline axes as follows:

- the 1 and 2 axes are in the basal plane of the crystallite;
- the 3 axis is parallel to the c-axis of the crystallite.

The relationship between the stresses and strains is found by rotating the stress and strain tensors through an angle $\phi$ about the 1 axis; i.e.

$$\varepsilon_{i'j'} = \sum_{i,j} R_{i'i}(\phi)\hat{\varepsilon}_{ij}R_{jj'}(-\phi) \quad , \tag{4.9}$$

where $\hat{\varepsilon}_{ij}$ refers to the crystallite coordinate system and $\varepsilon_{i'j'}$ refers to the fiber coordinate system, and

$$R(\phi) = \begin{pmatrix} 1 & 0 & 0 \\ 0 & \cos\phi & \sin\phi \\ 0 & -\sin\phi & \cos\phi \end{pmatrix} \quad . \tag{4.10}$$

(Note that the more general case of rotations to a completely arbitrary orientation of a crystallite must be used for some problems – such as understanding the effects of not having azimuthal symmetry in a fiber.)

Similarly, for the stress tensor we have the relation

$$\sigma_{\mathrm{T}i'j'} = \sum_{i,j} R_{i'i}(\phi)\hat{\sigma}_{\mathrm{T}ij}R_{jj'}(-\phi) \tag{4.11}$$

where $\hat{\sigma}_{\mathrm{T}ij}$ refers to the crystallite coordinate system. Thus it follows that

$$\hat{\varepsilon}_{ij} = \sum_{kl} S_{ijkl}\hat{\sigma}_{\mathrm{T}kl} = \sum_{klk'l'} S_{ijkl}R_{kk'}(-\phi)\sigma_{\mathrm{T}k'l'}R_{l'l}(\phi) \quad . \tag{4.12}$$

In tensile stress experiments, the components $\sigma_{\mathrm{T}ij}$ vanish ($\sigma_{\mathrm{T}ij} = 0$) except for $\sigma_{\mathrm{T}2'2'}$, so that in this case

$$\varepsilon_{i'j'} = \sigma_{\mathrm{T}2'2'} \sum_{ijkl} S_{ijkl}R_{i'i}(\phi)R_{jj'}(-\phi)R_{k2'}(-\phi)R_{2'l}(\phi) \quad . \tag{4.13}$$

Using this model, Young's modulus is given by

$$\begin{aligned} 1/E_{\mathrm{Y}} &= \langle\varepsilon_{2'2'}\rangle/\sigma_{\mathrm{T}2'2'} \\ &= \left\langle \sum_{ijkl} S_{ijkl}R_{2'i}(\phi)R_{j2'}(-\phi)R_{k2'}(-\phi)R_{2'l}(\phi) \right\rangle \quad . \end{aligned} \tag{4.14}$$

Recalling that $S_{44} = 4S_{3232}$, we obtain

$$\begin{aligned} 1/E_{\mathrm{Y}} &= S_{11} + (2S_{13} + S_{44} - 2S_{11})\langle\sin^2\phi\rangle \\ &\quad + (S_{11} + S_{33} - 2S_{13} - S_{44})\langle\sin^4\phi\rangle \quad . \end{aligned} \tag{4.15}$$

We can also derive an expression for the change in cross-sectional area of a fiber under uniaxial tension (i.e., the Poisson ratio $\nu$)

$$2\nu = \text{change in area/area} = \varepsilon_{1'1'} + \varepsilon_{3'3'} \tag{4.16}$$

to obtain

$$2\nu = E_Y[S_{12} + S_{13} + (S_{11} + S_{33} - S_{44} - S_{12} - S_{13})\langle \sin^2 \phi \rangle$$
$$+ (2S_{13} - S_{33} + S_{44} - S_{11})\langle \sin^4 \phi \rangle] \quad . \tag{4.17}$$

The torsional modulus will depend on the details of the azimuthal distribution of crystallites. When that distribution is completely random one obtains for the torsional modulus [Reynolds and Moreton 1980]

$$1/G = (0.5S_{44} + S_{11} - S_{12}) + (S_{11} + 2S_{33} + 12S_{13} - 1.5S_{44})\langle \sin^2 \phi \rangle$$
$$- (2S_{11} + 2S_{33} - 4S_{13} - 2S_{44})\langle \sin^4 \phi \rangle \quad . \tag{4.18}$$

The above solutions, i.e., (4.15) and (4.18), can be simplified for the case of perfectly aligned crystallites ($\phi \rightarrow 0°$), for which

$$E_Y = S_{11}^{-1} \quad \sim 1020 \text{ GPa}, \quad \nu = E_Y(S_{11} + S_{13}) \quad . \tag{4.19}$$

The angular averages in (4.15) and (4.18) are defined [Goggins and Reynolds 1967] simply in terms of the measured crystallite orientation functions $I(\phi)$ (Sect. 3.4.3)

$$\langle \sin^n \phi \rangle = \int_{-\pi/2}^{\pi/2} I(\phi) \sin^n \phi \cos \phi \, d\phi \left/ \int_{-\pi/2}^{\pi/2} I(\phi) \cos \phi \, d\phi \right. . \tag{4.20}$$

These angular averages are very sensitive to the tails of the orientation distribution function $I(\phi)$. The tails of the distribution are usually very difficult to measure accurately, and thus direct integration often leads to large uncertainties in the angular averages. Thus many authors have used analytic approximations for the distribution function and then calculated the integrals using those expressions. Once a method for determining the angular averages has been developed, and reasonable values for $S_{ij}$ have been selected, then comparison can be made between measured and predicted values of $E_Y$ and $G$.

Another approach to the calculation of $E_Y$ was considered by Ruland [1969a], who compared the predictions of the uniform stress and uniform strain models and suggested an alternative "elastic unwrinkling" model. According to Ruland's model [Ruland 1969a, 1969b], the graphite layers are linked in such a way that they form long wrinkled ribbons along the fiber axis. An applied stress causes the ribbons to straighten. The environment of the layers produces a resistance to the tilting of the layers. Under the action of an applied stress, the resulting Young's modulus can be written as

$$1/E_Y = \ell_z S_{11} + m_z k_e , \quad \text{where} \tag{4.21}$$

$$\ell_z = \int_0^\pi I(\theta) \sin^2 \theta \, d\theta \left/ \int_0^\pi I(\theta) \sin \theta \, d\theta \right. ,$$

$$m_z = \int_0^\pi I(\theta) \cos^2 \theta \, d\theta \left/ \int_0^\pi I(\theta) \sin \theta \, d\theta \right. \tag{4.22}$$

and where $k_e$ is an elastic compliance which Ruland [1969a] did not specify. The angle of integration used by Ruland is $\theta = \frac{\pi}{2} - \phi$ as shown in Fig. 3.15. The value of $k_e$ was shown to be roughly independent of fiber type and $T_{HT}$. Although the model is appealing because of its simplicity, the fact that there is no physical model for $k_e$ based on other parameters makes it difficult to justify. Both the elastic unwrinkling model and the uniform stress model can explain in a natural way the increase in modulus with applied stress. For instance, Curtis et al. [1968] were able to predict values for the strain-dependent misorientation $\phi(\varepsilon)$ and obtain good agreement with values measured by x-ray techniques (Sect. 3.4.3) with the fiber under strain. In view of the fact that the compliance, $k_e$, is unidentified, and that the uniform stress model can give a reasonable interpretation of experimental results, it is probably preferable to use the uniform stress model for $E_Y$. Further use of the uniform stress model is made below in understanding the piezoresistance properties of fibers as discussed in Sect. 8.7.

The torsional modulus is much more difficult to calculate, since it depends on the azimuthal arrangement of the carbon ribbons as well as the misorientation along the axis of the fiber. This preferred orientation can be explained with a simple analogy. Consider a hypothetical material consisting of four long, thin, relatively narrow strips of brass (e.g., $1000 \times 10 \times 1$ mm). The torsional modulus can be obtained by suspending these four strips, attaching a moment to the bottom, and timing the torsional motion. Suppose that the four strips were loosely packed on top of each other; then the torsional modulus would assume the value $G$, say. If, however, the four strips were brazed to form a solid piece of dimensions $1000 \times 10 \times 4$ mm, another, higher, value of $G$ would be obtained. Yet another value of $G$ would be obtained if the four pieces were brazed at their edges to form a hollow, square tube. Thus, the torsional moduli of fibers will depend on the way in which the fibrils are arranged.

In principle, it should be possible to calculate $G$ for each fiber morphology, but this does not appear to have been done. Some idea of the possible variation can be obtained by considering the extreme cases of fibers with radial structure ($G = C_{44} \sim 4$ GPa) and a highly ordered CCVD fiber ($T_{HT} = 3500°C$) with perfectly faceted graphite surfaces [$G = (C_{11} - C_{12})/2 \sim 44$ GPa]. Clearly, these ideas need to be quantified with detailed models.

Experimental data for $E_Y$ and $G$ are given in Sect. 6.1.2 and are compared with the above models in Sect. 6.1.4, while the effects of neutron irradiation are discussed in Sects. 6.1.2 and 12.5.

### 4.3.3 Rayleigh Waves in Thin Carbon Films

If the sample thickness $d$ is small ($<100$ Å), a Rayleigh wave with small sound velocity ($\sim 10^2$ m/s) can propagate in the film without damping. This wave is polarized along the $c$-axis. Carriers are strongly scattered by the phonons associated with this Rayleigh wave even at temperatures below 1 K, since typical phonon energies interacting with carriers are at most 1 K. Though the carrier relaxation rate that is related to the interaction with the Rayleigh wave phonons is one order of magnitude smaller than that for the impurity scattering, the electron-Rayleigh wave interaction plays an important role in the negative magnetoresistance of disordered carbons (Sect. 8.4.2) at low temperature [Sugihara and Dresselhaus 1986].

Pregraphitic carbons heat treated at low temperature have a turbostratic structure and the correlation length along the $c$-axis is small. Accordingly, it is reasonable to assume that the sample is composed of an assembly of many thin films which are weakly coupled to each other elastically. Based on (4.5) and imposing appropriate boundary conditions on the thin film surface, a solution of the Rayleigh wave equation has been obtained for small damping and small sound velocity. The solution shows that the conduction carriers are strongly scattered even at 4.2 K [Sugihara and Dresselhaus 1986, 1987]. This situation is very different from the case for bulk graphite samples.

To solve (4.5) for the case of a thin film, the following boundary conditions are imposed at $z = 0$ and $z = -d$, where $d$ is the film thickness along the $c$ axis. Strain-free and stress-free boundary conditions give rise to the same equations:

$$\frac{\partial u_x}{\partial z} + \frac{\partial u_z}{\partial x} = \frac{\partial u_y}{\partial z} + \frac{\partial u_z}{\partial y} = \frac{\partial u_z}{\partial z} = 0 \tag{4.23}$$

at $z = 0$ and $z = -d$. A damped plane wave solution of (4.5) is written in the form

$$u(r) = U \, e^{\kappa z} \, e^{i(qx - \omega t)} \tag{4.24}$$

for $q$ along the $x$-axis. The general solution $u(r) \propto \exp[\,i\,(q \cdot r - \omega t)]$ can be obtained from (4.24) by rotating the coordinate axis in the $xy$-plane. Inserting (4.24) into (4.5), yields $U_y = 0$ and equations for the $U_x$ and $U_z$ components:

$$(\omega^2 - v_l^2 q^2 + \zeta \kappa^2)U_x + i\zeta q\kappa U_z = 0 \,, \; i\zeta q\kappa U_x + (\omega^2 - \zeta q^2 + v_z^2 \kappa^2)U_z = 0 \,. \tag{4.25}$$

These equations can be solved for $\kappa^2(\omega)$ and have two positive roots $\kappa_1^2$ and $\kappa_2^2$, if the Rayleigh wave velocity $v_R = \omega/q$ satisfies the condition

$$v_R^2 < \zeta = C_{44}/\varrho_m \quad . \tag{4.26}$$

From these roots for $\kappa^2(\omega)$, an explicit expression for the displacement vector $u = (u_x, 0, u_z)$ can be found [Sugihara and Dresselhaus 1986, 1987]. By introducing the phonon creation and annihilation operators $b_q^+$ and $b_q$ the energy of the Rayleigh wave is quantized according to

$$2 \int dr \; \frac{\varrho_m}{2} \left\{ \left| \frac{\partial u_a}{\partial t} \right|^2 + \left| \frac{\partial u_z}{\partial t} \right|^2 \right\} = \sum_q \hbar \omega_q \left( b_q^+ b_q + \frac{1}{2} \right), \tag{4.27}$$

where $u_a = (u_x, u_y, 0)$. From these equations, explicit expressions for $u_a$ and $u_z$ are obtained [Sugihara 1986a; Sugihara and Dresselhaus 1987]. The solutions show that the displacements follow a weakly damped Rayleigh wave polarized along the $c$-axis where the in-plane displacement $u_a$ is two orders of magnitude smaller than $u_z$.

These Rayleigh waves are considered further (Sect. 8.4.2) in connection with the unusual low temperature negative magnetoresistance of pregraphitic carbon fibers.

## 4.4 Raman Effect for Single Crystal graphite

The unit cell of graphite has a symmetry consistent with the space group $D_{6h}^4$, and the point group homomorphic with this space group is $D_{6h}$. From the properties of this point group the zone center optic modes (at $k = 0$) can be decomposed into the following irreducible representations

$$\Gamma_{opt} \rightarrow 2E_{2g} + E_{1u} + A_{2u} + 2B_{1g} \quad . \tag{4.28}$$

Of these modes, the $E_{2g}$ modes are Raman active, the $E_{1u}$ and $A_{2u}$ modes are infrared active and the $B_{1g}$ modes are optically inactive. The lattice site

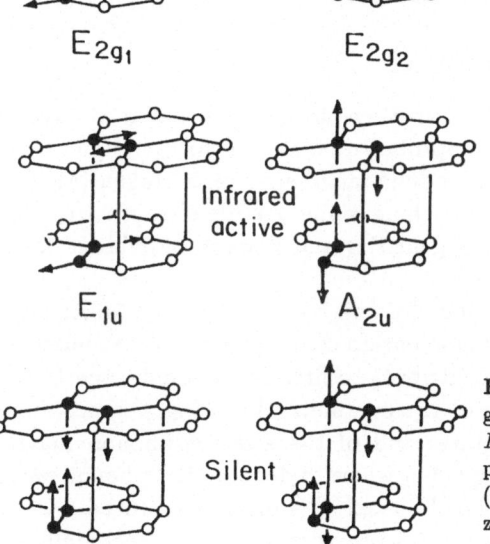

Fig. 4.3. Zone-center optical modes for graphite. For the in-plane modes ($E_{1u}$, $E_{2g_1}$, and $E_{2g_2}$), only one of the degenerate pair of modes is shown. The $c$-axis modes ($A_{2u}$, $B_{1g_1}$, $B_{1g_2}$) are non-degenerate. The zero frequency acoustic modes ($E_{1u}$, $A_{2u}$) correspond to pure translations and are not shown [Dresselhaus and Dresselhaus 1982]

displacements corresponding to these normal modes are shown in Fig. 4.3. The $E_{2g_1}$ mode is an in-plane rigid layer shear mode which occurs at a very low frequency because in this case the layers are rigidly displaced against the weak interlayer restoring force. On the other hand, $E_{2g_2}$ also represents an in-plane mode but occurs at a high frequency because the neighboring atoms in each of the layer planes are displaced against the strong in-plane restoring force. The first-order Raman spectrum of a large single crystal of graphite has a single high-frequency mode at about 1580 cm$^{-1}$, corresponding to the Raman-allowed $E_{2g_2}$ mode shown in Figs. 4.1 and 4.3. Under favorable circumstances, the Raman-active low frequency $E_{2g_1}$ mode can also be observed.

## 4.5   Raman Effect in Disordered Carbons

Raman spectra are very sensitive to changes that break the translational symmetry. Finite size effects, as occur in small-dimensional crystals, and lattice defects, as occur in carbon fibers, cause a breakdown of the translational symmetry. For this reason the Raman effect is commonly used to characterize disordered carbons [Dresselhaus and Dresselhaus 1982].

Because of the long wavelength of light relative to lattice dimensions, the first-order Raman-allowed modes are confined to the center of the Brillouin zone. The Raman-active modes correspond to in-plane vibrations. The $E_{2g}$ modes are uniquely identified by the polarization ($\parallel$, $\perp$), whereby the incident and scattered light are at right angles. The experiments are carried out in the back-scattering geometry because graphite is opaque to the laser light, and the high reflectivity of graphite is favorable for back-scattering.

### 4.5.1   Raman Spectroscopy of Fibers

The technique can be used for qualitative evaluation of a batch of carbon fibers, or, by employing a Raman microprobe and focusing the laser light down to 1 or 2 $\mu$m, the Raman spectra of individual fibers can be taken [McNeil et al. 1986, Lespade et al. 1982]. By probing along the fiber length with the Raman microprobe, structural inhomogeneities and flaws in individual fibers can be detected on a 1 $\mu$m scale.

The introduction of lattice disorder (or the reduction in crystallite size) relaxes the selection rules on crystal momentum conservation, so that phonons throughout the Brillouin zone can contribute to first-order Raman scattering in accordance with the magnitude of the symmetry breaking perturbation, the magnitude of the density of states and of the electron-phonon matrix element for each mode of frequency $\omega_s(q)$ where $q$ denotes the phonon wave vector and $s$ the phonon branch. The observation of disorder-induced lines in the first- and second-order Raman spectra provides a sensitive technique for the non-destructive observation of lattice disorder in carbon fibers.

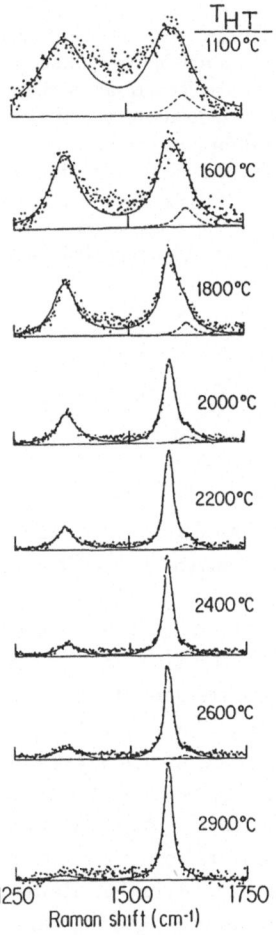

$T_{HT}$
1100°C

1600 °C

1800°C

2000°C

2200°C

2400°C

2600°C

2900°C

1250    1500    1750
Raman shift (cm$^{-1}$)

**Fig. 4.4.** First-order Raman spectra for benzene-derived carbon fibers heat treated at various temperatures ($T_{HT}$). As $T_{HT}$ is increased, the intensity of the disorder-induced line at ~1360 cm$^{-1}$ decreases and the linewidth of the Raman-allowed line decreases. At the highest $T_{HT}$ of 2900°C, the line at ~1360 cm$^{-1}$ can barely be detected. Solid lines represent a Lorentzian fit to the experimental points. Dashed lines represent a Lorentzian fit to a line at ~1620 cm$^{-1}$ [Chieu et al. 1982]

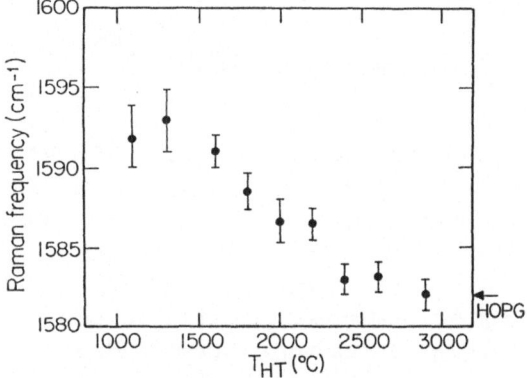

**Fig. 4.5.** Plot of the Raman frequency for the dominant peak in the first-order spectra near ~1580 cm$^{-1}$ vs heat-treatment temperature ($T_{HT}$). As $T_{HT}$ increases, the Raman-mode frequency approaches that of HOPG (highly ordered pyrolytic graphite) [Chieu et al. 1982]

The dominant disorder-induced Raman line in the first-order Raman spectrum occurs at ~1360 cm$^{-1}$ corresponding to a high density of states for phonon modes near the Brillouin zone boundary. As an example, we show in Fig. 4.4 the first-order Raman spectra for benzene-derived carbon fibers heat treated to various temperatures $T_{HT}$ [Chieu et al. 1982]. Here the dominant features are the Raman-allowed line at ~1580 cm$^{-1}$ and the disorder-induced line at ~1360 cm$^{-1}$. In addition, a small feature near ~1620 cm$^{-1}$ is seen in some of the traces; this feature is identified with the high phonon density of states for mid-zone phonons near the maximum phonon frequency (see Fig. 4.2). The upshift in frequency of the $E_{2g_2}$ mode with decreasing $T_{HT}$ (see Fig. 4.5) can be explained by this high density of phonon states away from the zone center. As $T_{HT}$ is increased, distinct differences in the spectra of Fig. 4.4 are observed with regard to the ratio of the integrated intensities of

99

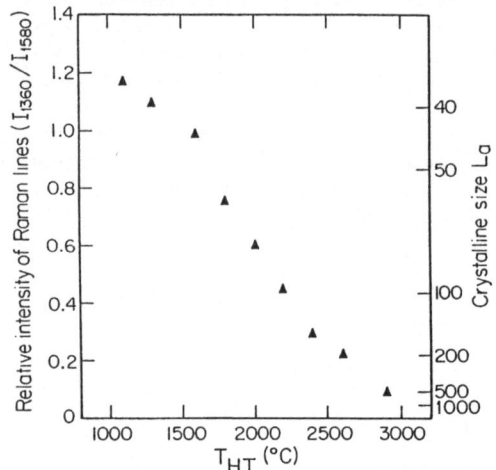

**Fig. 4.6.** Plot of the intensity ratio of the disorder-induced line at $\sim 1360\,\mathrm{cm}^{-1}$ to the Raman-allowed line at $\sim 1580\,\mathrm{cm}^{-1}$ vs heat-treatment temperature ($T_{\mathrm{HT}}$) for benzene-derived carbon fibers. On the right scale, the intensity ratio $R = I_{1360}/I_{1580}$ is related to the crystallite size $L_a$ using the results of Tuinstra and Koenig [1970] and Chieu et al. [1982]

the disorder-induced peak at $\sim 1360\,\mathrm{cm}^{-1}$ and the Raman-allowed peak at $\sim 1580\,\mathrm{cm}^{-1}$ ($R = I_{1360}/I_{1580}$). Also a dramatic increase in the linewidths of the Raman lines at $\sim 1360\,\mathrm{cm}^{-1}$ and $\sim 1580\,\mathrm{cm}^{-1}$ is observed as the amount of disorder is increased [Tuinstra and Koenig 1970; Wada et al. 1980; Vidno and Fischbach 1981; Chieu et al. 1982; Dillon et al. 1984]. In this context, Fig. 4.6 illustrates a plot of the intensity ratio $R$ vs heat treatment temperature $T_{\mathrm{HT}}$ for the benzene-derived fibers [Chieu et al. 1982]. The intensity ratio $R = I_{1360}/I_{1580}$ has been shown to be closely related to the in-plane crystallite size $L_a$ [Tuinstra and Koenig 1970], so that $L_a$ vs $T_{\mathrm{HT}}$ is also plotted in Fig. 4.6 (see right hand scale). The correlation of $R$ with $L_a$ makes Raman spectroscopy a powerful tool for characterizing the structural order of carbon fibers. Important correlations have also been established between the Raman intensity ratio $R$ and the resistivity [Chieu et al. 1982], the magnetoresistance [Chieu et al. 1983; Rahim et al. 1986] and the thermal conductivity [Heremans et al. 1985].

From Fig. 4.6, it follows that $R$ decreases with increasing $T_{\mathrm{HT}}$. For $T_{\mathrm{HT}} \leq$ 1700°C, the decrease in $R$ with $T_{\mathrm{HT}}$ is relatively slow, but becomes more rapid for $1700° \leq T_{\mathrm{HT}} \leq 2400$°C. Above $T_{\mathrm{HT}} = 2900$°C, the ratio $R$ is so small ($R <$ 0.1) that it is difficult to measure $R$ accurately. A value of $R = 0.1$ corresponds to an $L_a$ value of $\sim 500$ Å. The first-order Raman results show that for $T_{\mathrm{HT}} \geq$ 2900°C, the vapor-grown fibers are almost completely graphitized and three-dimensional graphite ordering is nearly fully established. Resistivity vs $T_{\mathrm{HT}}$ curves of benzene-derived fibers (Fig. 8.19) also show a similar behavior to Fig. 4.6 [Chieu et al. 1982]. As is seen in Figs. 4.4 and 4.7, the linewidth for the 1580 cm$^{-1}$ line decreases dramatically with increasing $T_{\mathrm{HT}}$, as is also the case for the linewidth of the disorder-induced line at $\sim 1360\,\mathrm{cm}^{-1}$.

To account for the first-order Raman spectra observed for carbon fibers heat treated at various $T_{\mathrm{HT}}$, a calculation of the Raman spectra for disordered

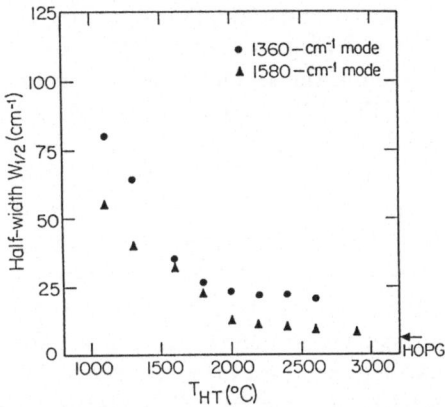

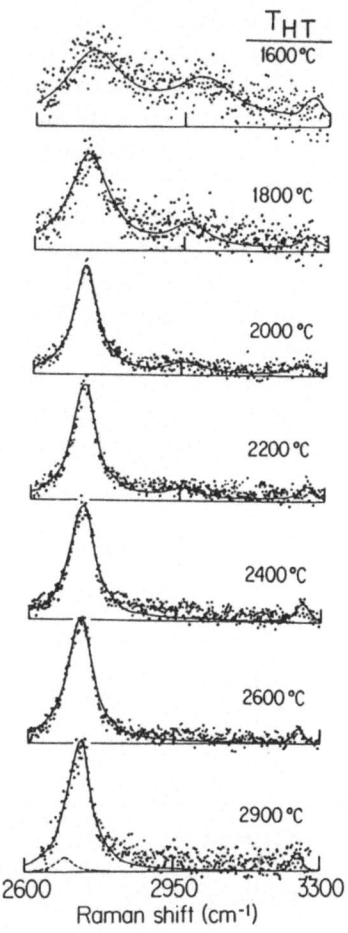

$T_{HT}$
1600°C

1800°C

2000°C

2200°C

2400°C

2600°C

2900°C

2600        2950        3300
Raman shift (cm$^{-1}$)

**Fig. 4.7.** Plot of half width at half maximum (HWHM) linewidth vs heat-treatment temperature ($T_{HT}$) for the 1360- and 1580-cm$^{-1}$ Raman lines for benzene-derived carbon fibers. For $T_{HT} < 1800°C$, the linewidth HWHM decreases rapidly with increasing $T_{HT}$. A constant value of HWHM is approached for $T_{HT} \geq 2000°C$ [Chieu et al. 1982]

**Fig. 4.8.** Second-order Raman spectra for benzene-derived graphite fibers heat treated at various temperatures ($T_{HT}$). Increasing $T_{HT}$ causes the dominant mode at $\sim2730\,cm^{-1}$ to sharpen and causes an attenuation of the disorder-induced mode at $\sim2970\,cm^{-1}$. For $T_{HT} = 2900°C$, a shoulder near 2700 cm$^{-1}$ in the 2730 cm$^{-1}$ line (characteristic of highly ordered graphite) is barely visible [Chieu et al. 1982]

graphite was carried out by Lespade et al. [1982]. The calculation of the Raman spectra for the disordered graphite followed the same formulation as was previously applied to crystalline graphite by Al-Jishi and Dresselhaus [1982a], except that the wave vector conservation rule was relaxed to take into account the finite coherence length of the lattice vibrations. The Lespade calculation was successful in providing a qualitative explanation of the spectra observed in pregraphitic carbon fibers.

Characterization studies of lattice disorder in carbon fibers [Chieu et al. 1982, 1983] and in ion-implanted graphite [Elman et al. 1982] have shown that the second-order Raman spectra are often more sensitive than the first-order spectra to small amounts of disorder in the graphite lattice. The dominant features in the second-order graphite Raman spectra, where phonons with wave vectors $q$ and $-q$ can contribute, are a strong line at $\sim2730\,cm^{-1}$ and a weak but sharp feature at $\sim3250\,cm^{-1}$. The lines at $\sim2730\,cm^{-1}$ and $\sim3250\,cm^{-1}$

respectively correspond to the second harmonics of the maxima in the phonon density of states at $\sim 1360\,\text{cm}^{-1}$ and $\sim 1620\,\text{cm}^{-1}$ (Figs. 4.1 and 4.2). In addition, a broad feature at $\sim 2970\,\text{cm}^{-1}$ is observed in the second-order spectrum of disordered graphite and disordered fibers (Fig. 4.8). This disorder-induced feature is associated with a combination mode from a high density of phonon states with different wave vectors (e.g., $[\sim 1360\,\text{cm}^{-1}] + [\sim 1620\,\text{cm}^{-1}]$). The mixing of modes with different wave vectors can occur through disorder or finite size effects [Nemanich and Solin 1979; Elman et al. 1982; Chieu et al. 1983]. Raman studies have also been carried out to characterize intercalated graphite and graphite fibers (see Sect. 10.3) and ion-implanted graphite samples (see Sect. 11.2). Second-order Raman spectra are shown in Fig. 4.8 for vapor-grown fibers for a series of heat treatment temperatures. Of particular interest is a comparison of these spectra with the corresponding first-order spectra in Fig. 4.4. Also for the second-order spectra, the linewidth decreases

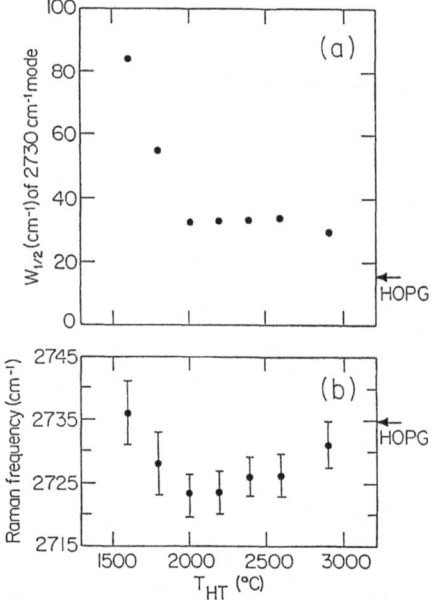

Fig. 4.9a,b. (a) Half width at half maximum (HWHM) linewidth and (b) peak mode frequency of the dominant line in the second-order Raman spectra at $\sim 2730\,\text{cm}^{-1}$ vs heat-treatment temperature ($T_{HT}$) for benzene-derived fibers. The corresponding linewidth and peak mode frequency for HOPG are also indicated [Chieu et al. 1982]

with increasing $T_{HT}$ and eventually approaches the value characteristic of single crystal graphite. Above $T_{HT} \simeq 2000°\text{C}$, the linewidth is close to that of highly ordered graphite, as seen in Fig. 4.9a. Small variations in the Raman frequency of the dominant second-order Raman line near $2730\,\text{cm}^{-1}$ are also observed as a function of $T_{HT}$ as shown in Fig. 4.9b [Chieu et al. 1982].

Raman studies on PAN and pitch-based fibers show the importance of the precursor material in the ultimate graphitizability of the fibers [Kwizera et al. 1982, 1983]. When heated to the same $T_{HT}$ for the same length of time, the vapor-grown fibers are most readily graphitized followed by mesophase pitch fibers and then by PAN fibers, in that order. Similar conclusions have been reached using other characterization techniques, such as temperature-dependent thermal conductivity measurements [Heremans et al. 1985].

The spatial distribution of the lattice imperfections can be sensitively monitored in a single fiber by the Raman microprobe technique [Lespade et al. 1982; McNeil et al. 1986]. With this technique, the laser light is focused down to a small spot ($\approx 1\text{--}2\,\mu$m) by a microscope attachment. By using a diode array detector, the Raman spectrum of an entire spectral region can be simultaneously recorded and a satisfactory signal-to-noise ratio can be achieved in a short time interval (e.g., 2 min). Thus the Raman microprobe is especially well suited for scanning the structural perfection of a fiber along its length. Results for the first- and second-order Raman microprobe spectra show significant variation from one fiber to another for fibers prepared in the same nominal way. As noted above, the Raman microprobe is also used to scan single fibers in order to identify structural inhomogeneities along the fiber length. The Raman microprobe can in a similar way be used to detect staging inhomogeneities along the length of an intercalated fiber, as discussed in Sect. 10.3.

## 4.6  Photoconductivity in graphite Fibers

The recent observation of photoconductivity in electrically conducting graphite fibers (Fig. 4.10) provides another non-destructive technique for the characterization of the structural ordering of the fibers [Steinbeck et al. 1986]. Since the photoconductivity phenomenon in carbon fibers is entirely due to the presence of defects, and since the photoconductivity technique has much in common with Raman microprobe spectroscopy, it is appropriate to discuss the photoconductivity phenomenon at this point in the text.

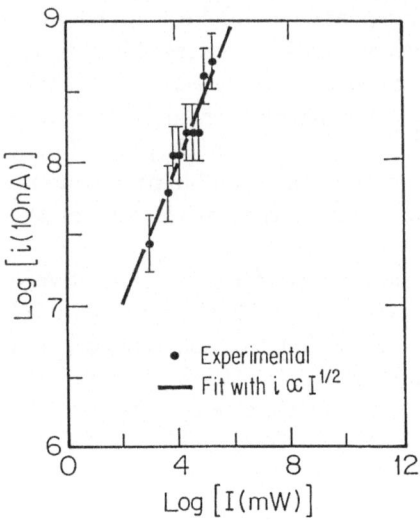

Fig. 4.10. Log-log plot of the magnitude of the photocurrent $i$ (in units of $10\,$nA) vs the incident laser power $I$ (in mW) for vapor grown carbon fibers. Note the good fit of the experimental data points to an $I^{1/2}$ dependence. [Steinbeck et al. 1986]

103

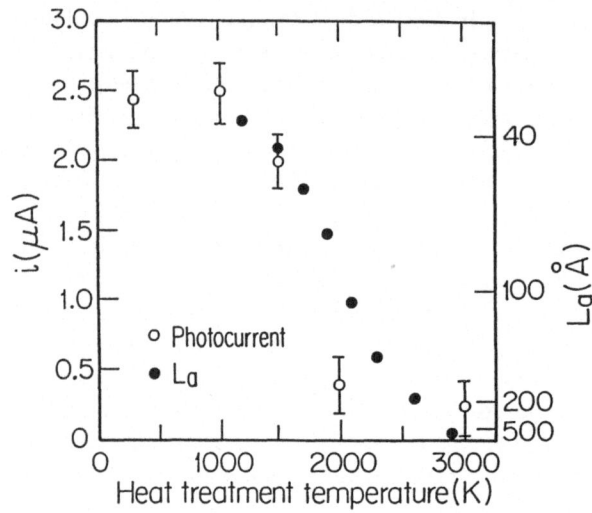

**Fig. 4.11.** Relationship between the magnitude of the photocurrent and the in-plane coherence length as a function of heat treatment temperature for vapor grown carbon fibers [Steinbeck et al. 1986]

The photoconductivity signal is found to decrease with increasing structural order, as is seen in Fig. 4.11, and for this reason the photoconductivity technique is suitable for studying structural defects in carbon fibers. Though a promising characterization technique, it has not yet been applied systematically to the characterization of defects in carbon fibers. The photocurrent $i$ is observed to vary with the illumination intensity, $I$, as $I^{1/2}$ (see Fig. 4.10). This dependence on $I$ implies that carrier trapping is responsible for the observed photocurrent. The $I^{1/2}$ dependence of the photocurrent on the laser intensity is the same relationship as was reported by MacFarlane et al. [1970] for the photoconductivity in amorphous carbon films. The $I^{1/2}$ relation may be explained on the basis of localized defect states lying in a band about the Fermi level and acting as carrier traps. Since carriers will be trapped close to the band edges, the density of carrier recombination centers will increase in proportion to the photoexcited carrier density. The increase in the density of the recombination centers will then increase the recombination rate for excited carriers, since the decay rate for excited carriers is proportional to both the excited carrier density and the density of recombination centers. The photocurrent will therefore be proportional to the square root of the photoexcitation rate.

The rise and fall times of the photocurrent are 50 and 100 ms, respectively, in as-grown benzene-derived fibers [Steinbeck et al. 1986]. The rise and fall times of the photocurrent are independent of the illumination intensity and fiber temperature over the temperature range $10 \leq T \leq 300$ K. These observations are consistent with rise and fall time data reported by MacFarlane et al. [1970] for amorphous carbon films, and give strong evidence that the photoconductivity is not a thermal effect.

The observation that the photoconductivity is a strong function of heat treatment temperature $T_{HT}$ (Fig. 4.11) shows that the magnitude of the photocurrent is strongly correlated with the in-plane coherence length $L_a$ in the fiber. The existence of photoconductivity in graphite fibers has been attributed to the population of localized defect states near the Fermi level which contribute to electrical conduction via a hopping mechanism. Also shown in Fig. 4.11 is the relationship between the in-plane coherence length $L_a$ and the annealing temperature for graphite fibers taken from Chieu et al. [1982]. A comparison of the behavior of the photocurrent and the in-plane coherence length indicates that the decrease in the photocurrent is associated with the annealing of defects responsible for the photoconduction process. To explain photoconductivity in as-grown graphite fibers, and graphite fibers which have been annealed at temperatures below 1000 K, a two-dimensional zero gap semiconductor model is used and a large density of localized states is introduced near the Fermi level. As the graphite fibers are heat treated, the density of localized states decreases. The addition of localized states to the band model picture for disordered graphites has been suggested by Mrozowski [1971], who used localized states to explain the electrical conduction properties of disordered graphites via hopping. Bright [1979] also used localized states to explain the negative magnetoresistance behavior of pregraphitic carbons. These localized states are also thought to be responsible for the measured photocurrent in graphite fibers [Steinbeck et al. 1986]. Further work is needed to develop the photoconductivity technique for quantitative characterization of structural defects in carbon fibers.

# 5. Thermal Properties

## 5.1 Specific Heat

To date, there have been no direct measurements of the specific heat of carbon fibers. The only heat capacity work on carbon fibers known to us is the differential scanning calorimetry study carried out on a number of ex-PAN, ex-pitch and vapor grown carbon fibers before and after bromination [Jaworske et al. 1987]. This work was directed toward relating the degree of structural order to the threshold for intercalation and is discussed in Sect. 10.1.

On the other hand, there has been a considerable amount of work on the heat capacity of various carbons as a function of heat treatment temperature. Thus, in this section we briefly review the information which has been obtained from heat capacity measurements on carbons which may have important implications for carbon fibers.

The subject of specific heat in carbons has been reviewed by Mrozowski [1979]. We will not repeat his extensive discussion here, but will only discuss the important qualitative results:

1. The lattice vibration contributions to the specific heat of graphite can be understood in terms of the anisotropic phonon spectrum of graphite (see Sect. 4.2 for further discussion). Detailed measurements of the heat capacity $C(T)$ in pristine graphite were carried out in the years 1953–1958 by several groups [DeSorbo and Tyler 1953; Bergenlid et al. 1954; Keesom and Pearlman 1955; DeSorbo and Nichols 1958]. A quadratic temperature dependence $(C \sim T^n, n = 2)$ has been verified from about 13 K to about 50 K while at higher temperatures a dependence of $n = 1.8$ (at 90 K) has been observed [Bergenlid et al. 1954]. At low temperatures (below 10 K) a gradual change in the temperature dependence of the heat capacity from $n = 2$ to $n = 3$ takes place and in addition a linear term must be added to $C(T)$ due to conduction carriers [Krumhansl and Brooks 1953]. At higher temperatures, the temperature dependence of the lattice contribution is determined by the details of the phonon spectrum, and has been reviewed in detail by B.T. Kelly and Taylor [1973].

2. In disordered carbons there is a significant linear term in the specific heat at low temperature. This term has contributions both from the conduction electrons and from the disorder in the carbon or graphite. For materials

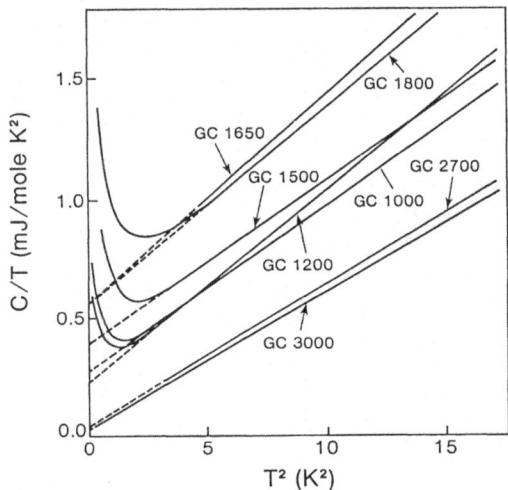

**Fig. 5.1.** Specific heat curves for variously heat-treated glassy carbons (GC) (original material GC10 from Tokai Electrode Mfg. Co. has $T_{HT} = 1000°C$). [Mrozowski 1979]

heat treated below about 2000°C the defects are more important than the electronic contribution to the linear term.

3. Below 1 K there are often one or two specific heat peaks that depend significantly on both the magnetic field and the degree of order in the carbon (see Figs. 5.1-3). Well-ordered graphite does not have this type of anomaly, although it can be induced by radiation damage [Delhaès et al. 1971; Mrozowski 1979]. The size of the specific heat peak is correlated with the size of the linear term in the specific heat (see Fig. 5.4) and it has been suggested [Mrozowski 1979] that the low temperature specific heat peaks are due to defect levels. Mrozowski [1979] also suggested that an antiferromagnetic ordering was associated with these defects, though there is no direct evidence for such magnetic ordering at low temperature. In the case of antiferromagnetic ordering, a linear contribution to the heat capacity would arise from the spin glass contribution due to the same magnetic defects. For $T$ below the peaks in the specific heat, there is no large linear $T$ term in the heat capacity. At present there is no explanation of the origin of these peaks in terms of a microscopic model.

As already mentioned, there has been no comparable study of the specific heat of carbon fibers. Fits of the vibrational contributions to the heat capacity would provide checks on models for the vibrational spectrum and elastic constants of fibers annealed at different heat treatment temperatures $T_{HT}$. The sensitivity of both the linear term and the low temperature peaks to the degree and nature of the disorder could provide new insights into the detailed microstructure of the commercially important high strength low modulus fibers. These fibers are heat treated in the temperature range where the magnitudes of the linear term and of the peak in the heat capacity are largest. At low temperatures these fibers also have an anomalous temperature

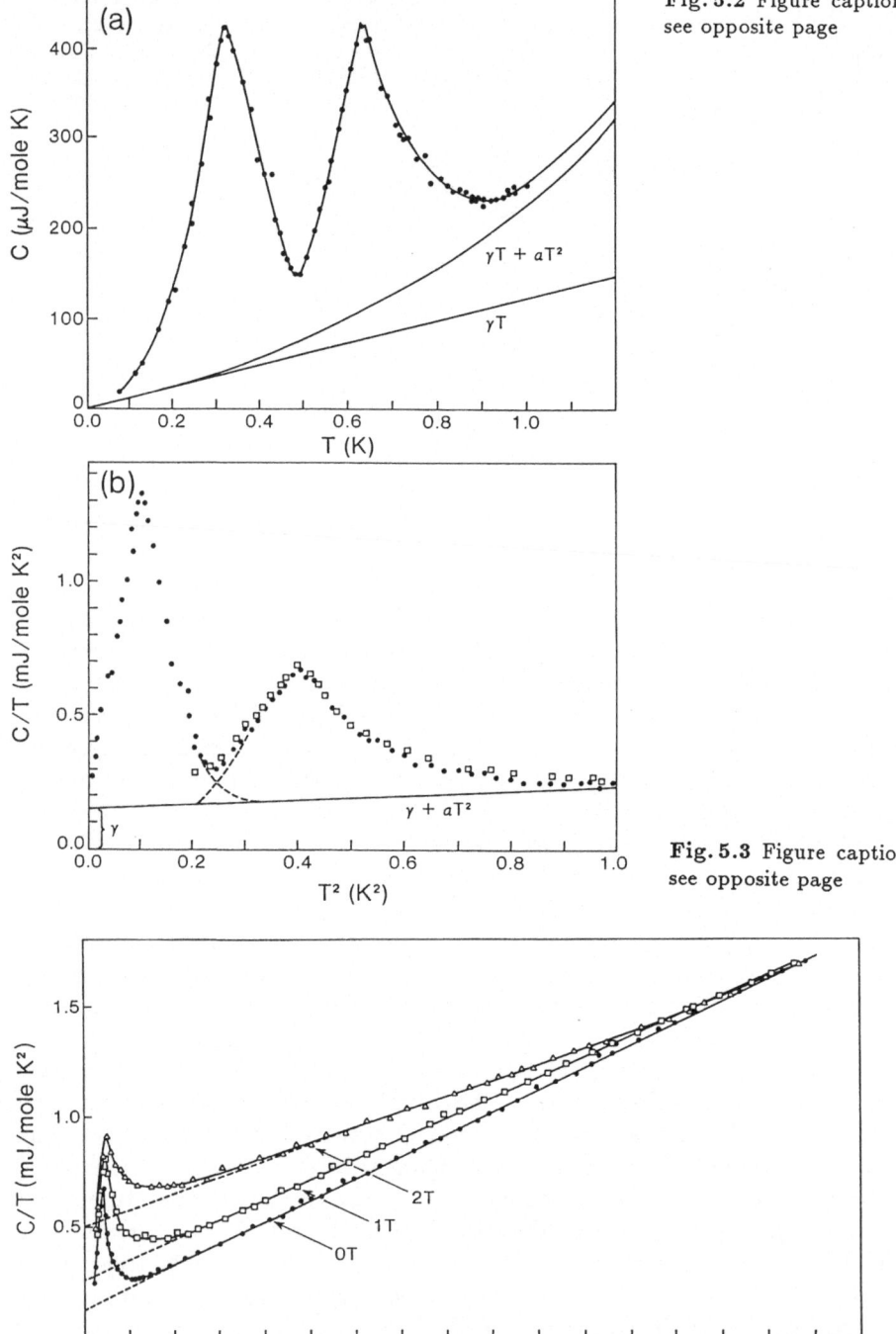

(a)

$\gamma T + aT^2$

$\gamma T$

**Fig. 5.2** Figure caption
see opposite page

(b)

$\gamma + aT^2$

$\gamma$

**Fig. 5.3** Figure caption
see opposite page

2T

1T

0T

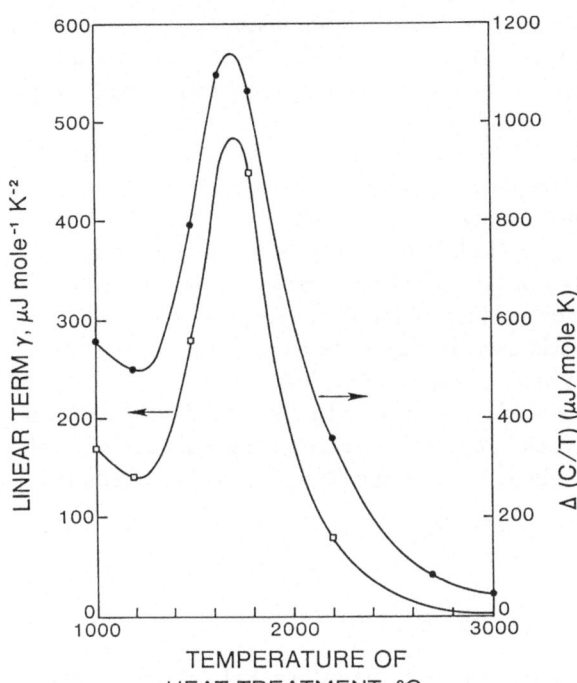

**Fig. 5.4.** Dependence on $T_{HT}$ of the linear $\gamma$ term in the heat capacity, and of the height of the peak tail in the $C/T$ vs $T^2$ curve at $0.63$ K above the $\gamma + \alpha T^2$ line. The observations were made on glassy carbon as a function of heat treatment temperature $T_{HT}$ [Mrozowski 1979]

dependence in the electrical resistivity [Spain et al. 1983b] (see Sect. 8.3), as well as a negative magnetoresistance (see Sect. 8.4) [Hambourger 1987].

## 5.2 Thermal Expansion

Because of the anomalously small room temperature thermal expansion coefficient of graphite in the $a$ direction $(\alpha_a \sim -1 \times 10^{-6}/°C)$ [Morgan 1972], the thermal expansion of individual carbon filaments is difficult to measure. Of particular interest is the negative value of $\alpha_a$ for graphite at room temperature and the large anisotropy of $\alpha$, with a large positive thermal expansion along the $c$-axis $(\alpha_c \sim 9 \times 10^{-6}/°C)$. Many of the thermal expansion measurements relevant to carbon fibers have been made on oriented fiber composites.

**Fig. 5.2a,b.** Specific heat curves for NCC carbon $(T_{HT} = 1250°C)$ obtained in the temperature range below 1 K showing both cubic and linear contributions as extrapolated from data obtained at higher temperature. (a) The plot of $C$ vs $T$ shows the sharpness of the two peaks in the specific heat and (b) the plot of $C/T$ vs $T^2$ shows the cubic decay of both peaks on the low temperature side. The broken curves show a possible separation into individual contributions for each peak [Mrozowski 1979]

**Fig. 5.3.** Specific heat curves $C/T$ vs $T^2$ for the NCC carbon of Fig. 5.2 $(T_{HT} = 1250°C)$ at magnetic fields of 0, 1 and 2 T [Mrozowski 1979]

Here the expansion parallel to the fiber axis is dominated by the fiber thermal expansion because of the high modulus of the fiber relative to the matrix [Morgan 1972]. In the transverse direction, however, the contributions from the matrix material are significant.

One expects that the thermal expansion of carbon fibers is dominated by the thermal expansion of the graphite crystallites which compose the fibers. Thus one should be able to use a polycrystalline model for the fiber to predict its thermal expansion. Polycrystalline models of bulk graphite have been reviewed by B.T. Kelly and Taylor [1973]. The method used for the determination of the thermal expansion coefficient for fibers employs essentially the same polycrystalline model as is used to calculate the internal stress distribution and the appropriate elastic constants in carbon fibers.

Using a theoretical model similar to that used to describe the elastic properties of fibers (Sect. 4.3.2), one can express the thermal expansion coefficients $\alpha_\parallel$ and $\alpha_\perp$, which are respectively $\parallel$ and $\perp$ to the fiber axis, as [Reynolds 1971]

$$\alpha_\parallel = A\langle \sin^2 \phi \rangle \alpha_c + (1 - A\langle \sin^2 \phi \rangle)\alpha_a \tag{5.1}$$

and

$$\alpha_\perp = A\langle \cos^2 \phi \rangle \alpha_c/2 + (1 - A\langle \cos^2 \phi \rangle/2)\alpha_a \quad, \tag{5.2}$$

where $A$ is a parameter which accounts for the internal stress and porosity, and is usually in the range 0.5–0.7 for polycrystalline graphite, while $\phi$ is the misorientation angle defined in Fig. 3.15. The angular averages $\langle \sin^2 \phi \rangle$ and $\langle \cos^2 \phi \rangle$ are defined and discussed in Sect. 4.3.2.

The thermal expansion coefficients $\alpha_a$ and $\alpha_c$ which enter (5.1) and (5.2) have been measured in bulk graphite over a wide temperature range, extending from low temperature up to 3000 K. This work is summarized by Morgan [1972] and is largely based on measurements by Bailey and Yates [1970]. The functional form of the temperature dependence of the in-plane (Fig. 5.5) and $c$-axis (Fig. 5.6) lattice constants are very different and the non-monotonic temperature dependence of the in-plane lattice constant $a_0(T)$ is highly anomalous. The corresponding temperature dependences of the thermal expansion coefficients $\alpha_a(T)$ (shown in Fig. 5.7) and $\alpha_c(T)$ (shown in Fig. 5.8) are highly anisotropic, with $\alpha_a(T)$ being unusually small in magnitude and $\alpha_c(T)$ being unusually large. The negative values for $\alpha_a$ below $\approx 700$ K are highly anomalous. In contrast $\alpha_c$ exhibits a very usual temperature dependence. A phenomenological model for $\alpha_a(T)$ and $\alpha_c(T)$ developed by Riley [1945] relates the thermal expansion to the heat capacity and the elastic constants of graphite. The model is quite successful in fitting $\alpha_a(T)$ and $\alpha_c(T)$ over a wide temperature range. (See the solid curves in Figs. 5.7 and 5.8.)

A few thermal expansion measurements on carbon fibers have been reported to date [Butler and Diefendorf 1969; Butler et al. 1971; Tanabe et

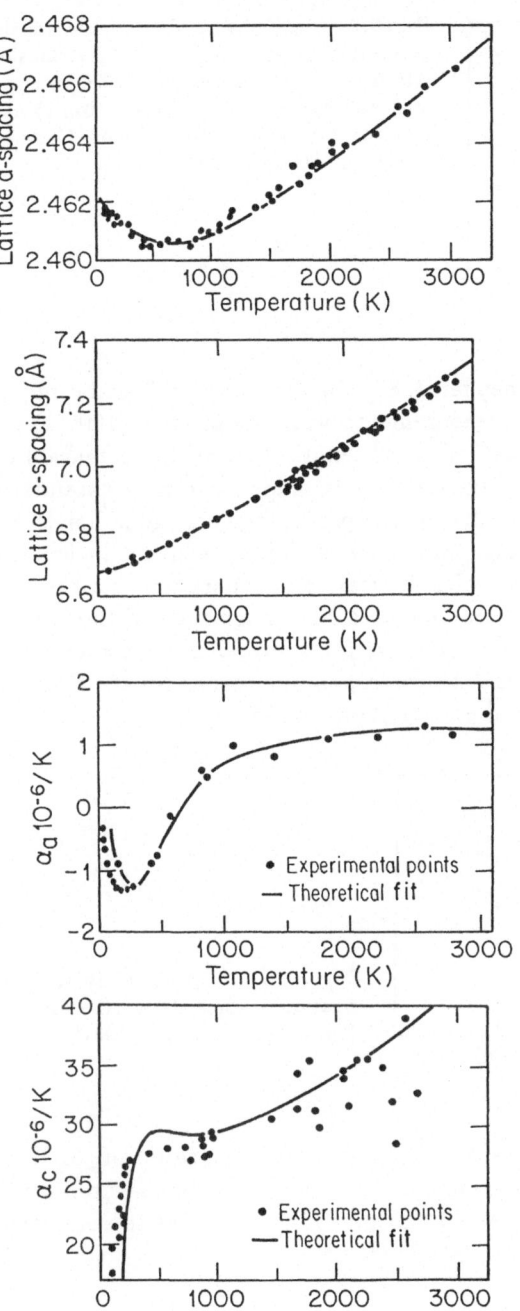

**Fig. 5.5.** Anomalous lattice thermal expansion for pristine graphite: the $a$-spacing as a function of temperature [Morgan 1972]. The solid line is from the phenomenological model of Riley [1945]

**Fig. 5.6.** Temperature dependence of the lattice $c$-spacing (twice the interlayer separation) [Morgan 1972]. The solid line is from the phenomenological model of Riley [1945]

**Fig. 5.7.** Thermal expansion coefficient $\alpha_a$ as a function of temperature [Morgan 1972]. The solid line is from the phenomenological model of Riley [1945]

**Fig. 5.8.** Thermal expansion coefficient $\alpha_c$ as a function of temperature [Morgan 1972]. The solid line is from the phenomenological model of Riley [1945]

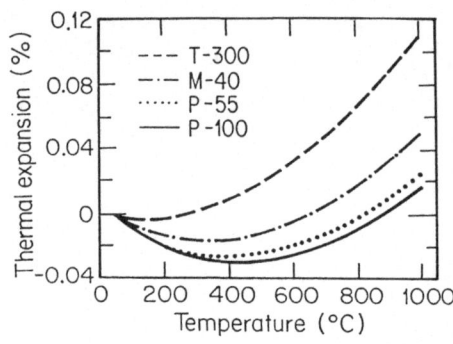

Fig. 5.9. Temperature dependence of the percentage longitudinal thermal expansion relative to room temperature (20°C) for various ex-PAN (T-300 and M-40 Toray) and ex-pitch (P-55 and P-100 Union Carbide) fibers [Tanabe et al. 1987]

al. 1987; Yasuda et al. 1987; Sheaffer 1987]. In the work of Tanabe et al. [1987] and Yasuda et al. [1987] measurements were made up to 1000°C on samples consisting of ~1000 fibers, ~5 cm in length, and held at a tension of ~5 g. Calibration of the thermal expansion was made in terms of standard quartz and tungsten rods. The results for the percentage change in length vs temperature for several fibers (Fig. 5.9) show an initial decrease in length, followed by an increase in length with increasing temperature. As the degree of structural order of the fibers increases, the magnitude of the length change decreases, and the temperature where the thermal expansion changes from negative to positive increases. The corresponding results for the thermal expansion coefficient $\alpha$ for these fibers are shown in Fig. 5.10.

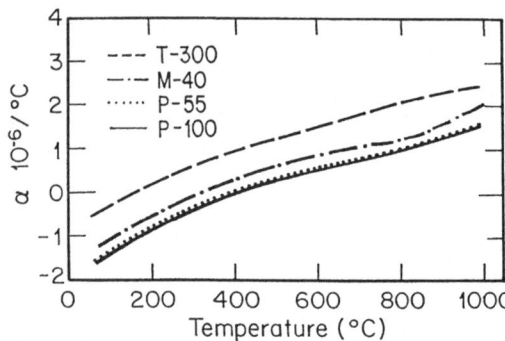

Fig. 5.10. Longitudinal thermal expansion coefficients for the fibers of Fig. 5.9 [Tanabe et al. 1987]

A good fit for the observed results for $\alpha_{\parallel}$ is obtained in terms of (5.1). By determining the crystalline orientation function $I(\phi)$ from the measured width of the (002) x-ray diffraction lines, $\alpha_{\parallel}$ can be evaluated using the empirical relation (see Sect. 3.4.3)

$$I_m(\phi) = I_0 \cos^m \phi \quad , \tag{5.3}$$

from which the average $\langle \sin^2 \phi \rangle$ can be computed, yielding

$$\langle \sin^2 \phi \rangle = 1/(m + 3) \quad . \tag{5.4}$$

As the degree of orientational alignment increases (see Figs. 3.16 and 3.17), the proper limits $m \to \infty$ and $\alpha_\parallel \to \alpha_a$ are obtained. These results show that the more ordered fibers exhibit a larger (more negative) longitudinal coefficient of thermal expansion at room temperature than fibers with a lower degree of structural order, in accordance with (5.1).

Reynolds [1971] has shown that $A$ is a constant which depends on the internal stress and on the pores and voids in the sample. Good agreement between the model given by (5.1, 5.3, and 5.4) and the experimental measurements for $\alpha$ in Fig. 5.10 were obtained up to 700°C for $A = 0.6$ and the thermal expansion coefficients $\alpha_a(T)$ and $\alpha_c(T)$ for graphite [Morgan 1972; Touloukian et al. 1979]. In the work of Tanabe et al. [1987] and Yasuda et al. [1987] the orientational alignment parameter $m$ in (5.3) and (5.4) was measured independently by x-ray diffraction for each of the fibers shown in Figs. 5.9 and 5.10, yielding data such as given in Fig. 3.16. From these data the values $m = 15$ for the T-300 fibers; $m = 56$ for M-40; $m = 73$ for P-55; and $m = 261$ for P-100 are determined. Above 700°C, discrepancies from the model were observed. These discrepancies were attributed to relaxation of the internal stress and to a change in the shape and size of the pores in the fibers at high temperatures, especially for the more disordered fibers with high pore density.

Using an adaptation of a laser light diffraction technique, Sheaffer [1987] was able to measure the *transverse* thermal expansion of individual carbon fibers up to 1400°C while the fibers were held under a preload tension of 1000 psi. The results were calibrated against a tungsten filament and were interpreted in terms of a linear (positive) thermal expansion of the fiber diameters. The results for several ex-PAN and ex-pitch fibers are summarized in Table 5.1. In this work, no attempt was made to interpret these results

Table 5.1. Transverse and longitudinal thermal expansion coefficient of various carbon fibers

| Fiber | Type | Fiber diameter [$\mu$m] | Average transverse thermal expansion coefficient [$10^{-6}/°C$] | Average longitudinal thermal expansion coefficient [$10^{-6}/°C$] |
|---|---|---|---|---|
| Hercules HM-3000[a] | PAN | 7.5 | 10.9 | |
| Celanese 950[a] | PAN | 5.7 | 13.1 | |
| Union Carbide P-55[a] | Pitch | 10.8 | 12.0 | |
| Hercules AS-1[a] | PAN | 7.1 | 12.5 | |
| Union Carbide P-55[b] | Pitch | | 11.8 | −1.37 |
| Union Carbide P-75[b] | Pitch | | 12.4 | −1.48 |
| Union Carbide P-100[b] | Pitch | | 9.4 | −1.48 |
| Toray T-50[b] | PAN | | 6.7 | −1.22 |

[a] Scheaffer [1987]
[b] Wagoner et al. [1987]

in terms of (5.2). This equation suggests that the less ordered fibers should exhibit a lower value for $\alpha_\perp$.

Values for the transverse and longitudinal thermal expansion coefficients at room temperature were determined for several ex-pitch fibers [Wagoner et al. 1987] from measurement of the transverse and longitudinal thermal expansion coefficients in both a unidirectional fiber/epoxy composite sample and in a similar sample of the composite material without the fibers. The analysis was done using a composite cylinder assemblage model. The same analysis also yielded the axial and transverse Young's moduli and the shear modulus from measurements of the dynamic resonant frequencies of the longitudinal and torsional vibrations (Sect. 6.1.2). The measurements were made on composites with two types of resins (Fiberite 934 and Union Carbide ERLX-1962) and over a temperature range between 30°C and 120°C. The results averaged over the two types of resin matrix material are summarized in Table 5.1. For the case of the P-55 fibers, which were also measured directly by the laser interference method [Sheaffer 1987], good agreement was obtained by the two methods. In general, the results on the thermal expansion coefficients of carbon fibers by Yasuda et al. [1987], Sheaffer [1987] and Wagoner et al. [1987] are consistent with each other and with the interpretation of the thermal expansion measurements in carbon fibers given by (5.1-4).

The very low thermal expansion along the fiber axis and the highly anisotropic thermal expansion of well-graphitized fibers presents a differential thermal expansion problem when carbon fibers are to be bonded to the matrix binder of a fiber composite material which has to withstand thermal cycling.


## 5.3   Thermal Conductivity

Measurements of the temperature variation of the thermal conductivity can be used as a tool to characterize carbon fibers prepared from various precursors under different conditions [B.T. Kelly and Taylor 1973]. One advantage of the thermal conductivity characterization technique is that it gives an overall view of the defect density of an entire fiber, in contrast to electron microscopy techniques which provide a detailed local view (see Sects. 3.7.1 and 3.7.2). The thermal conductivity is related to other measured quantities known to characterize the degree of disorder in the material, such as x-ray diffraction, Raman spectroscopy, resistivity and magnetoresistance measurements.

Since the heat carrying capacity of a single fiber is very small, most of the thermal conductivity measurements in fibers have been made on multifilamentary samples. In the case of the vapor grown fibers, however, thermal conductivity measurements on a single fiber have been possible [Piraux et al. 1984; Nysten et al. 1985a,b; Heremans et al. 1985].

114

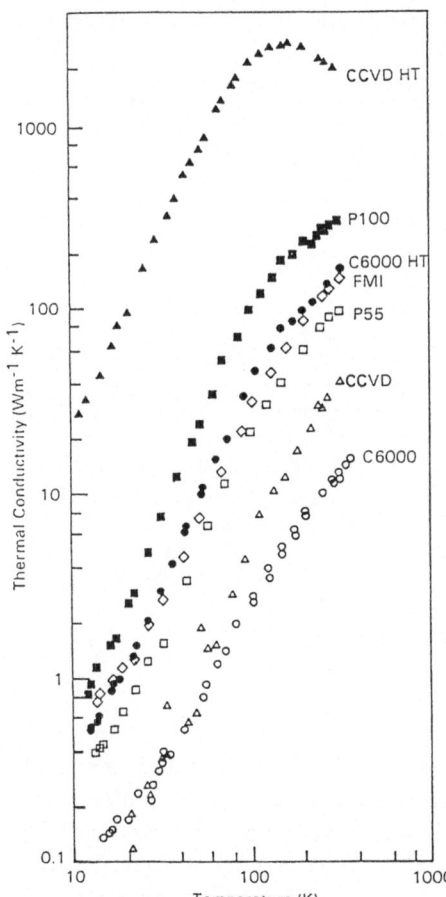

**Fig. 5.11.** Log-log plot of the temperature dependence of the thermal conductivity of various carbon fibers for a variety of precursor materials and heat treatment temperatures (see Table 5.2) [Heremans et al. 1985; Heremans and Beetz 1985]

Figure 5.11 shows the observed temperature dependence of the thermal conductivity for various carbons and graphite fibers [Heremans et al. 1985]. Similar results were previously obtained by Piraux et al. [1984] and by Nysten et al. [1985a, 1985b]. The results show a large increase in the magnitude of the thermal conductivity $\kappa$ and a shift of the peak in $\kappa(T)$ to lower temperatures as the crystalline perfection increases. Properties of the fibers in Fig. 5.11 are listed in Table 5.2, showing the interconnection of the various characterization parameters which determine the thermal conductivity and other properties of the carbon fibers.

The thermal conductivity of graphite has been extensively studied experimentally and theoretically [Hooker et al. 1963; Holland et al. 1966; Dreyfus and Maynard 1967; B.T. Kelly 1967, 1969b, 1981; Morelli and Uher 1985] and is of particular interest because of its very large magnitude. Of all materials, diamond and graphite (in-plane) exhibit the highest known room tempera-

**Table 5.2.** Correlation of the thermal conductivity with other properties related to the degree of graphitization of carbon fibers[a]

| Property[b] | Units | CCVD | CCVD-HT | P-55 | P-100 | C6000 | C6000-HT | FMI |
|---|---|---|---|---|---|---|---|---|
| $T_{HT}$ | °C | 1130 | 3000 | | | | 3000 | |
| $d_{002}$ | nm | 0.347 | 0.335 | 0.341 | 0.337 | 0.353 | 0.3395 | 0.341 |
| $L_c$ | nm | 3.8 | $\geq 100$ | 8 | 18 | 1.6 | 7.2 | 5.4 |
| $\varrho(300K)^c$ | $\mu\Omega$m | 10.2 | 0.71 | 9.6 | 3.6 | 18 | 5.9 | 3.0 |
| $\varrho(300K)^d$ | $\mu\Omega$m | | | 7.5 | 2.5 | 15.0 | | |
| $\kappa(300K)^c$ | Wm⁻¹K⁻¹ | 35 | 1950 | 100 | 300 | 12 | 155 | 140 |
| $\kappa(300K)^d$ | Wm⁻¹K⁻¹ | | | 100 | 520 | | | |
| $\sigma_T{}^d$ | GPa | | | 2.1 | 2.2 | 2.7 | | 3.3 |
| $E_Y{}^d$ | GPa | | | 380 | 720 | 230 | | 480 |
| $l_\phi$ | nm | 3.6 | 2900 | 15 | 130 | – | 28 | 22 |
| $I_{1360}/I_{1580}$ | – | 0.778 | ~0 | 0.316 | 0.109 | – | 0.196 | 0.337 |

[a] For the various fibers listed, CCVD denotes as-deposited fibers while CCVD-HT denotes catalytic chemical vapor deposited fibers subsequently heat treated. Data for pitch-based (P-55 and P-100), and PAN-based (FMI and C6000) fibers are also included.
[b] The various properties are $\varrho(300K)$, the room temperature electrical resistivity; $\kappa(300K)$, the room temperature thermal conductivity; $\sigma_T$, the tensile strength; $E_Y$, Young's modulus; $l_\phi$, the phonon mean free path; and $I_{1360}/I_{1580}$, the ratio of the disorder-induced intensity at ~1360 cm⁻¹ to the Raman allowed intensity at 1580 cm⁻¹.
[c] As measured by Heremans et al. [1985].
[d] As reported by the manufacturers.

ture thermal conductivity. The very high anisotropy of the room temperature thermal conductivity $(\kappa_a/\kappa_c \approx 200)$ in graphite is also of particular interest, allowing for very high in-plane thermal conductivity and poor conductivity in the c direction. The poor c-axis thermal conductivity is associated with the weak interplanar binding in the c direction. The thermal conductivity is less anisotropic than the electrical conductivity because phonons also contribute to $\kappa$, even if there is no free carrier contribution. Phonon processes dominate both the in-plane and c-axis thermal conductivities [Hooker et al. 1963; Holland et al. 1966; de Combarieu 1968; Chau and Liu 1974; B.T. Kelly 1981]. In the most elementary approach, the high temperature $(T \geq 100K)$ in-plane thermal conductance can be modeled as a superposition of the conductances of the individual layer planes, whereas the c-axis thermal resistance can be modeled as a superposition of the thermal resistances of the individual layer planes.

As shown in Fig. 5.11, a large variation in the magnitude of the thermal conductivity $\kappa$ is found for the various fibers. A strong correlation is however found between $\kappa$ and the crystalline perfection of the fiber. For example, the values of the thermal conductivity at room temperature for vapor grown fibers heat treated to 3000°C (CCVD-HT in Fig. 5.11) exceed that of copper by a factor of 4 and are indeed comparable to those obtained for HOPG prepared at the same heat treatment temperatures [Piraux et al. 1984; Heremans and Beetz 1985; Heremans et al. 1985; Woollam et al. 1986a]. The lowest $T_{HT}$ sample (C6000) in Fig. 5.11 has an almost three orders of magnitude smaller thermal conductivity at $\sim 100$ K than that for the most defect-free vapor grown fibers.

In the relaxation time approximation, the thermal conductivity for graphite along the basal plane is given by [Kelly 1981]

$$\kappa_a = \sum_{s,q} k_B \left[ \frac{\hbar \omega_s(q)}{k_B T} \right]^2 \tau_s(q) \frac{\exp[\hbar \omega_s(q)/k_B T]}{\{\exp[\hbar \omega_s(q)/k_B T] - 1\}^2} [v_{sa}(q)]^2 \quad . \qquad (5.5)$$

where $\tau_s(q)$ is the relaxation time of a phonon of wave vector $q$ in the $s^{th}$ mode propagating in the basal plane, $\omega_s(q)$ denotes the phonon dispersion relation of the $s^{th}$ mode at $q$ and $v_{sa}(q)$ denotes the phonon velocity component in the basal plane. The three acoustic modes (the in-plane longitudinal and transverse waves and the out-of-plane vibration) all contribute to the thermal conductivity [see (4.7) and (4.8)]. In applying this analysis, we assume that the phonon dispersion relations are not much affected by the lattice disorder present in fibers. Assuming $C_{44} = 0$, we obtain the simple dispersion relations

$$\omega_1 = v_1 q_a, \quad \omega_2 = v_t q_a, \quad \omega_3 = (\delta^2 q_a^4 + v_z^2 q_z^2)^{1/2} \quad , \qquad (5.6)$$

where $v_1$ and $v_t$ refer to the longitudinal and transverse velocity of sound, $q_a = \sqrt{q_x^2 + q_y^2}$ and $\delta$ is a constant related to the bending of the honeycomb planar graphite network (Sect. 4.2.1).

Several scattering mechanisms contribute to the relaxation rate $1/\tau_s(q)$: normal phonon-phonon scattering, umklapp phonon-phonon processes, isotope scattering, strain field scattering, defect scattering and crystallite boundary scattering. For carbons with an in-plane coherence length $L_a < 10^3$ Å, at low temperatures the phonon mean free path $l_\phi$ is mainly controlled by boundary scattering. Then,

$$l_\phi \simeq L_a = v_s(q)\tau_s(q) \quad \text{for} \quad s = 1, 2, 3, \qquad (5.7)$$

and (5.5) thus can be approximated by

$$\kappa_a \simeq l_\phi \sum_{s,q} C_s(q) v_s(q) \quad , \qquad (5.8)$$

where $C_s(q)$ is the specific heat component for phonon $q$ in the $s^{th}$ mode and the phonon mean free path $l_\phi$ is assumed to be independent of $T$, $s$, and $q$. This is probably a good approximation at low $T$ where phonon scattering is dominated by the crystal defects.

In practical applications of the thermal conductivity to the characterization of graphite fibers, it is often convenient to use the simplified Debye formula

$$\kappa = \tfrac{1}{3} C_v v l_\phi \quad , \qquad (5.9)$$

where $C_v$ is the specific heat at constant volume, $v$ is an average phonon group velocity (the velocity of sound) and the phonon mean free path $l_\phi$ is simply given by $l_\phi = v\tau$. For the case of graphite, heat is carried by phonons over the entire temperature range $(T < 300K)$ so that the electronic contribution $\kappa_e$ is negligible. The reason for this is the huge phonon contribution to $\kappa_a$ and the relatively low carrier concentration for pristine graphite. For intercalated graphite fibers, $\kappa_e$ is important at low temperatures and follows the Wiedemann-Franz law [Issi 1987]

$$\kappa_e = LT\sigma \quad , \tag{5.10}$$

where $L$ is the Lorenz number ($L_0 = 2.44 \times 10^{-8} V^2 K^{-2}$ for a free electron gas), $T$ is the temperature and $\sigma$ the electrical conductivity (see Sect. 10.5).

The temperature dependence of the thermal conductivity obeys a $T^{2.3}$ law for temperatures above 30 K, reaching a maximum of $\sim 2.5 \times 10^3$ Wm$^{-1}$K$^{-1}$ at $T \sim 140$ K and then decreasing with further increase in temperature as umklapp phonon-phonon scattering processes become important [Piraux et al. 1984; Heremans et al. 1985]. Below the maximum in $\kappa_a$, the specific heat is proportional to $T^2$, and the phonon mean free path $l_\phi$ is limited by boundary scattering due to the crystallite size, while phonon-phonon scattering (um-klapp processes) dominates above the maximum in $\kappa_a$. No significant fiber size effects due to boundary scattering are observed in fiber diameters down to 12 $\mu$m, the fiber diameters being much larger than the crystallite sizes $(< 0.1 \mu m)$. The in-plane vibrations (longitudinal and transverse modes) contribute a term to the thermal conductivity proportional to the square of the temperature and the out-of-plane vibrations contribute a term of similar magnitude proportional to $T^{2.6}$. The total thermal conductivity is proportional to $T^{2.3}$ [Kelly 1969b, 1981].

The phonon mean free path $l_\phi$ values obtained from (5.9) and the measurements in Fig. 5.11 are listed in Table 5.2. Single crystal graphite values for $C_v$ and $v$ are assumed for all fibers. This table shows that there is a consistent correlation between the various quantities that characterize the degree of graphitization of the fiber samples. For the vapor grown fiber category, a factor of 50 increase in $\kappa_a$ can be achieved by heat treatment to 3000°C. For the vapor grown fibers heat treated to 3000°C, $\kappa(T)$ is very similar to that for HOPG or even single crystal graphite [Heremans and Beetz 1985]. However, for PAN-based fibers an anomalous temperature dependence was reported [Piraux et al. 1984; Heremans et al. 1985]. By measurements on a variety of samples covering a wide range of $l_\phi$ and $\kappa_a$ values, Heremans et al. [1985] and Heremans and Beetz [1985a] showed that the thermal conductivity characterization technique, which measures $l_\phi$ in the bulk, correlates very well with the Raman microprobe technique, which measures $L_a$ within the optical skin depth [Rahim et al. 1986]. Good correlation was also obtained between the thermal conductivity, temperature dependent resistivity and x-ray (002) linewidth measurements. However, for samples that are not well ordered, the agreement between the thermal conductivity, electrical resistivity and x-ray diffraction characterization techniques is not as good [B.T. Kelly and Gilchrist 1969]. These discrepancies are attributed to point defects and a distribution of $L_a$ values which affect the various property measurements differently.

Finally, we note that (5.9) is not always valid. This equation is based on the assumption that the reciprocal relaxation time is given by $1/\tau \simeq v_s/l_\phi$. In some cases the simple formula leads to an erroneous conclusion, indicating the need for a more detailed calculation that takes into account a non-uniform phonon distribution perpendicular to the temperature gradient [Carruthers 1961].

# 6. Mechanical Properties

The most important applications of carbon fibers utilize their high strength-to-weight ratio, and therefore mechanical properties are of special technological interest (see Table 1.1). Some idea of the potential strength of carbon fibers can be gleaned from R. Bacon's [1960] study of carbon whiskers, in which he found extensional moduli of 800 GPa [120 Msi (megapounds per square inch)] and breaking stresses of 20 GPa (3 Msi). Since typical tensile strengths of steels are 1 GPa, and their densities are about four times that of carbon, the potential of carbon fibers as structural materials is clear (see Tables 1.1 and 6.1). Actual tensile strength values realized in commercial fibers are lower than those of carbon whiskers, but still offer significant advantages over conventional metals.

Table 6.1. Representative data[a] for the mechanical properties of selected carbon fibers

| | Company | Fiber | $E_Y$ [GPa] | $\varepsilon_T$ [%] | $\sigma_T$ [GPa] | Density [g/cm³] |
|---|---|---|---|---|---|---|
| Ex-PAN, low modulus, high density | Celanese | Celion 3000 | 235 | 1.5 | 3.6 | 1.77 |
| | Hercules | Magnamite ASAW | 235 | 1.5 | 3.6 | – |
| | Hysol Grafil | Apollo HS 38-750 | 260 | 1.9 | 5.00 | – |
| Ex-PAN, intermediate modulus | Hysol Grafil | Apollo IM43-600 | 300 | 1.3 | 4.0 | – |
| | Hercules | HMS-10 | 345 | 0.6 | 2.2 | – |
| | Toray | T800 | 300 | 1.9 | 5.7 | – |
| Ex-PAN, high modulus | Celanese | GY-70 | 520 | 0.4 | 1.86 | 1.96 |
| | Toray | M50 | 500 | 0.45 | 2.2 | – |
| Ex-mesophase-pitch (standard extrusion technology) | Union Carbide | P-25 | 140 | 1.0 | 1.4 | – |
| | Thornel | P-55 | 380 | 0.5 | 2.1 | 2.02 |
| | | P-75 | 500 | 0.4 | 2.0 | 2.06 |
| | | P-100 | 690 | 0.3 | 2.2 | 2.15 |
| | | P-120 | 820 | 0.2 | 2.2 | – |
| CCVD[b] | General Motors | As grown | 237 | 1.2 | 2.9 | 1.8 |
| | | $T_{HT} = 2700°C$ | 700–760 | 0.36 | 2.6–2.9 | 2.0 |

[a] Extensive lists of data are available in Lovell [1986] and Hughes [1987].
[b] Data for the as-grown CCVD fibers are from Tibbetts and Beetz [1987] and at 2700°C are from Beetz [1987].

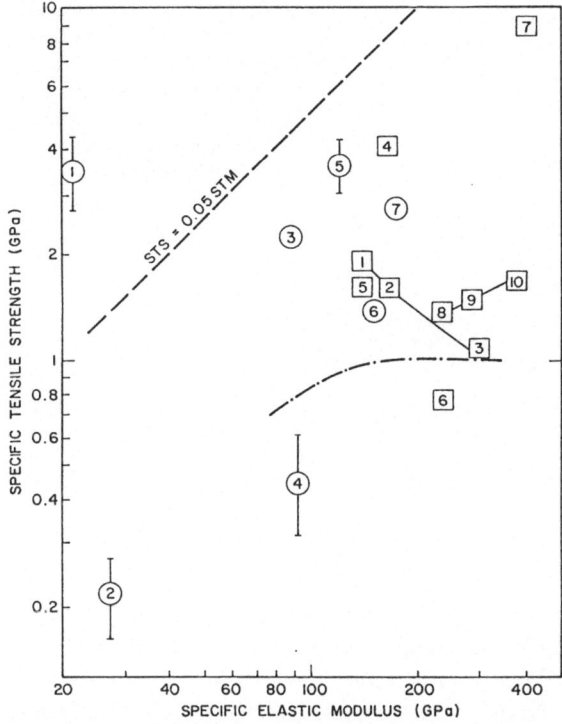

**Fig. 6.1.** A plot of the specific tensile stress vs specific elastic modulus for various carbon fibers and other filaments (adapted from Hughes [1987]). The legend for the various points is: ① silica (whiskers), ② steel, ③ Aramid, ④ alumina, ⑤ polyethylene (experimental), ⑥ boron, ⑦ silicon carbide; [1] high tensile carbon fibers (ex-PAN), [2] intermediate modulus (ex-PAN), [3] ultra-high modulus (ex-PAN), [4] ex-PAN (experimental), [5] CCVD as deposited, [6] CCVD with $T_{HT} = 3000°C$, [7] graphite whisker, [8] ex-pitch Carbonic HM50, [9] ex-pitch Carbonic HM60, [10] ex-pitch Carbonic HM80. The dot-dash curve refers to ex-mesophase-pitch fibers [Bright and Singer 1979] while the dashed line refers to the "theoretical limit" relating the specific tensile strength (STS) to the specific tensile modulus (STM)

The potential of carbon fibers can be seen from Fig. 6.1, in which the specific tensile breaking stress (breaking stress/specific gravity) is plotted against the specific tensile elastic modulus (Young's modulus/specific gravity) for various kinds of carbon fibers, and compared with filaments of other materials. It is to be noted, firstly, that some carbon materials have very high specific strengths and moduli. Strength and modulus are, of course, the parameters which are of greatest importance in composites for aircraft applications (Chap. 12).

Figure 6.2 plots the same type of data as Fig. 6.1, but is for the breaking strength vs tensile modulus for various commercial carbon fibers. In addition, lines of constant strain are included in Fig. 6.2 so that typical breaking strains can be estimated.

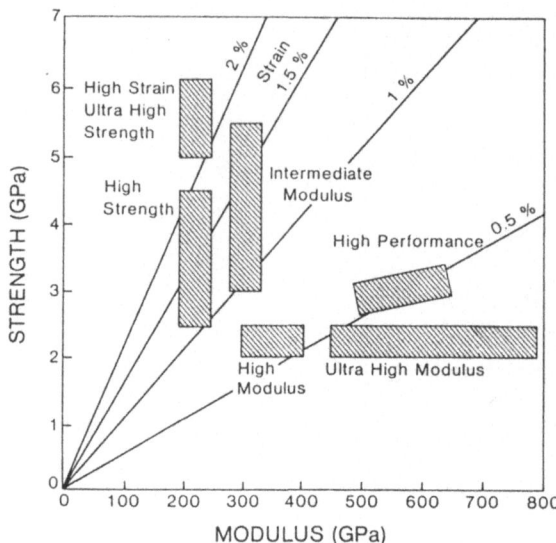

**Fig. 6.2.** The breaking strength of various types of carbon fibers plotted as a function of Young's modulus. Lines of constant strain can be used to estimate the breaking strains [Tansosen-i Konwakai 1986; Endo 1987]

Figures 6.1 and 6.2 illustrate many points which will be enlarged upon later, but it is worthwhile to introduce some concepts at this point to act as a guide. Firstly, carbon fibers have elastic moduli which lie between about 200 and 800 GPa ($1 \text{GPa} = 10^9 \text{ N/m}^2 = 10^{10} \text{ dynes/cm}^2$, and is about 150 000 psi). The theoretical modulus of perfectly aligned graphite fibers would be about 1000 GPa, which is nearly reached in carbon whiskers [R. Bacon 1960]. Section 4.3.2 discusses models which explain why the modulus is decreased by disorder, principally the misalignment of graphene planes along the fiber axis. Experimental data are presented in Sect. 6.1.2 and compared to these models in Sect. 6.1.4.

Commercial carbon fibers with elastic moduli of about 230 GPa are known as high-strength, or low-modulus, fibers. High-modulus fibers have values in the range of 480-690 GPa, with intermediate modulus fibers typically about 300 GPa. Other nomenclatures are sometimes used, such as ultra-high for high, and high for intermediate modulus. Significant improvements in the mechanical properties are still being realized in the research laboratory today, and may become incorporated in the commercial product of tomorrow (see Hughes [1986a, 1987]). Also, the newer high-modulus ex-pitch fibers (Carbonic HM50, HM60, and HM80) [Endo 1988a] show marked increases in tensile strength over the older Thornel fibers.

The strengths of carbon fibers also vary widely. The theoretical strength can be evaluated from an expression which is discussed in most textbooks on the strength of materials (see, for example, Cottrell [1964], A. Kelly [1966]):

$$\sigma_T = (E_Y \gamma_a / a_d)^{1/2} \quad , \tag{6.1}$$

122

where $\sigma_T$ is the breaking stress, $E_Y$ is Young's modulus, $\gamma_a$ the surface energy (i.e., the energy needed to separate two planes in the crystal, per unit area; $\gamma_a = 4.2$ J/m$^2$ for the prismatic bonds in graphite [Reynolds 1971]), while $a_d$ is the distance between the planes which are to be separated by the applied tensile stress. This expression, sometimes referred to as the Orowan-Polanyi equation [Polanyi 1921; Orowan 1949], generally overestimates the strength, since it assumes a defect-free crystal. However, certain defect-free whiskers were observed to have experimental strengths fairly close to the predictions of (6.1) [Griffith 1920]. The surface energy $\gamma_a$ in (6.1) is a measure of the bond strength, as is Young's modulus of elasticity, $E_Y$. Using a proportionality between these two quantities, (6.1) becomes

$$\sigma_T^{\max} = A_\sigma E_Y \tag{6.2}$$

with the constant, $A_\sigma$, taking a value between 0.05 and 0.1. This prediction (for $A_\sigma = 0.05$) is plotted as the dashed line in Fig. 6.1. It can be seen that certain carbon fibers are approaching this line as fiber technology becomes more sophisticated. Unfortunately, the fibers with highest modulus fall well below it, from which we conclude that the full potential of carbon fibers has not been realized, by any means.

It is reasonable to assume that the reason why the experimental strength lies well below the "theoretical" limit is because of cracks and similar flaws in the fibers. Then, (6.1) becomes [Griffith 1920]

$$\sigma_T = (2E_Y\gamma_a/\pi l_f)^{1/2} \quad , \tag{6.3}$$

where $l_f$ is the length of the critical flaw which initiates rupture. As an example, for a high-modulus fiber with $E_Y = 400$ GPa and $\sigma_T = 2.5$ GPa, (6.3) yields a critical flaw length of $l_f = 170$ nm. This model gives some qualitative insight into the fracture process, but gives quantitatively unreasonable results for carbon fibers, as discussed below.

Earlier reviews of the mechanical properties have appeared for rayon-based fibers [R. Bacon 1973] and PAN-based fibers [Reynolds 1973]. A summary of their elastic properties as well as the mechanical properties of mesophase pitch-based fibers is given in Sects. 6.1 and 6.2. The relatively limited work that has been carried out on CCVD (vapor grown) fibers is discussed in Sect. 6.3.

## 6.1 Elastic Parameters

### 6.1.1 Experimental Techniques

The extensional, or Young's, modulus $E_Y$ of individual fibers can be measured by a conventional extensiometer, in which the applied force is $\sim$0.1 N. Since the fibers are very delicate, great care must be taken in mounting them in the jaws of the apparatus. This is usually achieved by gluing the ends across a slit in a thin piece of cardboard. This piece of cardboard is then mounted in a tensile test machine with low load capability. The slit in the cardboard is completed across the full width of the cardboard, and an extension versus load curve is obtained. The measurements can also be made on fiber bundles. In this case the fibers are usually bonded in a polymer whose mechanical properties are such that the measurements of carbon fiber properties are not significantly affected (see Hughes et al. [1980]). Chi et al. [1985] have recently discussed the relationship between the single fiber strength and values obtained from fiber bundles. Other references can be found in their paper. Some practical considerations are discussed by Hughes [1986a]. Test methods for high modulus fibers are considered by McMahon [1973], while standard methods for single fibers and resin-impregnated tows are given in ASTM D 3379-75 and D 2343-67.

An interesting experimental technique was developed by Beetz and Budd [1983] in which a small amplitude oscillatory strain ($\varepsilon_{osc} \sim 10^{-3}$) was superimposed on a slowly increasing strain ($\sim 10^{-3}$/s). This sensitive technique allowed non-linear stress-strain behavior to be studied to high precision (Sect. 6.1.2) [Beetz and Budd 1983].

Modulus data have also been obtained on single fibers by an ultrasonic method [R.E. Smith 1972] (see Fig. 6.3), in which short pulses of 1 MHz ultrasound were tuned with a pair of transducers at either end [Curtis et al. 1968]. The transit time yields the propagation velocity $v = \sqrt{(E_Y/\varrho_m)}$, where $\varrho_m$ is the mass density.

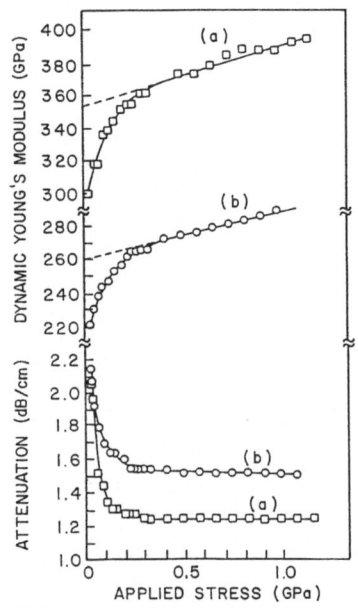

**Fig. 6.3.** Ultrasonically measured Young's modulus and ultrasonic attenuation vs applied stress for (a) a high modulus and (b) a high strength ex-PAN fiber [Curtis et al. 1968]

R.E. Smith [1972] developed a method for evaluating the elastic constants of long fibers aligned in a resin composite, using sonic or ultrasonic waves. The importance of the technique is that it allows the full set of elastic constants of the plate to be evaluated, from which both the longitudinal and transverse Young's modulus ($E_t$) can be evaluated. This approach was recently extended and amplified to determine not only the axial and transverse values for Young's modulus and the shear modulus, but also the longitudinal and transverse thermal expansion coefficients for carbon fibers [Wagoner et al. 1987]. The transverse modulus is related to the passage of a shear wave along the axis of the fiber, so that it contains information about the resistance to bending. Values of $E_Y$ can also be obtained from static measurements on composites. The values of the longitudinal Young's modulus obtained by the ultrasonic method appear to be several percent higher than those obtained at sonic frequencies, or statically. This is probably a consequence of the nonlinear elastic interaction between the fibers and the matrix.

The torsional rigidity can be measured by tuning the oscillations of a single fiber loaded with a body of known moment of inertia (see Reynolds [1973]) [Fischbach and Srinivasagopalan 1978; Srinivasagopalan 1979]. This technique also gives information about internal damping mechanisms. Young's modulus can also be measured by a similar resonance method [Jouquet and Schill 1971].

Raman spectroscopy has also been applied to characterize the effect of stress on the frequency and linewidth of the Raman-allowed $E_{2g_2}$ mode of single fibers using a Raman microprobe with 2 $\mu$m spatial resolution (Sect. 4.5.1). By using linearly polarized light, the stress-induced strain parallel and perpendicular to the fiber axis can be studied independently. Specific application of this technique has been made to CCVD fibers [Sakata et al. 1987, 1988].

### 6.1.2 Experimental Observations

The modulus of carbon fibers is dependent on the applied stress in an unusual manner, increasing as stress is increased until fracture occurs. This is opposite to that which occurs in most organic and metal fibers, where there is a yield stress at which the modulus drops significantly. This effect was demonstrated using conventional extensiometer techniques by Beetz [1982c] in ex-pitch fibers and by Voet and Morawski [1975] using dynamic techniques. Figure 6.4 shows stress-strain curves for an ex-PAN fiber (Union Carbide T-300) and a CCVD filament grown from natural gas [Tibbetts and Beetz 1987]. Hughes [1986a] suggested that extension data for fibers should not only be expressed in terms of the average Young's modulus, but in the form

$$E_Y(\varepsilon) = E_Y(0)(1 + f_\epsilon \varepsilon) + \dots \quad , \tag{6.4}$$

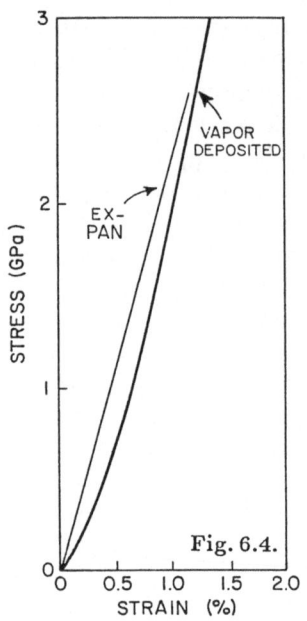

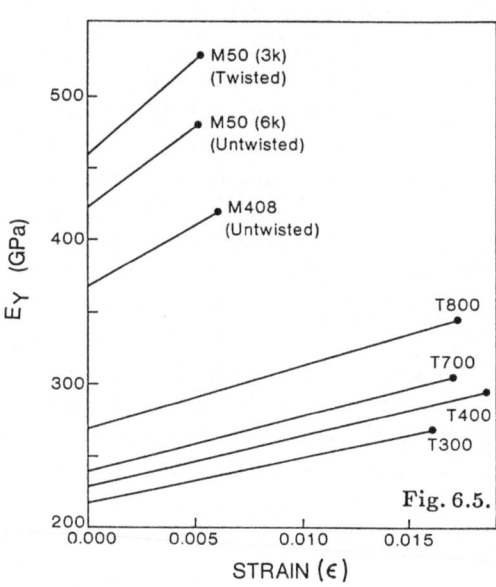

**Fig. 6.4.** Stress-strain curves for ex-PAN (230 GPa modulus) and vapor deposited fibers (as grown) [Tibbetts and Beetz 1987]

**Fig. 6.5.** Experimental data for the variation of elastic modulus $E_Y$ with strain for several ex-PAN fibers manufactured by Toray. The black dots indicate the breaking point of the various fibers and tows [Hughes 1986a]

where $E_Y(\varepsilon)$ is the modulus as a function of strain, $\varepsilon$, and $E_Y(0)$ is the zero strain value, while $f_\varepsilon$ is a constant. Figure 6.5 shows such experimental data for a set of fibers manufactured by Toray, and the comparative properties of various fibers are listed in Table 6.1. It can be seen in Fig. 6.5 that typical $f_\varepsilon$ values range from 15 to 28.

An even more striking example of the effect of increased fiber modulus with increased stress is found in the ultrasonically derived curves of Fig. 6.3 by Curtis et al. [1968], who also showed that the acoustic attenuation fell with applied stress. These effects are usually interpreted in terms of the "unwrinkling" of fibrils, and this type of model has been discussed in Sect. 4.3.2.

The elastic modulus increases strongly with heat treatment temperature, as shown in Fig. 6.6. This major trend can be related to a decrease in the mean misorientation angle $\langle \phi \rangle$ of the fibrils (see Fig. 3.15 and Sect. 3.4.3). A decrease in $\langle \phi \rangle$ can also be induced by stretching the fiber during the oxidation stage [Watt and W. Johnson 1969] or by stretching during testing, as discussed in the paragraph above.

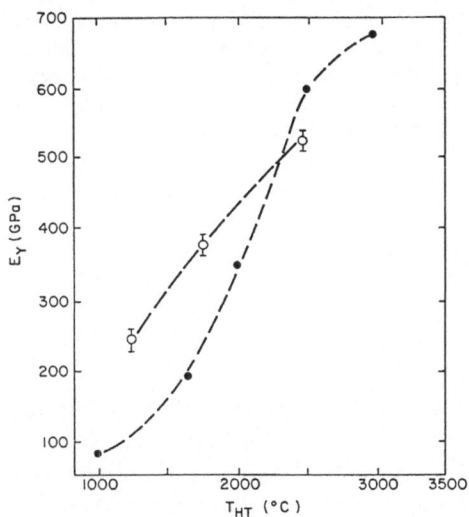

**Fig. 6.6.** Young's modulus of fibers as a function of heat treatment temperature ($T_{HT}$) for ex-mesophase-pitch fibers shown as open circles (Bright and Singer [1979], averaged from their Table 1) and for ex-PAN fibers shown as solid circles (data from Table 2.2 of Riggs [1985]). The dashed lines are drawn to guide the eye

The elastic properties of fibers as a function of measurement temperature are reviewed in Sect. 9.3. It is noted there that the Young's modulus can fall by ~50% at temperatures of 2400°C.

Within this framework, the major trends of the modulus for different fibers can be understood on the basis of their structural order. Ex-rayon fibers have relatively poor alignment, unless they are hot-stretched during the carbonization stage. Ex-PAN fibers retain the alignment obtained during stretching in the (low-temperature) oxidation stage, and their Young's modulus increases as the alignment in the oxidized, extruded, polymer increases [Moreton 1976]. These fibers have correspondingly higher moduli than (unstretched) ex-rayon fibers. Even higher moduli are realized in ex-pitch fibers, which align during the extrusion of the mesophase. High moduli are also found in CCVD filaments, to be discussed in Sect. 6.3.

In many fibers the modulus also decreases with increasing diameter of the fiber, falling about 20% for increasing diameters of PAN-based fibers between 7 and 9 $\mu$m [B.F. Jones and Duncan 1971], with a more rapid variation for ex-rayon fibers. These data are consistent with a multiphase structure, in particular a layer of surface material with properties differing from the interior. Evidence for such structures is discussed in Sect. 3.7.2.

Some empirical relations have also been used to relate Young's modulus $E_Y$ to the in-plane crystallite size $L_a$ for ex-PAN fibers [Guigon et al. 1984b]:

$$E_Y = 492 - 43.5 \times 10^5 / L_a \qquad (6.5)$$

where $E_Y$ is in gigapascals (GPa) and $L_a$ is in angstroms (Å), and the fit that was obtained is given in Fig. 6.7, where $R = 0.93$ denotes a correlation factor [a perfect fit to (6.5) would have $R = 1$].

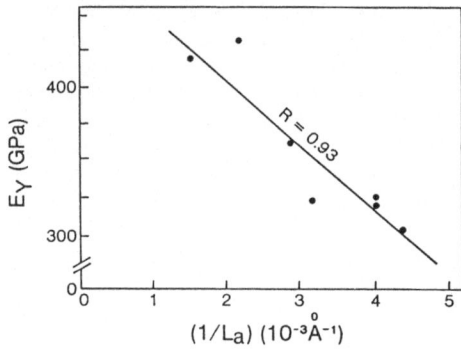

**Fig. 6.7.** Plot of Young's modulus $E_Y$ vs $1/L_a$ for high modulus, ex-PAN fibers. Here $R = 0.93$ is the correlation factor signifying the goodness of the fit to the data [Guigon et al. 1984b]

Touzain and Bagga [1984] have studied the correlation between the interlayer spacing $\tilde{c}$ as measured by x-ray diffraction ($d_{002}$) and the Young's modulus and the tensile strength for various ex-PAN and ex-pitch fibers and have found a good correlation between the increase in the graphitization index $g_p$ (see Sect. 3.1.7) and the increase in $E_Y$, but little correlation was found between $\sigma_T$ and $g_p$.

A clear connection between the cross-sectional microstructure and the mechanical properties of mesophase pitch-based carbon fibers has recently been demonstrated [Endo 1988a]. Hamada et al. [1987a,b] have further shown that the cross sectional microstructure can be controlled by the extrusion conditions, such as the temperature (viscosity) of the pitch, whether or not the pitch is stirred during extrusion, the shape of the stirrer as well as the subsequent heat treatment conditions (Sect. 2.1.3). These authors further characterized the microstructure by x-ray diffraction spectra and linewidths, electron microscopy (both SEM and TEM), electrical resistivity and magnetoresistance, and then correlated these properties with the tensile strength and bulk modulus. Their results show that the basic planar structural unit of the microstructure of fibers such as Thornel P-120 yields a higher modulus, associated with the higher degree of graphitization and 3D interlayer correlation, but a lower tensile strength relative to fibers such as Carbonic HM80 which are turbostratic with a wavy ribbon microstructure and show no 3D correlations (see Fig. 3.30). These authors were able to show that the wavy ribbon microstructure of these fibers increased their tensile strength. They further showed (Sect. 2.1.3) that this change of microstructure could be implemented by stirring the mesophase pitch while spinning the fibers. This conjecture was confirmed by showing that the magnetoresistance ($\Delta\varrho/\varrho_0$) for the fibers spun with stirring was negative even after heat treatment to 2700°C, while fibers spun without stirring and prepared under otherwise identical conditions show a positive magnetoresistance of much larger magnitude (Sect. 8.4.2). They found that by raising the temperature of the mesophase pitch during spinning, they could decrease the viscosity and decrease the degree of structural ordering.

128

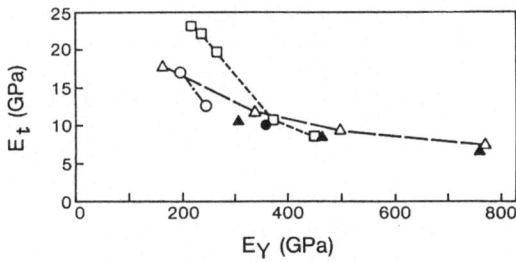

**Fig. 6.8.** Experimental data for the transverse modulus $E_t$ vs Young's modulus $E_Y$ for ex-PAN and ex-pitch fibers. The following notation is used: for ex-PAN: (□) static, (•) sonic, (o) ultrasonic; and for ex-pitch: (△) static, (▲) sonic [from Wagoner et al. 1987]

Values of the transverse modulus, $E_t$, of ex-PAN and ex-pitch fibers are plotted in Fig. 6.8 as a function of Young's modulus using several experimental techniques [Wagoner et al. 1987]. The lower values of the ultrasonically derived values of these quantities are to be noted. The major trend of $E_t$ is to lower values at higher $E_Y$, and values of $E_t$ are generally higher for ex-PAN fibers than ex-pitch.

The theory and experimental results on the torsion modulus of carbon fibers has been reviewed by Srinivasagopalan [1979]. The torsion modulus is found to vary significantly from fiber to fiber, having little correlation with Young's modulus and with the preferred orientation. The values of the torsion modulus are however correlated with the fiber precursor material. Ex-PAN fibers have torsion moduli $G$ between 17 and 28 GPa, ex-rayon fibers between 10 and 15 GPa, and ex-mesophase pitch fibers between 9 and 13 GPa [Srinivasagopalan 1979; Fischbach and Srinivasagopalan 1978; R.E. Smith 1972] (see Fig. 6.9). The results are also sensitive to the shape of the fiber cross section [Jouquet and Schill 1971].

The torsional modulus increases significantly with increasing tensile stress. The increase is linear and can be more than 10% before the fiber breaks. This effect is usually attributed to an inhomogeneous microstructure, such as skin-core heterogeneity (Sect. 3.7.2) [Srinivasagopalan 1979, and references therein]. Both the extensional and torsional moduli for carbon fibers change less markedly with temperature than comparable values for polycrystalline graphite [Jouquet and Schill 1971]. These authors concluded that dislocations play a less important role in the mechanical properties of fibers. This is just what one might expect for a turbostratic material.

Boron doping changes the mechanical and structural properties in a similar way to increases in $T_{HT}$ [Allen et al. 1971]. Boron doping appears to enhance the alignment of the crystallites during the processing of the polymer precursor. The fact that boron doping acts to pin the dislocations in graphite, thereby increasing the low-frequency value of $C_{44}$ [Seldin and Nezbeda 1970], may also be significant.

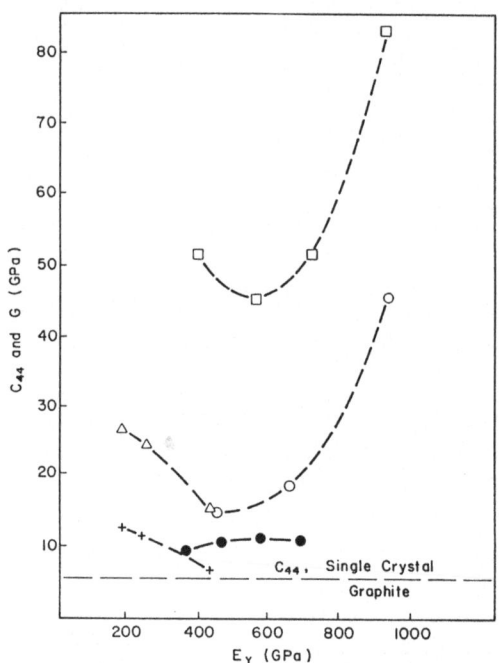

**Fig. 6.9.** Summary of experimental and theoretical data for the torsional modulus $G$ and the reciprocal elastic compliances for the crystallites, ($C_{44} = 1/S_{44}$). ($\triangle$) experimental data for $G$ [Reynolds and Moreton 1980], ($\bullet-\bullet$) experimental data for $G$, ex-pitch fibers [Srinivasagopalan 1979]; ($+$) calculated values for $C_{44}$, ex-PAN [Reynolds and Moreton 1980]; calculated values for $C_{44}$, (o) ex-pitch, ($\square$) ex-PAN fibers [Goldberg 1985]. Curves are drawn as guides to the eye

This effect may also be of importance in neutron irradiation-induced changes in the elastic moduli of ex-PAN fibers [Allen et al. 1969, 1971; G.A. Cooper and Mayer 1971; Bullock 1972a; B.F. Jones and Wilkins 1972] and ex-pitch fibers [Price et al. 1985]. Irradiation is usually carried out with distributions of neutron energies, and fluences are given for those neutrons with energies greater than a fixed value. This makes a direct comparison of effects observed by different workers somewhat difficult. However, the results appear to substantiate an increase in modulus of ~10% occurring for both high strength and high modulus ex-PAN fibers after neutron irradiation, with similar increases for ex-pitch fibers. Typical fluences used in this work are ~$10^{22}$ neutrons/m$^2$ ($E > 1$ MeV). Higher doses (e.g., $4 \times 10^{22}$ neutrons/m$^2$, and $E > 1$ MeV) result in a decrease in modulus.

Bullock et al. [1973] showed that the x-ray $d_{002}$ spacings increased from ~3.44 Å to 3.50 Å for doses of $10^{23}$ neutrons/m$^2$ and $E > 1$ MeV. Also, the apparent crystallite misorientation angle, $\langle \phi \rangle$ *decreased* from ~16° to 12° in this dose range. This decrease is somewhat surprising, since neutrons knock atoms out of lattice sites, increasing disorder. The increase in $d_{002}$ can be attributed to the production of interstitials, for instance. The decrease in $\langle \phi \rangle$ can be correlated with the increase in $E_Y$ (see next section), but it must be remarked that a decrease of $\langle \phi \rangle$ of the magnitude observed would increase $E_Y$ by more than 10% (see Fig. 6.11 and discussion). Therefore, other factors are involved. Irradiation at 77 K resulted in no change in $E_Y$ [Bullock 1972a].

### 6.1.3  Internal Friction

Although the mechanical properties of carbon fibers have been studied extensively, internal friction measurements have not been available until very recently [Taborek 1987]. Structure in the internal friction as a function of temperature provides information about defect or impurity mobility which is difficult to obtain by other means.

Both the Young's modulus and the internal friction of graphite fibers can be measured by monitoring the resonant frequency and damping of a mechanical oscillator made of a short length of fiber clamped at both ends [Taborek 1987]. Bending mode resonant vibrations of typical frequency 1 kHz and amplitude $10^{-6}$ cm can be driven and detected electromagnetically when exciting a fiber of 1 cm length and $\sim 10 \mu$m diameter. The internal friction of the fiber is measured by the width of the resonance curve $Q = \omega_0/\Delta\omega$, which can be measured to one part in $10^2$. Measurements of $Q$ show a characteristic temperature where $Q$ has a minimum value (or $Q^{-1}$ has a maximum value) (Fig. 6.10a). At temperatures above room temperature, the internal friction is high, presumably due to thermally activated sliding of grain boundaries. As a function of temperature, $Q^{-1}$ shows a broad Bordoni-type peak centered at 115 K associated with the movement of dislocations, with two sharper peaks superimposed at $T = 160$ K and 220 K. At helium temperatures, the $Q$ can be greater than $10^5$. The sharp peaks in $Q^{-1}$, which occur at the same temperatures as elastic anomalies in bulk graphite, are due to the thermally activated motion of defects in the graphite crystal lattice. Since a perfect

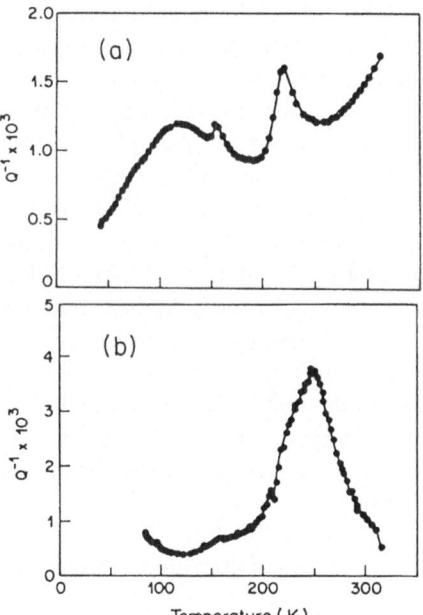

Fig. 6.10a,b. $Q^{-1}$ as a function of temperature for (a) a pristine Union Carbide P-25 fiber and (b) a bromine residue compound made from a Union Carbide P-120 fiber. Note the sharp features at 160 K and 220 K for the pristine fiber and the large internal friction peak at 250 K caused by intercalation [Taborek 1987]

harmonic crystal has no dissipation, the $Q$ is determined by the defects in the fiber. A defect such as an impurity, a vacancy, a dislocation or grain boundary moves in response to an imposed stress, but the motion typically requires hopping over a potential barrier, which requires a certain relaxation time $\tau$. If the relaxation time is either very long or very short compared to the frequency of the applied stress, the defect motion does not dissipate energy. When the hopping time for the defect is comparable to the period of oscillation, the defects can efficiently extract energy from the macroscopic motion; since the hopping time $\tau$ is an exponential function of temperature, $\tau \sim \exp(E/k_B T)$, the dissipation is sharply peaked at a characteristic temperature. The details of the internal friction depend on the structure of the fiber. For example, Fig. 6.10b shows the temperature dependence of the internal friction of a fiber intercalated with bromine and then allowed to outgas in air to form a residue compound (see Chap. 10). Although the graphite peak in $Q^{-1}$ at 160 K is essentially unaffected by bromine intercalation, it is dwarfed by a much larger peak at 250 K which is introduced by the intercalant. Studies of the internal friction may provide a promising method of studying the mechanical properties and defects in graphite fibers.

### 6.1.4 Comparison of Elastic Moduli with Theory

Models for the moduli of carbon fibers are introduced in Sect. 4.3.2. Experimental results for ex-PAN fibers are compared to the uniform stress model in this section, and results for CCVD filaments with Ruland's unwrinkling model in Sect. 6.3.

Goldberg [1985] used a Gaussian expression to fit the misorientation intensity function $I(\phi)$ for a set of ex-PAN and ex-pitch fibers, as discussed in Sects. 3.4.3 and 5.2. Experimental data for $E_Y$ were then analyzed by using single crystal values for the compliances $S_{11}$, $S_{12}$, and $S_{13}$ (see Table 4.1) then adjusting $S_{44}$ to obtain the best fit to the data. Fortunately, Young's modulus for the fiber is not sensitive to the choice of $S_{13}$, but its variation with crystallite orientation depends strongly on the value of $S_{44}$ (see Fig. 6.11). Table 6.2 compares the best fit values of $S_{44}$ to the measured values of Young's modulus for a wide range of ex-PAN and ex-mesophase pitch fibers [Goldberg 1985]. The Young's moduli used in these estimates are significantly higher than are usually quoted for these fibers because they have been corrected for two effects: the compliance of the tensile testing fixture and the porosity of the fibers (this was not done in the data of Reynolds and Moreton discussed below). The porosity correction assumes that the pores do not interfere with the stress, i.e., the pores lie roughly parallel to the basal planes. This is consistent with the small angle scattering results described in Sect. 3.5.

Figure 6.11 illustrates curves of the variation of $E_Y$ with the mean misorientation angle $\langle \phi \rangle$, using (4.15). The value of $E_Y$ falls rapidly with $\langle \phi \rangle$,

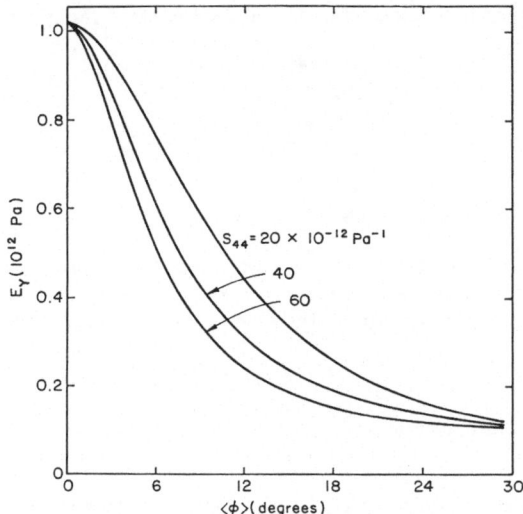

**Fig. 6.11.** Calculated curves of Young's modulus as a function of average misorientation angle $\langle\phi\rangle$ for various values of $S_{44}$ according to the uniform stress model [Goldberg 1985]

**Table 6.2.** Measured values[a,b] of Young's modulus $E_Y$, orientational misalignment $\langle\phi\rangle$, and calculated values of shear compliance $S_{44}$

| Fiber type | $E_Y$ [GPa] | $\langle\phi\rangle$ [°] | $S_{44}$ [$10^{-12}$Pa$^{-1}$] |
|---|---|---|---|
| Celion[c] | 370 | 20.6 | 19 |
| G-50[c] | 560 | 11.4 | 22 |
| GY-70[c] | 730 | 7.5 | 19 |
| GR-21[c] | 930 | 5.8 | 12 |
| VSB[d] | 450 | 8.0 | 66 |
| VSC[d] | 660 | 5.9 | 54 |
| P-100[d] | 960 | 3.6 | 21 |

[a] From Goldberg [1985].
[b] $E_Y$: Young's modulus corrected for compliance of test fixture and porosity. The orientational misalignment angle $\langle\phi\rangle$ is obtained from the measured full width at half integrated intensity of the (002) x-ray reflection.

[c] Ex-PAN fibers (Celanese).
[d] Ex-pitch fibers (Union Carbide).

and the curves are sensitive to the value of $S_{44}$. A 50% reduction in $E_Y$ is obtained for a value of $\langle\phi\rangle$ of only about $10°$, indicating the importance of good alignment in obtaining a high modulus.

Reynolds and Moreton [1980] analyzed data for ex-PAN fibers similar to those of Goldberg [1985] over a limited range of $E_Y$ values using a $\cos^m \phi$ form for $I(\phi)$ and experimental data for $E_Y$ to obtain the exponent $m$ using (4.15), (4.20) and (5.3). These values of $m$ were then compared to x-ray or electron microscopy data. Their results are shown in Table 6.3. It is clear from these

**Table 6.3.** Calculated and observed orientation parameters for ex-PAN fibers[a,b] (adapted from Reynolds and Moreton [1980])

| $T_{HT}$ [°C] | $E_Y$ [GPa] | $G$ [GPa] | $m_{calc}$ | $\langle\phi\rangle_{calc}$ [°] | $\langle\phi\rangle_{obs}$ [°] | $S_{44}$ $[10^{-12}$ Pa] |
|---|---|---|---|---|---|---|
| 1000 | 197 | 26 | 13 | 18 | 14–19 | 77 |
| 1500 | 245 | 24 | 21 | 15 | 12–17 | 84 |
| 2500 | 428 | 14 | 107 | 7 | 5–8 | 143 |

[a] The crystal orientation function $I_m(\phi) = \cos^m(\phi)$ is used in performing the averages in (4.15), (4.18) and (5.3). For perfectly aligned crystallites $m \to \infty$.
$\langle\phi\rangle$ is the average HWHM (half width at half maximum) of the orientational misalignment angle.
[b] $E_Y$ is Young's modulus.
$G$ is the torsional modulus.
$S_{44}$ values are from Reynolds, private communication.

and the above data that reasonable agreement can be obtained between the uniform stress model and experimental data for Young's modulus.

As discussed in Sect. 4.3.2, no satisfactory models have been developed to relate the compliances for the crystallites to the torsional modulus $G$. Although values of $S_{44}$ can be found to fit the experimental data for $E_Y$, there are difficulties in relating these $S_{44}$ values to physical processes in the fibers. The model values for $S_{44} = 1/C_{44}$ are summarized in Fig. 6.9. Also included are selected experimental data on the torsional modulus $G$.

As noted in Sect. 4.3.2, the torsional modulus $G$ depends sensitively on the interactions between the crystallites or ribbons, and $G$ is a function of the elastic moduli $C_{11}$, $C_{12}$, $C_{44}$. Values of $G$ which are only 5% of the value of $S_{44}$ for single crystal graphite ($S_{44} = 240 \times 10^{-12}\text{Pa}^{-1}$) are not difficult to understand in principle. Models need to be developed to explain this quantitatively.

It is more difficult to understand the "stiff" (i.e., lower) values of $S_{44}$ obtained by the models and their trends with $E_Y$. These values are describing the local shear response of the individual ribbons or crystallites, and would be expected at first sight to be approximately those of graphite. As discussed in Sect. 4.3.1, experimental data for $C_{44}$ on bulk graphite based on ultrasonic measurements give low frequency values of $S_{44}$ that are higher than the single crystal value, with values as high as $5000 \times 10^{-12}\text{Pa}^{-1}$, due to dislocation softening. This softening is not expected within the ribbons of ex-pitch or ex-PAN fibers.

Two possible explanations for "stiffer" values of $S_{44}$ in fibers than for single crystal graphite are:

1. Covalent bonds are linking the planes. These bonds could be due to high concentrations of non-basal dislocations, such as those in Fig. 3.4, or to $sp^3$ bond linkages, or to chemical bond linkages involving impurity

134

atoms, particularly at the edges of the ribbons, or at vacancies and vacancy clusters.

2. Ribbon bending stiffens the shear behavior. This can be imagined simply by shearing a stack of papers. If the papers are corrugated, their shear deformation is achieved with greater difficulty. The complicated bending patterns of the ribbons (see for example Fig. 3.30) ensure that shear deformation would be more difficult with respect to stress in any direction.

Although these explanations are reasonable, it is difficult to understand why the model values of $(1/S_{44})$ *increase* with increasing modulus (Fig. 6.9) in the calculations of Goldberg [1985]. It would be anticipated that interlayer bending and layer bending would decrease with increase in $E_Y$. This trend, leading to an increase in $S_{44}$, is consistent with the model values of $(1/S_{44})$ obtained by Reynolds and Moreton [1980]. It would be interesting to obtain data for the elastic behavior of the newer ex-pitch fibers (Carbonic HM50, HM60, HM80) which are spun so as to increase the corrugation of the layers. A decreased value of $S_{44}$ [or an increased value of $(1/S_{44})$] would be expected, compared to results for the earlier ex-pitch fibers used by Goldberg [1985].

There are also no models for the transverse modulus $E_t$, which is another parameter which should depend sensitively on the folding of the graphene ribbons. Again, it would be interesting to compare the values of $E_t$ for the older Thornel and the newer Carbonic ex-pitch fibers.

Another test of the uniform stress model was carried out by Curtis et al. [1968], who measured the change in $\langle \phi \rangle$ induced by stretching ex-PAN filaments with an applied load, and compared the increase in $E_Y$ with the value predicted by the model. It may be concluded that the uniform stress model can satisfactorily fit data for the Young's modulus $E_Y$, but models are badly needed to relate the values of $S_{44}$ used to the torsional modulus $G$. With such a model it should be possible to check on the reasonableness or otherwise of the values of reported compliances, particularly $S_{44}$.

To summarize, fiber microstructure influences the mechanical properties in several ways. First of all, the degree of orientation determines how close the elastic properties (such as Young's modulus) are to those of single crystal graphite or HOPG. Second, the details of the nature of defects in the crystallites determine the appropriate elastic constants to use in any model for the properties. These effects and the stacking arrangement of the fibrils are particularly important for the shear moduli of the crystallites. Finally, the degree of validity of the uniform stress, uniform strain, or elastic unwrinkling models will depend upon the details of fiber type, i.e., CCVD onion skin, ex-pitch radial, ex-pitch random, ex-PAN, etc. The degree to which the stress is not uniform within the fiber will depend on the details of the microstructure. Wu and McCullough [1977] have developed variational techniques which enable one to utilize detailed information about the microstructure of composites to accurately calculate the properties of composites. Application of these techniques to polycrystalline models of carbon fibers should provide additional insight into the degree of validity of these models. It would be particularly interesting to study the effects of ribbon bending in these models.

## 6.2 Fracture, Stress and Strain

Carbon fibers fracture in a brittle manner when subjected to tensile or binding stresses. Thus, the stress-strain curves shown in Fig. 6.4 do not have a plastic-strain region, but terminate abruptly at the breaking point. Typical strain-to-failure (breaking strain) values, $\varepsilon_T$, lie between 1 and 2% for low modulus, high strength, ex-PAN fibers ($E_Y \sim 250$ GPa) so that tensile strength $\sigma_T$ (stress at failure or breaking stress) values are usually between 3 and 4.5 GPa. When these ex-PAN fibers are heat-treated further, to form high modulus types (e.g., $E_Y \sim 350$–500 GPa), the ultimate tensile strength *falls* to values $\sigma_T$ between 2.5 and 1.8 GPa as the modulus increases, corresponding to a *reduction* in the strain-to-failure. This trend to lower $\varepsilon_T$ values for higher $E_Y$ also is observed for ex-rayon and ex-pitch fibers, so that the potential of carbon fibers to achieve single crystal values ($E_Y \sim 1000$ GPa and $\sigma_T \sim 20$ GPa) cannot be realized in commercial fibers at the present time.

Reviews of the mechanical properties of carbon fibers have been written by R. Bacon [1973] (ex-rayon) and Reynolds [1973] and Reynolds and Moreton [1980] (ex-PAN) and by D.J. Johnson [1987b] (ex-PAN). Rather than trying to duplicate this material, the aim here will be to give an overview of the area, attempting to understand the reasons for the decrease in strain-to-failure at higher modulus, and looking for the intrinsic strength of these materials. Again, the discussion relating to CCVD filaments will be dealt with in Sect. 6.3.

### 6.2.1 Experimental Observations

As mentioned above, one of the most striking features of the results on fracture stress is the decrease in strain-to-failure at higher modulus values. Figure 6.12 illustrates the tensile strength of ex-polymer fibers from PAN and pitch precursors as a function of $T_{HT}$. The dashed curve shows data of Watt and W. Johnson [1971] on ex-PAN fibers, which reach a maximum strength at $T_{HT} \sim 1500°$C and then decrease. Thus, the fracture stress initially increases with Young's modulus, and then decreases [Moreton et al. 1967]. More recent data for Celanese ex-PAN fibers are also included in Fig. 6.12. The higher values of the tensile stresses result from improved processing procedures, but show a similar decrease in tensile stress with $T_{HT}$ above 1500°C. These data are from Riggs [1985]. Values for ex-pitch fibers are taken from Bright and Singer [1979] (their Table 1) and are presented as an average of four types

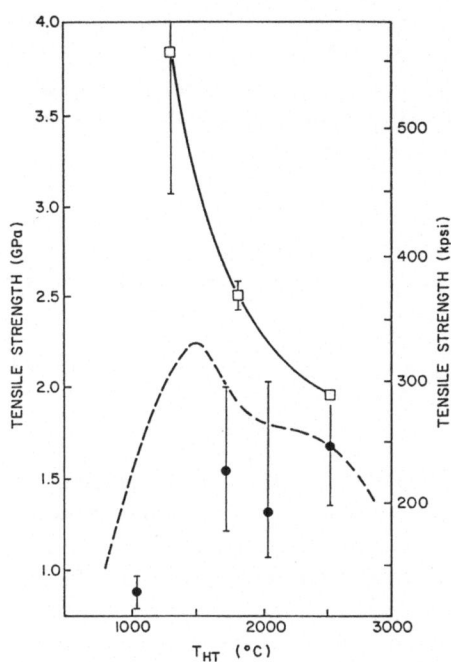

Fig. 6.12. Tensile strength of fibers as a function of heat treatment temperature: (---) early data on ex-PAN fibers [adapted from Watt and W. Johnson 1971]; (□) modern ex-PAN fibers [Riggs 1985]; (•) ex-mesophase-pitch fibers [Bright and Singer 1979]

of filaments, with the error bars representing the range of values. More recent ex-pitch fibers have considerably higher values of tensile stress, and are discussed later.

These data can be plotted as a function of elastic modulus, but Figs. 6.1 and 6.2 already summarize this information, since the density does not change dramatically with $T_{HT}$ (see Sect. 3.3 and Table 2.2). The tensile strength of ex-PAN fibers falls with increasing $E_Y$, reaching a plateau for ex-pitch, and increases linearly with $E_Y$ for ex-rayon fibers, following the approximate relationship [Bacon and Schalamon 1969]

$$\sigma_T = 0.005 E_Y + 0.51 \text{ [GPa]} \quad . \tag{6.6}$$

It is to be noted that $\sigma_T$ values for ex-rayon fibers lie below those for ex-pitch, which lie below those for ex-PAN fibers.

It is emphasized that the results summarized in Fig. 6.12 are for particular processing conditions. There are many variables. For instance, the amount of fiber stretching during the oxidation stage of ex-PAN fibers is very important in aligning the graphene planes. Manocha and Bahl [1982] studied the changes in mechanical properties occurring as a function of stretching, concluding that excessive stretching led to the introduction of a large number of defects, which decreased the tensile strength. Clearly, the stretch used in commercial fibers is optimized, subject to other preparation parameters. Also, the modulus and strength of both ex-PAN fibers stretched during the carbonization stage increase monotonically with Young's modulus (see Reynolds [1973] for a review), in contrast with the earlier result for unstretched fibers. This hot-stretching is an expensive process and is not used commercially. Instead, improvements have been made to mechanical properties through careful control of precursors, additives, surface treatments, etc.

The importance of certain details of the microstructure on the fracture process has been emphasized by Guigon et al. [1984b]. These authors found an empirical relation between the tensile strength $\sigma_T$ and the longitudinal $(r_l)$ and transverse $(r_t)$ radii of curvature of the graphene ribbons (see Fig. 6.13) [Guigon et al. 1984b]:

$$\sigma_T = 3.07 - 8.52/S \quad , \tag{6.7}$$

where

$$S = L_a(r_t^{-1} + r_l^{-1}) \quad . \tag{6.8}$$

Further support for this concept comes from the work of Endo [1988a] on Carbonic ex-pitch fibers. These fibers have improved fracture properties compared to Thornel fibers, consistent with Fig. 6.13. The increased mechanical strength of the Carbonic fibers was attributed by Endo [1988a] to a folded ribbon structure and the arrest of cracks at the folds. These folds reduce the elastic modulus somewhat, and reduce the electrical and thermal conductivities compared to the Thornel fibers. These data are discussed further in Sect. 6.2.2.

As noted in Sect. 4.1, the in-plane elastic constants are not sensitive to changes in temperature near room temperature, so that a fracture process depending on Young's modulus is not expected to vary strongly with temperature. No effect of temperature on breaking stress was seen between 20° and 200°C [Zureck et al. 1981]. However, measurements to high temperature indicate that the breaking stress can be increased by ~10% upon increasing $T$ to 2400°C. These results are reviewed in Sect. 9.3.

Figure 6.14 summarizes the strain-to-failure versus Young's modulus of several types of carbon fibers. The data represented by lines are those reviewed earlier by Reynolds [1973] on ex-rayon and ex-PAN fibers, while more recent data are included on ex-pitch [Endo 1988a], commercial ex-PAN [Riggs 1985] and CCVD fibers [Tibbetts et al. 1986; Tibbetts and Beetz 1987]. Important improvements are found in the performance of high strength ex-PAN carbon fibers, as noted before, with some strain-to-failure values as high as 1.8% being recorded. There have also been marked improvements in the strain-to-failure of high modulus ex-pitch fibers [Hughes 1986a, 1987; Endo 1988a]. Some recent data for Carbonic ex-pitch fibers types HM50, HM60 and HM80 [Endo 1988] are included in Fig. 6.14. Similarly, Hughes reports strain-to-failure values as high as 2% (see Fig. 6.14) for research ex-PAN fibers with $E_Y \sim$ 290 GPa (high strength type), which again illustrates recent developments that continue to improve the mechanical properties of carbon fibers.

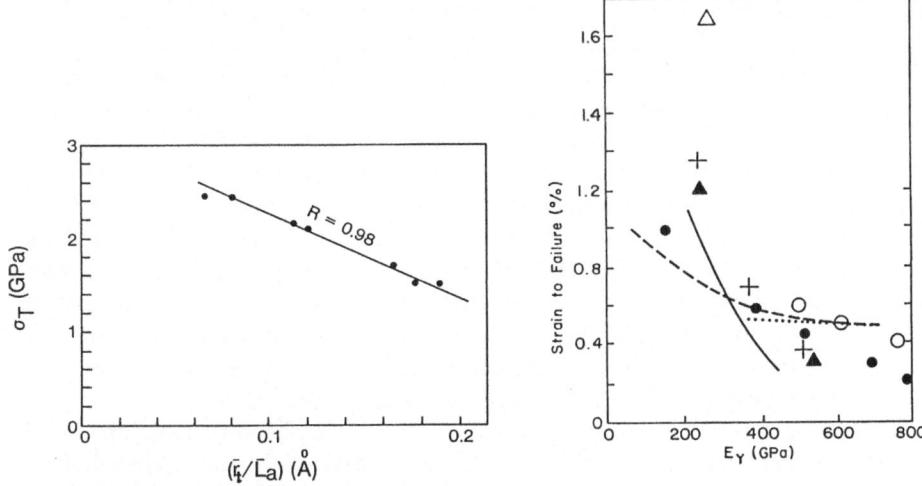

**Fig. 6.13.** A plot of the fracture stress $\sigma_T$ vs $\bar{r}_t/\bar{L}_a$, where $\bar{r}_t$ is the average transverse radius of curvature, $\bar{L}_a$ is the average in-plane crystallite dimension, and $R$ is the correlation factor signifying the goodness of the fit to the data [Guigon et al. 1984b]

**Fig. 6.14.** Strain-to-failure of various types of fibers as a function of modulus;
(–)     ex-PAN unstretched;
(· · ·)   ex-PAN hot-stretched;
(- - -)  ex-rayon hot-stretched [adapted from Reynolds 1973];
(•)     ex-pitch (commercial) [D.J. Johnson 1987a];
(▲)     CCVD [Tibbetts and Beetz 1987];
(+)     Commercial ex-PAN [Riggs 1985];
(o)     Carbonic ex-pitch [Endo 1988a];
(△)     ex-PAN (experimental) [Hughes 1987]

The most important feature of the curves is the trend to lower values for the strain-to-failure, $\varepsilon_T$, as the modulus increases. It is this decrease in $\varepsilon_T$ that prevents the breaking strength from increasing as the modulus increases. Unless $\varepsilon_T$ can be increased, the full potential of carbon fibers cannot be realized. One of the goals of experimental and theoretical research has been the understanding of this phenomenon.

It was mentioned in the introduction to this chapter that brittle materials usually fracture via the presence of flaws (cracks, pores, impurity particles, etc.). The Griffith theory [Griffith 1920] postulates that stresses near the tips of these flaws can be much higher than average, as calculated by dividing the applied stress by the area of the fiber. When the stress is sufficiently high, the stored elastic energy near the crack tip is higher than that needed to create new surfaces (or break bonds), so that the crack elongates, and the solid ruptures. Equation (6.3) expresses the breaking stress $\sigma_T$ in terms of the modulus and the critical flaw length $l_f$. Unfortunately, this relationship is not useful quantitatively for carbon fibers for several reasons [Reynolds 1973].

Firstly, measurements of $\sigma_T$ versus defect size do not follow (6.3). The data are badly scattered [Williams et al. 1970; Sharp and Burney 1971; Whitney and Kimmel 1972] about the theoretical line with the surface energy taken as $\gamma_a = 4.2$ J/m$^2$ [Reynolds 1971]. Secondly, $\sigma_T$ does not depend on $E_Y^{1/2}$, but rather, in many cases $\sigma_T \approx E_Y$, as for ex-rayon fibers (6.6). Thirdly, the flaw lengths calculated from (6.3) are unrealistically large. For instance, a value $l_f = 170$ nm was calculated at the beginning of this section for a fiber with $\sigma_T = 2.5$ GPa, and $E_Y = 400$ GPa. As discussed in Sect. 3.5, large voids are present in carbon fibers, but their lenticular shape (see Figs. 1.3 and 1.4), aligned roughly along the fiber axis, does not lead to critical stresses being developed under longitudinal loads. Also, the carbon ribbons are not so wide that they can accommodate a lateral flaw of these dimensions. The mechanisms proposed for fracture are discussed further in Sect. 6.2.2.

The fracture of brittle materials is a probabilistic phenomenon, depending on the chance that a flaw of a critical size is present in the material. It is found that individual fibers fracture at varying stress levels, even though the fibers are prepared by the same process. The scatter in Young's modulus, for instance, from a given batch of ex-PAN fibers is typically ~10% for $E_Y$, but 25% for $\sigma_T$ [Reynolds 1973]. Also, the fracture stress $\sigma_T$ falls as the length (gauge length) of the fiber being tested decreases (Moreton [1969], Beetz [1982a, 1982b] for ex-PAN; Chwastiak et al. [1979] for ex-pitch fibers). This is simply related to the increased probability of finding a critical flaw in a longer fiber.

It is conventional to test many fibers to fracture, and then plot the cumulative probability of fracture versus $\sigma_T$ on a log-log basis, as is done in Fig. 6.15. This plot is important from a technological point of view. It would

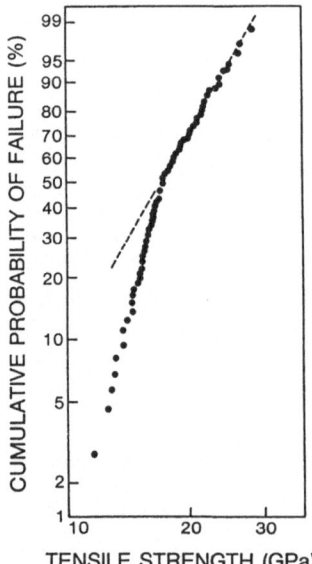

**Fig. 6.15.** A Weibull plot [Weibull 1951] for ex-pitch fibers [data from Beetz 1982a]

be desirable that all fibers fracture over a relatively small range of values, so that reliable engineering estimates could be made of fracture stress in practical situations. Where a wide range occurs, one is forced to accept a relatively low value of $\sigma_T$ at which, say, over 90% of fibers tested would not fracture. One version of such a plot is the Weibull plot [Weibull 1951] for carbon fibers (Fig. 6.15), but this plot is not a standard plot, since it consists of two regions [Moreton 1969; Hitchon and Phillips 1979; Beetz 1982a, 1982b]. These two regions of Fig. 6.15 were interpreted on the basis of a bimodal fracture process. Beetz generalized the Weibull analysis for this case [Beetz 1982a], and investigated the self-consistency of his analysis in a subsequent paper [Beetz 1982b], using data from ex-pitch fibers for four different gauge lengths. The different fracture processes leading to this bimodal plot have not been identified.

The above statistical data may be interpreted to give an estimate of the "intrinsic strength" $\sigma_i$ of the fibers [J.W. Johnson and Thorne 1969; Chwastiak et al. 1979; J.B. Jones et al. 1980; Beetz 1982a, 1982b]. When the fiber is embedded in a matrix, there is a critical length, called the "critical load-transfer length" $l_c$ (see A. Kelly [1966], for example), which provides a scale with which to estimate $\sigma_i$. In practical terms $l_c$ is the minimum length of fiber which must be embedded in the matrix in order to ensure that applied stresses lead to fiber breakage, rather than pull-out (which is the process whereby the fiber-matrix interface breaks down before the fiber breaks). An alternative way to interpret this effect is to identify $l_c/2$ with the length over which the longitudinal stress in the fiber rises to a maximum from a broken fiber end. Typical $l_c$ values are 0.1–0.3 mm (see Hughes et al. [1980] for example). Using this $l_c$ length value, $\sigma_T$ has been variously estimated as 5.3 GPa [Chwastiak et al. 1979], 3.8 GPa [J.B. Jones et al. 1980], and 3.5 GPa [Beetz 1982b, 1982c]. At the length scale of $l_c$, the probability of a large flaw is considerably reduced.

Several workers have attempted to measure the breaking stress over a length scale comparable to $l_c$. This can be done using the elastic loop test [Sinclair 1950]. In this test, a fiber is formed into a loop, or knot, and pulled until fracture occurs. In this test the regions of maximum strain and failure are restricted to a small area, so that the intrinsic strength of the fiber can be obtained in principle [Williams et al. 1970; W.R. Jones and J.W. Johnson 1971; Da Silva and D.J. Johnson 1984]. Deformation becomes non-Hookean at higher stresses as the layers begin to buckle, elastically at first on the compression side of the fiber, at a strain of ~0.5%. The stress distribution is complex, and a buckling instability on the compressive side (inner side of the loop) of the fiber occurs for high modulus fibers, so that the data are difficult to interpret.

The ultimate failure appears to start on the tensile side, and much higher strain-to-failure values $\varepsilon_T$ (2%–3% after careful etching to avoid surface flaws) can be obtained than with classical tensile tests. This suggests that the intrinsic tensile strength may be as high as 7 GPa for ex-PAN fibers. The fracture surfaces after failure are quite different after tensile and flexural failure for ex-PAN fibers, but similar for ex-pitch fibers [Da Silva and D.J. Johnson 1984], because ex-PAN fibers have a random structure compared to the more sheet-like structure of the particular fibers made from mesophase pitch used in their study.

Da Silva and D.J. Johnson [1984] obtained intrinsic strength, $\sigma_i$, values of $\sim$2.8–3.6 GPa for various ex-PAN and ex-pitch fibers. They concluded that fibers with random structure (ex-PAN) have greater flexibility and less brittleness than those with a more planar structure (ex-pitch). Since fracture surfaces rarely showed the presence of gross flaws, it was stressed that differences in mechanical behavior between these types of fibers must be sought in their fine structure. It was concluded that $\sigma_i$ values must be treated with caution, since the flaw distribution is not fully understood, particularly in the surface regions, which are often critical in fracture. However, the results were consistent with failure in both tension and compression by the Reynolds-Sharp mechanism, to be discussed in the next section.

There have been a large number of studies in which electron microscopy has been used to study the flaws in fibers (see also Sect. 3.7). For instance, J.W. Johnson [1969] and J.W. Johnson and Thorne [1969] used SEM techniques to identify flaws on fracture surfaces of ex-PAN fibers, and examples of these flaws are given in Fig. 3.24. Both surface and interior flaws were found to be responsible for fracture.

A TEM technique was used by Sharp and Burney [1971] to characterize cavities and inclusions in the interior of ex-PAN fibers. These authors speculated that cavities were formed during heat treatment above 1500°C when volatile constituents escaped. An interesting observation by these authors was that the largest flaws were often not those that led to failure. Sharp and Burney [1971] also showed that the walls of some cavities in heat-treated carbon fibers contained graphitic material. This may well be related to the catalytic activity of certain impurities [Moreton and Watt 1974]. Reynolds and Moreton [1980] deliberately contaminated ex-PAN fibers grown under clean-room conditions, and concluded that small concentrations ($\sim$0.1% by mass) of silicon and iron oxide could reduce fracture stresses by as much as a factor of 3. J.B. Jones et al. [1980] also investigated the fracture surfaces of ex-pitch fibers which had been deliberately contaminated. They identified the flaws that led to failure, then carried out ESCA analysis of the regions near the flaws, showing the presence of contaminants such as transition metals. These elements are known to catalyze graphitization. Further work on ex-PAN fibers by Bennett and D.J. Johnson [1979] and D.J. Johnson [1980] showed the presence of large, misoriented crystallites, in the walls of the cavities (see Fig. 1.5c). These will be discussed in the next section, in connection with the Reynolds-Sharp model of fracture.

These studies were very important in the recent development of commercial fibers with enhanced mechanical properties (see for example Izuka et al. [1986], Holleyman [1986]). It was realized that contaminants such as transition metals were reducing the tensile strength, and that it was essential to remove them as far as possible from manufacturing equipment, especially by filtering the polymer down to small particle sizes.

Changes in tensile strength can result from processing conditions, such as incomplete oxidation before carbonization, producing the sheath-core structure discussed in Sect. 3.7.2 [Watt and W. Johnson 1971]; Wicks and Coyle 1976]. This effect presumably plays a role in the dependence of the tensile strength on fiber diameter [Perry et al. 1971; B.F. Jones and Duncan 1971]. However, as discussed in Sect. 3.7.2 there is some controversy over the nature and extent of the sheath-core structure. It is to be stressed that it is difficult to generalize about mechanical properties, since details of the microstructure can be extremely important in modifying such properties.

An example of property modification which depends sensitively on processing conditions is that of boron doping, whose effect on mechanical properties has been reviewed recently by Agrawal et al. [1986]. Boron is known to catalyze the graphitization of carbons, and Ezekiel [1973], reported that boron additions of between 0.1% and 1% produced fibers with a more uniform structure and higher graphitization. Dramatic increases in modulus ($\sim$300%) and in strength ($\sim$50%) were obtained. Allen et al. [1969] and Pepper et al. [1978] also obtained improved mechanical properties. However, these latter researchers failed to detect the presence of boron in their fiber using chemical analysis.

Agrawal et al. [1986] carried out structural and chemical analysis on the fibers of Pepper et al. [1978] and concluded that the important structural properties of the fibers were not affected by the presence of boron in the furnace atmosphere. The added boron was found to be in the near-surface regions, and a large fraction was present as boron carbide. They concluded that it was unlikely that boron addition would eliminate the large voids present in the fibers, and that processing variables other than boron-doping should be examined to eliminate them.

Although these reports appear to contradict each other, it is possible that processing conditions led to different effects on structural and mechanical properties. Again, care must be taken in generalizing from specific experiments.

Allen et al. [1969], Cooper and Mayer [1971], and Bullock [1971, 1972b] showed that relatively low doses of fast neutron irradiation were sufficient to increase the $E_Y$ and $\sigma_T$ values of ex-PAN fibers. Fast neutrons knock atoms out of lattice sites, creating vacancies and interstitials, the latter of which diffuse to form interstitial loops, as described in Sect. 3.1.3 (for a review, see B.T. Kelly [1981]). A dose of $10^{22}$ neutrons/m$^2$ ($E > 1$ MeV) used in the above study displaces only about 1 atom in 10 000.

Bullock [1971, 1972b] showed that the tensile strengths of high-strength ex-PAN fibers irradiated at room temperature increased by ~17% from doses of $8.5 \times 10^{21}$ neutrons/m² (E > 1MeV) then fell to 25% less than the starting values for $4.5 \times 10^{22}$ neutrons/m². If the same fibers were irradiated to $4.5 \times 10^{22}$ neutrons/m² at liquid nitrogen (LN₂) temperatures, then increases in $\sigma_T$ of 30% were recorded. Somewhat smaller increases occurred in high modulus ex-PAN fibers at room temperature (4% at $4.5 \times 10^{22}$ neutrons/m²) but the strength *decreased* for the same dose in LN₂ (−10%). These results were interpreted in terms of three effects: 1) The effect of neutron damage on the bulk fiber increases the strength. 2) Radiation-enhanced oxidation can pit the surface, leading to a reduction in $\sigma_T$ at high fluences in air [Bullock 1974]. 3) Irradiations at 77 K can store "Wigner" energy (see B.T. Kelly [1981]), which is released on warming to room temperature, damaging the fiber and reducing $\sigma_T$. Similar increases in the modulus and strength of ex-pitch fibers were found by Price et al. [1985] irradiated at room temperature, and their data are considered further in Sect. 12.5.

The beneficial effects of this neutron irradiation are possibly related to the increase in $C_{44}$ (or decrease in $S_{44}$) associated with the creation of covalent links between the planes, as discussed in Sect. 6.1.2. The Reynolds and Sharp model (Sect. 6.2.2) for the fracture process, discussed in the next section, supposes that highly graphitized crystallites which are badly aligned with respect to the fiber axis are responsible for the initiation of fracture. Neutron irradiation damages all the crystallites but the effect is more important for the very large crystallites. Radiation damage would therefore convert the most highly graphitic ribbons into microstructures more like those found in a majority of the ribbons within the fiber. Accordingly, the stress needed to initiate fracture would increase, as observed.


### 6.2.2   Models of Fracture Processes in Ex-polymer Fibers

The experimental results outlined in the previous section can be summarized as follows:

1. Carbon fibers fracture at stresses well below the theoretical limit.
2. The low values of the experimental stress-to-failure $\sigma_{exp}$, the dependence of $\sigma_{exp}$ on gauge length, and the observation that fracture often occurs near the surface or interior flaws, all point to the importance of flaws in aiding the fracture process. However, the data do not fit the Griffith formula (6.3).
3. Certain experiments point to the importance of catalytic impurities near fracture surfaces, and of well-graphitized crystallites near these impurities. When these large crystallites are badly aligned with respect to the fiber axis, they promote fracture.

144

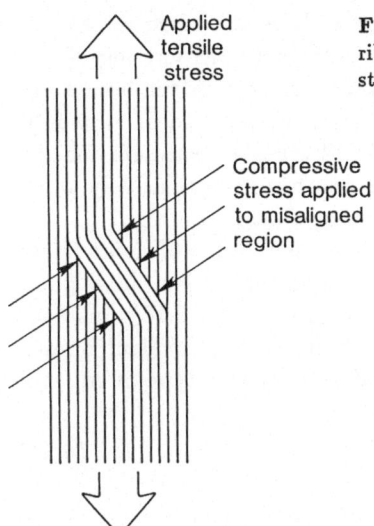

Applied tensile stress

Compressive stress applied to misaligned region

**Fig. 6.16.** Illustration of a misaligned section of a ribbon subjected to a tensile stress, which tends to straighten it, thereby producing a compressive stress

A model which accounts for these observations is that of Reynolds and Sharp [1974], which is of sufficient importance to discuss in greater detail, particularly since it illuminates ideas presented in the introductory paragraphs of this chapter.

These authors consider that fracture occurs in regions where the planes are misoriented at relatively large angles to the fiber axis, as illustrated in Fig. 6.16. If the "crystallites" have a mean misorientation angle $\langle \phi \rangle$, then statistically, some will be aligned at $2\langle \phi \rangle$. Alternately, there may be regions of the fiber which are misaligned around large flaws. In a polycrystalline graphite, failure would occur via shear of the planes over each other, and the critical shear stress $\sigma_S$ would be

$$\sigma_S = \left( \frac{2\gamma_c}{\pi \bar{c} S_{44}} \right)^{1/2} \quad , \tag{6.9}$$

where $\gamma_c$ is the surface energy for $\pi$-bonds between the basal planes, and $\bar{c}$ is the interplanar separation. This is essentially (6.1) rewritten for the case of basal shear. By contrast, the critical tensile stress for the case of perfectly aligned crystallites would be

$$\sigma_T = \left( \frac{2\gamma_a}{\pi a S_{11}} \right)^{1/2} > \sigma_S \quad , \tag{6.10}$$

where $a$ is the interatomic spacing in the basal plane, and $\gamma_a$ the surface energy for the covalent bonds within the basal plane. (The difference between the surface energies $\gamma_a$ and $\gamma_c$ may be easily understood by considering a model for graphite crystallites in which the graphene planes are represented by the pages of a book. $\gamma_c$ is the surface energy related to separating the

145

book into two halves by removing half the pages. It is therefore related to the rupture of the weak interlayer bonds. $\gamma_a$ is related to the cutting of the book into two across the pages, so that relatively strong covalent bonds are broken. As a consequence, $\gamma_a \gg \gamma_c$.) The fracture in the case of $\gamma_a$ would not involve shear but tensile stress. Equation (6.9) gives an estimate for the critical stress that is too small (polycrystalline graphites are relatively weak materials, see Fig. 12.10), while (6.10) gives a value that is too high.

Reynolds and Sharp [1974] argue that the above, low, value of shear stress is not applicable to fibers because neighboring crystallites subject misaligned regions to compressive stresses which inhibit this shear (Fig. 6.16). Instead, the planes rupture by tearing across the planes, breaking the covalent bonds, as illustrated in Figs. 6.17b and 6.17c. The condition for this type of rupture is

$$\sigma_{TS} = \left( \frac{2\gamma_a}{\pi a S_{44}} \right)^{1/2} \tag{6.11}$$

The value of $\sigma_{TS}$ lies between $\sigma_T$ and $\sigma_S$, since $S_{44} > S_{11}$, and $\gamma_a > \gamma_c$. Therefore, fracture would occur at a lower value of shear stress (6.11) than tensile stress (6.10).

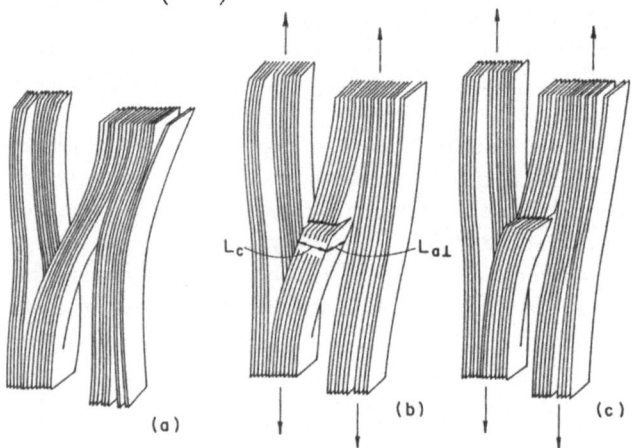

Fig. 6.17a-c. (a) Sketch of a region of a fiber with a badly misoriented crystallite. (b,c) Sketches as of a region of a fiber where fracture occurs at a badly misoriented crystallite due to shear rupture of the basal planes [D.J. Johnson 1987b]

If the crystallite is misoriented at an angle of $\phi$ to the fiber axis, then the resolved shear stress on the plane is

$$\sigma_S = \sigma \sin \phi \cos \phi \quad . \tag{6.12}$$

The shear strain-to-failure, $\varepsilon_S$, can be written as

$$\varepsilon_S = \frac{\varepsilon_L E_Y}{C_{44}} \sin \phi \cos \phi \quad , \tag{6.13}$$

146

where $\varepsilon_L$ is the longitudinal strain-to-failure. Using these equations with $C_{44} = S_{44}^{-1} = 4$ GPa and $\gamma_a = 4.2$ J/m$^2$, $a = 0.142$ nm, then $\varepsilon_S$ would be $\sim$200%. Reynolds and Sharp [1974] suggest that this would be associated with a perfect crystal, and that a more realistic value of $\varepsilon_S$ is 10% of this, i.e., $\varepsilon_S \sim 20\%$, as suggested by analysis of ruptured fibers.

The rupture of one crystallite by this failure mode would not lead to fiber rupture, unless the crack continued to propagate across neighboring crystallites. This can occur if the critical region is near a major flaw, so that many crystallites in the region fail under shear stress. Alternatively, fiber rupture could occur if an impurity were in the vicinity of the misaligned region (and may in fact be the cause of the misalignment). Heat treatment could lead to catalytic growth processes, which produce relatively large, perfect, crystallites. This is possibly the reason why heat-treated fibers have reduced fracture strains.

This model was considered further by Reynolds and Moreton [1980]. These authors used the uniform stress model, discussed in Sect. 4.3.2, to obtain elastic constants. They then assumed that a quantitative criterion for the failure of fibers could be obtained by requiring the stored energy to be constant:

$$\tfrac{1}{2}\sigma_T^2/E_Y = K_E \ , \tag{6.14}$$

where $K_E$ is a constant. They then searched for the highest value of this quantity, $K_E$, as a representative of a value for the intrinsic strength. A maximum in the strain energy $\sigma_T^2/2E_Y$ was found for a fiber with modulus of 210 GPa, prepared at 1300°C. Its elastic energy of 38 MJ/m$^3$ at fracture was taken as a minimum estimate for the intrinsic value. Then, using the uniform stress model (discussed for these same fibers in Sect. 4.3.2), the corresponding value of the tensile stress and strain were computed for this same stored energy (Table 6.4). The authors pointed out that this empirical value of the critical stored energy may need to be modified in the future. However, the calculated values are considerably higher than those observed for high modulus fibers.

Table 6.4. Elastic properties[a] and maximum fracture properties[b] observed and calculated for ex-PAN fibers (adapted from Reynolds and Moreton [1980])

| $T_{HT}$ [°C] | $E_Y$ [GPa] | $\langle\phi\rangle_{calc}$ [°] | $\langle\phi\rangle_{obs}$ [°] | $\sigma_T^{max}$(calc) [GPa] | $\sigma_T^{max}$(obs) [GPa] | $\varepsilon_T^{max}$(calc) [%] | $\varepsilon_T^{max}$(obs) [%] |
|---|---|---|---|---|---|---|---|
| 1000 | 197 | 18 | 14–19 | 3.9 | 2.45 | 2.0 | 1.36 |
| 1400 | 245 | 15 | 12–17 | 4.3 | 3.41 | 1.8 | 1.42 |
| 2500 | 428 | 7 | 5–8 | 5.7 | 3.13 | 1.3 | 0.83 |

[a] $\langle\phi\rangle$ is defined in Table 6.2.
[b] $\sigma_T^{max}$ is the maximum tensile strength at fracture and $\varepsilon_T^{max}$ is the maximum elongation at fracture.

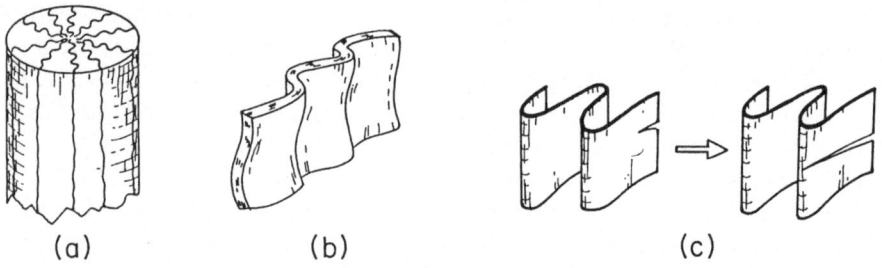

**Fig. 6.18a-c.** Inhibition of crack propagation by the folded ribbon structure: (a) Sketch of the folded ribbon microstructure for ex-pitch fibers with a radial organization of graphene layers; (b) expanded view of the folded ribbons; and (c) arrest of crack propagation by the folded ribbons [Endo 1988a]

As discussed above, Guigon et al. [1984a, 1984b] related the fracture stress to $L_a$ and the radii of folds $\bar{r}_t$ in the graphene ribbons [Guigon et al. 1984a, 1984b]. These authors suggested that cracks are arrested at folds in the ribbons, as illustrated in Fig. 6.18. As mentioned in the previous section, Endo [1988a] was able to show that the strength of Carbonic HM50, HM60, and HM80 fibers is improved due to the folding of the ribbons. These data are consistent with the Reynolds and Sharp model, which emphasizes the role of a critical crystallite (see Fig. 6.16) in the fracture process and identifies this crystallite as an extensive, well-graphitized, but misoriented crystallite. The work of Guigon et al. [1984a, 1984b] relates the fracture process to average crystallite properties. However, the critical and average crystallites are related. As the average crystallite dimension $L_a$ increases, so does the dimension of the critical flaw, growing and graphitizing under the catalytic influence of an impurity particle. Even if such a crystallite were extensive, and more fully graphitized than an average region of the fiber, fracture would be arrested by the folds. Again, it is the presence of neighboring folded ribbons that prevent the critical crystallite from growing under the favorable growth conditions provided by the catalyst, and from fracturing until a higher stress is applied than would be the case for unfolded ribbons.

## 6.3 Mechanical Properties of CCVD Fibers

Measurements of the mechanical properties of CCVD fibers were first published by Koyama [1972], who reported a range of values for Young's modulus (180-400 GPa) and values of tensile strength between 1 and 3 GPa. These strengths depended strongly on fiber diameter, decreasing for fibers with larger diameters. More recent work of Kandani et al. [1984] showed that the mechanical properties were independent of whether benzene or methane was used for growth, and again showed that the mechanical properties varied with fiber diameter. For instance, the Young's modulus fell from 300 to 160 GPa as the diameter increased from 8 to 30 $\mu$m while the tensile strength fell from ~2.6 to ~1.0 GPa for the same increase in fiber diameter. A similar trend was observed by Yetter et al. [1985] and by Tibbetts and Beetz [1986], as discussed below.

A more extensive study of as-grown CCVD fibers was carried out by Tibbetts and Beetz [1987]. They showed that the stress-strain curve was non-linear, corresponding to an increase in modulus of 28% as the fibers were strained to failure. This strain stiffening effect was more pronounced than that observed in ex-PAN (see Fig. 6.4) and ex-pitch fibers, and could be repeated in the CCVD fibers with little hysteresis (as it could also for ex-PAN and ex-pitch fibers) as long as the fiber did not fracture. The effect was ascribed to the increase in preferred orientation with applied stress, as was found by Curtis et al. [1968] for ex-PAN fibers, and Beetz [1982a, 1982b] for ex-pitch fibers (Sect. 6.1.2).

Typical stress-strain curves for a CCVD fiber and an ex-PAN fiber are compared in Fig. 6.4. Note that the as-grown CCVD fibers (diameter 8 $\mu$m) fracture at a strain of 1.3%, with an average tensile stress of 2.9 GPa. This stress value is similar to that for ex-PAN fibers ($\sigma_T$ = 3.1–4.6 GPa). However, the significant increase in modulus with strain for the CCVD fibers leads to a lower strain-to-failure for the CCVD fibers (1.3% versus 1.5%–2.2% for ex-PAN fibers).

Tibbetts and Beetz [1987] measured the tensile strength and Young's modulus of the CCVD fibers as a function of diameter (Fig. 6.19). They

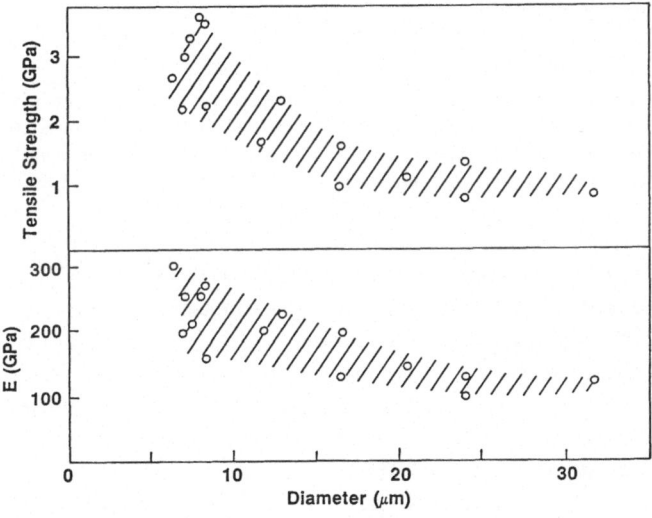

**Fig. 6.19.** Tensile strength and Young's modulus ($E_Y$) of fibers of different diameter grown from the vapor phase [Tibbetts and Beetz 1987]

showed, however, that the thicker fibers were obtained from a region of the furnace where growth rates were higher, and the resulting material had less orientational alignment of the crystallites. These lower values of modulus are likely due to the higher growth rate. The lower strength of the thicker fibers was thought to be due to the higher probability of finding critical flaws to initiate fracture, as was also found in other types of carbon fibers (Sect. 6.2.2).

These results were discussed in terms of Ruland's "elastic unwrinkling" model discussed in Sect. 4.3.2. Data for ex-PAN, ex-pitch, and CCVD fibers of Young's modulus versus $\langle \cos^2 \theta \rangle$ (note $\theta = \pi/2 - \phi$) lie close to a line in which the compliance constant $k_e$ is 0.03 GPa$^{-1}$ (Fig. 6.20). The same data can be fit quite well by the uniform stress model with a reasonable value for $S_{44}$ and all other compliances taken equal to single crystal values given in Table 4.1 (see the discussion in Sect. 6.1.4).

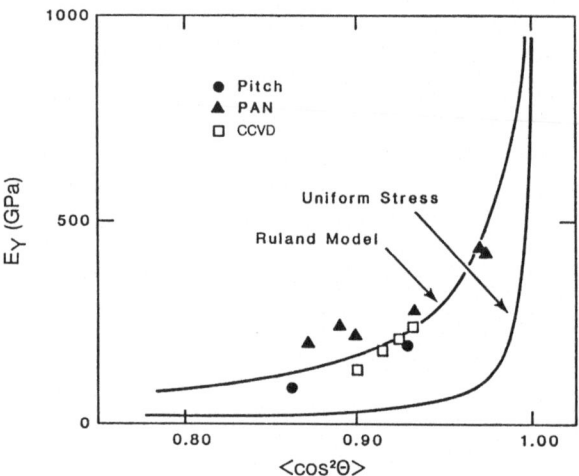

Fig. 6.20. Young's modulus as a function of the ordering parameter $\langle \cos^2 \theta \rangle$. The plot shows curves for the uniform stress and Ruland model ($k_e = 0.03$ GPa$^{-1}$) compared to several experimental points. The uniform stress model is also shown where single crystal values of the elastic constants (including $S_{44}$) have been used [Tibbetts and Beetz 1987]

A similar CCVD fiber of 13 $\mu$m diameter heat treated to 2880°C was shown to have an increased modulus (510 GPa) and decreased strength (1.7 GPa) and no strain stiffening was observed [Tibbetts et al. 1986a; Tibbetts 1987]. (Considerably higher values of the strength, up to 7 GPa, were reported by Endo et al. [1976].) The increased modulus is consistent with a decreased value of the mean misorientation angle, while the decreased strength can be related to the reduced fracture strain (0.35%).

The failure mode of the heat-treated fibers was found to be via a "sword-in-sheath" mode, as illustrated in Fig. 6.21 [Tibbetts and Beetz 1986]. Frac-

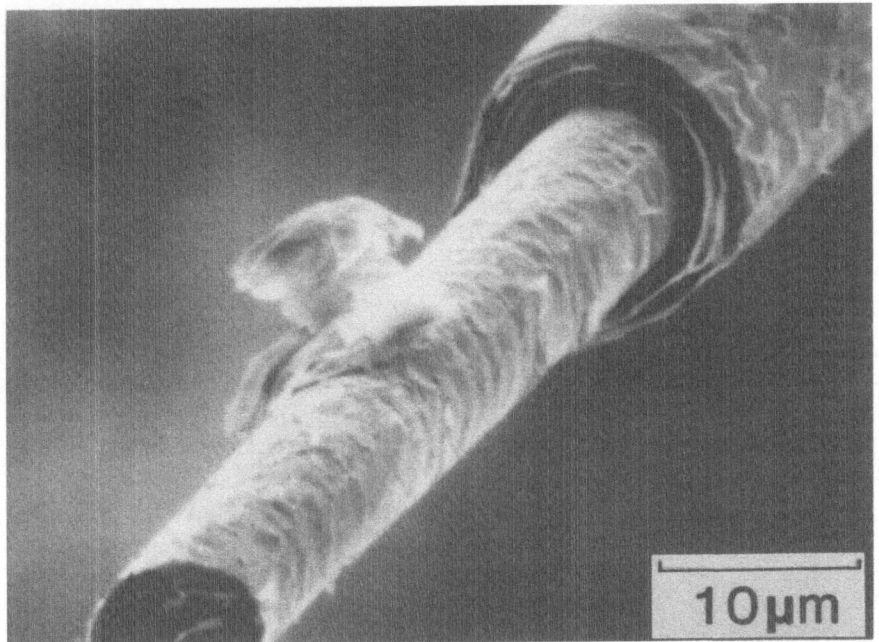

**Fig. 6.21.** Scanning electron micrograph of a fiber which has suffered a "sword-in-sheath" failure [Tibbetts and Beetz 1986]

ture was conjectured to occur by initiation of a crack at the surface, propagating inwards for a short distance, then propagating rapidly around the fiber. The corresponding reduction in fiber cross-sectional area then results in increased stress, leading to crack initiation and propagation at another flaw further along the fiber, similar to the first fracture. This process can continue, with cracks occurring over a length of several millimeters along the fiber. This brittle failure is then followed by shearing of the graphite layers, so that strength is reduced, and the "sword-in-sheath" phenomenon occurs. Nesting fragments were postulated to wedge together to allow strength to redevelop. This redevelopment of strength was observed in the tensile tests.

Raman microprobe measurements (see Sect. 4.5.1) were made on vapor-grown fibers heat treated to 3000°C to study the effect of tensile stress $\sigma_T$ on the Raman-allowed $E_{2g_2}$ mode at ~1580 cm$^{-1}$ in order to develop a sensitive characterization tool for the mechanical properties [Sakata et al. 1987, 1988]. The results show a linear downshift in the $E_{2g_2}$ mode with increasing $\sigma_T$ until failure (i.e., breakage of the fiber at ~930 MPa) for both the polarizations ($\sigma \parallel E$) and ($\sigma \perp E$) of the laser light. The splitting of the $E_{2g_2}$ mode by the applied stress is indicative of different strain magnitudes induced parallel and perpendicular to the applied stress, with a 3.5 times larger shift of the

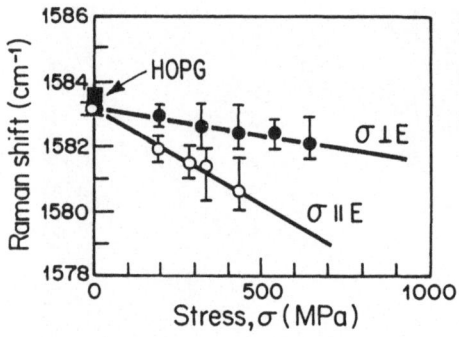

Fig. 6.22. Plot of the Raman shift for the Raman-allowed peak near 1580 cm$^{-1}$ vs the applied uniaxial tensile stress $\sigma$ for a single CCVD fiber and polarizations $\parallel$ and $\perp$ to the applied stress (HOPG is highly oriented pyrolytic graphite) [Sakata et al. 1987, 1988]

Raman frequency for the longitudinal polarization ($\sigma \parallel \boldsymbol{E}$) as compared with the transverse polarization ($\sigma \perp \boldsymbol{E}$) as shown in Fig. 6.22. A small linear stress-induced increase in the Raman linewidth $\Delta\Gamma$ is also observed, and is attributed to an increased inhomogeneity in the fiber; the linewidth $\Delta\Gamma$ is larger by ~20% for the polarization ($\sigma \parallel \boldsymbol{E}$) as compared with ($\sigma \perp \boldsymbol{E}$). By calibrating the applied stress with the induced strain, the Raman microprobe characterization technique can be used to monitor stresses and strains in individual graphite fibers non-destructively. Since the stress-induced Raman shifts are small, this technique can be applied quantitatively only to well-graphitized fibers [Sakata et al. 1987, 1988].

The various studies that have been carried out to date indicate that CCVD fibers have potentially useful properties for applications where chopped high strength ex-PAN fibers are currently used. It may be possible to increase the stress-to-failure value by higher temperature heat treatments and other processes, as well as to utilize the unique mode of fracture of the CCVD fibers.

# 7. Electronic Structure

## 7.1 Introduction and Overview

The wide range of structures that occur in carbon filaments is described in Chap. 3, in terms of microstructures, lattice defects and impurities. This range of structures in turn implies that a wide range of electronic structures can occur. At one extreme, where the filaments are graphitic, the electronic structure approximates that of single crystal graphite. At the other extreme, where the filaments can be modeled as an inhomogeneous mixture of partially aligned polymeric molecules, a model based on the electronic structure of macromolecules is more appropriate. In between these two extremes, disorder of various kinds modifies the electronic structure in several ways.

Some of these changes are unique to carbon, and result from the fact that graphite is a narrow band overlap semimetal. In the absence of interplanar interactions, graphite would be a zero gap semiconductor. Thus the small band overlap in graphite results from interactions between planes. These interplanar interactions can be modified, and effectively reduced to zero, by introducing stacking disorder, resulting in turbostratic carbon. At this point, the planar sheets of carbon atoms (graphene planes) can be viewed as a zero-band-gap semiconductor, in which the degeneracy between the conduction and valence bands results from the two-dimensional lattice symmetry. When long range order in the planes is disrupted, this degeneracy is lifted, and a narrow band gap semiconductor results, in which localized states lie in the gap. This disorder can be extended continuously to describe amorphous carbon as a limiting case.

It has already been remarked in Chap. 3 that there are close parallels between the structure of carbon/graphite filaments and bulk, heat-treated carbons. In both cases the heat treatment temperature ($T_{HT}$) at fixed residence time can be used as a rough guide to specimen perfection (parameters more appropriate than $T_{HT}$ will be discussed in Chap. 8). To account for the electronic properties of carbon, Mrozowski [1971] developed a simple model to explain the major changes in the electronic structure which occur as $T_{HT}$ is increased in graphitizable carbons, as sketched in Fig. 7.1. Superimposed on this sketch are resistivity, magnetoresistance and Hall effect data for polymer-based fibers and CCVD filaments. These properties are only sketched roughly

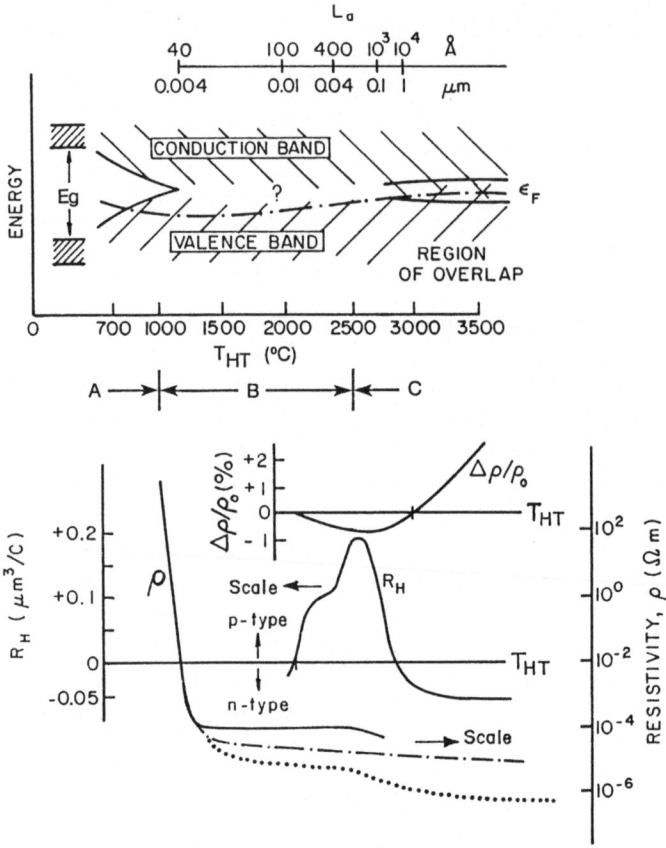

**Fig. 7.1.** Schematic variation of the electrical resistivity, Hall coefficient, and magnetoresistance of heat-treated bulk carbons, pyrocarbons, and PAN fibers, as a function of $T_{HT}$. The in-plane crystallite size for bulk heat-treated carbons is indicated. Also included is a sketch of a simple band model proposed by Mrozowski [1971] to qualitatively explain these observations. Room temperature data are from Robson et al. [1973] for PAN fibers (*dot-dashed line*), from Mrozowski [1971] for bulk carbons (*full line*) and from Chieu et al. [1983] for benzene-derived fibers (*dotted line*)

and will be considered in greater detail later. There are several versions of this sketch, and an early one comparing bulk properties to those of ex-PAN fibers was discussed by Robson et al. [1971, 1972].

Several distinct regions can be identified from Fig. 7.1, and these regions are discussed separately:

1) $T_{HT} \leq 1000°C$ *(Region A in Fig. 7.1)*. In the low $T_{HT}$ region ($\leq$ 1000°C), an ex-polymer fiber consists of a set of partially aligned macromolecules, with H, N, and O as the principal constituents aside from carbon.

Electronically, an energy gap exists between the occupied and unoccupied states. In order to explain the positive Hall coefficient, it must be concluded that the predominant defects produce holes, so that the electrochemical potential lies near the valence band edge. Coulson et al. [1957] considered the evolution of the electronic energy levels of aromatic molecules (planar molecules with carbon hexagons and edge hydrogen atoms), and showed that the density of electronic states near the bonding-anti-bonding energy gap began to approximate that of an infinite sheet of carbon atoms when the number of carbon atoms exceeded $\sim 50$ (e.g., $m$ and $n$ in Fig. 7.2a are equal to about 5). As $T_{HT}$ increases, the excitation gap $\Delta E$ therefore decreases (Fig. 7.2b), reaching about 0.5 eV at this molecular size.

It is important to understand that this excitation gap is an average value, and that a statistical distribution of molecular sizes produces a range of values of $\Delta E$. Accordingly, the transition from an insulator to a metal at $T_{HT} \sim 1000°C$ occurs by percolation of electrons and holes via those regions of the material that have a smaller value of $\Delta E$, or a higher value of the highest

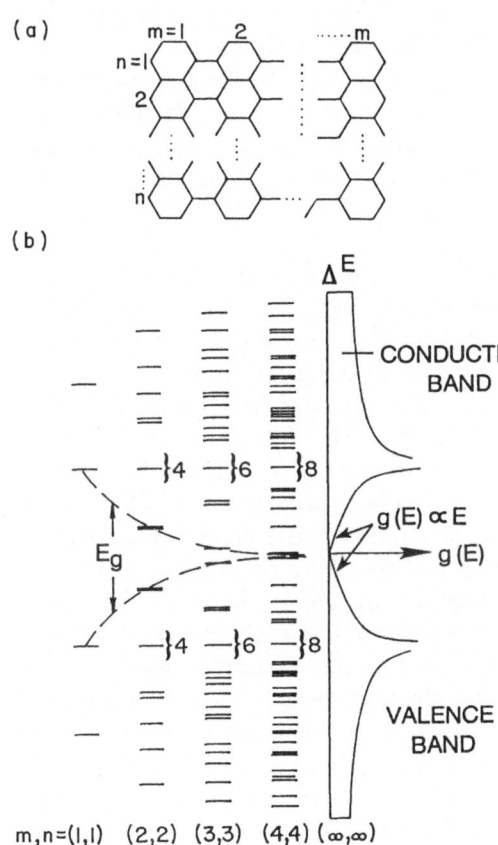

Fig. 7.2a,b. The energy levels of aromatic (hexagonal) molecules; (a) the model for the molecule and (b) the energy levels for symmetric ($m = n$) planar molecules, with a graphene plane ($m = n = \infty$) as the limiting case. The density of states $g(E)$ for the graphene plane (2D graphite) is proportional to $E$ near the band overlap. $E_g$ is the excitation gap (adapted from Coulson et al. [1957])

155

valence states. This insulator-to-metal transition has been reviewed in detail by Delhaès and Carmona [1981] for bulk carbons.

The region of $T_{HT}$ labeled A in Fig. 7.1 will not be considered here further, since ex-polymer fibers have only been studied extensively for $T_{HT} \geq 1300°C$. At these low $T_{HT}$ values ($\leq 1300°C$), the ex-polymer fibers are not really carbon fibers, but rather partially pyrolyzed fibers. In contrast, CCVD filaments (discussed further below) also have formation temperatures of $\sim 1100°C$, but the action of the catalyst enhances structural perfection so that their physical properties are more representative of a bulk carbon with $T_{HT} \geq 1500°C$.

2) $1000°C \leq T_{HT} \leq 2500°C$ *(Region B in Fig. 7.1)*. This is the region which Mrozowski [1971] marked with a question mark in Fig. 7.1. The most striking observation for this range of $T_{HT}$ is that of negative magnetoresistance. For example, for mesophase pitch fibers, a negative magnetoresistance is observed for $700°C < T_{HT} < 3000°C$ and for benzene-derived fibers in the range $1400°C < T_{HT} < 2100°C$. This phenomenon is so characteristic of carbons with poorly developed structural ordering that there is a unique explanation for the negative magnetoresistance based on the two-dimensional nature of these materials [Bright 1979]. This explanation is based on an increase in carrier density when a magnetic field is applied. For a magnetic field $B$ applied in a direction perpendicular to the graphitic sheets (graphene layers), the magnetic energy levels for two-dimensional (or turbostratic) carbon are given by McClure [1956] and by Uemura and Inoue [1958] as

$$E_N = \pm\sqrt{3}\gamma_0 a_0 \sqrt{2Ns}/2 \quad , \tag{7.1}$$

where $N$ is the quantum number for the magnetic energy level $E_N$, $\gamma_0$ is the nearest neighbor in-plane overlap integral and $a_0$ is the in-plane lattice constant ($a_0 = 2.46Å$). (Here $s = eB/\hbar$ in SI units, and $s = eB/\hbar c$ in cgs units.) The relation (7.1) implies that a large energy separation occurs between magnetic energy levels with low quantum numbers ($N = 0, 1, 2$) at modest magnetic fields $B$ (see Sect. 7.3.2 and Fig. 7.3). Thus the quantum limit, in which all carriers are in the ground state ($N = 0$), can be achieved at only $B = 1.0$ T for a hypothetical case where acceptors shift the electrochemical potential to 37.5 meV below the band degeneracy [Bright 1979]. In this case, an increasing magnetic field can result in an increase in the free carrier density, which gives rise to a negative magnetoresistance (Sect. 8.4.2).

In region B of $T_{HT}$ (Fig. 7.1), the simple two-band model cannot be used, since the fundamental relationship $kl > 1$ is not satisfied, where $l$ is the electron mean free path and $k$ is the electron wave vector. However, if a decrease of structural perfection accompanies an increase in carrier density as $T_{HT}$ decreases, an approximate constancy of the electrical resistivity in this region of $T_{HT}$ is expected (see Fig. 7.1). An increase in carrier density (hole density) could be achieved by an increase of $|E_F|$ for $E_F < 0$. However,

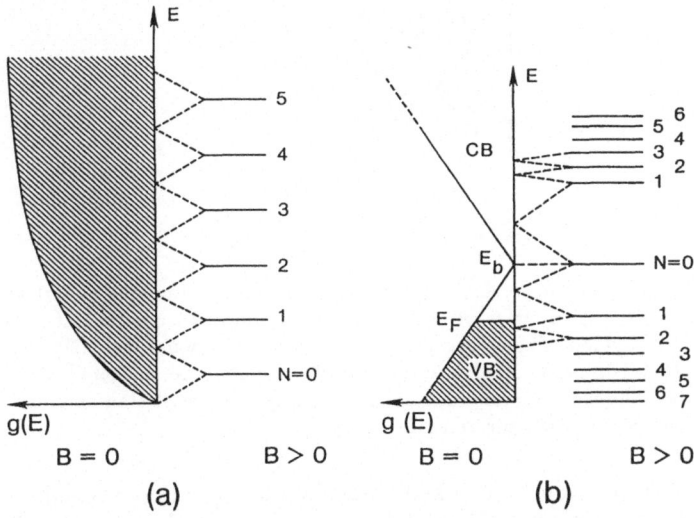

Fig. 7.3a,b. Density of states for (a) 3-dimensional simple parabolic band and (b) 2-dimensional graphite (a graphene plane of infinite extent). In each case, the density of states for zero magnetic field ($B = 0$) is shown to the left of the energy axis. The coalescence of the zero field levels to form the Landau levels for $B > 0$ and $k_B = 0$ is shown to the right of the energy axis where $k_B$ is the wave vector along the magnetic field. The Fermi level in (b) is depressed by the addition of acceptors

the sign change of the Hall coefficient from $p$-type for $T_{HT} \geq 1500°C$ to $n$-type below this $T_{HT}$, indicates that $E_F$ moves from the valence band (for $T_{HT} \geq 1500°C$) to the conduction band (for $T_{HT} \leq 1500°C$).

The problem is compounded when the effects of finite size of the hexagonal carbon planar networks are taken into account. Using a model for a polymer-derived fiber, in which the planar networks are assumed to be of width $W$ and of infinite length, the electronic structure has been calculated by Chausse and Hoarau [1969] and by Hoarau and Volpilhac [1976]. The essential results of their calculations are sketched in Fig. 7.4. One of the models they considered is a ribbon of width W which has edge atoms for the lowest row of atoms as shown in Fig. 7.4a. For this model, there are localized states at $E = 0$, so that a semiconductor with a gap of $2\Delta$ results (Fig. 7.4b), where the band gap $E_g = 2\Delta$ is given by

$$E_g = \frac{2.1}{W} \quad , \tag{7.2}$$

in which $E_g$ is in eV and $W$ in nanometers (see Fig. 7.4). A typical ribbon of width 20 nm would be predicted to have a band gap of approximately 0.1 eV. Different models for the edge states predict either a zero or a finite energy gap rather than a band overlap, in contrast to phenomenological models for 3D graphite.

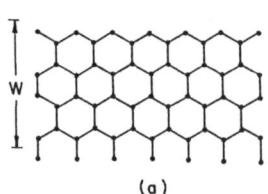 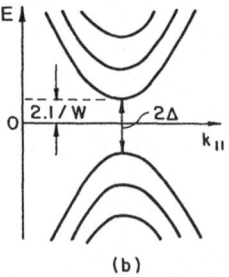

(a)                                                    (b)

Fig. 7.4a,b. The electronic structure calculated by Chausse and Hoarau [1969] and by Hoarau and Volpilhac [1976]. Two types of edge atom configurations were assumed, as indicated in (a) (top and bottom edges), resulting in the one-dimensional dispersion relationships shown in (b) for the case of edge atoms. Localized states are predicted and a band gap of $2\Delta$ is noted. The wave vector along the ribbon length is designated as $k_\parallel$ and the energy extrema occur at the Brillouin zone boundary.

Three types of disorder can be invoked to explain an apparent overlap of conduction and valence bands. Firstly, atomic disorder can result in band tailing of the electronic energy bands (see Mott [1967] for a discussion). However, the tail states are localized and no evidence for hopping-type conductivity has been found from the temperature dependence of the conductivity below 300 K (Sect. 8.3). Secondly, the relaxation time, $\tau$, for the inelastic scattering of electrons in fibers can become short compared to that in single crystal graphite. For example, $\tau$ for an ex-PAN fiber heat treated to 2000°C is of the order of $10^{-15}$ s. Accordingly, the lifetime broadening was estimated to be $\Delta E \sim \hbar/\tau \sim 0.3$ eV [Goldberg 1985]. This is discussed further in Sect. 8.3. Thirdly, the individual macromolecules making up the fiber have a distribution of sizes, configurations, impurities, etc. The work function for each macromolecule varies (see for example Coulson et al. [1957]) so that charge transfer has to occur between the macromolecules of the filament in order to maintain a constant electrochemical potential. This again results in an apparent band overlap when the properties of the aggregate are considered [Spain et al. 1983a], and allows for the percolation of carriers to explain the insulator-to-metal transition, discussed above.

3) $T_{HT} > 2500°C$ *(Region C in Fig. 7.1)*. This is the region in which the carbon develops from a turbostratic interlayer ordering to 3-dimensional graphite. The electronic structure of turbostratic carbon (2200°C $\leq T_{HT} \leq$ 2500°C) can be modeled as a collection of two-dimensional layer planes, since all interlayer interactions are essentially washed out when the layer planes are stacked in a random (turbostratic) fashion, as was shown by McClure and Ruvalds [1964]. For $T_{HT} = 3500°C$, the structure of bulk carbons can approach that of single crystal graphite, in which both the in-plane perfection and the interlayer site correlations become well-developed. The electronic structure of single crystal graphite is described in detail in Sect. 7.2.

158

It is to be noted that the dispersion relationship $E(k)$ for two-dimensional graphite (i.e., a single graphene sheet of hexagonal honeycomb carbon) is unusual [Lomer 1958; Johnston 1955]. Using a two-dimensional planar hexagonal Brillouin zone in the $(k_x, k_y)$ plane, where $k$ is the electron wave vector, the valence and conduction bands are degenerate at the zone corners, with an energy dispersion relation that is linear in $\kappa$:

$$E(\kappa) = \pm \frac{\sqrt{3}}{2} a_0 \gamma_0 \kappa \qquad (7.3)$$

in which the + and − signs, respectively, denote the conduction and valence bands, and the electron energy $E(\kappa)$ and wave vector $\kappa = \sqrt{\kappa_x^2 + \kappa_y^2}$ are measured from the zone corner, $a_0$ is the lattice constant (see Sect. 3.1.1) and $\gamma_0$ is the in-plane energy overlap parameter for neighboring atoms (see Table 7.1). The electronic density of states, $g(E)$ for a single plane is then given by

$$g(E) = \frac{8|E|}{3\pi a_0^2 \gamma_0^2} \quad , \qquad (7.4)$$

where the spin degeneracy and the two independent Brillouin zone corners have been included (see Fig. 7.3). If $E$ and $\gamma_0$ are expressed in eV and the lattice constant is expressed in meters, then the units for the two-dimensional density of states $g(E)$ are eV$^{-1}$m$^{-2}$. Equations (7.3) and (7.4) hold only for energies which are close (e.g., < 1 eV) to the band touching point [i.e., where the conduction and valence bands are degenerate and $g(E) = 0$]. The linear $E(\kappa)$ relation in both the conduction and valence bands results in a logarithmic singularity in the density of states (see Fig. 7.2b). In molecular terms, this is the energy where the molecular energy levels are $2n$-fold degenerate for an $n \times n$ "molecule". In band terms, it is the energy of the van Hove singularity where the energy surfaces touch the sides of the planar Brillouin zone (point $M$ in Fig. 7.5).

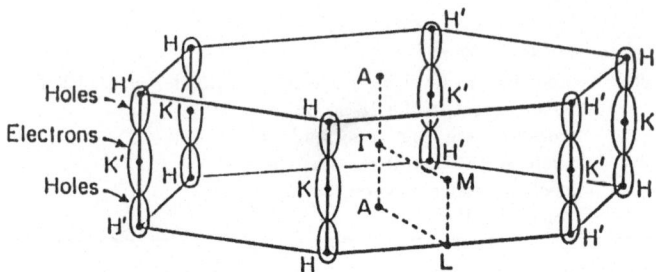

Fig. 7.5. Graphite Brillouin zone showing several high symmetry points and a schematic version of the graphite electron and hole Fermi surfaces located along the $HK$ axes

**Table 7.1.** Slonczewski-Weiss-McClure band parameters for graphite – their magnitudes and physical significance

| Band parameter | Order of magnitude [eV] | Physical origin | Reference[a] |
|---|---|---|---|
| $\gamma_0$ | $3.16 \pm 0.05$ | Overlap of neighboring atoms in a single layer plane. Determines 2D properties. | [Toy et al. 1977] |
| $\gamma_1$ | $0.39 \pm 0.01$ | Overlap of orbitals associated with carbon atoms located one above the other in adjacent layer planes. Width of $\pi$-bands at point $K$ is $4\gamma_1$. | [Misu et al. 1979] |
| $\gamma_2$ | $-0.020 \pm 0.002$ | Interactions between atoms in the next nearest layers and from coupling between $\pi$- and $\sigma$-bands. Band overlap is $2\gamma_2$. The Fermi surface is most sensitive to $\gamma_2$. | [Soule et al. 1964] |
| $\gamma_3$ | $0.315 \pm 0.015$ | Coupling of the two $E_3$ bands by a momentum matrix element. Trigonal warping of the Fermi energy is determined by $\gamma_3$. | [Doezema et al. 1979] |
| $\gamma_4$ | $\sim 0.044 \pm 0.024$ | Coupling of $E_3$ bands by a momentum matrix element. Determines inequality of $K$-point effective masses in valence and conduction bands. | [Mendez et al. 1980] |
| $\gamma_5$ | $0.038 \pm 0.005$ | Interactions between second nearest layer planes. Introduced in $E_1$ and $E_2$ to be consistent with $E_3$ in the order of the Fourier expansion of $E(k)$. | [Misu et al. 1979] |
| $\Delta$ | $-0.008 \pm 0.002$ | Difference in crystalline fields experienced by inequivalent carbon sites in layer planes. Volume of minority hole carrier pocket is sensitive to $\Delta$. | [Toy et al. 1977] |
| $E_F$ | $-0.024 \pm 0.002$ | The Fermi level is measured with respect to the $H$-point extremum (see Fig. 7.7) and is fixed by the condition that the electron concentration = hole concentration. | [Mendez et al. 1980] |

[a] References are to experimental determinations of band overlap parameters.

From the linear dispersion relation, the corresponding energy-independent electron velocity is obtained:

$$v(\kappa) = \frac{1}{\hbar}\frac{\partial E}{\partial \kappa} = \frac{\sqrt{3}}{2\hbar}a_0\gamma_0 \sim 10^6 \text{m/s} \quad . \tag{7.5}$$

In this region, the gradual increase of interlayer correlations is thought to produce an increase in overlap between the conduction and valence bands from zero for the 2D turbostratic graphite, to $\sim$40 meV for 3D graphite. However, no theoretical models have been developed for the partially graphitic case to substantiate this.

The electrical conductivity can be interpreted in simple terms using the Drude formula

$$\sigma = \frac{ne^2\tau}{m^*} = ne\mu = \frac{ne^2l}{m^*v_F} \quad , \tag{7.6}$$

where $n$ is the density of carriers, $e$ the charge on the electron, $\tau$ the relaxation time for scattering, $m^*$ the effective mass, $\mu$ the mobility, $l$ the mean-free path for scattering, and $v_F$ the Fermi velocity. The formula (7.6) is valid because in this range of $T_{HT}$ values (region C of Fig. 7.1) the fundamental relationship $kl > 1$ is satisfied, where $k$ is the wave-vector magnitude (for a discussion see Ziman [1972]). In this regime, the transport properties are usually interpreted in terms of a two-band model (electrons in the conduction band and holes in the valence band), as discussed in Sect. 8.3.3, and the magnetoresistance is positive (Sect. 8.4.1) while the Hall coefficient is negative.

CCVD fibers develop into a fully graphitized structural state, as discussed in Sect. 3.7.3, but ex-polymer fibers do not, so that ex-PAN fibers heat treated to 3500°C, for example, have structural and electronic characteristics more like those of bulk, graphitizable carbons with $T_{HT} \sim 2500$°C.

It is clear that a complete description of the electronic structure of carbons still awaits development. At present only a qualitative description can be given. However, the band structure of fully graphitic material is extremely well developed, and will now be discussed in detail.

## 7.2 The Slonczewski-Weiss Model for Graphite

Because of the large anisotropy of the crystal structure, most models for the electronic structure of graphite start from a two-dimensional approximation, treating the intraplanar interaction between the $2s, 2p_x, 2p_y$ atomic orbitals to form strongly coupled bonding and antibonding trigonal orbitals. These trigonal orbitals give rise to three bonding and three antibonding $\sigma$-bands separated by $\sim$10 eV in the two-dimensional graphite band structure. In these models, the weakly coupled $p_z$ atomic wave functions correspond to two $\pi$-bands which are degenerate by symmetry at the six Brillouin zone corners at

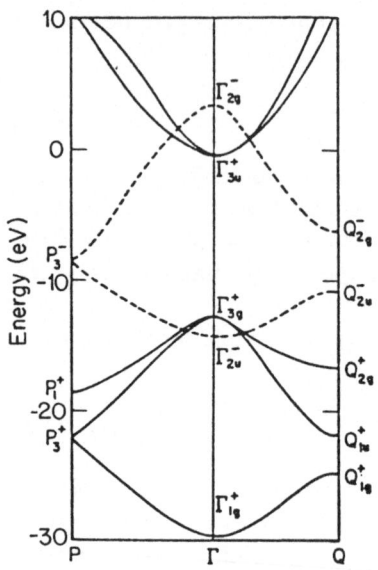

Fig. 7.6. Results of band structure calculations for two-dimensional graphite along the $P$–$\Gamma$–$Q$ directions of the two-dimensional zone, in which $P$ corresponds to the $HK$ axis and $Q$ corresponds to the $LM$ axis of the three-dimensional zone [from Zunger 1978]. The solid curves are for the $\sigma$-bands and the dashed curves for the $\pi$-bands. The Fermi level passes through the degenerate $P_3^-$ levels, giving rise to a dispersion relation linear in $\kappa$ for the $P_3^{\pm}$ levels

point $P$ forming the degenerate $P_3^-$ levels (Fig. 7.6), through which the Fermi level passes. The points $P$, $Q$ and $\Gamma$ of the two-dimensional zone correspond respectively to $HKH$, $LML$ and $A\Gamma A$ of the three-dimensional zone shown in Fig. 7.5. A large number of calculations of the two-dimensional graphite electronic structure have been made [Wallace 1947; Corbató 1956, 1957; Bassani and Pastori Parravicini 1967; Doni and Pastori Parravicini 1969; Painter and Ellis 1970]. The dispersion relations for a two-dimensional band structure for graphite have also been calculated [Painter and Ellis 1970; Zunger 1978] and the results of Zunger [1978] are shown in Fig. 7.6. The degenerate $\pi$-bands at point $P$ are of particular interest since the Fermi level passes through this $P$-point degeneracy, and the $E(\kappa)$ relations away from the $P$-point are linear in $\kappa$ for the $\pi$-bands [see (7.3)]. We further note that at the $P$-point the three antibonding $\sigma$-bands lie far above the $\pi$-bands in energy, and the three bonding $\sigma$-bands lie far below. Three-dimensional band models confirm the following features: (1) The Fermi surface is located near the Brillouin zone edges $HKH$ and $H'K'H'$ (shown in Fig. 7.5). (2) The widths of the $\pi$-bands in the vicinity of the Brillouin zone edges are much less than the separation between the $\pi$-bands and the bonding and antibonding $\sigma$-bands. Therefore two-dimensional band models have been applied extensively to the qualitative interpretation of much experimental data on the electronic structure of graphite.

Graphite, however, is a three-dimensional solid with $ABAB\dots$ stacking of the graphite layers, giving rise to four carbon atoms per unit cell as shown in Fig. 3.1. Although the interlayer interaction is small, it has a profound effect on the four $\pi$-bands near the Brillouin zone edges, causing a band overlap that is responsible for the semimetallic properties of graphite, whereas the two-dimensional band model gives a zero gap semiconductor for the graphene plane. Detailed phenomenological models for the dispersion relations for the four $\pi$-bands have been developed by Slonczewski and Weiss [1958], and by McClure [1957, 1960]. This SWMcC model has been applied extensively to explain the transport, diamagnetic, quantum oscillatory, optical and magneto-optical properties dependent on the electronic structure of bulk crystalline graphite near the Fermi level. A comprehensive summary of the electronic properties of graphite has been given by McClure [1971] and by Spain [1973].

Whereas the two-dimensional band models give the in-plane dispersion relations for the two $\pi$-bands for a planar graphite unit cell, the three-dimensional SWMcC model gives three-dimensional dispersion relations for the four $\pi$-bands corresponding to the full graphite unit cell containing four crystallographically distinct atoms based on the $ABAB$ stacking sequence (see Fig. 3.1). The effect of different stacking arrangements of the graphite layers ($AAA$, $ABAB$, $ABCABC$) on the electronic structure has been considered by McClure [1969] and by Samuelson et al. [1980], who showed that each stacking arrangement results in slightly different energy levels near $E_{\mathrm{F}}$.

The SWMcC model gives a phenomenological treatment of the electronic structure based on crystal symmetry. The most general form of the Hamiltonian consistent with crystal symmetry is developed for $k$ values in the vicinity of the Brillouin zone edges. In the $k_z$ direction, a Fourier expansion is made, and rapid convergence is obtained because of the weak interplanar binding. In the layer planes, a $\kappa \cdot p$ expansion is made since the extent of the Fermi surface in the basal planes is small compared with Brillouin zone dimensions. The SWMcC model is commonly written in terms of the $(4 \times 4)$ Hamiltonian for the $\pi$-bands

$$
H = \begin{pmatrix} E_1 & 0 & H_{13} & H_{13}^* \\ 0 & E_2 & H_{23} & -H_{23}^* \\ H_{13}^* & H_{23}^* & E_3 & H_{33} \\ H_{13} & -H_{23} & H_{33}^* & E_3 \end{pmatrix} \quad ,
\tag{7.7}
$$

where the band edge energies are given by

$$
\begin{aligned}
E_1 &= \Delta + \gamma_1 \Gamma + \tfrac{1}{2}\gamma_5 \Gamma^2 \\
E_2 &= \Delta - \gamma_1 \Gamma + \tfrac{1}{2}\gamma_5 \Gamma^2 \\
E_3 &= \tfrac{1}{2}\gamma_2 \Gamma^2
\end{aligned}
\tag{7.8}
$$

and the interaction terms are

$$
\begin{aligned}
H_{13} &= (-\gamma_0 + \gamma_4 \Gamma)\sigma \exp(i\alpha)/\sqrt{2} \\
H_{23} &= (\gamma_0 + \gamma_4 \Gamma)\sigma \exp(i\alpha)/\sqrt{2} \\
H_{33} &= \gamma_3 \Gamma \sigma \exp(i\alpha) \quad ,
\end{aligned}
\tag{7.9}
$$

163

in which $\kappa$ is the in-plane wave vector measured from the $HKH$ Brillouin zone edge, $\alpha$ is the angle between $\kappa$ and the $K\Gamma$ direction, and

$$\Gamma = 2\cos\pi\xi \quad . \tag{7.10}$$

The dimensionless wave vectors $\xi$ along the $k_z$ direction and $\sigma$ in the basal plane are respectively given by

$$\xi = k_z \bar{c}_0 / \pi \quad \text{and} \tag{7.11}$$

$$\sigma = \frac{\sqrt{3}}{2} a_0 \kappa \quad , \tag{7.12}$$

where $\bar{c}_0 = 3.35\,\text{Å}$ and $a_0 = 2.46\,\text{Å}$. Each of the seven parameters $(\gamma_0, \ldots, \gamma_5, \Delta)$ of the SWMcC model can be identified with overlap and transfer integrals within the framework of the tight binding approximation, but in practice they are evaluated experimentally. The tight binding identification of the seven SWMcC band parameters together with their numerical values are summarized in Table 7.1.

The eigenvalues of the SWMcC Hamiltonian (7.7) yield the energy dispersion relations which are schematically illustrated in Fig. 7.7. Along the Brillouin zone edge $HKH$, two of the four solutions are doubly degenerate and are labeled by $E_3$. The remaining two solutions are non-degenerate and are denoted in this figure by $E_1$ and $E_2$. The degeneracy of the two $E_3$ levels is lifted as we move away from the zone edge, and this is indicated on the left-hand side of the figure with reference to the plane defined by $\xi = 0$. At the $H$ point ($\xi = 1/2$), the levels $E_1$ and $E_2$ are degenerate, and the double degeneracy of these levels is maintained throughout the planes $\xi = \pm 1/2$, as shown on the right-hand side of Fig. 7.7.

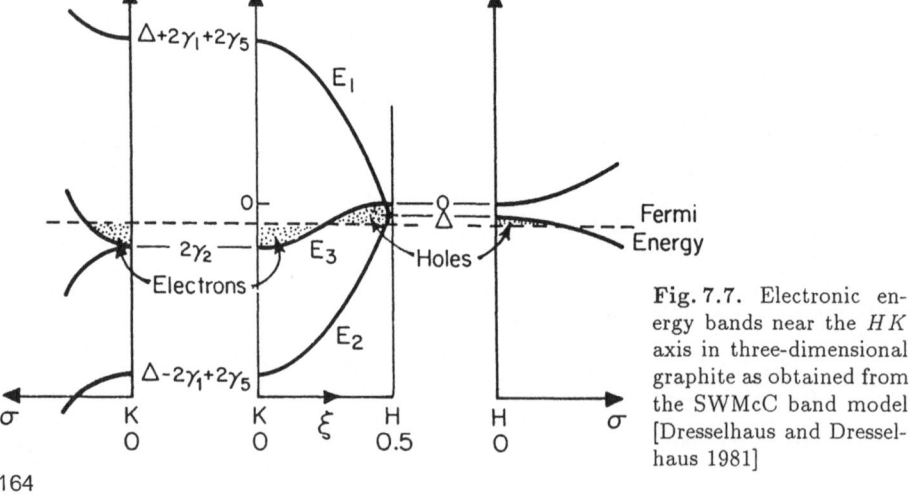

Fig. 7.7. Electronic energy bands near the $HK$ axis in three-dimensional graphite as obtained from the SWMcC band model [Dresselhaus and Dresselhaus 1981]

The SWMcC Hamiltonian has simple solutions in certain special cases. If $\gamma_3$ is neglected, the four solutions

$$E = \tfrac{1}{2}(E_1 + E_3) \pm [\tfrac{1}{4}(E_1 - E_3)^2 + (\gamma_0 - \gamma_4 \Gamma)^2 \sigma^2]^{\frac{1}{2}} \quad , \tag{7.13}$$

$$E = \tfrac{1}{2}(E_2 + E_3) \pm [\tfrac{1}{4}(E_2 - E_3)^2 + (\gamma_0 + \gamma_4 \Gamma)^2 \sigma^2]^{\frac{1}{2}} \tag{7.14}$$

are obtained. These solutions have been applied to the interpretation of a large variety of experiments relevant to the electronic structure and Fermi surface of graphite [McClure 1971; Spain 1973]. It is of interest to note that whereas two-dimensional graphite is a zero-gap semiconductor, three-dimensional graphite is semimetallic with a band overlap of $2\gamma_2$ ($\sim 0.040\,\mathrm{eV}$) and a bandwidth along the Brillouin zone edge of $4\gamma_1$ ($\sim 1.56\,\mathrm{eV}$). In the two-dimensional limit, the only non-vanishing band parameter in Table 7.1 is $\gamma_0$, so that in this limit (7.13) and (7.14) yield (7.3) which exhibits the celebrated linear $\kappa$ relation of the two-dimensional graphite $\pi$-bands.

The density of electronic states associated with two-dimensional graphite and given by (7.4) is linear in $|E|$, as is shown schematically in Fig. 7.3. In three-dimensional graphite the density of states assumes a finite value at $E = 0$ as shown in Fig. 7.8, where the densities of states for electrons $g_e(E)$ and for holes $g_h(E)$ are plotted as a function of energy. It is noted that $g_e(E)$ and $g_h(E)$ become approximately linear in $E$ for energies below the $K$ point minimum or above the $H$ point maximum in the $E_3$-band. The singularities in $g_e(E)$ and $g_h(E)$ at these $K$ and $H$ point extrema are denoted by $E_x$ in Fig. 7.8. If, by acceptor doping, the Fermi level falls below $E_x$ (Acceptors) in Fig. 7.8, only holes are present, or if $E_F$ rises above $E_x$ (Donors), only electrons are present.

The SWMcC model was generalized by L.G. Johnson and Dresselhaus [1973] to yield dispersion relations for the $\pi$-bands throughout the Brillouin zone. Using symmetry requirements to specify the form of the Hamiltonian,

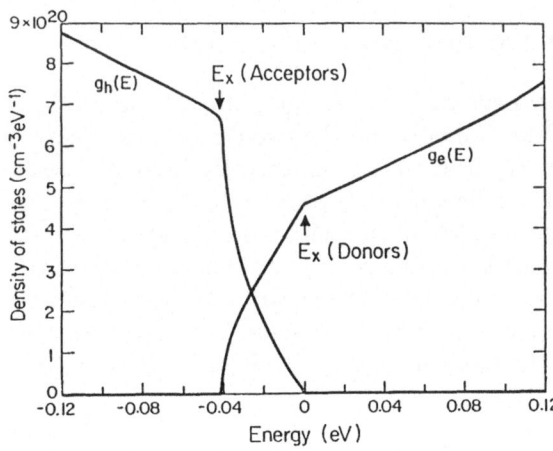

Fig. 7.8. Energy dependence of the density of states for electrons and holes for three-dimensional graphite [Dresselhaus et al. 1977]

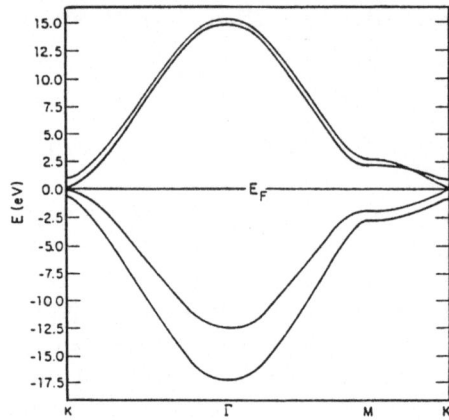

Fig. 7.9. Graphite $\pi$-bands along several high symmetry directions calculated by L.G. Johnson and Dresselhaus [1973] on the basis of a three-dimensional Fourier expansion of $E(\mathbf{k})$ for the SWMcC model along the $HK$ axis and by comparing the calculated frequency-dependent dielectric function $\varepsilon(\omega)$ to the results from optical reflectivity measurements

a three-dimensional Fourier expansion was used for the basis functions. The band parameters of the model were evaluated using (1) Fermi surface data in the vicinity of the Brillouin zone edges $HKH$ and $H'K'H'$, (2) fits to the optical data below 6 eV, and (3) the requirement that the dispersion relations reduce to those of the SWMcC model in the vicinity of the Brillouin zone edges. The resulting dispersion relations for the $\pi$-bands along several high symmetry directions are shown in Fig. 7.9.

First principles three-dimensional calculations of the electronic band structure of graphite have also been carried out using a variety of techniques [Van Haeringen and Junginger 1969; Nagayoshi et al. 1973; Zunger 1978], yielding good agreement near the Fermi level with the widely used phenomenological SWMcC model. These first principles calculations have also been extended to consider the effect of pressure on the graphite electronic structure [Nagayoshi 1977].

## 7.3 Electronic Structure in a Magnetic Field

In this section, results for the Landau levels for graphite on the basis of both the three-dimensional SWMcC model and the two-dimensional Wallace model are summarized. The energy levels do not agree with the semiclassical energy levels for low quantum numbers because of interband coupling. This interband coupling is especially important for the Landau levels with small quantum number ($N \leq 1$). In particular, the $N = 0$ level plays a decisive role in two-dimensional diamagnetism (Sect. 8.1). The negative magnetoresistance observed in pregraphitic carbons is also a direct consequence of the two-dimensional characteristics of the Landau levels (Sect. 8.4.2).

### 7.3.1 Landau Levels in Three-Dimensional Graphite

In the absence of $H_{33}$ (i.e., $\gamma_3 = 0$), analytical solutions of the Landau levels are obtained from the Hamiltonian in (7.7). Suppose that the magnetic field is applied parallel to the c-axis, then the wave vector components $\kappa_x$, $\kappa_y$, which are perpendicular to the magnetic field $B$, should satisfy the commutation relation

$$[\kappa_x, \kappa_y] = \frac{s}{i} \quad , \tag{7.15}$$

where $s$ is proportional to the magnetic field $B$ and is defined in Sect. 7.1. We define the raising and lowering operators $\kappa_+$ and $\kappa_-$ in terms of $\kappa_x$ and $\kappa_y$:

$$\kappa_\pm = \frac{1}{\sqrt{2}}(\kappa_x \pm i\kappa_y) \quad , \tag{7.16}$$

and the operation of $\kappa_\pm$ on the harmonic oscillator wave function $U_N$ is given by

$$\begin{aligned}
\kappa_+ U_N &= [(N+1)s]^{\frac{1}{2}} U_{N+1} \\
\kappa_- U_N &= (Ns)^{\frac{1}{2}} U_{N-1}
\end{aligned} \quad , \quad N = 0, 1, 2, \ldots \quad . \tag{7.17}$$

Then, the SWMcC-Hamiltonian (7.7) in a magnetic field becomes

$$\mathcal{H} = \begin{bmatrix} E_1 & 0 & ip_1\kappa_+ & -ip_1\kappa_- \\ 0 & E_2 & -ip_2\kappa_+ & -ip_2\kappa_- \\ -ip_1\kappa_- & ip_2\kappa_- & E_3 & 0 \\ ip_1\kappa_+ & ip_2\kappa_+ & 0 & E_3 \end{bmatrix} \quad , \tag{7.18}$$

where

$$p_{1,2} = p_0(1 \mp \nu), \quad p_0 = \frac{\sqrt{3}}{2}\gamma_0 a_0, \quad \nu = (2\gamma_4/\gamma_0)\cos(\tilde{c}_0 k_z) \tag{7.19}$$

and $\tilde{c}_0 = 3.35\,\text{Å}$ is the graphite interlayer spacing and in (7.19) $p_1$ refers to the minus sign and $p_2$ refers to the plus sign.

The Schrödinger equation $\mathcal{H}\Psi = E\Psi$ in a magnetic field reduces to the following finite algebraic equation [McClure 1960; Inoue 1962]:

$$\begin{bmatrix} E_1 & 0 & ip_1(Ns)^{1/2} & -ip_1[(N+1)s]^{1/2} \\ 0 & E_2 & -ip_2(Ns)^{1/2} & -ip_2[(N+1)s]^{1/2} \\ -ip_1(Ns)^{1/2} & ip_2(Ns)^{1/2} & E_3 & 0 \\ ip_1[(N+1)s]^{1/2} & ip_2[(N+1)s]^{1/2} & 0 & E_3 \end{bmatrix} \begin{bmatrix} C_1 \\ C_2 \\ C_3 \\ C_4 \end{bmatrix}$$

$$= E_N \begin{bmatrix} C_1 \\ C_2 \\ C_3 \\ C_4 \end{bmatrix} \quad , \tag{7.20}$$

where the wave function $\Psi_N$ is assumed to be of the form

167

$$\Psi_N = \begin{bmatrix} C_1 U_N \\ C_2 U_N \\ C_3 U_{N-1} \\ C_4 U_{N+1} \end{bmatrix} \quad , \tag{7.21}$$

and the relations (7.17) are employed. The secular equation for $N \geq 1$ is thus given by

$$[(E_1 - E_N)(E_3 - E_N) - p_1^2(2N+1)s][(E_2 - E_N)(E_3 - E_N) - p_2^2(2N+1)s]$$
$$- (p_1 p_2 s)^2 = 0 \quad , \tag{7.22}$$

where $E_N$ denotes the energy eigenvalues for Landau level $N$. We use the superscript $j$ to denote one of the four levels corresponding to each $N$. Special cases occur for $N = 0$ where $C_3 = 0$, and for $N = -1$ where $C_1 = C_2 = C_3 = 0$. For these special cases the energy levels $E_0^j$ and $E_{-1}^j$ are given by

$$(E_1 - E_0^j)(E_2 - E_0^j)(E_3 - E_0^j) - [(E_1 - E_0^j)p_2^2 + (E_2 - E_0^j)p_1^2]s = 0 \tag{7.23}$$

for $N = 0$, and by

$$E_{-1}^j = E_3 \tag{7.24}$$

for $N = -1$. It should be noted that the secular equation cannot be factored in the presence of a magnetic field. This implies that the magnetic field mixes states associated with different bands. If the last terms in (7.22) and (7.23) are neglected, we obtain the SWMcC result, see (7.13, 7.14), by a replacement

$$(2N + 1)s \rightarrow \kappa^2 \quad . \tag{7.25}$$

This corresponds to the Bohr-Sommerfeld-Onsager relation:

$$\oint d\kappa_x d\kappa_y = 2\pi s(N + \tfrac{1}{2}) , \quad (N \gg 1) \quad . \tag{7.26}$$

Figure 7.10 illustrates the three-dimensional Landau levels in the quantum limit ($B = 7.5 \,\mathrm{T}$), where $E_0^-(1) = E_3$ corresponds to the level for $N = -1$ and $E_0^+(2)$ to one of the three levels for $N = 0$, see (7.23), and the other two levels are not shown in Fig. 7.10. In this figure, the energies $E_1^-(1)$ and $E_1^+(2)$ denote the two levels for $N = +1$ located in the vicinity of $E_3$ that are obtained from (7.23) [Sugihara and Woollam 1978]. In Fig. 7.10, the subscript on the energy levels refers to the quantum number $N$, the $+$ sign refers to the conduction band levels (the $-$ sign to the valence band levels) and the integer in parenthesis refers to one of the four solutions to (7.22). In this notation the subscript 0 is used to label both the $N = 0$ and $N = -1$ levels.

### 7.3.2 Landau Levels in Two-Dimensional Graphite

The band structure for pregraphitic carbons is approximated by the two-dimensional simplification of the Wallace model [Wallace 1947]. Neglecting

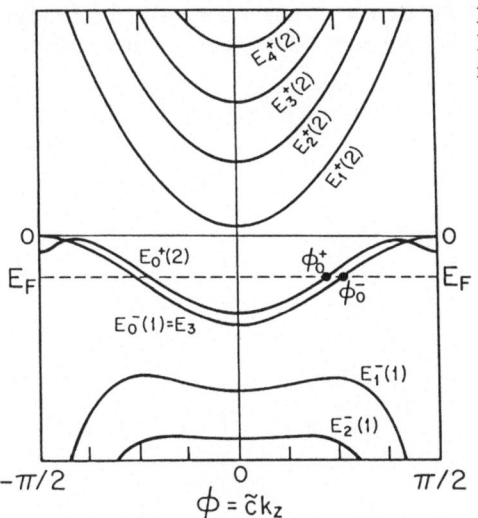

Fig. 7.10. Sketch of the three-dimensional Landau levels for pristine graphite in a high magnetic field ($\simeq 7.5$ T)

all the band parameters except $\gamma_0$ (Sect. 7.2), we obtain the following Hamiltonian matrix:

$$\mathcal{H} - E\mathcal{I} = \begin{vmatrix} -E & p_0\kappa e^{i\alpha} \\ p_0\kappa e^{-i\alpha} & -E \end{vmatrix} \quad , \tag{7.27}$$

where $\mathcal{I}$ is the unit matrix, $\kappa^2 = \kappa_x^2 + \kappa_y^2$ and $p_0 = \sqrt{3}\gamma_0 a_0/2$. From (7.27) we obtain $E(\kappa) = \pm p_0\kappa$, which is the celebrated linear dispersion relation for zero field ($B = 0$), where the nearest neighbor in-plane overlap integral is $\gamma_0 = 3.16$ eV, the in-plane lattice constant is $a_0 = 2.46$ Å and

$$\tan\alpha = \kappa_x/\kappa_y \quad . \tag{7.28}$$

In the presence of a magnetic field, the wave vector components $\kappa_x$, $\kappa_y$ which are perpendicular to the magnetic field $B$ satisfy the commutation relation (7.15), while the raising and lowering operators $\kappa_+$ and $\kappa_-$ satisfy the commutation relation

$$[\kappa_-, \kappa_+] = s \quad . \tag{7.29}$$

The action of the raising and lowering operators on the harmonic oscillator states $U_N$ yields

$$\kappa_- U_N = \sqrt{Ns}\, U_{N-1}, \quad \kappa_+ U_N = \sqrt{(N+1)s}\, U_{N+1} \tag{7.30}$$

for $N = 0, 1, 2,\ldots$ . Therefore, the secular equation

$$(\mathcal{H} - E\mathcal{I})\begin{pmatrix} U_N \\ U_{N+1} \end{pmatrix} = 0 \tag{7.31}$$

leads to

169

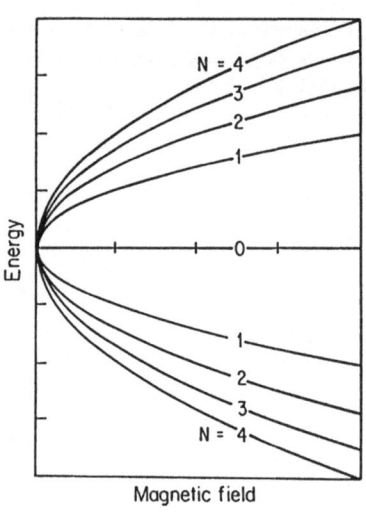

**Fig. 7.11.** Sketch of the two-dimensional Landau levels for turbostratic (2D) graphite in a magnetic field

$$\begin{vmatrix} -E & ip_0\sqrt{2Ns} \\ -ip_0\sqrt{2Ns} & -E \end{vmatrix} = 0 \quad , \tag{7.32}$$

which has solutions

$$E_N = \pm p_0\sqrt{2Ns} \quad , \tag{7.33}$$

for $N = 0, 1, 2,...$ , which has already been quoted as (7.1). A sketch of the two-dimensional magnetic energy levels is shown in Fig. 7.11. The density of states for each magnetic energy level is

$$g_N = \frac{4}{(2\pi)^2}\left(\frac{2\pi}{\tilde{c}_0}\right)s = \frac{2s}{\pi\tilde{c}_0} \quad , \tag{7.34}$$

taking account of the spin degeneracy and the two independent corners of the two-dimensional Brillouin zone. Since $s$ is proportional to $B$ (Sect. 7.1), the density of states in a magnetic field is proportional to the magnetic field $B$, but independent of energy and quantum number $N$. Of particular interest is the $N = 0$ level, which has a finite density of states given by (7.34), while, in the absence of the magnetic field, the density of states is zero for $E = 0$, the energy where the valence and conduction bands of two-dimensional graphite are degenerate (see Fig. 7.3). A sketch showing the relation between the density of states in a magnetic field and in zero field for two-dimensional graphite is shown in Fig. 7.3. In this figure it is seen that when the magnetic field is turned on, a group of states which were continuously distributed in $k$-space in the absence of a field coalesce to form the $N = 0$ Landau level. The unique structure and spacing of the Landau levels in two-dimensional graphite is responsible for the large diamagnetism (Sect. 8.1) and for the negative magnetoresistance in pregraphitic carbons (Sect. 8.4.2).

As shown in Figs. 7.3 and 7.11, the energy of the $N = 0$ Landau level is zero and the number of coalescing states increases as the magnetic field increases. The energy of the $N = 0$ Landau level is given by

$$E = \int_{-\lambda}^{\lambda} g_0(E')f(E')E'dE' \quad , \tag{7.35}$$

where $g_0(E) = 2|E|/(\pi \tilde{c} p_0^2)$ is the density of states in the absence of a field, $f(E)$ is the Fermi function and $\lambda$ is defined by

$$\frac{1}{2} \int_0^{p_0\sqrt{2s}} dE' g_0(E') = \int_0^{\lambda} dE' g_0(E') \quad , \tag{7.36}$$

so that the Bloch states in the interval $-\lambda < E < \lambda$ coalesce into the $N = 0$ Landau level which lies at $E = 0$. In zero magnetic field, the valence and conduction bands are degenerate at $E = 0$, where $g_0(0) = 0$. For small magnetic fields, $\lambda$ is small compared to the thermal energy. Then, we have

$$E \simeq \frac{4}{\pi \tilde{c} p_0^2} \int_0^{\lambda} dE' E'^3 (\partial f/\partial E')_{E'=0} = [p_0^2 s^2/(\pi \tilde{c})](\partial f/\partial E')_{E'=0} \quad . \tag{7.37}$$

These expressions are used in the calculation of the diamagnetism and negative magnetoresistance for pregraphitic carbons (the Bright model [1979]) discussed in Sect. 8.4.2.

# 8. Electronic and Magnetic Properties

The electronic structure of carbon fibers was introduced in Chap. 7. This chapter explores the electronic and magnetic properties of carbon fibers, beginning with equilibrium, and then considering steady state properties. It is quite useful to consider similarities in the properties of fibers to those of bulk carbons, with perturbations derived from the special structural features of the fibers. This can produce effects which are unique to carbon fibers. In turn, this allows information to be deduced about their electronic structure, and in some cases, their microstructure. In fact, some electronic effects are particularly sensitive to certain structural features, and provide valuable probes of structural defects.

The use of Système International (SI) units has been adhered to in all other sections, but most researchers have used Gaussian units for magnetic properties, and this will be followed here, with a brief discussion of engineering units where appropriate.

## 8.1 Diamagnetism

The degree of graphitization can be characterized very sensitively through measurements of the diamagnetic susceptibility $\chi$. It must be pointed out, however, that the diamagnetism reaches its upper limit long before the complete transformation of a carbon into graphite is achieved. For more highly ordered carbons, measurement of the anisotropy of $\chi$ provides a sensitive characterization technique. In particular, the establishment of three-dimensional order results in a decrease of the diamagnetic anisotropy. This sensitivity to crystalline order can be seen clearly in Table 8.1 where the diamagnetic susceptibility (also denoted by $\chi_{dia}$) of vapor-grown carbon fibers with different heat treatment temperatures $T_{HT}$ are compared with other physical properties [Ohhashi et al. 1983].

The dependence of the absolute value of the diamagnetic susceptibility $|\chi|$ on $T_{HT}$ is shown schematically in carbon fibers in Fig. 8.1. For samples with $T_{HT}$ below 1000°C, the aromatic networks attain coherence over only a small distance so that $|\chi|$ is very small and also temperature independent. Heat treatment of fibers above 1000°C results in two-dimensional growth of the aromatic networks, so that the diamagnetism due to conduction carriers starts

**Table 8.1.** Room temperature properties[a] of vapor-grown carbon fibers with different heat treatment temperatures [Ohhashi et al. 1983]

| $T_{HT}$ [°C] | $d_{002}$ [Å] | $L_c$ [Å] | $L_a$ [Å] | $I_{1350}/I_{1580}$ | $\chi_{dia} \times 10^6$ [emu CGS/g] | $\chi_{ESR} \times 10^8$ [emu CGS/g] | $\varrho \times 10^8$ [Ωm] |
|---|---|---|---|---|---|---|---|
| 1100[b] | 3.484 | 40 | 30 | 2.9 | −1.0 | +0.6 | 950 |
| 1500 | 3.449 | 55 | 30 | 1.1 | −2.1 | +2.6 | 1000 |
| 1800 | 3.431 | 170 | 100 | 0.7 | −3.9 | +1.5 | 1000 |
| 2000 | 3.426 | 330 | 200 | 0.6 | −5.7 | +1.6 | 700 |
| 2500 | 3.368 | 600 | 570 | ∼ 0 | −6.5 | +2.2 | 140 |
| 2800 | 3.357 | 1000 | – | ∼ 0 | −6.8 | +0.3 | 80 |

[a] The properties in this table include:
$d_{002}$, the interplanar spacing between graphite layers as measured by x-ray diffraction ($d_{002} = \tilde{c}$);
$L_c$, $L_a$, the c-axis and in-plane crystallite sizes;
$I_{1350}/I_{1580}$, the ratio of the disorder-induced to Raman-allowed integrated Raman line intensities;
$\chi_{dia}, \chi_{ESR}$, the magnetic susceptibilities from diamagnetic and ESR measurements, where 1 emu CGS/g = $(10^3/4\pi)$ SI units;
$\varrho$, the electrical resistivity.
[b] As-grown CCVD fiber.

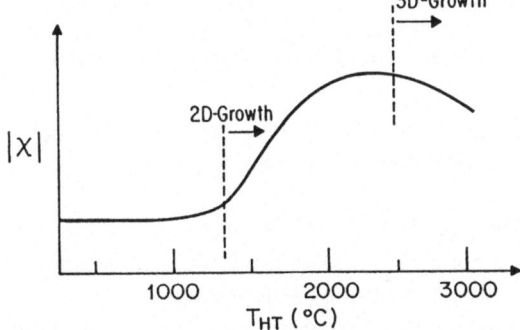

**Fig. 8.1.** Schematic curve of the absolute value of the diamagnetic susceptibility $\chi$ vs $T_{HT}$. The regions of 2D and 3D growth are indicated

to increase and $|\chi|$ reaches a maximum before three-dimensional ordering is established. In the $T_{HT}$ range corresponding to 3D ordering, $|\chi|$ decreases weakly. Figure 8.2 shows plots of the measured $\chi_T$ (total susceptibility) vs $1/T$ for benzene-derived fibers heat treated at various temperatures [Matsubara et al. 1986]. (In order to convert the susceptibility to SI units, which are dimensionless, the susceptibility in emu/g should be multiplied by $1000/4\pi$.) The total susceptibility includes contributions from the core electrons, though the dominant contribution is the unusually large diamagnetic term discussed above. In Fig. 8.2, saturation behavior in $\chi_T$ is observed at low temperatures. The dependence of $\chi_T$ at saturation on $T_{HT}$ is seen to be consistent with the behavior of $\chi(T_{HT})$ shown in Fig. 8.1. The corresponding $\chi_T$ vs $1/T$ plot for ex-PAN fibers heat treated to similar temperatures (Fig. 8.3) shows a behavior similar to that for benzene-derived fibers (Fig. 8.2) except that the magnitude of $\chi_T$ is larger for the vapor-grown fibers (consistent with the

173

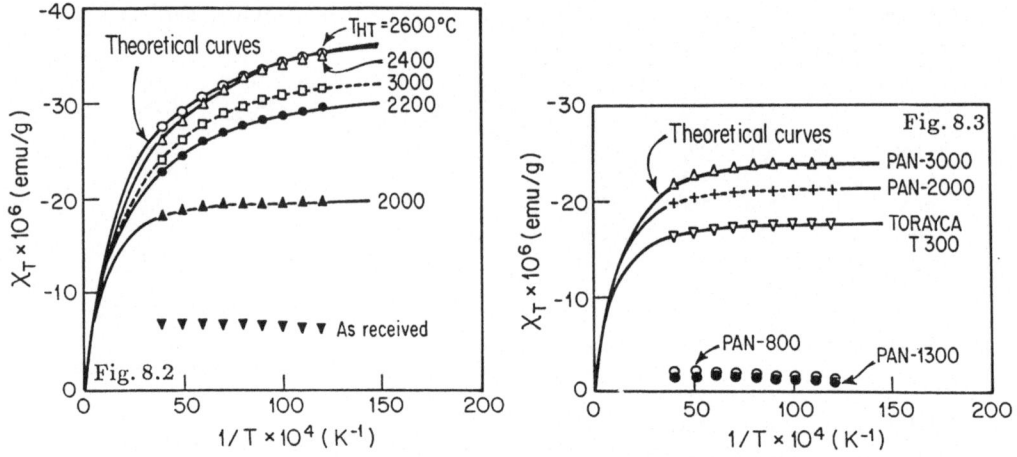

**Fig. 8.2.** Total susceptibility $\chi_T$ vs reciprocal temperature $(1/T)$ plots for benzene-derived fibers with various heat treatment temperatures (°C) [Matsubara et al. 1986]; solid lines are based on the McClure-Hickman theory [McClure and Hickman 1982], while the broken line is the theory of Sharma et al. [1974]

**Fig. 8.3.** Plot of the total susceptibility $\chi_T$ vs reciprocal temperature $(1/T)$ for ex-PAN (PAN-based) fibers with various heat treatment temperatures (°C) [Matsubara et al. 1986]. Solid lines are the calculations based on the McClure-Hickman theory [Matsubara et al. 1986]

higher degree of ordering of vapor-grown fibers) and the saturation effect is more pronounced for PAN fibers (consistent with their turbostratic planar arrangement).

The theory of the diamagnetic susceptibility of graphite shows that the most important contribution to $\chi_T$ comes from interband coupling terms. As is shown below, the diamagnetic susceptibility has a dependence $\chi \sim 1/T$ at high temperatures where in-plane interactions are dominant, while $\chi$ is essentially temperature independent at low $T$, especially for turbostratic carbons. For single crystal graphite, the high-temperature diamagnetic susceptibility provides an accurate determination of the nearest neighbor in-plane interaction energy $\gamma_0$ (Sect. 7.2), while the temperature dependence of $\chi$ is highly sensitive to several other graphite band parameters (namely $\gamma_1$, $\gamma_2$, $\gamma_3$). For disordered carbons, the diamagnetic susceptibility provides valuable information on the defect structure, especially the crystallite size and the parameters describing the ribbon structure. Of particular interest are the theoretical calculations for carbon fibers [McClure and Hickman 1982], which show how the temperature dependence of $\chi_T$ can be exploited to provide detailed information on the overall width of the graphitic ribbons and the folding of these ribbons.

Extensive theoretical investigations of the diamagnetism in graphite and carbon fibers were carried out by McClure and his group [McClure 1956, 1960;

Sharma et al. 1974; McClure and Hickman 1982; Sun and McClure 1983].
The Landau-Peierls contribution to the diamagnetism accounts for only a
few percent of the observed value. In particular, a vanishing value for the diamagnetic susceptibility is expected for two-dimensional graphite, according to
the Landau-Peierls formula. This difficulty was removed by considering the
two-dimensional Landau levels of graphene layers. In the following, the Mc-
Clure theory for two-dimensional graphite bands which resolves this problem
is briefly sketched, showing how the unusual electronic structure of graphite
gives rise to an especially large diamagnetic contribution to $\chi$.

### 8.1.1 Diamagnetism in Two-Dimensional Graphite

When a magnetic field is applied in a direction perpendicular to the carbon
planes, the magnetic energy levels for two-dimensional carbon are given in
Sect. 7.3.2 by [McClure 1956; Uemura and Inoue 1958]

$$E_N = \pm p_0 \sqrt{2Ns}, \quad N = 0, 1, 2, \ldots \quad , \tag{8.1}$$

where

$$p_0 = \frac{\sqrt{3}}{2} \gamma_0 a_0 \quad , \tag{8.2}$$

in which for graphite $\gamma_0 = 3.16$ eV, $a_0 = 2.46$ Å, and $s = eB/\hbar c$ in cgs
units and $s = eB/\hbar$ in SI units. The corresponding density of states for each
magnetic level is linear in $B$ from (7.34)

$$g_N = \frac{2s}{\pi \tilde{c}_0} \quad , \tag{8.3}$$

taking into account the spin degeneracy and the two independent corners
of the two-dimensional Brillouin zone. Since pregraphitic carbons are tur-
bostratic, $\tilde{c}_0$ is replaced by $\tilde{c}$ for fibers in (8.3), where $\tilde{c}$ denotes an interlayer
distance somewhat larger than for graphite ($\tilde{c}_0 = 3.35\,\text{Å} \leq \tilde{c} \leq 3.45\,\text{Å}$).

The physical reason for the diamagnetism is qualitatively explained as
follows. When the magnetic field is turned on, a group of states whose energies are continuously distributed in $k$-space coalesce to the $N = 0$ level (see
Fig. 7.11 and Sect. 7.3.2).

Since the total energy of the group of energy levels which forms the $N = 0$
Landau level is zero, the change in energy with magnetic field is negative.
Neglecting the difference between the free energy and the total energy, the
contribution to the susceptibility is given by $(1/B)(\partial E/\partial B)$, yielding

$$\chi \simeq -\frac{p_0^2}{2\pi \tilde{c}} \left( \frac{e}{\hbar c} \right)^2 \frac{\operatorname{sech}^2(E_F/2k_B T)}{k_B T} \quad , \tag{8.4}$$

where $E_F$ denotes the Fermi energy. The contribution to the susceptibility
from the other levels ($N \neq 0$) is negligible. Putting $a_0 = 2.46$ Å and $\tilde{c} =$

3.35 Å, we obtain the diamagnetic susceptibility per gram,

$$\chi \simeq -0.0093\gamma_0^2 \, \frac{\text{sech}^2(E_\text{F}/2k_\text{B}T)}{T\varrho_\text{m}} \quad \text{emu/g} \quad , \tag{8.5}$$

where the prefactor is 0.74 in SI units, $\varrho_\text{m}$ is the density of graphite ($\varrho_\text{m} =$ 2.26 g/cm$^3$ for single crystal graphite) and the nearest neighbor overlap energy $\gamma_0$ is in eV units. The magnitude $|\chi|$ given by (8.5) is about three times larger than the value calculated exactly [McClure 1956]. This discrepancy is due to neglect of the entropy term. The exact calculation yields

$$\chi = -0.0032\gamma_0^2 \, \frac{\text{sech}^2(E_\text{F}/2k_\text{B}T)}{T\varrho_\text{m}} \quad \text{emu/g} \quad . \tag{8.6}$$

At high temperature, (8.6) becomes $-0.014/T$ emu/g (or $-1.11/T$ in SI units), when $\gamma_0 = 3.16$ eV and $\varrho_\text{m} = 2.26$ g/cm$^3$ are inserted. This is in good agreement with the observed temperature dependence of the susceptibility for pure graphite measured by Ganguli and Krishnan [1941]. The magnetic susceptibility for graphite is unusually large and negative, and the temperature dependence of (8.6) is unique to graphite. The calculated diamagnetic susceptibility based on the 2D-model explains the high temperature behavior well. However, a satisfactory account of the behavior observed at low temperature (e.g., $\chi \to$ constant value) requires use of a more detailed band model, including interaction parameters such as $\gamma_1$, $\gamma_2$ and $\gamma_3$ [McClure 1960; Sharma et al. 1974].

### 8.1.2 Magnetic Susceptibility of Carbon Fibers

The structure of carbon fibers can be regarded as an aggregate of two-dimensional graphitic ribbons preferentially oriented in the direction of the fiber axis. The ribbons are a few layer planes thick, and the stacking is usually turbostratic [Oberlin 1979, 1984]. The graphene layers in turbostratically stacked graphite behave like isolated monolayers and do not exhibit $ABAB$ stacking as discussed in Chaps. 1 and 3. In the calculations, the graphitic ribbons are considered to be approximately infinite along their lengths (i.e., very long along the fiber axis), while their transverse sizes are assumed to be small compared to the fiber diameter. The magnetic susceptibility of flat ribbons of graphite has been calculated by Hoarau and colleagues [Chausse and Hoarau 1969; Hoarau and Volpilhac 1976; Volpilhac and Hoarau 1978] in order to understand the strong size dependence of the susceptibility in partially graphitized carbon materials. However, their calculations were based on an oversimplified model. McClure and Hickman [1982] (hereafter MH) subsequently proposed a theory based on the folded-ribbon model that explains the observed susceptibility quite well with a reasonable set of parameters.

According to the MH model [McClure and Hickman 1982], each ribbon is assumed to consist of periodically folded graphitic (graphene) monolayers

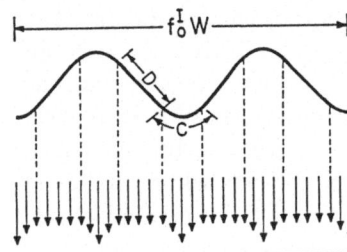

**Fig. 8.4.** Cross section of the folded ribbon model. The infinite length of the ribbon ($x$ direction) is perpendicular to the plane of the diagram. The figure corresponds to the case of $n = W/L = 4$. The in-plane length of a fold is given by $L$ and $f_0^I$ gives the ratio of the projected width to the actual folded lengths. The vertical arrows represent the effective component of a magnetic field applied perpendicular to the ribbon (from top to bottom)

which have very large lengths (along the fiber axis) but finite widths in the ribbon cross sections. Interactions between graphitic layers are neglected. Figure 8.4 shows the cross section perpendicular to the ribbon length according to the folded ribbon model. The electrons moving on a graphite layer plane interact with the component of the magnetic field perpendicular to the layer planes (indicated by the arrows in Fig. 8.4). This is equivalent to the case where a spatially varying magnetic field is applied to a flat layer if the effect of bond-bending upon the interaction between neighboring atoms is neglected.

The true width $W$ of the folded ribbon follows the curved trajectory and is assumed to be given by an integral number ($n$) times one-half of the folding wavelength ($2L$), so that $W = nL$, where $L = C + D$ in the figure and is a measure of half of the ribbon period. The projected width of the ribbon is represented by $f_0^I W$, where the folding ratio $f_0^I$ is smaller than unity and $f_0^I = 1$ corresponds to a completely stretched out ribbon. According to MH, the total susceptibility $\chi_T$ is expressed by

$$\chi_T = (f_0^I)\chi_{00}^I + [1 - (f_0^I)^2]\chi_{11}^{II} + \chi_b \quad , \tag{8.7}$$

where $\chi_{00}^I$ is the susceptibility of a flat ribbon, $\chi_{11}^{II}$ is the susceptibility for a sinusoidal variation of the field with wave vector $q = \pi/L$ and $\chi_b(= 0.5 \times 10^{-6}$ emu/g$)$ is a paramagnetic contribution resulting from a correction due to the susceptibility anisotropy and a term representing the isotropic part of the susceptibility. Here $\chi_{00}^I$ and $\chi_{11}^{II}$ are complicated functions of $L$ and $W$.

Figure 8.3 shows the temperature dependence of $\chi_T$ of PAN-based carbon fibers with various $T_{HT}$ values [Matsubara et al. 1986] and the solid lines are for calculations based on the MH theory. In this figure, $\chi_T$ for PAN-800 and PAN-1300 are very small and decrease gradually with decreasing $T$. In these samples the aromatic networks do not grow very much during the heat treatment, and therefore there is negligible diamagnetism of the conduction carriers. The slight negative dependence of $|\chi_T|$ on $1/T$ for PAN-800 and PAN-1300 fibers can be ascribed to the small value of the paramagnetic component described by the Curie law. For Torayca-T 300, PAN-2000 and PAN-3000 fibers, $|\chi_T|$ increases stepwise with each increase in $T_{HT}$, and each curve of $\chi_T$ vs $1/T$ saturates at low temperatures. For these three samples, the diamagnetism is mainly due to the contribution of conduction carriers, despite the small size of the crystallites. In fact, $\chi_T$ for PAN-3000 at room temperature is $-21.5 \times 10^{-6}$ emu/g ($-1.71 \times 10^{-3}$ SI), which is just the same as $\chi_T$ for graphite. However, as $T$ is lowered, the discrepancy between PAN-3000 and graphite becomes measurable. As shown in Fig. 8.2 (see broken lines), $\chi_T$ of graphite at 77 K is $-30 \times 10^{-6}$ emu/g ($-2.4 \times 10^{-3}$ SI), while PAN-3000 shows a definite leveling off in $\chi_T(T)$ that is caused by the crystallite size effect.

**Table 8.2.** Parameter sets[a] which fit experimental susceptibility data for ex-PAN fibers [Matsubara et al 1986]

| Sample | $L$[Å] | $W$[Å] | $f_0^I$ [b] | $W f_0^I$[Å] |
|---|---|---|---|---|
| Torayca-T 300 | 59.6 | 217.3 | 0.500 | 108.7 |
| PAN-2000 | 74.6 | 249.2 | 0.510 | 127.1 |
| PAN-3000 | 81.0 | 272.6 | 0.560 | 152.7 |

[a] $L, W$ are the length and width of graphitic ribbons.
[b] $f_0^I$ is the folding ratio (see Fig. 8.4).

There is negligible contribution by the conduction carriers to the diamagnetism in PAN-800 and PAN-1300. Therefore, the three parameters $L, W$ and $f_0^I$ have only been evaluated for Torayca-T 300, PAN-2000 and PAN-3000 in Table 8.2; the experimental plots are satisfactorily reproduced by the theoretical curves (solid lines) in Fig. 8.3. As $T_{HT}$ is increased, all three parameters $L, W$ and $f_0^I$ tend to increase. The increase in ribbon width $W$ implies a growth of the two-dimensional ribbon. A gradual increase of $f_0^I$ indicates a weakening of the folding by annealing or a gradual stretching out of the ribbon. The average crystallite size $L_a$ defined by x-ray diffraction should be compared with the folding wave length $2L$ rather than with $W$ as shown in Fig. 8.4. Similar calculations have been performed for the diamagnetism of benzene-derived fibers and a good fit is obtained (see Fig. 8.2). However, the evaluated $f_0^I$ values do not appear to be consistent with a monotonic dependence on heat treatment temperature $T_{HT}$. This suggests that the folded-ribbon model should be modified in the case of benzene-derived fibers. It should be noted that $\chi_T$ of BDF-3000 is smaller than those of BDF-2600 and BDF-2400. This corresponds to a 2D to 3D transformation in benzene-derived fibers (see Fig. 8.1). Thus analysis of the temperature dependence of $\chi_T$ for carbon fibers provides unique information on the width of the graphene ribbons and on their degree of folding.

## 8.2 Electron Spin Resonance

Electron spin resonance (ESR) absorption in graphite and in carbons was discovered independently by Castle [1953] and by Hennig et al. [1954]. In the following few years, various explanations for the observed ESR behavior were proposed until Wagoner [1960] demonstrated convincingly that the resonance in natural graphite and in highly heat treated carbons is due to conduction carriers. However, the analysis of the ESR data in pregraphitic carbons is very complicated because a large number of localized spins contribute to the resonance absorption in addition to the ESR signal from the conduction carriers. Furthermore, the mixing of the $g$-values for conduction and localized holes and electrons and the various contributions of each to the linewidths by the exchange interaction make it difficult to separate each contribution out from the total absorption intensity. In spite of this situation, the ESR study of carbons has provided a useful monitor of the carbonization process and has been reviewed by Mrozowski [1979].

ESR measurements provide four important quantities: (1) absorption intensity, (2) line shape, (3) $g$-shifts and (4) linewidth. ESR absorption in disordered carbons is composed of two contributions: a) conduction carriers and b) localized spin centers. The significant point is that the conduction carriers and the localized spins exhibit different behaviors with regard to these four quantities. By analyzing the observed spectra, useful information can be obtained on crystallite size, mean deviation of the $a$-axis of the crystallites from the fiber axis, carrier density, localized spin density, and Fermi level.

The ESR properties of polycrystalline graphite or carbon fibers with high $T_{HT}$ can be analyzed by introducing an appropriate averaging procedure for the ESR line for single crystal graphite. This analysis provides useful information on the crystallite size and layer plane alignment relative to the fiber axis (Sect. 8.2.2).

As a result of the high anisotropy of graphite, the conduction carrier absorption varies strongly with the alignment of the crystal in the magnetic field. The $g$-value and the ESR line width $\Delta H$ vary according to the following expressions:

$$g = g_\perp + (g_\parallel - g_\perp)\cos^2\theta$$

$$\Delta H = \Delta H_\perp + (\Delta H_\parallel - \Delta H_\perp)\cos^2\theta \quad , \tag{8.8}$$

where $\theta$ is the angle between the magnetic field and the $c$-axis. The $g$-value anisotropy is $(g_\parallel - g_\perp) = 0.047$ at room temperature in pure graphite, and $g_\perp$ is equal to $2.0026 \pm 0.0002$, close to the free electron value of 2.0023 [Wagoner 1960]. Here $g_\parallel$ and $g_\perp$ refer to $g$-factors for the magnetic field parallel and perpendicular to the $c$-axis. As is shown in Fig. 8.5, $g_\perp$ is independent of temperature, whereas $g_\parallel$ is large and strongly dependent on temperature and changes drastically upon the addition of a small concentration of electrically

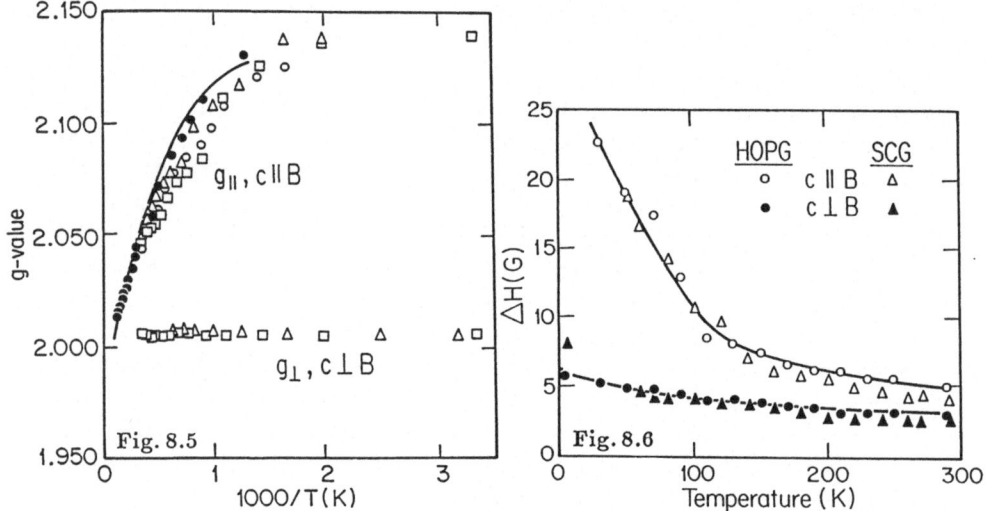

**Fig. 8.5.** Temperature dependence of the $g$-value of graphite for the magnetic field $\parallel$ and $\perp$ to the $c$-axis, where the solid line is the McClure-Yafet theory [McClure and Yafet 1963] as discussed in the text. Experimental data of Murata [1982] are: ($\square$) highly oriented pyrolytic graphite; ($\triangle$, o) single crystal graphite; and of Wagoner [1960]: (•) single crystal graphite

**Fig. 8.6.** Temperature dependence of the ESR linewidth $\Delta H$ of graphite [Murata 1982] showing motional narrowing. Results for $\Delta H$ are shown for the magnetic field both $\parallel$ and $\perp$ to the $c$-axis. Solid lines are aids to the eye. (Note $1\,\mathrm{G} = 10^{-4}\,\mathrm{T}$)

active impurities such as boron atoms. A similar behavior is observed for the linewidth $\Delta H$, see Fig. 8.6, where it is seen that $\Delta H_\perp$ is essentially temperature independent while $\Delta H_\parallel$ increases rapidly below $\sim 100\,\mathrm{K}$. The anisotropy component $(g_\parallel - g_\perp)$, which is caused by the spin-orbit interaction, is approximately proportional to $1/T$ above $150\,\mathrm{K}$ and saturates at low temperatures similar to the diamagnetic susceptibility (see Fig. 8.2). These behaviors are a direct consequence of the unique band structure of graphite.

The observed $g$-shift $\Delta g$ for graphite is an average of $\Delta g(k)$ over the Fermi surface and $\Delta g(k)$ is a sensitive function of the energy dispersion relation $E_k$ and of the wavevector $k_z$ normal to the basal planes. The dominant contribution comes from the region in $k_z$ where $E_k$ crosses the $E_3$ level (see Fig. 7.7) because of the factor $[E_k - E_3(k_z)]^{-1}$ in the resonant denominator. With increasing temperature, spin diffusion becomes very rapid, leading to an averaging of the $\Delta g(k)$ contributions and to a small $g$-shift (this is referred to as motional narrowing), as is seen in Fig. 8.5.

Another important ESR property of graphite is that the line shape is of the Dysonian form which is characteristic of conduction electron spin resonance in metals. According to Dyson's theory [Dyson 1955; Feher and Kip

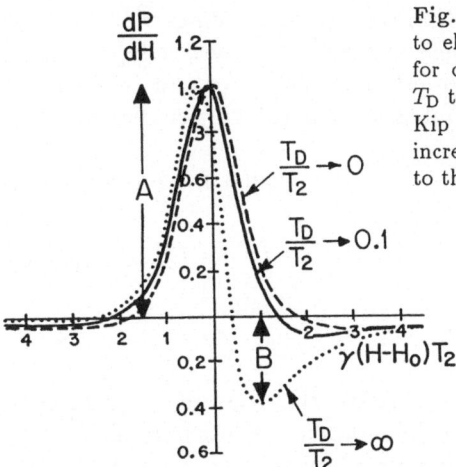

Fig. 8.7. Derivative of the power absorption due to electron spin resonance in thick metal plates for different ratios $T_D/T_2$ of the diffusion time $T_D$ to the spin-spin relaxation time $T_2$ [Feher and Kip 1955]. The increase in peak height $B$ with increase in $T_D/T_2$ is associated with power loss to the conduction electron spins

1955], the ratio of peak heights $A/B$ lies between 19.0 and 2.5 (see Fig. 8.7), where 19.0 corresponds to the rapid diffusion case $T_D \ll T_2$. Here $T_2$ is related to the ESR linewidth by $g\mu_B \Delta H \simeq \hbar/T_2$ and $T_D$ is the time required for a spin to diffuse across the skin depth $\delta$, and is given by

$$T_D = \frac{3}{2} \frac{\delta^2}{vl} \quad , \tag{8.9}$$

where $v$ is the mean electron velocity and $l$ is the mean free path. For single crystal graphite (and $H \parallel c$-axis) at room temperature, $A/B = 3.0$ [Wagoner 1960]. However, (8.9), which is valid for isotropic conductors, should be modified since graphite is highly anisotropic. This was done by Wagoner [1960].

As mentioned at the beginning of this chapter, the ESR absorption in disordered carbons is composed of contributions from conduction carriers and from localized spin centers. The localized spin centers exhibit a specific heat anomaly below 1 K whose magnitude increases with magnetic field [Mrozowski 1979] (Sect. 5.1). Highly ordered graphite has only conduction carrier contributions.

The exchange interaction between the conduction electrons and localized spins affects the susceptibility, $g$-factor and linewidth. The observed behaviors were explained by the exchange coupling model which has been employed in the problem of Cu-Mn dilute alloys [Pifer and Longo 1971]. According to this model, the total susceptibility is given by contributions from the conduction electrons $\chi_{\text{cond}}$ and the localized spin centers $\chi_{\text{loc}}$:

$$\chi_{\text{tot}} = \chi_{\text{cond}} + \chi_{\text{loc}} , \quad \text{where} \tag{8.10}$$

$$\chi_{\text{cond}} = \frac{\chi_{\text{cond}}^0 (1 + \alpha \chi_{\text{loc}}^0)}{1 - \alpha^2 \chi_{\text{cond}}^0 \chi_{\text{loc}}^0} , \qquad \chi_{\text{loc}} = \frac{\chi_{\text{loc}}^0 (1 + \alpha \chi_{\text{cond}}^0)}{1 - \alpha^2 \chi_{\text{cond}}^0 \chi_{\text{loc}}^0} \quad . \tag{8.11}$$

181

Here, $\chi^0_{\mathrm{cond}}$ and $\chi^0_{\mathrm{loc}}$ refer to the susceptibilities of conduction carriers and localized spins in the absence of exchange interaction, respectively; $\alpha$ is the exchange coupling constant and to lowest order in $J$ we can write $\alpha = Jg(E_{\mathrm{F}})/2n$, where $J$ is the exchange integral, $n$ is the carrier density, and $g(E_{\mathrm{F}})$ is the density of states at $E_{\mathrm{F}}$.

In the limit of strong exchange interaction, the $g$-value and linewidth are given by

$$g_{\parallel} = \frac{1}{1+\chi_{\mathrm{r}}} g_{\parallel\mathrm{cond}} + \frac{\chi_{\mathrm{r}}}{1+\chi_{\mathrm{r}}} g_{\parallel\mathrm{loc}} \quad \text{and} \tag{8.12}$$

$$\Delta H = \frac{1}{1+\chi_{\mathrm{r}}} \Delta H_{\mathrm{cond}} + \frac{\chi_{\mathrm{r}}}{1+\chi_{\mathrm{r}}} \Delta H_{\mathrm{loc}} \quad . \tag{8.13}$$

Here, $\chi_{\mathrm{r}} = \chi_{\mathrm{loc}}/\chi_{\mathrm{cond}}$, and $g_{\parallel\mathrm{cond}}$ and $g_{\parallel\mathrm{loc}}$ denote the $g$-values of conduction carriers and localized spins, respectively. The linewidth contributions $\Delta H_{\mathrm{cond}}$ and $\Delta H_{\mathrm{loc}}$ have similar meanings. Using these relations, Kazumata [1983] and Kazumata et al. [1986] carried out a detailed analysis of the ESR spectra in neutron-irradiated graphite.

The procedure outlined above is applicable to ESR measurements on carbon fibers. The analysis for carbon fibers is more complex than in the case of graphite. From the analysis of the ESR absorption intensity, lineshape, $g$-shift and linewidth, the concentration of localized spins can be determined as well as estimates for the crystallite size, orientation of the crystallites with respect to the fiber axis, the carrier density and the Fermi level for the carbon fibers.

### 8.2.1  The $g$-Shift in Graphite

First we briefly sketch the theory of the $g$-shift in 3D and 2D graphite. The ESR line shape is of the Dysonian form, which is characteristic of conduction electron spin resonance in metals. The $g$-value of the resonance shows a remarkably large anisotropy. At room temperature in pure graphite, the $g$-value varies from $2.0026 \pm 0.0002$ to $2.0495 \pm 0.0002$ as the magnetic field is shifted from perpendicular to parallel to the $c$-axis. The $g$-value depends on $T$ and on the position of the Fermi level with respect to the band edge [Wagoner 1960]. Also, the width of the ESR line decreases with increasing temperature (motional narrowing) (see Figs. 8.5 and 8.6).

McClure and Yafet [1963] presented a theory of the $g$-factor based on a semiclassical approximation. Here, the McClure-Yafet theory is expressed in a different form by employing the Dresselhaus-Dresselhaus Hamiltonian [1965]. To derive the $g$-shift, we write the Hamiltonian for the four $\pi$-bands as

$$\mathcal{H} \begin{pmatrix} U_N \\ U_N \\ U_{N-1} \\ U_{N+1} \end{pmatrix} = E \begin{pmatrix} U_N \\ U_N \\ U_{N-1} \\ U_{N+1} \end{pmatrix} \quad , \tag{8.14}$$

where $U_N$ denotes the pertinent harmonic oscillator wave functions (Sect. 7.3.1) and $\mathcal{H}$ is given by

$$\mathcal{H} = \begin{pmatrix} E_1 & \lambda_{12}^z \sigma_z & ip_1 \kappa_+ & -ip_1 \kappa_- \\ \lambda_{12}^z \sigma_z & E_2 & -ip_2 \kappa_+ & -ip_2 \kappa_- \\ -ip_1 \kappa_- & ip_2 \kappa_- & E_3 + \lambda_{33}^z \sigma_z & 0 \\ ip_1 \kappa_+ & ip_2 \kappa_+ & 0 & E_3 - \lambda_{33}^z \sigma_z \end{pmatrix} , \qquad (8.15)$$

where $\lambda_{12}^z$ and $\lambda_{33}^z$ are the spin-orbit interaction matrix elements and $\sigma_z$ denotes the Pauli spin operator. Other quantities in (8.15) are defined in Sect. 7.3.1. To obtain the $g$-shift $\Delta g(E_k)$ from (8.14) and (8.15), terms of order $\lambda^2$ are neglected and $s(2N+1)$ is replaced by $\kappa^2$, see (7.25) and (7.26), while terms of order $(1/N)$ are disregarded compared with unity. The total $g$-shift observed in actual measurements is thus represented by an average over the Fermi surface

$$\Delta g = \sum_k \Delta g(E_k) \frac{\partial f_0}{\partial E_k} \bigg/ \sum_k \frac{\partial f_0}{\partial E_k} \qquad (8.16)$$

and contributions from both electrons and holes must be considered. However, $\Delta g$ diverges since $\Delta g(E_k)$ includes the term $[E_k - E_3(k_z)]^{-1}$, which is obtained from the spin-dependent term by solving the secular equation related to (8.15). This divergence comes from the region of small $N$ or a small $k_z$ range where $E_k$ crosses $E_3(k_z)$. In this region, (8.14) must be solved exactly without neglecting terms $\mathcal{O}(1/N)$ compared with unity. Although a finite $\Delta g$-value is obtained with this refinement, the sign of $\Delta g$ is negative [McClure and Ruvalds 1964], in contradiction with the observed result. Although the McClure-Yafet [1963] theory is incorrect for the reasons stated above, most experimental papers still interpret their data in terms of this model, as for example in Fig. 8.5. A reliable theory for the $g$-factor of graphite remains to be developed.

There is no corresponding difficulty in calculating the $g$-factor for graphite intercalation compounds, which have a large carrier density and where it is necessary to consider many Landau levels $N \gg 1$ [Sugihara et al. 1984].

The $g$-shift of two-dimensional graphite can also be calculated, using the procedure outlined above [McClure and Yafet 1963]. For two-dimensional graphite the same analysis yields [McClure 1980]

$$\Delta g = \frac{\eta \lambda x}{E_F \ln(2 \cosh x)} \left( \frac{3ma_0^2 \gamma^2}{2\hbar^2 E_F} \frac{x}{\cosh^2 x} \pm \eta \tanh x \right) , \qquad (8.17)$$

where

$$x = \frac{E_F}{2k_B T} , \qquad (8.18)$$

$\lambda$ is the spin-orbit coupling parameter and $\eta$ is the fraction of 3D-states admixed in the $\pi$-band. The + and − signs in (8.17) depend on the sign of the

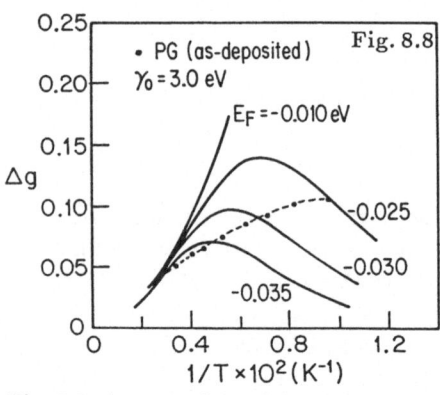

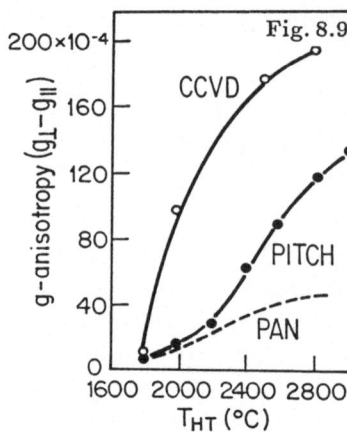

**Fig. 8.8.** A comparison of the observed $g$-shift (points) of as-deposited pyrographite with the theory (solid curves) for the $T$ dependence of $\Delta g$ for two-dimensional graphite for various values of the Fermi level [Kawamura et al. 1983]

**Fig. 8.9.** Plot of the $g$-factor anisotropy vs heat treatment temperature for ex-PAN, ex-pitch and CCVD fibers where $g_{\parallel}$ and $g_{\perp}$ refer to the $g$-factors $\parallel$ and $\perp$ to the fiber axis (see text) [Robson et al. 1971; Bright 1979; Ohhashi et al. 1986]

Fermi level $E_F$. The first term in (8.17) comes from the $N = 0$ Landau level and the second term is related to the $N \neq 0$ levels. Figure 8.8 illustrates a comparison of the observed $g$-shift vs $1/T$ plotted for as-deposited pyrographite with theory (for different Fermi energies) given by (8.17) [Kawamura et al. 1983]. However, the agreement between theory and experiment in Fig. 8.8 is poor. The discrepancy possibly stems from the fact that the material is not a well-developed 2D graphite. In summary, the theory of the $g$-factor in carbons and also in 3D graphite still remains in an unsatisfactory state and an improved theory is needed to explain the experimental results in detail. Nevertheless, interpretation of the experiments in terms of the McClure-Yafet theory provides estimates for the various fiber parameters discussed above.

### 8.2.2 Anisotropy Effects

Robson et al. [1971] reported results of ESR studies on ex-PAN carbon fibers for various $T_{HT}$ (see Fig. 8.9). Although the observed ESR lines were often quite broad, the $g$-anisotropy behavior could be explained on the basis of an increase in the size of the graphite crystallites with increasing heat treatment temperature and motional averaging by the conduction electrons [Singer and Wagoner 1962]. With regard to heat treatment temperature, Fig. 8.9 shows that the anisotropy in the $g$-factor for the fibers increases dramatically with increasing $T_{HT}$ (i.e., increasing crystalline order), with increasingly larger anisotropies found in going from ex-PAN fibers [Robson et al. 1971] to ex-pitch fibers [Bright 1979] to CCVD fibers [Ohhashi et al. 1986] for a given value of $T_{HT}$. In Fig. 8.9 $(g_{3'} - g_{2'})$ is plotted vs $T_{HT}$, where $g_{3'}$ and $g_{2'}$ are respectively the $g$-factors $\perp$ and $\parallel$ to the fiber axis (see notation in Sect. 4.3.2) and

$$g_{2'} = g_1 + (g_3 - g_1)\langle \sin^2 \phi \rangle \quad ,$$
$$g_{3'} = [(g_1 + g_3) - (g_3 - g_1)\langle \sin^2 \phi \rangle]/2 \,\,,$$

(8.19)

where $g_1$ and $g_3$ are the single crystal $g$-factor values for $H \perp c$-axis and $H \parallel c$-axis, respectively, and $\phi$ is the misorientation angle given by Fig. 3.15 [i.e., $(\pi/2 - \phi)$ is the angle between the fiber axis and the $c$-axis of the crystallite]. Thus Fig. 8.9 clearly demonstrates the power of the ESR technique for characterization of the degree of crystalline alignment in carbon fibers, and shows that the CCVD fibers heat treated to 2800°C approach the $[(g_3 - g_1)/2] = 0.023$ value for graphite.

J.B. Jones and Singer [1982] carried out more detailed ESR studies of ex-PAN and mesophase pitch fibers heat treated at 3000°C. Figures 8.10a and 8.11a show the experimental ESR spectra of their pitch fibers, where the

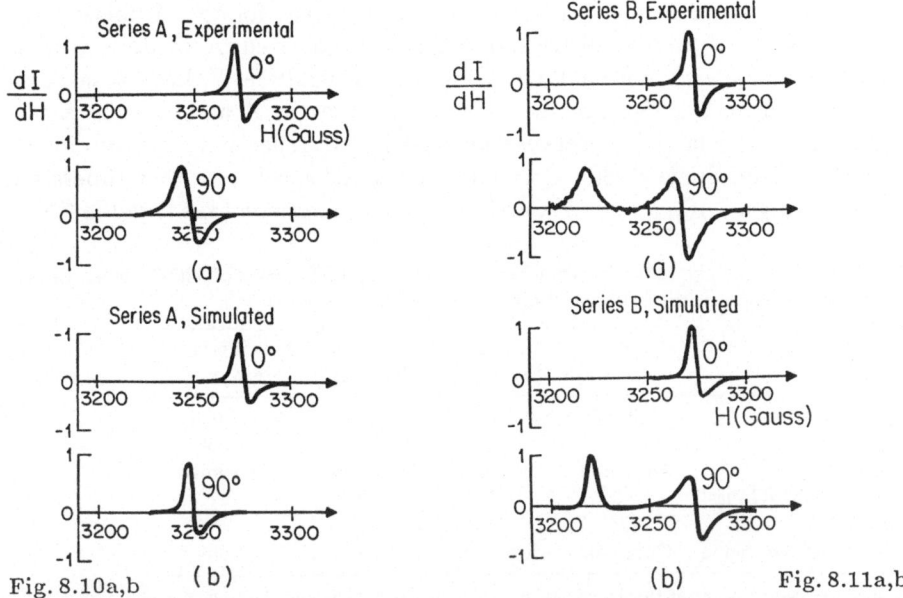

Fig. 8.10a,b (b)                                    (b)                    Fig. 8.11a,b

Fig. 8.10a,b. (a) Experimental ESR spectra of series $A$ pitch-based fibers with the fiber axis at 0° and 90° to the magnetic field. Curves are normalized for comparison with calculated spectra. Areas under the corresponding absorption curves are equal. (b) Normalized simulations of ESR spectra of series $A$ fibers for $\beta = 0°$ and $\beta = 90°$. Input parameters are listed in Table 8.3 [Jones and Singer 1982]

Fig. 8.11a,b. (a) Normalized experimental ESR spectra of series $B$ pitch-based fibers. The signal-to-noise ratio of the 90° spectrum was very poor due to the extreme anisotropy broadening. (b) Normalized simulations of ESR spectra of series $B$ fibers for $\beta = 0°$ and $\beta = 90°$. Input parameters are listed in Table 8.3 [Jones and Singer 1982]

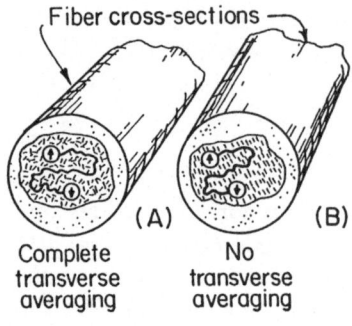

Fiber cross-sections

(A) Complete transverse averaging

(B) No transverse averaging

**Fig. 8.12.** Illustration of the effect of domain size for ordered spins on the transverse motional narrowing. Note that for (B), the spin alignment is very good within a single domain so that transverse averaging does not occur. The random arrangement of the spins in (A) results in complete transverse averaging [Jones and Singer 1982]

symbols 0° and 90° signify that the magnetic field is applied parallel and perpendicular to the fiber axis, respectively, and $\beta$ denotes the angle between the fiber axis and the magnetic field. It is of special interest that the spectra for $\beta = 0°$ and 90° are so different for a given fiber and that the spectra labeled 90° are completely different for fibers $A$ and $B$. These different properties are attributed to different structural features, as illustrated in Fig. 8.12. For the $B$ fibers, there are relatively large structurally coherent "domains" giving rise to the broad resonance for the $B$ series. In contrast, for the $A$ fibers, relatively small structurally coherent domains are indicated. Accordingly, in $A$, complete transverse motional averaging occurs, while in $B$, little transverse averaging occurs. Differences in the domain sizes in $A$ and $B$ were ascertained from the large differences observed in the reflectivity measurements [J.B. Jones and Singer 1982]. Computer-simulated spectra for these fibers are shown in Figs. 8.10b and 8.11b, and the fitting parameters are listed in Table

**Table 8.3.** Single "crystallite" parameters for calculated ESR curves of 3000°C heat treated PAN- and mesophase pitch-based fibers [Jones and Singer 1982]

|  | PAN | Pitch (Series $A$) | Pitch (Series $B$) | Single crystal[d] graphite |
|---|---|---|---|---|
| $g_1 = g_2$ [a] | 2.0027 | 2.0026 | 2.0026 | 2.0026 |
| $g_3$ | 2.028 | 2.035 | 2.036 | 2.050 |
| Line shape | Dysonian | Dysonian | Dysonian | Dysonian |
| Linewidth[b] [Gauss] | 4.2 | 4.5 | 4.8 | 3.7 |
| FWHM[c] | 12° | 7° | 7° | 0° |
| Motional averaging | Complete | Complete | None | None |

[a] $g_1 = g_2$ denotes the $g$-tensor in the layer plane and $g_3$ is the value along the $c$-axis.

[b] For the Dysonian shape, the peak-to-peak linewidth is that of a Lorentzian curve in the limit of infinite skin depth. For these spectra, the linewidth was assumed to be independent of crystallite orientation.

[c] FWHM (the full width half maximum intensity of the ESR line) is a parameter sensitive to the preferred orientation. It is expressed in terms of the orientation distribution of the crystallites with respect to the fiber axis.

[d] These values are from Wagoner [1960]. The experimental curves for the fibers are similar to the "slow diffusion" Dysonian line shape at 9 GHz. The entry of FWHM = 0° for single crystal graphite should be interpreted as FWHM $\ll 1°$.

8.3. Contributions from localized spins to the ESR absorption are negligible for the pitch-based fibers ($A$ and $B$) and for the 3000°C heat treated PAN-based fibers since the ratio of the ESR intensities at the temperatures 124 K and 300 K is almost unity (the resonance absorption due to localized spins is proportional to $1/T$). The computed curves (Figs. 8.10b and 8.11b) reproduce the experimental spectra well, indicating that ESR lineshapes can be explained for fibers using theoretical models.

### 8.2.3 Application of ESR to Vapor-Grown Carbon Fibers

Following the discussion of the use of ESR to characterize ex-polymer carbon fibers in Sect. 8.2.2, an application of the ESR technique has recently been made to vapor-grown fibers [Marshik et al. 1986]. The ESR spectra of the as-grown fiber are shown in Fig. 8.13, where the curves are identified by $\beta$, the angle between the fiber axis and the magnetic field. The positions of both resonances are very close to the free electron spin value, i.e., $g = 2.0023$, and the spectra exhibit a Lorentzian line shape. This implies that this resonance is due to localized spin centers and not due to conduction carriers (see Sect. 8.2 and Fig. 8.7). Figure 8.14 shows the ESR spectra of vapor-grown graphite fibers heat treated to $T_{HT} = 2250$°C. For $\beta = 0$ (Fig. 8.14a) the resonance is a typical Dysonian line shape, which is characteristic of conduction carriers (see Fig. 8.7), while for $\beta = 90°$ (Fig. 8.14b) a broad inhomogeneous line is observed with two resonances corresponding to $g_\parallel$ and $g_\perp$ (see Figs. 8.11a and 8.11b). These spectra are to be compared with Fig. 8.15, which shows the ESR spectra for fibers heat treated to high temperature $T_{HT} \geq 2900$°C. Figure 8.15 illustrates the high sensitivity of ESR to crystalline perfection for the small range of $T_{HT}$ at these very high temperatures. The narrowing of the spectra with increasing $T_{HT}$ is indicative of an increase in structural order. By analyzing these spectra, information is obtained on the order parameter which

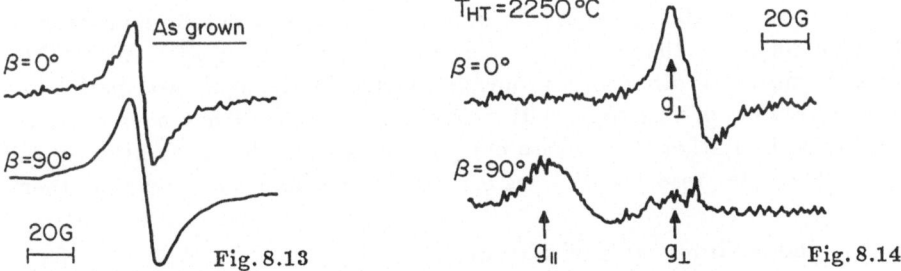

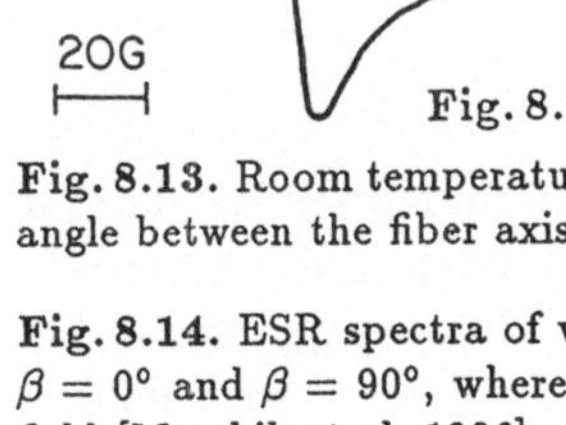

Fig. 8.13. Room temperature ESR spectra of as-grown CCVD fibers, where $\beta$ denotes the angle between the fiber axis and the magnetic field [Marshik et al. 1986]

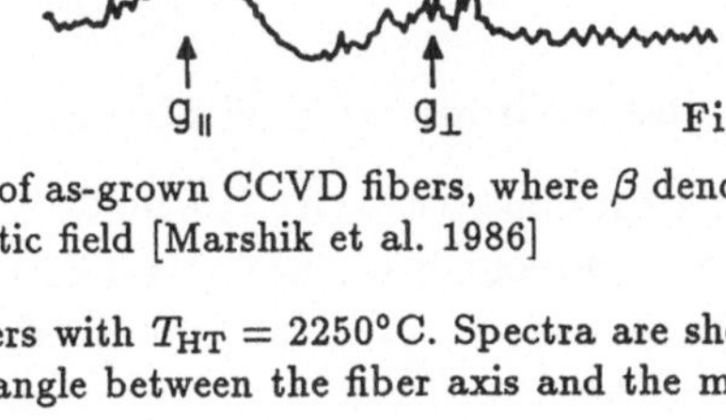

Fig. 8.14. ESR spectra of vapor-grown fibers with $T_{HT} = 2250$°C. Spectra are shown for $\beta = 0°$ and $\beta = 90°$, where $\beta$ denotes the angle between the fiber axis and the magnetic field [Marshik et al. 1986]

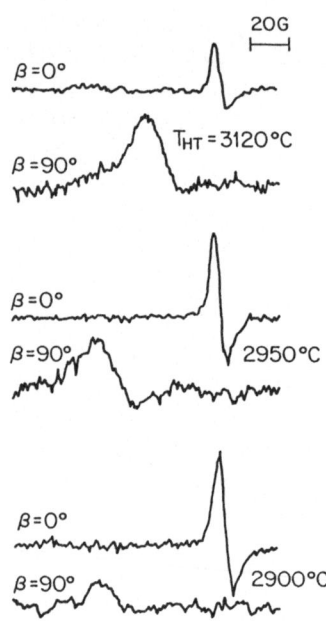

Fig. 8.15. ESR spectra of various vapor-grown fibers with $T_{HT}$ near 3000°C. Spectra are shown for both $\beta = 0°$ and $\beta = 90°$ and cover a relatively small $T_{HT}$ range [Marshik et al. 1986]

specifies the mean deviation of the $a$-axis of the crystallites from the fiber axis [Marshik et al. 1986]. Although the theory for the $g$-factor is unsatisfactory, a great deal of useful information can nevertheless be extracted from ESR measurements. This technique is especially useful when combined with other independent characterization techniques.

## 8.3 Electrical Resistivity

The electrical resistivity of carbon fibers has been studied extensively because of the usefulness of employing parameters such as the room temperature resistivity, $\varrho(300\,\text{K})$, or the resistivity ratios $\varrho(300\,\text{K})/\varrho(77\,\text{K})$ or $\varrho(300\,\text{K})/\varrho(4.2\,\text{K})$ as a diagnostic of other fiber properties, including the structural perfection. A brief review of this topic with references up to 1979 has been given by Spain [1981]. Rather than review every publication in detail, an attempt will be made in the present work to assess the major trends and to explain them.

### 8.3.1 Experimental Techniques

The longitudinal resistivity is usually measured using a four-point method, in which two current leads are attached to the ends of a single fiber, or bundle of fibers, and two potential probes are introduced at about a quarter of the length of the fiber from each end. Contacts are usually applied with gold or silver epoxy. Great care must be taken to ensure that the fiber is not damaged by the application of leads, nor put into a state of strain, either by applying contacts, or through thermal expansion when the fiber is locally heated in applying leads.

Fiber diameters can be measured with an optical microscope for thick fibers, and a scanning electron microscope for thinner ones. Caution must be exercised, since fibers are often non-circular in cross section.

Noise is generated both in the fiber and at the contacts. This noise is orders of magnitude above the level of Johnson noise (Sect. 8.9). This noise is more problematical when small changes arising from magnetoresistance or pressure, for example, are being studied. The noise level recorded by the measuring instrument can often be reduced by using an ac bridge technique with small bandwidth.

Very high frequencies are required to couple electromagnetic radiation to small fibers. A technique for contactless resistivity measurements on 10 $\mu$m diameter fibers at 10 GHz has been described by Azzeer et al. [1985].

Fibers can be heated readily by electrical currents, and this can lead to erroneous resistivity values. Even pulsed currents can heat fibers appreciably, since their heat capacity is low. It is thus advisable to immerse the fiber in a boiling fluid to ensure good thermal contact for measurements at fixed temperature, or in an exchange gas when varying the temperature. Even in this case, excess currents can heat the fiber by hundreds of degrees [Spain et al. 1983a]. Such temperature excursions can be achieved with much lower power levels in a vacuum [Chung and Wong 1986], and result in large changes in electrical resistance.

### 8.3.2  Electrical Resistivity Data

Curves of the electrical resistivity as a function of temperature are very useful for categorizing the lattice perfection of the fiber. Such resistivity curves are very similar for widely different classes of carbon-based materials (see Spain 1981 for a review.) Figure 8.16 illustrates a set of $\varrho(T)$ curves for ex-PAN fibers and bulk heat treated carbons derived from anthracene, while Fig. 8.17 illustrates similar curves for ex-mesophase pitch fibers, arc-grown whiskers, and single crystal graphite. Data for the whiskers are corrected by a factor of 10 from the original publication [R. Bacon 1960]. Figure 8.18 compares data for a set of heat treated benzene-derived CCVD fibers with single crystal graphite. These results can be discussed in terms of three distinct regimes:

1. Fibers of the highest perfection have resistivity curves approaching that of single crystal graphite. It will be shown below that these results can be interpreted using the graphite band model. The effects of electron scattering by defects must be considered in addition to the intrinsic scattering by phonons. Curves in Figs. 8.16-8.18 with room temperature resistivities lying below $\sim 5 \times 10^{-6}$ $\Omega$m fall into this category.

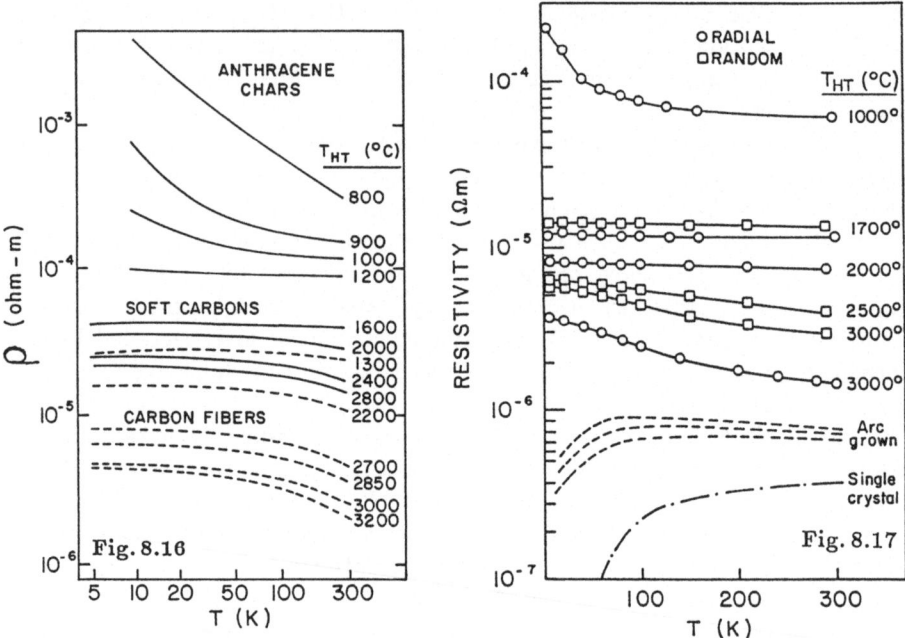

**Fig. 8.16.** Temperature dependence of the resistivity of ex-PAN fibers [Spain et al. 1983a] compared to bulk and heat treated carbons (ex-anthracene) [Mrozowski 1952; Griffith and Gayley 1966]

**Fig. 8.17.** Plots of the resistivity of ex-mesophase pitch fibers (labeled by their $T_{HT}$ values in °C) as a function of temperature [Bright 1979], compared to arc-grown whiskers [R. Bacon 1960], and single crystal graphite [Soule 1958]

2. Partially carbonized carbons and fibers have resistivity curves which show increasing resistivity with decreasing $T$ at lower temperatures. Such material is exemplified in Fig. 8.16 by anthracene chars heat treated below ~1000°C, and by the ex-pitch fiber with $T_{HT} \leq 1000$°C in Fig. 8.17. Resistivity values are above ~$10^{-4}$ Ωm. An excitation gap between bonding and anti-bonding states produces the decrease of conducting carriers as the temperature is lowered. Since this effect arises from the presence of hydrocarbons and other molecular impurities, these phenomena will not be considered further.

3. In between these two extremes ($5 \times 10^{-6} < \varrho < 10^{-4}$ Ωm) lie curves which are nearly independent of temperature, and which shift only weakly with heat treatment temperature. The fundamental condition $kl > 1$ for using the Drude formula (see Sect. 7.1) is not satisfied for these fibers. Also, the disorder in the fiber producing the short mean free path, $l$, does not result in a hopping-type conductivity fitting the Mott [1967] form:

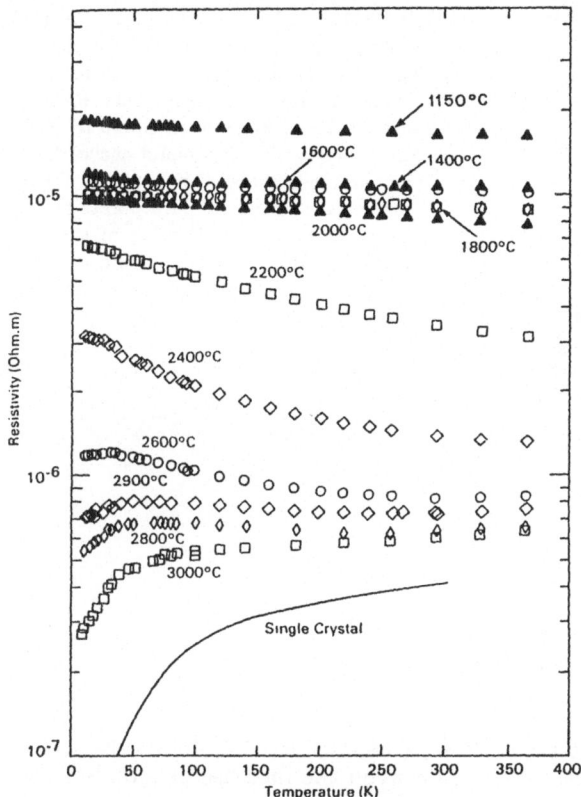

**Fig. 8.18.** The resistivity of benzene-derived fibers plotted as a function of temperature. Curves are drawn for fibers with several heat treatment temperatures as indicated in the figure (including the as-grown fiber labeled 1150°C), and the results are compared to single crystal graphite [Heremans 1985]

$$\ln \sigma \propto T^{-1/4} \quad . \tag{8.20}$$

This remains a region in which only qualitative models for the transport properties have been developed.

It is to be noted that none of the resistivity curves completely attain the low values for single crystal graphite. This results from the residual defects remaining in the fibers. For the same value of heat treatment temperature, the curves lie lower for CCVD fibers than for ex-pitch, and ex-PAN fibers, in that order, reflecting the relative degree of difficulty to graphitize these carbons. This is represented in Fig. 8.19, where room temperature resistivities are plotted as a function of heat treatment temperature $T_{HT}$. The curve for ex-rayon fibers lies above that of ex-PAN mainly because of the higher preferred orientation in the latter (see Sect. 3.4.3). However, the lower values for ex-mesophase pitch and CCVD fibers are related to their enhanced tendency to

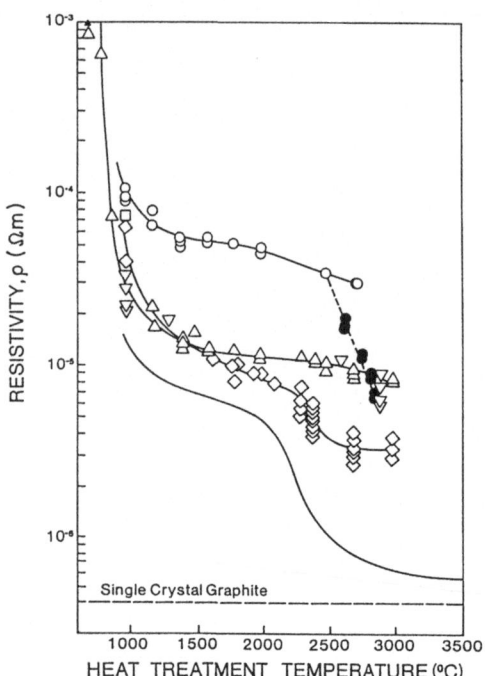

**Fig. 8.19.** Schematic variation of the room temperature electrical resistivity vs $T_{HT}$ for: (○) ex-rayon; (●) hot stretched rayon; (△ and ▽) ex-PAN fibers [Robson et al. 1972]; (◇) ex-pitch fibers [Bright and Singer 1979] and (solid curve) benzene-derived fibers [Chieu et al. 1983]. The scatter of typical data points about the mean give an indication of the uncertainty. The dashed line indicates the decrease in resistivity produced by hot stretching the ex-rayon fibers

graphitize. The relatively flat portions of the curves occurring around $T_{HT} = 1500°$C arise because of a near balance between the increase of mobility as $l$ increases with $T_{HT}$, and a decrease in the carrier density. This latter decrease will be related to collision broadening of the energy levels in the next section. The plateau in $\varrho(T)$ which is found in Figs. 8.16-8.18 is also observed in bulk heat treated carbons [Mrozowski 1952] and pyrocarbons [Klein 1964].

Another feature worthy of note in Fig. 8.18 is the presence of two regions of $T_{HT}$ for which the resistivity vs $T$ curves of CCVD fibers are almost flat. The curves above 2800°C are nearly independent of temperature because the carrier density increases, and the mobility decreases with $T$, so that the two effects nearly cancel each other out. This is discussed further in the next section. The curves for $T_{HT}$ below 2000°C are flat because both the carrier density and mobility are approximately independent of temperature in this range. However, our lack of understanding of the transport processes in this region is underscored by the observation of an unusual resistivity maximum near ~20 K for ex-PAN fibers with $T_{HT} \sim 1300°$C, as shown in Fig. 8.20 [Spain et al. 1983b; Goldberg et al. 1983]. Further data on the corresponding anomalous magnetoresistance of these fibers is presented in Sect. 8.4. No explanation for this maximum in the resistivity has been presented so far.

Endo et al. [1983a] have obtained data on graphitic CCVD fibers for measurement temperatures up to to 1100 K, as illustrated in Fig. 8.21. In this

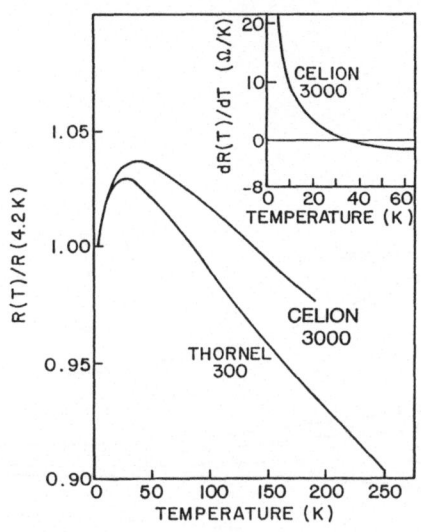

◄ **Fig. 8.20.** The variation with temperature of the reduced resistivity of ex-PAN fibers showing anomalous behavior at low temperature. The inset shows the temperature dependence of $dR/dT$ below 60 K [Spain et al. 1983b]

▼ **Fig. 8.21.** Linear plot of the temperature dependence of the resistivity for benzene-derived graphite fibers ($T_{HT} = 2900°C$). Results are shown both for a pristine fiber in the range $T < 1100$ K, and for the intercalated fibers (lower right-hand inset) in the range $T < 600$ K [Endo et al. 1983a]

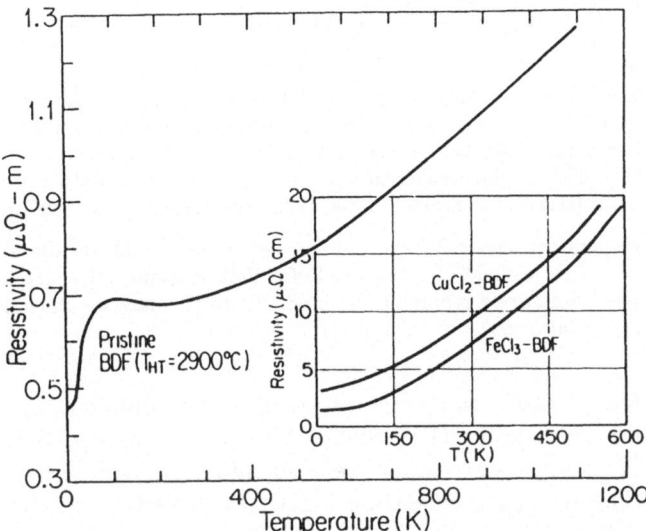

figure, the resistivity reaches a very shallow minimum at 225 K, remaining almost temperature independent between 100 K and 400 K and then increasing at higher temperatures. This increase of the resistivity at higher temperature is a feature of many types of graphitic carbons (see Spain 1981), and is attributable to a more rapid decrease of the phonon-dominated mobility with temperature than the approximately linear increase of carrier density.

The room temperature resistivity, $\varrho(300\,\text{K})$, the temperature coefficient of resistivity or the ratio $\varrho(300\,\text{K})/\varrho(77\,\text{K})$ are valuable parameters for correlating a variety of fiber properties. This was first pointed out by Ezekiel

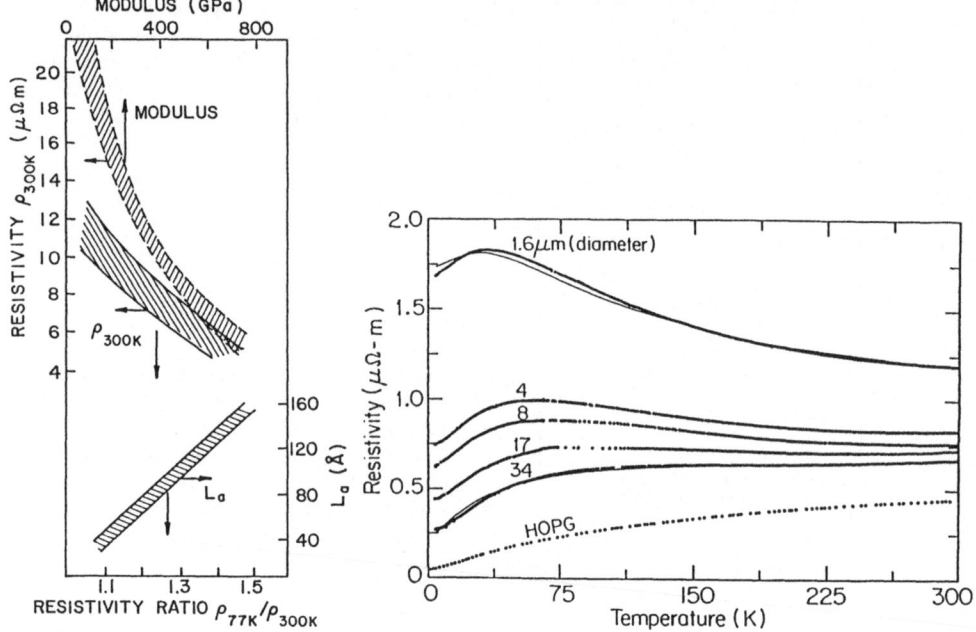

**Fig. 8.22.** Sketch of the variation of fiber resistivity $\varrho(300\,\text{K})$ with elastic modulus. Data from ex-polymer fibers prepared from several precursors fall within the shaded area. Also shown are correlations between the room temperature resistivity $\varrho(300\,\text{K})$ and the resistivity ratio $\varrho(77\,\text{K})/\varrho(300\,\text{K})$. The in-plane coherence length $L_a$ vs the resistivity ratio $\varrho(77\,\text{K})/\varrho(300\,\text{K})$ is also shown [Data from Ezekiel (1969, 1970) and Robson et al. (1973)]

**Fig. 8.23.** Resistivity vs temperature curves for five vapor-grown fibers of different diameters 1.6, 4.8, 17 and 34 $\mu$m ($T_{\text{HT}} = 3500°\text{C}$). The result for bulk graphite (HOPG) is included for comparison. The solid curves represent the model fit to the data for the 1.6 and 34 $\mu$m fibers (from Tahar et al. 1986)

[1969, 1970] who plotted $\varrho(300\,\text{K})$ from several precursors of different $T_{\text{HT}}$ as a function of the elastic modulus (Fig. 8.22). Also plotted in Fig. 8.22 for the same set of fibers is the correlation between $\varrho(300\,\text{K})$ and the ratio $\varrho(77\,\text{K})/\varrho(300\,\text{K})$. The in-plane correlation length $L_a$ is further plotted vs $\varrho(77\,\text{K})/\varrho(300\,\text{K})$ in Fig. 8.22. Robson et al. [1973] developed this idea by presenting correlations between $\varrho(77\,\text{K})/\varrho(300\,\text{K})$ and the ESR $g$-factor anisotropy, thermoelectric power, transverse magnetoresistance, and interlayer spacing. Correlations of these properties with the resistivity were statistically superior (by nearly an order of magnitude) than those using only $T_{\text{HT}}$. From this, it is clear that the resistivity and its temperature dependence are probing a fundamental structural property of the fibers.

One of the reasons that the magnitude of $T_{\text{HT}}$ is not a good parameter with which to correlate other properties is that other variables such as the residence time at $T_{\text{HT}}$, the nature of the precursor material and the size of the fibers need to be taken into account. Figure 8.23, for example, illustrates resistivity curves for benzene-derived fibers heat treated at 3500°C for one hour, with diameters ranging from 1.6 to 34.0 $\mu$m [Tahar et al. 1986], showing that for the same heat treatment conditions, a variety of $\varrho(T)$ behaviors are observed, depending on the fiber diameter. These results are considered in greater detail in Sect. 8.3.4.

### 8.3.3 Simple Two-Band Model

The simple two-band (STB) model was first used by Klein [1962, 1964, 1966] to explain galvanomagnetic data for pyrolytic carbons. There are several approximations which can be used, but all start with four parameters: the electron and hole mobilities, $\mu_n$ and $\mu_p$, and the corresponding carrier densities $n$ and $p$. For simplicity the approximation $\mu_n = \mu_p = \bar{\mu}$ is employed. In the STB model, the electron and hole energies are represented by

$$E_e = \frac{\hbar^2}{2m_n^*}k^2 , \quad E_h = \Delta - \frac{\hbar^2}{2m_p^*}k^2 , \qquad (8.21)$$

where $k^2 = k_x^2 + k_y^2$ while $\Delta$ denotes the band overlap, and the zero of energy is taken at the conduction band minimum. Then $n$ and $p$ take the form

$$n = C_n k_B T \ln[1 + \exp(E_F/k_B T)] ,$$
$$p = C_p k_B T \ln\{1 + \exp[(\Delta - E_F)/k_B T]\} , \qquad (8.22)$$

where

$$C_n = (2m_n^*)/[\pi \tilde{c} \hbar^2] , \quad C_p = (2m_p^*)/[\pi \tilde{c} \hbar^2] , \qquad (8.23)$$

$E_F$ is the Fermi energy and $\tilde{c}$ is the interplanar spacing (3.35 Å in graphite). If we assume that $n = p$, and $C_n = C_p$, then it follows that $E_F$ is at midgap; i.e., $E_F = \Delta/2$. In this case the total carrier concentration is given by

$$N = n + p = C k_B T \ln[1 + \exp(E_F/k_B T)] , \qquad (8.24)$$

where $C = 2C_n = 2C_p$.

Next, we consider in detail the temperature dependence of the mobility and the zero field resistivity of the graphite fiber. Usually the total mobility $\mu(T)$ is represented by Matthiessen's rule

$$\frac{1}{\mu(T)} = \frac{1}{\mu_{BS}} + \frac{1}{\mu_{ph}(T)} , \qquad (8.25)$$

where $\mu_{BS}$ is the temperature-independent boundary scattering term and $\mu_{ph}(T)$ is the temperature-dependent phonon scattering term. Figure 8.24 shows that a good fit to the experimental zero field resistivity curve of $\varrho$ vs $T$ for the vapor-grown fibers (circles) with $T_{HT} \sim 3000°$C is obtained with the two band model (solid curve) using the parameters $N(T = 0) = 3 \times 10^{24}/\text{m}^3$,

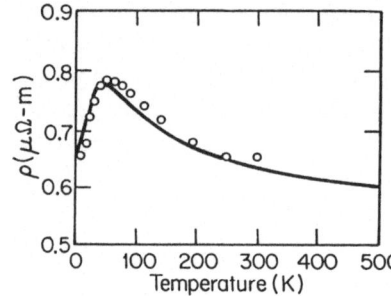

Fig. 8.24. The zero field resistivity $\varrho$ vs $T$ for the vapor-grown fibers (*circles*) with $T_{HT} = 3000°C$ [Chieu et al. 1983] and the theoretical curve (*solid line*) due to Woolf et al. [1984]

Table 8.4. Parameters obtained by fitting resistivity data to a simple two-band model for vapor-grown fibers[a]

| $T_{HT}$ | $t_R$[min] | $\varrho_0[\mu\Omega m]$ | $l_{BS}$ [Å] | $\sqrt{\mu_e\mu_h}$ | $(n+p)[10^{24}/m^3]$[b] |
|---|---|---|---|---|---|
| 3000 | 3 | 0.26 | 6200 | 1.5 | 5.5 |
| 2900 | 14 | 0.68 | 2440 | – | 5.4 |
| 2800 | 16 | 0.53 | 3040 | 1.3 | 4.95 |
| 2600 | 42 | 1.15 | 1400 | 0.66 | 5.5 |
| 2400 | 36 | 3.2 | 500 | 0.33 | 5.6 |
| 2200 | 30 | 6.6 | 240 | 0.09 | 5.6 |
| 2000 | 30 | 9.7 | 180 | – | 5.1 |
| 1800 | 30 | 10.1 | 160 | – | 5.5 |
| 1600 | 25 | 11.2 | 140 | – | 5.7 |
| 1400 | 20 | 11.5 | 140 | – | 5.5 |
| As grown | – | 18.4 | 90 | – | 5.4 |

[a] [Heremans 1985]. Here $t_R$ denotes the residence time of the fiber at temperature $T_{HT}$, while $l_{BS}$ denotes the mean free path due to boundary scattering deduced from x-ray linewidths. The electron and hole mobilities $\mu_e$ and $\mu_h$ are obtained by fitting to a two-band model.
[b] The carrier density $(n + p)$ is estimated from the Drude model relation $(1/\varrho_0) = (n+p)e^2 l_{BS}/(m^* v_F)$ where $m^* = 0.025 m_0$ and $v_F = 10^6$ cm/s are assumed.

$E_F = 0.0053$ eV, $\mu_{BS} = 3.3$ m$^2$/Vs, and $\mu_{ph}(T) = 3.4 \times 10^2/T$ m$^2$/Vs [Woolf et al. 1984] and the mean free path $l_{BS}$ of the carriers was identified with boundary scattering and evaluated from x-ray linewidth data (see Table 8.4). The band overlap $\Delta = 2E_F \approx 0.01$ eV may be compared to $2|\gamma_2| = 0.04$ eV for single crystal graphite [Dresselhaus and Dresselhaus 1981]. Reasonable values for $N(T = 0)$ and $\mu_{ph}(T)$ are obtained from the simple two-band model, compared with those for single crystal graphite [Spain 1973; Ono and Sugihara 1966].

Another approach was used by Spain et al. [1983a] who fitted resistivity data for ex-PAN fibers. These authors obtained least-squares fit values for $E_F$ and $\Delta$ from the temperature dependence of the resistivity assuming equal electron and hole mobilities. Then, a value for the effective mass

**Table 8.5.** Parameters obtained by fitting resistivity data[a] for ex-PAN fibers to a two-band model, and comparing with lifetime-broadening [Spain et al. 1983a]

| $T_{HT}$ [°C] | $\Delta$ [eV] | $E_F$ [eV] | $p$ [$10^{24}$ m$^{-3}$] | $v_F$ [$10^6$m/s] | $l$ [nm] | $\tau$ [$10^{-14}$s] | $\Delta E$ [eV] | $k_F l$ |
|---|---|---|---|---|---|---|---|---|
| >3000 | 0.025 | −0.002 | 2.7 | 0.87 | 30 | 3.4 | 0.019 | 2.70 |
| 3000 | 0.029 | −0.002 | 3.2 | 0.94 | 23 | 2.45 | 0.027 | 2.23 |
| 2850 | 0.043 | −0.003 | 4.7 | 1.14 | 13 | 1.14 | 0.057 | 1.53 |
| 2700 | 0.052 | −0.001 | 5.4 | 1.23 | 10 | 0.81 | 0.080 | 1.27 |
| 2200 | 0.134 | −0.002 | 13.6 | 1.95 | 4 | 0.2 | 0.32 | 0.81 |

[a] Residence time at $T_{HT}$ is $t_R = 30$ min. A value of $m^*/m_0 = 0.012$ is used in this analysis. The various parameters are defined by:
$\Delta$ = conduction-valence band overlap,
$E_F$ = Fermi level at 0 K measured with respect to bottom of conduction band,
$p$ = hole density at 0 K (the electron density $n = 0$ in all cases at 0 K),
$v_F$ = Fermi velocity,
$l$ = carrier mean free path,
$\tau$ = carrier relaxation time ($= l/v_F$),
$\Delta E = \hbar/\tau$ = lifetime energy broadening,
$k_F$ = wave vector at the Fermi level.

($m^* = 0.012m_0$) was deduced from the magnitude of the resistivity of the fiber with $T_{HT} = 2200°C$, using the value of 4 nm for the mean free path due to boundary scattering ($l_{BS}$) deduced from x-ray measurements. Data for a variety of ex-PAN fibers are given in Table 8.5 from which it can be seen that the band overlap $\Delta$ increases with decrease of $T_{HT}$. The procedure used in this analysis, like all adaptations of the STB, is somewhat arbitrary. However, these authors used several different approximations to fit the data, and showed that the same trends occurred for all approximations. Although the effective mass is about half that of graphite, the Fermi velocity is close to the value of $10^6$ m/s characteristic of turbostratic carbon, see (7.5).

An explanation was forwarded to explain the increase in the band overlap with decreasing $T_{HT}$, based on the inhomogeneity of the material [Spain et al. 1983a]. However, Goldberg [1985] showed that good agreement could be obtained by assuming that the band overlap arose from lifetime-broadening of the energy levels (see Sect. 7.1). Values of $\Delta E = \hbar/\tau$ are also listed in Table 8.5, and can be compared to $\Delta$. Both lifetime broadening and inhomogeneity effects are probably important in producing the observed trend in the apparent band overlap with decreasing $T_{HT}$.

Although the STB model successfully accounts for the observed resistivity behavior of the vapor-grown fibers annealed at 3000°C using an appropriate set of the parameters, we should, however, bear in mind that the STB model is not applicable in all cases. The STB model cannot account for the size effect of the resistivity in CCVD fibers [Tahar et al. 1986] discussed below. Also, the model cannot be used to describe the resistivity of low-$T_{HT}$ samples. This is illustrated in Table 8.5, where values of the dimensionless parameter $k_F l$ are listed for each fiber. The values of $k_F l$ tend to less than unity for lower $T_{HT}$ values, which indicates that band models are inadequate to explain transport phenomena (see Sect. 7.1).

### 8.3.4 Size Effects in the Resistivity of Vapor-Grown Fibers

The resistivity of vapor-grown graphite fibers exhibits a striking size dependence [Tahar et al. 1986]. Resistivity versus temperature curves for vapor-grown fibers ($T_{HT} = 3500°C$) with different fiber diameters are shown in Fig. 8.23. Below 40 K the resistivity data are well fit by the functional form

$$\varrho(T) = \varrho_0 + \varrho_1 T \quad , \tag{8.26}$$

where $\varrho_0$ and $\varrho_1$ are temperature-independent constants. Whereas $\varrho_1$ shows no observable diameter dependence, $\varrho_0$ increases linearly with $1/d$ for $d > 5\,\mu m$ according to the relation

$$\varrho_0(d) = \alpha_0 + \beta_0/d \quad . \tag{8.27}$$

A fit to the measured temperature dependence of the resistivity was made in Fig. 8.23 for each of the fiber diameters, using the simple two-band model. The results of the fit are shown explicitly by the solid curves for the smallest and largest diameter fibers. From the analysis of these results, the value of the in-plane mean free path $l$ is found to decrease with decreasing fiber diameter $d$, becoming approximately equal to $d$ for fibers with $d \leq 5\,\mu m$. The fitting procedure confirmed that the mobility for holes and electrons is essentially the same, $\mu_e = \mu_h = \mu$, for each of the fiber diameters, though $\mu$ itself decreases significantly with decreasing $d$. The analysis also indicated that the Fermi level for these fibers is approximately a factor of 2 less than for bulk graphite, indicating a corresponding decrease in carrier density.

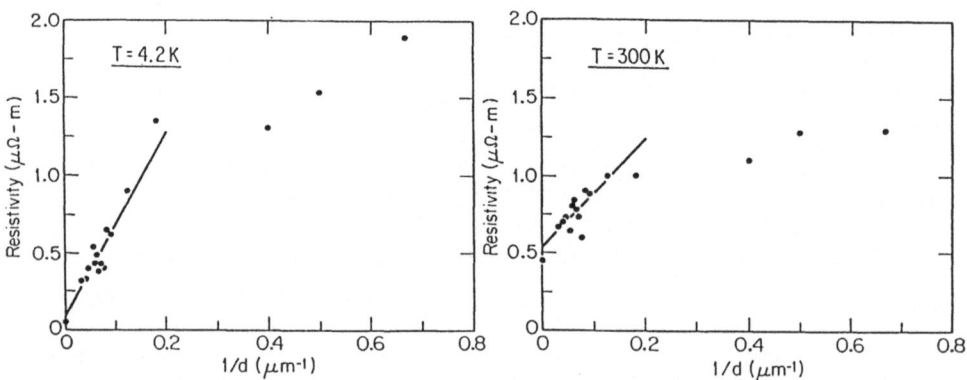

**Fig. 8.25.** Plot of resistivity $\varrho$ vs inverse fiber diameter for $T = 4.2$ K and $T = 300$ K for vapor-grown fibers heat treated at $T_{HT} = 3500°C$ (from Tahar et al. 1986)

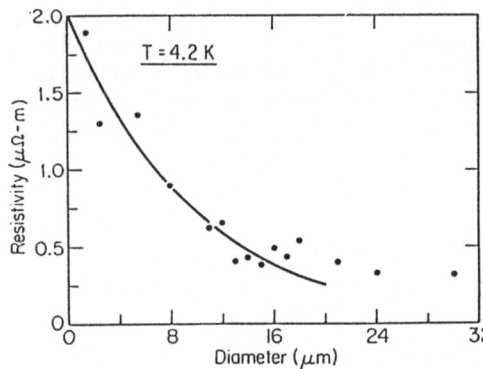

Fig. 8.26. Observed dependence of the resistivity on fiber diameter at 4.2 K for vapor-grown fibers with $T_{HT} = 3500°C$ [Tahar et al. 1986]. The solid line is a guide to the eye and shows an approximately exponential dependence

For larger $1/d$ values (or $d < 5\,\mu$m) it was found that $\varrho(d)$ increases more slowly with $1/d$ (see Fig. 8.25) and the size dependence can be represented by an exponential factor

$$\varrho(d) = \varrho(0)e^{-d/\cdot} \quad , \tag{8.28}$$

as shown in Fig. 8.26, where $d_0$ exhibits a weak temperature dependence

$$d_0 = 10 + 6.455 \times 10^{-3}T \tag{8.29}$$

for $d_0$ in units of micrometers, thus accounting for the difference in behavior observed at $T = 4.2$ K and $T = 300$ K as shown in Fig. 8.25 [Tahar 1985; Tahar et al. 1986].

It is reasonable to assume that for very thin fibers, the crystallite size, which is responsible for the boundary scattering of carriers, is proportional to the fiber diameter. Accordingly, Matthiessen's law for this case implies that

$$\frac{1}{\mu} = \frac{1}{\mu_I} + \frac{1}{\mu_{BS}} + \frac{1}{\mu_{ph}} \tag{8.30}$$

or in terms of the mean free path

$$\frac{1}{l} = \frac{1}{l_I} + \frac{1}{l_{BS}} + \frac{1}{l_{ph}} \quad , \tag{8.31}$$

where the subscript I denotes the impurity and/or defect scattering. Here $1/\mu_{BS}$ or $1/l_{BS}$ correspond to the boundary scattering term $\beta_0/d$ in (8.27) and $l_{BS}$ is given by the crystallite dimension perpendicular to the fiber axis [Tahar et al. 1986].

Analysis of the experimental $\varrho(T)$ curves on the basis of the simple two-band model indicates that $l_{BS}$ is approximately equal to $L_a$. As will be shown in the following, (8.31) is valid only if the condition

$$\frac{1}{l_I} + \frac{1}{l_{ph}} \gg \frac{1}{l_{BS}} \tag{8.32}$$

is satisfied. As is shown in the following, Fig. 8.25 indicates that for $d < 5\,\mu$m,

199

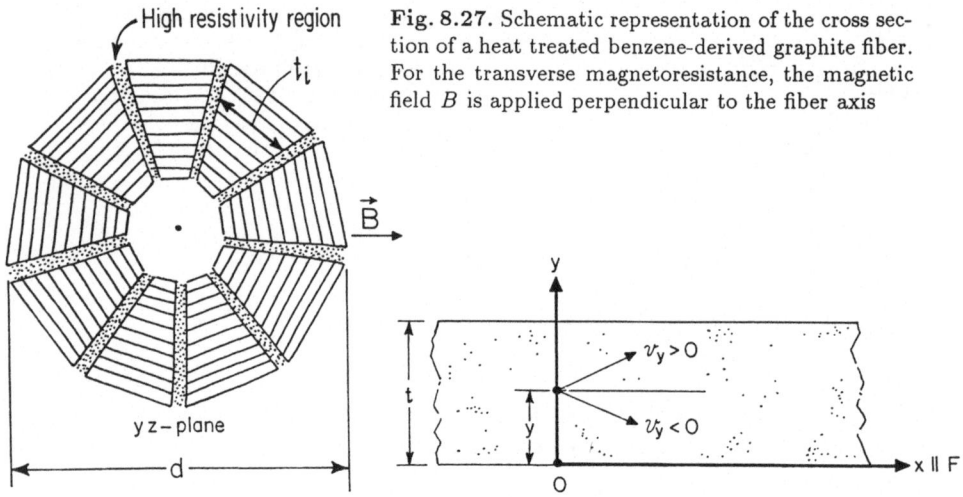

High resistivity region

Fig. 8.27. Schematic representation of the cross section of a heat treated benzene-derived graphite fiber. For the transverse magnetoresistance, the magnetic field $B$ is applied perpendicular to the fiber axis

Fig. 8.28. Geometry of the size effect calculations. The $xy$-plane is the carbon layer plane and the $x$-axis is parallel to the fiber axis and the electric field $F$

the inequality in (8.32) does not hold. To understand the case where boundary scattering is important, we must solve the Boltzmann equation, correctly taking into account the boundary conditions [Fuchs 1938; Chambers 1950; Sondheimer 1952].

The following calculation summarizes an application of these theories to the case of a two-dimensional conductor, which approximates conduction along the fiber axis of a highly graphitized vapor-grown fiber (see Fig. 8.27). An electric field is applied along the $x$-axis which is chosen parallel to the fiber axis. The total conductivity $\sigma$ is assumed to be given by a sum over the contributions from each layer $\sigma = \sum_i \sigma_i t_i / \sum_i t_i$. Consider a carbon layer with a thickness $t$ (see Figs. 8.27 and 8.28). The Boltzmann equation for transport in an electric field $F$

$$-eFv_x\frac{\partial f_0}{\partial E} + v_y\frac{\partial f}{\partial y} = -\frac{f - f_0}{\tau} \cdot \tag{8.33}$$

has a solution

$$f_1 \equiv f - f_0 = eF\tau v_x \frac{\partial f_0}{\partial E}\left[1 + \phi(v)\exp(-y/\tau v_y)\right] \quad , \tag{8.34}$$

where the function $\phi(v)$ is determined by the boundary conditions at the surface $(y = \pm t/2)$. These equations have two different solutions $\phi^+$ and $\phi^-$ according to whether $v_y$ is positive or negative:

$$\phi^+(v) = -(1-p)/(1 - pe^{-t/v_y\tau}), \qquad (v_y > 0)$$

$$\phi^-(v) = -e^{t/\tau v_y}(1-p)/(1 - pe^{t/\tau v_y}), \qquad (v_y < 0) \tag{8.35}$$

where $p$ is the coefficient of specular reflection at the surface and $(1-p)$ is the coefficient of diffuse scattering. For simplicity, the electron and hole bands are assumed to be represented by the simple two-band model of (8.21) where $k^2 = k_x^2 + k_y^2$ and the Fermi energy is $E_F = \Delta/2$ (see Sect. 8.3.3). At low temperature $(E_F \gg k_B T)$, the conductivity associated with one layer takes the form

$$\frac{\sigma}{\sigma_0} = 1 - \frac{4(1-p)}{\pi\lambda} \int_0^1 dx\; x\sqrt{1-x^2}\;\frac{1-e^{-\lambda/x}}{1-pe^{-\lambda/x}} \quad , \tag{8.36}$$

where $\lambda = t/l$ and $l = v\tau$ while $\sigma_0$ is the conductivity for $\lambda \to \infty$. Clearly, (8.36) implies that $(\sigma/\sigma_0)$ is equal to unity for $p = 1$ (specular reflection) and is equal to zero for $\lambda \to 0$. Figure 8.29(a) illustrates the calculated $\varrho = 1/\sigma$ versus $1/\lambda$ for $p = 0.2$.

From Fig. 8.27, the simplest assumption for the functional form for the width of a carbon layer in the fiber $t(d)$ is

$$t = \gamma d \quad , \tag{8.37}$$

where $\gamma = $ constant. Thus, comparison of the model calculation given in Fig. 8.29(a) qualitatively explains the observed size-dependent measurements (Fig. 8.25), especially the deviation of the measurements from the curve $\varrho(d) = \alpha_0 + \beta_0/d$ for large $1/d$ values. Figure 8.29(b) shows the corresponding $\varrho$ versus $\lambda$ curves for $p = 0.2$, which is appropriate to the experimental data shown in Fig. 8.26.

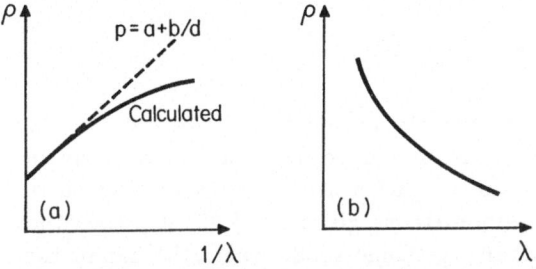

**Fig. 8.29a,b.** (a) Calculated resistivity $\varrho$ vs $1/\lambda$, where $\lambda = t/l$ and $t$ is the length of a planar segment (see Fig. 8.27), $l$ being the mean free path, and (b) calculated resistivity $\varrho$ vs $\lambda$ for the coefficient of specular reflection $p = 0.2$ [Sugihara 1986b]

So far our discussions have been limited to a single layer with width $t$. Since $t$ is different from layer to layer (see Fig. 8.27) in the vapor-grown fibers, we must sum up the contributions from the various layers

$$\frac{\varrho}{\varrho_0} = \frac{\sigma_0}{\sum_i \sigma_i r_i} \quad , \tag{8.38}$$

where $r_i$ is the reduced thickness $r_i = t_i/\sum_j t_j$, and $\sigma_i$ in (8.38) is given by $\sigma$ in (8.36). In spite of the crudeness of this model, it provides a qualitative explanation for the observed size-dependent resistivity data in CCVD filaments [Tahar et al. 1986]. The calculation further shows that deviations from the simple two band (STB) formula (8.31) become appreciable for $1/l_\mathrm{I} + 1/l_\mathrm{ph} \sim 1/l_\mathrm{BS} \sim 1/t$ or $\lambda \sim 1$, and therefore the more detailed solution (8.36) must be used for $\lambda < 1$.

Of particular interest is the behavior of submicrometer fibers. According to the schematic model of Fig. 8.27, the graphitization process associated with high heat treatment temperatures $T_\mathrm{HT}$ is modeled by the formation of graphitic segments of width $t$. For low $T_\mathrm{HT}$, the segment width $t$ is very small and an approximate circular cross section results. For high $T_\mathrm{HT}$, the segment width $t$ becomes relatively large. As the fiber diameter decreases, the approximation of segmenting the circle by graphitic segments (see Fig. 8.27) must eventually break down. Thus the core region of a vapor-grown fiber can at best approximate a 2D turbostratic graphite. See Sect. 2.2.4.

## 8.4 Magnetoresistance

Low field magnetoresistance measurements have been used extensively to monitor carrier mobilities in various forms of graphite and also in graphite intercalation compounds [Dresselhaus and Dresselhaus 1981]. Typically for metallic conductors, the Hall effect is sensitive to the carrier density, the magnetoresistance to the carrier mobility, and the resistivity to a product of the carrier density and carrier mobility. Thus, the magnetoresistance, which for simple metals is written as

$$\frac{\Delta\varrho}{\varrho_0} = \frac{\varrho(B) - \varrho(0)}{\varrho(0)} = (\mu B)^2 \quad , \tag{8.39}$$

is expected to be especially sensitive to the presence of lattice defects. The magnetoresistance of crystalline conductors is positive, indicative of an increase in cyclotron frequency as the magnetic field is increased, which increases the resistivity. The magnetoresistance given by (8.39) is positive for single crystal graphite, as for other crystalline solids where the conduction mechanism is associated with Bloch states; also $\Delta\varrho/\varrho_0$ is very large in graphite because of the high in-plane carrier mobility.

Magnetoresistance measurements thus provide a highly sensitive measure of the degree of crystalline order. Fibers with the highest degree of structural perfection show the largest positive magnetoresistance $(\Delta\varrho/\varrho_0) = [R(B) - R(0)]/R(0)$, with smaller values of $\Delta\varrho/\varrho_0$ corresponding to more disordered graphite. As more disorder is introduced, the 3D ordering disappears and the fibers assume 2D or turbostratic ordering, where there is no interlayer site correlation between adjacent graphene planes. The magnetoresistance for the turbostratic carbon is negative, with the magnitude of the magnetoresistance decreasing as the disorder increases. In both the turbostratic and 3D

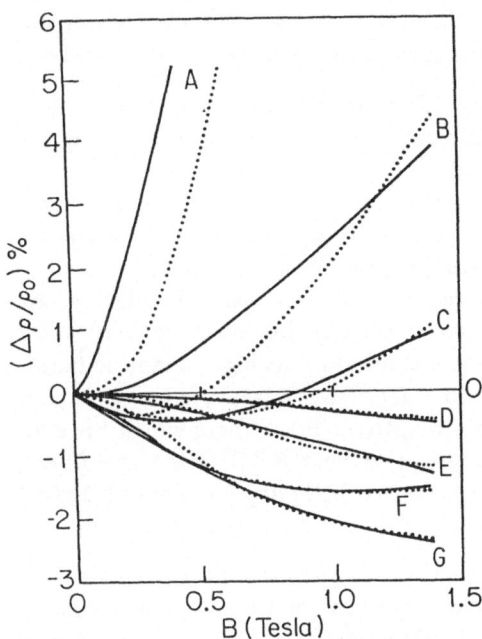

regime, the angular dependence of the magnetoresistance provides a sensitive characterization tool for the degree of preferred orientation of the graphene planes, with very small values of $(\Delta\varrho/\varrho_0)$ occurring for the magnetic field $B$ along the fiber axis, and the maximum $(\Delta\varrho/\varrho_0)$ for $B$ normal to the fiber axis. For highly disordered carbons, the magnetoresistance is small and positive and shows little anisotropy.

Magnetoresistance measurements on carbon fibers, however, show that $\Delta\varrho/\varrho_0$ is often negative (Fig. 8.30) which is characteristic of a wide range of carbons [see Bright 1979; Delhaès 1971]. Measurements on ex-PAN fibers [Robson et al. 1973], ex-mesophase pitch fibers and CCVD fibers [Endo et al.

**Table 8.6.** Values of parameters of Bright's model for the negative magnetoresistance of various fiber samples in Fig. 8.30 [Bright 1979]

|  | $T_{\mathrm{HT}}$ [°C] | $\varrho_{\mathrm{expt}}(4.2\ K)$ [$\mu\Omega$m] | $\varrho_{\mathrm{calc}}$ [$\mu\Omega$m] | $N_0$ [$10^{24}$m$^{-3}$] | $N_a$ [$10^{24}$m$^{-3}$] | $E_a$ [eV] | $\mu$ [m$^2$/Vs] |
|---|---|---|---|---|---|---|---|
| $(A)$ | 3000 | 3.79 | 3.92 | 19.2 | 1.1 | $-0.013$ | 1.110 |
| $(B)$ | 3000 | 5.08 | 4.88 | 9.1 | 1.1 | $-0.0058$ | 1.150 |
| $(C)$ | 3000 | 7.00 | 6.56 | 6.0 | 0.9 | $-0.0043$ | 1.040 |
| $(F)$ | 3000 | 6.62 | 6.80 | 4.5 | 1.6 | $-0.0075$ | 0.567 |
| $(G)$ | 2500 | 6.77 | 6.72 | 5.8 | 2.6 | $-0.0118$ | 0.371 |
| $(E)$ | 2000 | 9.24 | 9.18 | 5.4 | 3.2 | $-0.0182$ | 0.213 |
| $(D)$ | 1700 | 12.45 | 11.80 | 5.5 | 3.5 | $-0.0193$ | 0.152 |

1982a] show that the magnetoresistance behavior can be characterized in the same way as in the case of pyrolytic carbons [Delhaès et al. 1974], namely:

1. Highly graphitic fibers exhibiting 3D interplanar ordering have a positive magnetoresistance, which increases in magnitude as the x-ray in-plane coherence length, $L_a$, increases.
2. Turbostratic carbons with $\varrho < 10^{-5}$ $\Omega$m have a negative magnetoresistance. The magnitude of the negative magnetoresistance increases as the in-plane ordering increases [Bright 1979], until a maximum is reached, corresponding to the onset of three-dimensional ordering. As three-dimensional ordering sets in, the magnetoresistance may show negative values at low fields and positive values at high magnetic fields.
3. Fibers with higher disorder $(\varrho \geq 10^{-5}$ $\Omega$m$)$ have a weak, positive magnetoresistance. These include high strength ex-PAN fibers which have an anomalous $\varrho(T)$ at low temperatures (see Sect. 8.3.2). For these fibers, Hambourger [1986] has shown that there is little dependence of $(\Delta\varrho/\varrho_0)$ on the direction of $\boldsymbol{B}$.

The negative magnetoresistance observed in turbostratic carbons is due to a slight increase in the density of states with magnetic field, which results from the unusual Landau-level structure of two-dimensional graphene planes, discussed in Sect. 7.3.2. The resulting increase in the carrier density leads to a decreased resistance (see Sect. 8.3.3). This phenomenon will be discussed in Sect. 8.4.2, after a discussion of positive magnetoresistance (see Sect. 8.4.1). The discussion of the negative magnetoresistance is limited to graphitic fibers, since models have yet to be developed for highly disordered fibers.

### 8.4.1 Positive Magnetoresistance

Ex-PAN fibers $(T_{HT} > 2200°C)$ exhibit a negative magnetoresistance to the highest $T_{HT}$, while many ex-mesophase pitch fibers with high $T_{HT}$ values retain their negative magnetoresistance at weak magnetic field, but show positive values at high fields; others show a large positive magnetoresistance at the highest $T_{HT}$ values (see Fig. 8.30). However, measurements on benzene-derived fibers show that by increasing $T_{HT}$ to 2200°C, and above, the magnetoresistance at liquid-nitrogen temperature changes from negative to positive [Endo et al. 1982a]. The positive sign of the magnetoresistance is indicative of band conduction and provides direct evidence for the high degree of crystalline order that can be attained with the vapor-grown fibers. For $T_{HT} > 2200°C$, the magnitude of the low-field (positive) magnetoresistance increases with increasing $T_{HT}$ up to the highest $T_{HT}$ values (above 3000°C)

For highly ordered graphite fibers where the magnetoresistance is positive, the low temperature ($\sim$4.2 K) magnetoresistance provides a powerful characterization tool. Because of the very high degree of ordering that can be achieved by heat treatment of the vapor-grown fibers to temperatures above 2600°C, many characterization tools, such as resistivity, first-order Raman scattering, and the width of (002) x-ray diffraction peaks, become relatively insensitive to increases in crystalline ordering as the heat treatment temperature is further increased [Chieu et al. 1982]. The high sensitivity of the magnetoresistance technique is in part due to the very high value of $\Delta\varrho/\varrho(0) \sim 10^5$ which has been reported for single crystal graphite at low $T$. The magnetoresistance tool has also been widely used for pyrocarbons [e.g., by Klein 1962, 1964, 1966 and by Delhaès et al. 1974] and for highly oriented pyrolytic graphite by Spain et al. [1967].

The angular dependence of the high field magnetoresistance provides additional information on the organization of the layer planes in the fiber and is consistent with the observation that in benzene-derived fibers the graphite layer planes form facets with a polygonal organization about the fiber axis [Chieu et al. 1983], indicated by scanning electron micrographs and transmission electron microscope images (see Sects. 3.7.1 and 3.7.2). Also of interest is the observation of Shubnikov-de Haas (SdH) oscillations in the intercalated fibers, indicative of the long mean free path that can be achieved in these materials [Chieu et al. 1983, 1984; Oshima et al. 1983].

In practical measurements of the magnetoresistance of carbon fibers, the magnetic field usually makes a distribution of angles $\Theta$ with respect to the $c$-axis normal to the layer planes. Because of the high anisotropy of the magnetoresistance of graphite, the angular dependence of the magnetoresistance must be taken into account in the analysis of measurements on polycrystalline materials. This is done by first considering the angular dependence of $\Delta\varrho/\varrho_0$ and then carrying out a suitable angular average [Woolf et al. 1984], as summarized below.

Consider a graphite sample where the magnetic field makes an angle $\Theta$ with the $c$-axis. In nearly compensated specimens, the condition $n \simeq p$ is satisfied. On the basis of the simple two-band model, the magnetoresistance for well-graphitized samples is given by

$$\frac{\Delta\varrho(B)}{\varrho(0)} = \frac{(\bar{\mu}B)^2[m_a/m(\Theta)]^2}{1 + [(n-p)/(n+p)]^2(\bar{\mu}B)^2[m_a/m(\Theta)]^2} \quad , \qquad (8.40)$$

where $|n - p| \ll (n + p)$ and

$$\left(\frac{1}{m(\Theta)}\right)^2 = \frac{\cos^2\Theta}{m_a^2} + \frac{\sin^2\Theta}{m_a m_c} \qquad (8.41)$$

in which $m_a$ and $m_c$ are respectively the transverse and longitudinal effective mass components. Since $m_a/m_c \ll 1$, Eq. (8.40) becomes

$$\frac{\Delta\varrho(B)}{\varrho(0)} = \frac{(\bar{\mu}B)^2\cos^2\Theta}{1 + [(n-p)/(n+p)]^2(\bar{\mu}B)^2\cos^2\Theta} \quad , \qquad (8.42)$$

which can be rewritten as

$$\frac{B^2 \cos^2 \Theta}{\Delta\varrho/\varrho(0)} = \frac{1}{\bar{\mu}^2} + \left(\frac{n-p}{n+p}\right)^2 B^2 \cos^2 \Theta \quad . \tag{8.43}$$

Here we see that in the limit $n = p$, the simple formula

$$\frac{B^2 \cos^2 \Theta}{\Delta\varrho/\varrho(0)} = \frac{1}{\bar{\mu}^2} \tag{8.44}$$

is obtained.

To analyze experimental magnetoresistance results on fibers, it is necessary to carry out a suitable average over $\Theta$. As shown in Fig. 8.27, the cross section of benzene-derived fibers has an annular ring structure where each layer experiences a different field strength $B_i$. Averaging over $\Theta$ yields

$$\langle B^2 \cos^2 \Theta \rangle = \tfrac{1}{2} B^2 \tag{8.45}$$

and (8.43) then becomes

$$\frac{B^2}{\Delta\varrho/\varrho(0)} = \frac{2}{\bar{\mu}^2} + \left(\frac{n-p}{n+p}\right)^2 B^2 \tag{8.46}$$

for slightly uncompensated graphite fibers. In the low field limit, the second term in (8.46) is unimportant and the average mobility $\bar{\mu}$ can be evaluated directly. At higher fields the second term becomes important and by plotting $[B^2/(\Delta\varrho/\varrho_0)]$ vs $B^2$, the charge carrier imbalance factor $(n-p)/(n+p)$ can be determined as shown in Fig. 8.31. By carrying out magnetoresistance measurements as a function of temperature, the temperature dependence of the charge carrier imbalance can be found. Figure 8.31 shows a plot of $[B^2/(\Delta\varrho/\varrho_0)]$ vs $B^2$ curves for different measurement temperatures taken on vapor-grown fibers annealed at 3000°C. From this analysis the temperature dependence of the average mobility and of the charge carrier imbalance $(n-p)/(n+p)$ is ob-

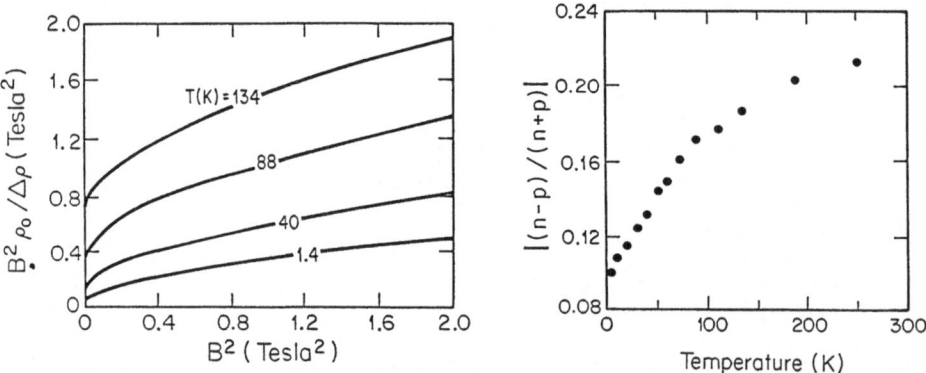

**Fig. 8.31.** Plot of $B^2/(\Delta\varrho/\varrho_0)$ vs $B^2$ for various temperatures for vapor-grown fibers annealed at 3000°C [Woolf et al. 1984]

**Fig. 8.32.** Plot of $|n-p|/|n+p|$ vs temperature obtained by fitting the magnetoresistance data of Fig. 8.31 to (8.46) [Woolf et al. 1984]

tained [Woolf et al. 1984] and the results are shown in Fig. 8.32. This sensitive determination of $(n-p)/(n+p)$ is useful for many transport measurements carried out on well-graphitized carbon fibers.

### 8.4.2 Negative Magnetoresistance in Pregraphitic Carbons and Fibers

As noted above, magnetoresistance measurements provide a sensitive indicator for the degree of graphitization of carbons, especially in the negative magnetoresistance regime. The magnetoresistance is much more sensitive, especially in the case of graphitized materials, than is for example the x-ray diffraction method. This arises from the sensitivity of the negative magnetoresistance to out-of-plane correlations, whereas the resistivity is more sensitive to in-plane order.

Thus, to provide a detailed characterization of carbon fibers, several characterization techniques are simultaneously employed, such as measurement of: (a) the interlayer distance $\tilde{c}$, (b) the $c$-axis crystallite size $L_c$ from the width of the (002) diffraction line, (c) the room temperature resistivity, (d) the tensile strength, (e) the Young's modulus and (f) the low temperature magnetoresistance $(\Delta\varrho/\varrho_0)$. Such an analysis has recently been carried out for ex-pitch fibers by Hamada et al. [1987b].

The negative magnetoresistance observed in carbon fibers is similar to that in bulk carbons, as is evident by comparing Fig. 8.30 with Fig. 8.33,

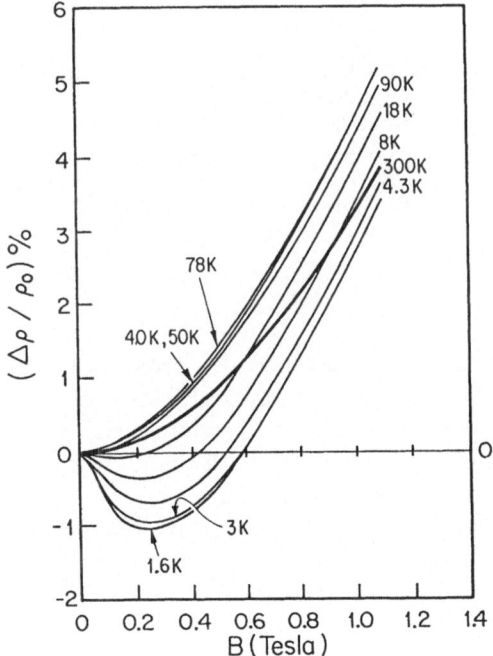

Fig. 8.33. Transverse magnetoresistance at different temperatures for pyrocarbons heat treated at 2560°C [Delhaès et al. 1974]

where data for bulk pyrocarbons are presented [Delhaès et al. 1974]. The data of Fig. 8.33 show that the positive magnetoresistance contribution (at $T > 50$ K) has a weak temperature dependence since the mobility depends only weakly on temperature. However, the negative component increases strongly at low temperature, which is consistent with attributing the negative magnetoresistance to the magnetic quantization of the energy levels as discussed below. Since the effect can change from positive to negative as the temperature is lowered, some caution is needed in assigning the negative effect to a particular range of $T_{HT}$ values. However, Fig. 8.34 shows magnetoresistance data at 77 K for several benzene-derived fibers heat treated to different temperatures, illustrating the trend from positive magnetoresistance at $T_{HT} = 2200°$C to negative values at 2000°C, then to smaller negative values at reduced $T_{HT}$. The same trend has already been shown in Fig. 8.30 for ex-mesophase pitch fibers [Bright 1979]. The curves for the CCVD fiber in Fig. 8.34 (e.g., $T_{HT} = 2200°$C) are consistent with a more pronounced structural ordering for the benzene-derived fibers. This is also evident in the comparison between CCVD and pitch carbon fiber data (Figs. 8.30 and 8.34), and is consistent with structural data for various types of fibers (see Chap. 3).

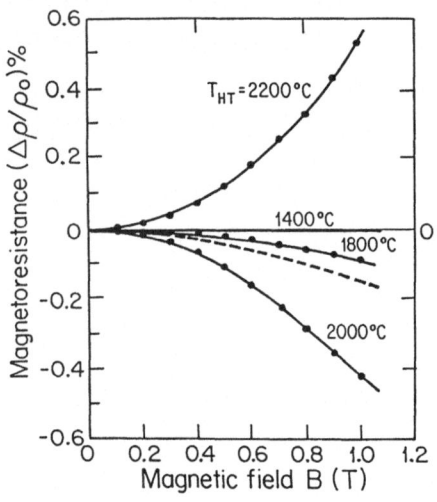

Fig. 8.34. Transverse magnetoresistance for vapor-grown carbon fibers heat treated between 1400°C and 2200°C, plotted as a function of magnetic field. Measurements were carried out at 77 K. The dotted curve is for PAN-based fibers heat treated at 2800°C [Endo et al. 1982a]

The angular dependence for the magnetoresistance also provides useful information about the structure of fibers [Endo et al. 1982a; Hishiyama et al. 1984]. The anisotropy of the magnetoresistance can be correlated with x-ray data sensitive to the preferred orientation (Sect. 3.4.3), with the more highly ordered fibers showing more pronounced anisotropy effects.

Several theories have been proposed to explain the negative magnetoresistance of disordered carbons [Fujita 1968; Yazawa 1969; Delhaès et al. 1974; Bright 1979]. Among them, only the Bright theory can account for the observed temperature and magnetic field dependence of the negative magnetoresistance. In view of its importance, the Bright theory is briefly reviewed.

In attempting to find an explicit solution for the negative magnetoresistance problem, Bright used the density of states derived in Sect. 7.3.2 for two-dimensional graphite, and introduced the following assumptions and approximations [Bright 1979]:

1. The effect of disorder on the density of states is considered by taking into account the Gaussian level broadening caused by collision processes and thermal broadening.
2. The small three-dimensional overlap effect is included by adding an excess density of states $N_0$ to the $N = 0$ Landau level.
3. Since the partially graphitized carbons exhibit a positive thermopower and Hall effect, an acceptor concentration $N_a \sim 10^{24}/m^3$ located at $E_a \sim -10^{-2}$ eV is assumed.

A simple two-carrier model was employed by Bright (1979) to calculate the resistivity, and four adjustable parameters $N_0$, $N_a$, $E_a$ and the carrier mobility $\mu = e\tau/m^*$ were chosen to fit the observed magnetoresistance results. According to the above assumptions, the density of states is given by

$$
\begin{aligned}
g(E) = \frac{AB\lambda}{\sqrt{\pi}} \bigg[ &\left(1 + \frac{N_0}{AB}\right) \exp(-\lambda^2 E^2) \\
&+ \sum_{N=1}^{\infty} \left\{ \exp[-\lambda^2(E-E_N)^2] + \exp[-\lambda^2(E+E_N)^2] \right\} \bigg] \ ,
\end{aligned}
\tag{8.47}
$$

where $1/\lambda$ represents the width of each level $N$. The zero of energy is taken at the conduction band minimum and $1/\lambda$ is related to the relaxation time $\tau$ by

$$
\frac{1}{\lambda} = \frac{1}{\sqrt{\ln 2}} \sqrt{(\hbar/\tau)^2 + (k_B T)^2} \ .
\tag{8.48}
$$

The electron and hole concentrations $n$ and $p$ are given by

$$
\begin{aligned}
n = \int_0^{\infty} g(E)dE/\{1 + \exp[(E - E_F)/k_B T]\} \\
p = \int_0^{\infty} g(E)dE/\{1 + \exp[(E + E_F)/k_B T]\}
\end{aligned}
\tag{8.49}
$$

and the Fermi energy $E_F$ is determined by the neutrality condition

$$
p = n + \frac{N_a}{\{1 + \exp[(E_a - E_F)/k_B T]\}} \ .
\tag{8.50}
$$

Assuming that $\mu_n = \mu_p = \mu$ for the electron and hole mobilities and noting that $\mu = e\tau/m^*$, the following expression for the field dependence of the resistivity $\varrho(B)$ is obtained:

$$
\varrho(B) = \frac{1 + \mu^2 B^2}{(n + p)e\mu[1 + \mu^2 B^2 (p - n)^2/(p + n)^2]}
\tag{8.51}
$$

and the magnetoresistance $[\varrho(B) - \varrho(0)]/\varrho(0) = \Delta\varrho/\varrho_0$ becomes

$$\frac{\Delta\varrho}{\varrho_0} = \frac{-1 + \dfrac{n_0 + p_0}{n + p}\left[1 + \mu^2 B^2\left(1 - \dfrac{(p - n)^2}{(p + n)(p_0 + n_0)}\right)\right]}{1 + \mu^2 B^2\dfrac{(p - n)^2}{(p + n)^2}} \quad , \qquad (8.52)$$

where the subscript zero indicates the zero-field value. In deriving (8.52) the mobility $\mu$ is assumed to be independent of $B$. In the limit that the electron carrier concentration $n = 0$, Eq. (8.52) reduces to

$$\frac{\Delta\varrho}{\varrho_0} = \frac{p_0}{p} - 1, \qquad (8.53)$$

corresponding to the well-known result that a single-carrier system has no magnetoresistance if the carrier concentration and mobility are field independent. To apply (8.52) to carbon fibers, an appropriate angular average is needed. However, to understand the physics of the negative magnetoresistance, the precise details employed in performing the angular average are not important.

Of particular importance is the case of low mobility samples at low fields. In this case the density of states in a magnetic field $g(E, B)$ is expanded in a power series in $B$, yielding a constant term, a term in $B^2$ and higher power terms [Bright 1979; Sugihara 1986b]. For low mobility fibers ($\mu \sim 0.1\,\mathrm{m^2/Vs}$), the expansion up to terms in $B^2$ provides a good approximation for magnetic fields $B \leq 1.5$ T. As is shown in Fig. 8.35, the density of states in a magnetic field $g(E, B)$ is almost independent of energy $E$ for $E \leq 0.01\,\mathrm{eV} \simeq E_\mathrm{F}$. At higher energies $g(E, B)$ increases monotonically by an amount proportional to $B^2$. At low temperatures ($k_\mathrm{B}T \ll E_\mathrm{F}$), the density of states $g(E, B)$ can then be approximated by

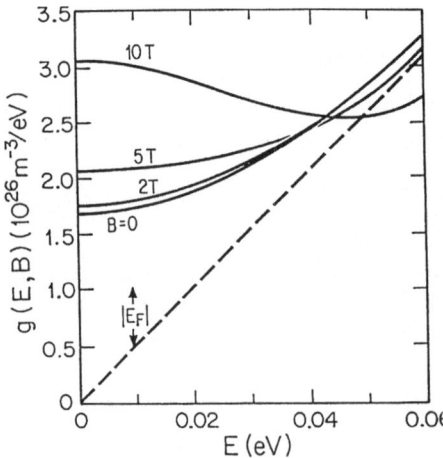

Fig. 8.35. Density of states as a function of energy for various magnetic field strengths for a carrier mobility of 0.1 m$^2$/Vs. The dashed line denotes the zero-field density of states in the absence of collision broadening. The Fermi energy $E_\mathrm{F}$ is determined by the charge neutrality condition ($E_\mathrm{F} < 0$) on the diagram [Bright 1979]

210

$$g(E, B) \simeq g(0, B) \simeq a + bB^2 \quad , \tag{8.54}$$

in which the factors $a$ and $b$ are given by

$$a = \frac{1}{\sqrt{\pi}}(N_0 + 2A/\lambda^2 D)\lambda \,, \quad b = \frac{1}{6\sqrt{\pi}}AD\lambda^3 \quad , \tag{8.55}$$

where the field-dependent coefficients $A$ and $D$ are given by

$$A = 2(s/B)/(\pi \bar{c}), \quad D = \tfrac{3}{2}(\gamma_0 a_0)^2 (s/B), \tag{8.56}$$

and $\lambda$ is the inverse level width, $\bar{c}$ denotes the interlayer separation, $a_0$ is the in-plane lattice constant and $\gamma_0$ is the nearest neighbor overlap energy for graphite. We note that $(s/B)$ is independent of magnetic field.

In the limit of $n \ll p$ (this is realized in pregraphitic carbons with low $T_{\mathrm{HT}}$), the magnetoresistance is given by (8.53), and since the hole concentration $p$ is proportional to the density of states $g(E, B)$, we obtain

$$\frac{\Delta\varrho}{\varrho_0} \simeq \frac{g(0, 0)}{g(0, B)} - 1 \simeq -\frac{b}{a}B^2 \sim -157\mu^4 B^2 \quad , \tag{8.57}$$

where $\mu$ is measured in m$^2$/Vs and $B$ is in tesla. In deriving (8.57), $m^*$ is assumed to be equal to 0.025 $m_0$ for pregraphitic carbons.

In the quantum limit, only the lowest quantum level ($N = 0$) is occupied. Since the density of states for each 2D magnetic level is proportional to $B$, see (7.34), the density of charged carriers increases roughly linearly with $B$. As is shown in the above calculation using the simple two-band model, the density of states $g(E, B)$ increases roughly proportional to $B^2$, which results in a quadratic field dependence of the carrier concentration and a quadratic negative magnetoresistance. However, in well-graphitized samples, the relation ($|p - n| \ll p + n$) holds, and the $\mu^2 B^2$ term in the numerator of (8.52) makes the dominant contribution to $(\Delta\varrho/\varrho_0)$ at high fields. Thus, the magnetoresistance eventually becomes positive as $B$ increases since the increase in carrier concentration is unimportant in this limit.

Typical results for the transverse magnetoresistance of various ex-pitch carbon fibers at 4.2 K are shown in Fig. 8.30 [Bright 1979]. The values of the adjustable parameters $N_0$, $N_a$, $E_a$ and the zero field mobility $\mu$ are listed in Table 8.6. The numerical values are reasonable and consistent with the assumptions of the model. It should be noted in particular that $N_a$ is of order $10^{24}$/m$^3$, consistent with localized defect states, whereas Yazawa [1969] found much higher values of $N_a \sim 10^{27}$/m$^3$. We note that the Bright calculation is insensitive to the value of $E_a$, since the defect level is far below the Fermi level.

Though the Bright theory can successfully account for the qualitative behavior of the negative magnetoresistance, it fails to explain the anomalous temperature dependence of the negative magnetoresistance observed at low temperatures [see Fig. 8.33 reported by Delhaès et al. (1974) and Figs. 8.36 and 8.37 reported by Hambourger (1986a, 1986b)].

211

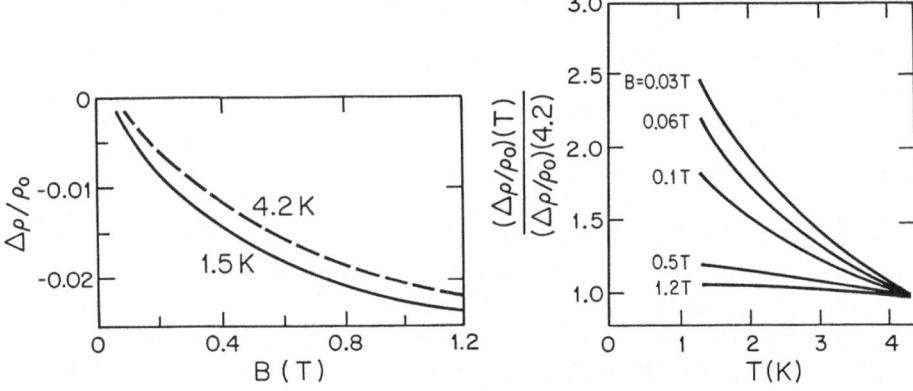

**Fig. 8.36.** Magnetoresistance vs magnetic field strength ($B$) for pitch-based P-100 fibers at very low temperatures [Hambourger 1986a, 1986b]

**Fig. 8.37.** Magnetoresistance (normalized to 4.2 K) vs temperature for several field strengths for pitch-based P-100 fibers [Hambourger 1986a, 1986b]

The anomalous low temperature magnetoresistance behavior can be explained by introducing two concepts which were not taken into account in the Bright theory. One is the introduction of a field-dependent scattering process [Rahim et al. 1986]. As was pointed out by McClure and Spry [1968], the force range of an ionized scattering center is proportional to $[g(E_F, B)]^{-1/2}$, and the force range decreases with increasing field intensity. This effect also gives rise to a negative magnetoresistance. Combining this effect with the electron-Rayleigh wave interaction, which is important in disordered carbons, the anomalous temperature dependent negative magnetoresistance can be explained as discussed below [Sugihara and Dresselhaus 1986].

For simplicity we focus our attention on the negative magnetoresistance region. Assuming $p \gg n$ and $\mu^2 B^2 \ll 1$, Eq. (8.52) for $\Delta\varrho(B)/\varrho(0)$ results in the following expression if $\Delta\varrho(B)/\varrho(0)$ is sufficiently small:

$$\frac{\Delta\varrho(B)}{\varrho(0)} \simeq -\frac{\Delta p(B)}{p(0)} - \frac{\Delta\mu(B)}{\mu(0)} \quad , \tag{8.58}$$

in which $p$ indicates the hole concentration, $\mu$ is the mobility, and (0) refers to the zero-field value of the parameters. In deriving (8.58), the approximation $p \gg n$ is assumed in (8.52). Since the density of states and carrier concentrations increase with $B$ [Bright 1979], the first term in (8.58) provides a negative contribution to $\Delta\varrho/\varrho_0$. Bright's model disregards the second term, but this is important in explaining the temperature dependent negative magnetoresistance. In accordance with Matthiessen's law, the inverse mobility $1/\mu$ is composed of several terms, as given by (8.30). Among these terms, only $1/\mu_I$ is field dependent. The relaxation rate $1/\tau_I$ is related to $1/\mu_I$ by $\mu_I = e\tau_I/m^*$ and is given by

$$\frac{1}{\tau_{\mathrm{I}}} \simeq \pi N_{\mathrm{I}}\, g(E_{\mathrm{F}})\, |\, v(0)\, |^2 \, / 2\hbar \quad , \tag{8.59}$$

where $N_{\mathrm{I}}$ denotes the ionized impurity concentration, and $v(0)$ is the Fourier component of the ionized impurity potential for $q = 0$. Since according to (8.54), the density of states at the Fermi level $g(E_{\mathrm{F}}, B)$ in the weak field region takes the form $a + bB^2$, and since the Thomas-Fermi approximation implies that $v(q)$ is almost independent of $q$ so that $v(0) \sim [g(E_{\mathrm{F}})]^{-2}$ [Sugihara and Dresselhaus 1986], it then follows that

$$\tau_{\mathrm{I}} \simeq 2\hbar \frac{g(E_{\mathrm{F}}, B)}{\pi N_{\mathrm{I}}} \propto a + bB^2 \quad . \tag{8.60}$$

Equation (8.60) indicates that $\tau_{\mathrm{I}}$ and $\mu_{\mathrm{I}}$ increase with magnetic field so that the second term in (8.58) also gives rise to a negative magnetoresistance since $\Delta\mu(B)/\mu(0)$ becomes

$$-\frac{\Delta\mu(B)}{\mu(0)} = -\frac{\mu(0)}{\mu_{\mathrm{I}}(0)}\frac{\Delta\mu_{\mathrm{I}}(B)}{\mu_{\mathrm{I}}(0)} = -\frac{\beta\mu^6(0)}{(1+K)^2}B^2 \quad , \tag{8.61}$$

where

$$K = \frac{N_0 \lambda^2 D}{2A} \propto N_0 \mu^2(0) \tag{8.62}$$

and $\beta$ is a numerical constant defined in (8.71). Here $N_0$ is the excess density of states added to the $N = 0$ Landau level in the Bright model, and is due to the small 3D-overlap effect. The first term $-\Delta p(B)/p(0)$ in (8.58) is given by

$$-\frac{\Delta p(B)}{p(0)} = -\frac{bB^2}{a} = -\frac{\alpha\mu^4(0)}{1+K}B^2 \quad , \tag{8.63}$$

where $\alpha$ is a numerical constant defined in (8.70) below. Explicitly $K$ is given by $K = 7.37 N_0 \mu^2$, where $m^* = 0.025 m_0$ is assumed and $N_0$ and $\mu$ are measured in units of $10^{24}/\mathrm{m}^3$ and $\mathrm{m}^2/\mathrm{Vs}$, respectively. Combining (8.58) and (8.61) with (8.63), the following result is obtained:

$$\frac{\Delta\varrho(B)}{\varrho(0)} \simeq -\left(\frac{\alpha\mu^4(0)}{1+K} + \frac{\beta\mu^6(0)}{(1+K)^2}\right)B^2 \quad . \tag{8.64}$$

Equation (8.64) would be $T$ independent at low temperatures if $1/\mu_{\mathrm{ph}}$ in (8.30) made a negligible contribution to $1/\mu$, as is the case for most conductors. However, this is not the case for carbon films, where the Rayleigh phonons contribute significantly to scattering processes and to the magnetoresistance down to 1 K and below.

At low temperatures, it is the Rayleigh phonons which contribute significantly to the magnetoresistance. Following the discussion in Sect. 4.3.3, an expression for the $k$-dependent relaxation rate due to the scattering by the Rayleigh wave phonons is obtained [Sugihara and Sato 1963; Sugihara 1986b]:

$$1/\tau_{\mathrm{R}}(E_k) \simeq \frac{2\pi k_{\mathrm{B}} T D^2}{\hbar \varrho_m v_{\mathrm{R}}^2 d^2 \Omega} \sum_{k'} \frac{1}{q^2} \left(1 - \frac{k_x'}{k_x}\right) \delta(E_k - E_{k'}) \quad , \tag{8.65}$$

where $\boldsymbol{q} = \boldsymbol{k}_a' - \boldsymbol{k}_a$, and $\boldsymbol{k}_a = (k_x, k_y, 0)$. Here $D$ denotes the electron-phonon coupling constant associated with the out-of-plane vibrations, and in bulk graphite $D \sim 3.7$ eV [Ono and Sugihara 1966]. In deriving (8.65), the introduction of the high temperature approximation for phonons $N_q \sim N_q + 1 \sim k_{\mathrm{B}} T/\hbar \omega_q$ is employed, since $\hbar \omega_q/k_{\mathrm{B}} = \hbar v_{\mathrm{R}} q/k_{\mathrm{B}} \simeq 5 \times 10^{-7} q < 1$ K for $q \sim 10^8$ m$^{-1}$. From (8.65), we then obtain

$$\frac{1}{\tau_{\mathrm{R}}} \simeq \frac{k_{\mathrm{B}} T}{4\hbar \varrho_m \tilde{c}_0 E_{\mathrm{F}}} \left(\frac{D}{v_{\mathrm{R}} d}\right)^2 \quad . \tag{8.66}$$

In evaluating (8.66), the following values of the parameters are employed: $v_{\mathrm{R}} = 6.3 \times 10^2$ m/s, $d = 70$ Å, and $E_{\mathrm{F}} = 0.01$ eV, yielding a value of $1/\tau_{\mathrm{R}} \sim 4 \times 10^{11} T/\mathrm{s}$ K. The magnitude of $1/\tau_{\mathrm{R}}$ at 4.2 K is larger than the 77 K relaxation rate associated with the in-plane vibration in bulk graphite. Though $1/\tau_{\mathrm{R}}$ is one order of magnitude smaller than the scattering rates for impurity scattering and for boundary scattering, $1/\tau_{\mathrm{R}}$ is responsible for the anomalous temperature-dependent negative magnetoresistance of pre-graphitic carbons at low temperatures. Furthermore, the anomalous increase below 30 K of the resistivity with temperature of ex-PAN fibers with low $T_{\mathrm{HT}}$ [Spain et al. 1983b] may be explained in terms of the electron-Rayleigh wave interaction (see Fig. 8.20).

To obtain the temperature dependence of $\Delta \varrho(B)/\varrho(0)$ at low $T$, assume that the sample is composed of an aggregation of many thin films which are weakly coupled to each other elastically. If the film thickness $d$ is small enough ($< 100$ Å), Rayleigh waves with small damping and small sound velocity scatter carriers even at $T \leq 1$ K, since the typical phonon energies associated with the wave are $\hbar \omega_q/k_{\mathrm{B}} \sim 1$ K. Though $1/\mu_{\mathrm{ph}}$ is one or two orders of magnitude smaller than $1/\mu_{\mathrm{BS}}$ and $1/\mu_{\mathrm{I}}$ for the boundary scattering and charged impurity scattering processes, the strong dependence of $\Delta \varrho(B)/\varrho(0)$ on $\mu(0)$ in (8.64) leads to an appreciable $T$-dependence of $\Delta \varrho(B)/\varrho(0)$ even at very low temperatures ($T \leq 1$ K) [Sugihara and Dresselhaus 1986].

An explicit expression for $\Delta \varrho(B)/\varrho(0)$ is obtained [Sugihara and Dresselhaus 1986] by combining (8.58, 8.61, 8.62) and (8.63):

$$\frac{\Delta \varrho(B)}{\varrho(0)} \simeq -\frac{(\ln 2)^2}{3} \left(\frac{\tau}{\tau_B}\right)^4 \left[\frac{1}{1+K} + \frac{1}{2} \left(\frac{\pi^3}{\ln 2}\right)^{1/2} \frac{N_{\mathrm{I}}}{N_0} \frac{K}{(1+K)^2}\right] \quad , \tag{8.67}$$

where $\tau_B$ is a characteristic time in a magnetic field $\tau_B = l_B/v$, corresponding to the characteristic magnetic length $l_B$ is $l_B = \sqrt{1/s}$. Here the electron velocity $v$ at the Fermi level is $v = \sqrt{3}\gamma_0 a_0/(2\hbar) = 1.02 \times 10^6$ m/s by (7.5), $N_{\mathrm{I}}$ is the ionized impurity concentration, and $K$ is defined by (8.62).

Evaluation of the dimensionless parameter $K$ of (8.62) in the weak field region where the density of states $g(E, B)$ is approximated by $g(0, B) \simeq a + bB^2$ yields

$$K = \frac{N_0 \lambda^2 D}{2A} = \frac{\pi \ln 2}{4} N_0 \tilde{c} (v\tau)^2 \tag{8.68}$$

in which $D$ and $A$ are given by (8.55) and (8.56), $\tilde{c}$ is the interlayer separation and $\lambda^{-1}$ represents the width of each Landau level. At low temperatures this width is given by $\lambda^{-1} \simeq (\hbar/\tau)/(\ln 2)^{1/2}$ [Bright 1979].

The explicit dependence of $\Delta\varrho(B)/\varrho(0)$ on $B$ and $T$ is sensitive to whether $K$ is small or large:

i) $K \ll 1$: This is realized for small values of $N_0$ and $\mu$, for example $N_0 \simeq 10^{24} \mathrm{m}^{-3}$ and $\mu \simeq 0.1 \ \mathrm{m}^2/\mathrm{Vs}$. In this case, (8.64) becomes

$$\frac{\Delta\varrho(B)}{\varrho(0)} \simeq -[\alpha\mu^4(0) + \beta\mu^6(0)]B^2 \ , \quad \text{where} \tag{8.69}$$

$$\alpha = \frac{(\ln 2)^2}{3} \left( \frac{v^2 m^{*2} s}{e^2 B} \right)^2 \quad \text{and} \tag{8.70}$$

$$\frac{\alpha}{\beta} = \frac{8}{N_{\mathrm{I}}\tilde{c}} (\ln 2\pi^5)^{-1/2} \left( \frac{e}{m^* v} \right)^2 \ . \tag{8.71}$$

Though $\alpha\mu^4(0)$ and $\beta\mu^6(0)$ are comparable in magnitude, they have different temperature dependences, since

$$\mu(0) = \frac{\bar{\mu}}{1 + R} \ , \tag{8.72}$$

where $R = \varepsilon T$ and $\varepsilon = \bar{\mu}/\mu_{\mathrm{R}}T$. Here $\mu_{\mathrm{R}}$ is the mobility associated with the electron-Rayleigh wave interaction process (see Sect. 4.3.3) and $\bar{\mu}$ is given by $1/\bar{\mu} = (1/\mu_{\mathrm{BS}}) + [1/\mu_{\mathrm{I}}(0)]$. Though $\varepsilon$ is small ($\sim 10^{-2}$) and is itself nearly temperature independent, the strong dependence of $\Delta\varrho(B)/\varrho(0)$ on $\mu(0)$ leads to an appreciable $T$ dependence for $\Delta\varrho(B)/\varrho(0)$. To evaluate (8.69), we insert the value $\varepsilon = 2 \times 10^{-2}$, which gives the result $(\Delta\varrho/\varrho_0)_{1.5\mathrm{K}}/(\Delta\varrho/\varrho_0)_{4.2\mathrm{K}} \simeq 1.27$, which is in qualitative agreement with Fig. 8.36. But the present theory cannot quantitatively account for the high field, low temperature region shown in Figs. 8.36 and 8.37.

ii) $K \gg 1$: In this case $\Delta\varrho(B)/\varrho(0)$ becomes

$$\frac{\Delta\varrho(B)}{\varrho(0)} \simeq -\frac{(v\tau)^2}{N_0\tilde{c}} s^2 \left( 0.15 + 0.39\frac{N_{\mathrm{I}}}{N_0} \right) \propto \frac{B^2}{(1 + \varepsilon\tau)^2}. \tag{8.73}$$

This gives a weaker temperature dependence than that of (8.69). However, $\Delta\varrho(B)/\varrho(0)$ itself is not small. To distinguish between the two cases $K \ll 1$ and $K \gg 1$ in (8.64), it is desirable to carry out more detailed measurements for various kinds of samples at low temperatures.

Although these models can qualitatively account for an increase in the magnetoresistance at low temperature for ex-mesophase pitch fibers (see Figs.

8.36 and 8.37), the models cannot account for the even steeper magnetoresistance effect with temperature found with ex-PAN fibers [Hambourger 1987]. This anomalous behavior may be related to the resistivity anomaly observed at low temperature in ex-PAN fibers [Spain et al. 1983b] (see Fig. 8.20). An explanation of this result awaits a suitable model.

## 8.5 Hall Effect

In general, measurements of the Hall effect provide information on the sign of the dominant carrier type and on the carrier density. However, the unfavorable geometry of carbon fibers makes it difficult to carry out Hall effect measurements. Nevertheless, Hall effect measurements on carbon fiber bundles have recently been reported [Espelette and Marchand 1987] using pressure contacts and a three-contact method. A mobile contact allowed cancellation of the voltage drop between the Hall probes in zero magnetic field, thereby properly separating the Hall voltage from magnetoresistance contributions. Corrections were also made for the voids between fibers in the fiber bundle. The same technique was applied to Hall effect measurements on coke and glassy carbon samples, giving good agreement with previous measurements on similar samples. Results for Hall coefficients for Toray M40 ex-PAN fibers ($T_{HT} \sim 2800°$C) and ex-mesophase pitch Union Carbide (UC) fibers ($T_{HT} \sim 2500°$C) are given in Table 8.7. The results show that although the ex-PAN fibers were heat treated to $2800°$C, their structure remained essentially 2D (i.e., turbostratic), as confirmed by magnetoresistance measurements, showing a negative magnetoresistance: $\Delta \varrho / \varrho_0 \sim -8 \times 10^{-4}$ at 77 K and $-6 \times 10^{-4}$ at 300 K. Hall measurements for the two ex-pitch fiber samples implied a higher degree of structural order, but still incomplete graphitization. This was confirmed by $\Delta \varrho / \varrho_0$ measurements, showing a much larger negative magnetoresistance ($-4.8 \times 10^{-3}$) at 77 K and a positive magnetoresistance ($\Delta \varrho / \varrho_0$) $\sim 2.3 \times 10^{-3}$ at 300 K. These arguments are based on the strong correlation between the positive Hall coefficient and the negative magnetoresistance in disordered carbons (see Fig. 7.1).

Table 8.7. Hall coefficient of carbon fibers[a] and carbons[b]

| Sample | $T_{HT}$ [°C] | $R_H$ [×10⁸m³/°C] 300 K | $R_H$ [×10⁸m³/°C] 77 K |
|---|---|---|---|
| Toray M40 (ex-PAN) | $\sim 2800$ | +1.9 | +2.4 |
| UC (sample 1) | $\sim 2500$ | −0.61 | −1.3 |
| UC (sample 2) | $\sim 2500$ | −0.635 | −1.05 |
| Coke | $\sim 2100$ | +1.6 | +2.0 |
| Coke | $\sim 2300$ | −2.8 | −4.0 |

[a] [Espelette and Marchand 1987].　[b] [Cherville et al. 1963].

# 8.6 Thermoelectric Power of Benzene-Derived Carbon Fibers

The fiber geometry is highly favorable for thermopower measurements. Since thermopower measurements also provide information on the carrier sign and carrier density, this technique has been more widely used than the Hall effect (see Sect. 8.5) for obtaining such information. On the other hand, the thermopower measurements are much more difficult to interpret quantitatively. Since most of the thermopower measurements to date have been made on vapor-grown fibers, the present review will emphasize this work.

The thermoelectric power (TEP) of carbon fibers prepared by thermal decomposition of benzene was extensively measured by Endo et al. [1977b] as a function of $T_{HT}$ and temperature. The original carbon fibers were prepared at 1080°C on a substrate, then heat treated stepwise for 30 minutes (residence time) at various temperatures from 1300°C to 3000°C using a graphite resistance furnace in a high purity argon atmosphere. TEP results obtained from benzene or methane-derived fibers are similar for the same heat treatment conditions, but the results vary significantly from fiber to fiber.

The TEP for graphite is small and exhibits an unusual temperature dependence (Fig. 8.38). The pronounced negative dip in Fig. 8.38 can be ascribed to the phonon drag effect [Tamarin et al. 1969; Jay-Gerin and Maynard 1970; Sugihara 1970; Sugihara et al. 1972]. The temperature dependence of the TEP in vapor-grown fibers heat treated at 3000°C [Endo et al. 1977b] almost coincides with that of single crystal graphite. Vapor-grown fibers heat treated at 3000°C by Heremans and Beetz [1985] showed a similar temperature dependence to that of Fig. 8.38 except that their data are shifted downward (by $\sim -10$ $\mu$V/K) so that no positive region between 60 K and 200 K was observed. Since the thermopower is very sensitive to $T_{HT}$ (see Fig. 8.39), study of the temperature dependence of the TEP provides a useful characterization tool for carbon fibers, in addition to yielding information on the dominant carrier type.

Comparison of Figs. 8.39 and 8.38 shows that the TEP is predominantly negative for three-dimensionally ordered fibers ($T_{HT} = 3000$°C) and positive for the 2D turbostratic fibers, in agreement with Hall effect measurements (see Sect. 8.5). It is further seen that the magnitude of the TEP decreases as the structural ordering decreases (Fig. 8.39), with a very small TEP found for the as-grown CCVD fibers. The behavior of the as-prepared vapor-grown carbon fibers [Endo et al. 1977b] shown in Fig. 8.39 is similar to that observed in glassy carbons heat treated at 900°–1000°C [Hishiyama and Kaburagi 1986]. Similar results for the temperature dependence of the TEP for as-grown CCVD fibers were reported by Heremans and Beetz [1985].

One might expect that the linear temperature region of the TEP for the less graphitized fibers in Fig. 8.39 (fibers $A$, $B$ for $T > 100$ K and fiber $C$ for

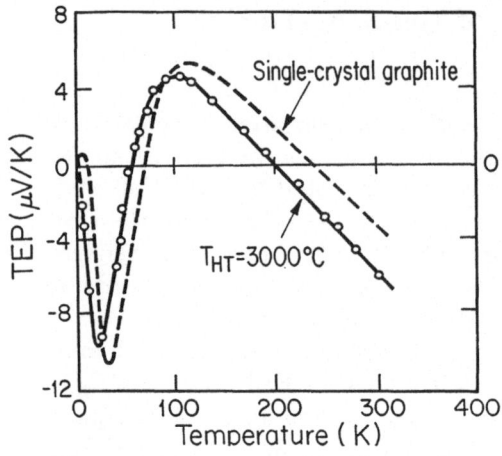

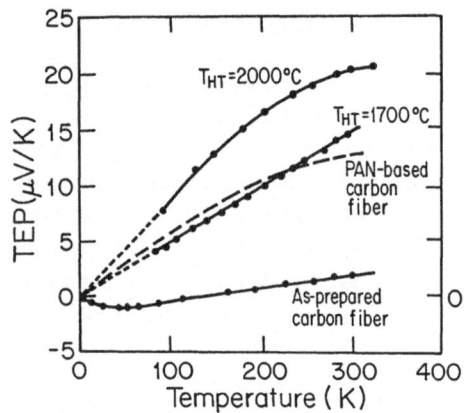

**Fig. 8.38.** Temperature dependence of the thermopower for a CCVD fiber heat treated at 3000°C [Endo et al. 1977b]. The dashed line is for the corresponding curve for kish single crystal graphite [Tsuzuku and Sugihara 1975; Takezawa et al. 1969]

**Fig. 8.39.** Plot of the thermopower vs temperature for various CCVD fibers [Endo et al. 1977b]: As-prepared vapor-grown carbon fibers; $T_{HT} = 1700°C$; $T_{HT} = 2000°C$. The dashed curve corresponds to ex-PAN carbon fibers [Arai and Tsuzuku 1972]

$T < 150$ K) and for the poorly graphitized fibers [Heremans and Beetz 1985] could be explained in terms of a diffusion effect. However, as will be shown in the following, this is not the case. The diffusion contribution is given by

$$S_d = \frac{\sigma_h S_h + \sigma_e S_e}{\sigma_h + \sigma_e} \quad , \tag{8.74}$$

where $S_i$ (in which $i =$ h or e for holes and electrons) and $\sigma_i$ denote the partial thermopower and conductivity of the $i^{th}$ carrier and the sign of $S_i$ is $S_h > 0$ and $S_e < 0$. The fiber with $T_{HT} = 2000°C$ in Fig. 8.39 shows a steeper initial slope in $S(T)$ than those of the other fibers, and for this fiber, $S(T)$ levels off at $T > 250$ K. This behavior indicates an elevation of the Fermi level and the thermal activation of electrons into the conduction band [Endo et al. 1977b]. The large positive value of the TEP implies that holes dominate the conduction in this fiber so that (8.74) becomes

$$S_d \simeq S_h \quad . \tag{8.75}$$

For a simple metal, $S_h$ at low temperatures is given by

$$S_h = \frac{\pi^2 k_B}{3e} \left( \frac{T}{T_F} \right) \quad , \tag{8.76}$$

where $T_F$ is the Fermi temperature. Using $T_F = 300$ K as an approximation for graphite and $T = 100$ K, a value of $S_h = 95$ $\mu$V/K is obtained, which is one order of magnitude larger than the observed value. To reconcile this estimate

218

with the experimental data, an unrealistically high $T_F$ is needed. Introduction of an electron contribution does not improve the situation. Similarly, the behaviors of the curves for $T_{HT} = 1700°C$ and $T_{HT} = 2000°C$ in Fig. 8.39 cannot be explained by the diffusion term alone. As was already pointed out in Sect. 8.3 (see Fig. 8.19), the conduction processes for the fibers with $T_{HT} \leq 2000°C$ cannot be described by a band conduction model. A new mechanism is needed to explain the transport properties of the samples with low $T_{HT}$ (below 2000°C).

The effect of a magnetic field on the thermopower (magneto-Seebeck effect) of carbon fibers heat treated at 3000°C was also measured by Endo et al. [1977b] and the results are shown in Fig. 8.40. Unlike the results for graphite fibers, the TEP peak in single crystal graphite shifts downward in temperature with increasing magnetic field [Sugihara et al. 1972; Tsuzuku and Sugihara 1975; Takezawa et al. 1971] and the negative dip tends to disappear.

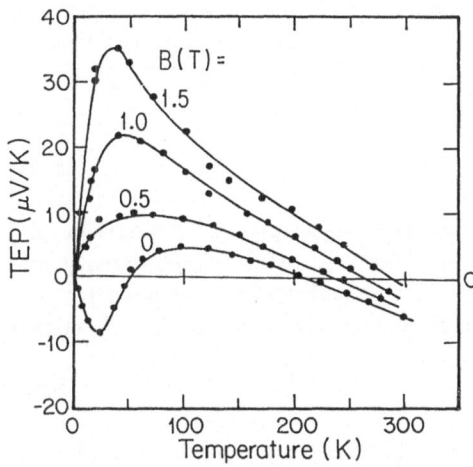

Fig. 8.40. Magneto-Seebeck coefficient (TEP) at various magnetic fields (in units of Tesla) of carbon fibers heat treated at 3000°C. The magnetic field is applied perpendicular to the fiber axis [Endo et al. 1977b]

To explain this anomalous magnetic field behavior, Endo et al. [1977b] suggested that the upward shift of the TEP with increasing magnetic field is a result of the cylindrical geometry of the graphite planes of the fiber. It has more recently been shown [Sugihara 1987] that this anomalous behavior can be explained better within the framework of the previous theory of the phonon drag effect in the thermopower [Sugihara et al. 1972; Tsuzuku and Sugihara 1975]. If the electric current density $j$ and a heat flux $w$ are expressed in terms of the temperature gradient $\nabla T$ and the electromotive force $F$ as

$$j = \sigma F - \beta \nabla T , \quad w = \chi F - \kappa \nabla T , \tag{8.77}$$

the TEP (magneto-Seebeck coefficient) is expressed by

$$S(B) = \frac{F_x}{\nabla_x T} = \frac{\sigma_{yy}\chi_{xx}(-B) - \sigma_{xy}\chi_{xy}(-B)}{T(\sigma_{xx}\sigma_{yy} - \sigma_{xy}\sigma_{yx})}, \tag{8.78}$$

219

where the Onsager reciprocal relation is employed. Here the temperature gradient is parallel to the fiber axis ($x$-axis) and the external magnetic field is perpendicular to the fiber axis ($z$-axis). Figure 8.38 indicates that in zero magnetic field the thermopower for vapor-grown carbon fibers heat treated to 3000°C is very similar to that for well-compensated single crystal graphite. It is then expected that

$$\frac{|n-p|}{n+p} \ll 1 \quad , \tag{8.79}$$

(see Fig. 8.32). Since $\sigma_{xx}\sigma_{yy} \gg \sigma_{xy}\sigma_{yx}$, Eq. (8.78) is approximated by

$$S = \frac{\chi_{xx}(B)}{T\sigma_{xx}} + \frac{\sigma_{xy}\chi_{xy}(B)}{T\sigma_{xx}^2} \quad , \tag{8.80}$$

where the relations $\chi_{xx}(-B) = \chi_{xx}(B)$, and $\chi_{xy}(-B) = -\chi_{xy}(B)$ are employed. (We note that these relations are not satisfied in a low symmetry crystal.) From the resistivity curve for vapor-grown carbon fibers heat treated at 3000°C (Fig. 8.18), the relaxation time $\tau$ can be estimated on the basis of the Drude formula. From the measured electrical resistivity at 50 K, which is $\varrho(50\ \text{K}) \simeq 5 \times 10^{-7}\Omega\text{m}$, an estimate for $\tau$ is obtained:

$$\varrho = \frac{m^*}{(n+p)e^2\tau} \quad . \tag{8.81}$$

Assuming reasonable values of the parameters appropriate for single crystal graphite, namely $n + p = 6 \times 10^{-24}$ m$^{-3}$ for the total carrier density and $m_c^* = 0.05m_0$ for the cyclotron effective mass, a value of $\tau = 0.65 \times 10^{-12}$ s is obtained. For $B = 1.5$ T the cyclotron frequency (in SI units) then becomes

$$\omega_c = \frac{eB}{m_c^*} = 5.28 \times 10^{12}\text{s}^{-1} \quad , \tag{8.82}$$

so that $\omega_c\tau = 3.42$ and $(\omega_c\tau)^2 = 12 \gg 1$. Though each layer sees a different field strength (see Fig. 8.27), most of the layers in the fiber experience a field strength that satisfies $(\omega_c\tau)^2 \gg 1$. According to previous theoretical work [Sugihara et al. 1972; Tsuzuku and Sugihara 1975], if $\omega_c\tau \gg 1$, then the TEP given by (8.80) takes the form

$$S(B) \simeq A_\text{d}T + \frac{A_\text{p}}{T^{2.7}} + B^2(\frac{D_\text{d}}{T} + \frac{D_\text{p}}{T^{4.7}}) \tag{8.83}$$

for $T_\text{d} < T < 60$ K, where $T_\text{d} \simeq 25$ K is the dip temperature corresponding to the minimum in the TEP (see Fig. 8.38). Here $A_\text{d}$ and $D_\text{d}$ are related to the diffusion term and $A_\text{p}$ and $D_\text{p}$ describe the phonon drag contribution. The phonon drag effect makes a dominant contribution below $\sim 60$ K and the coefficients $D_\text{d}$ and $D_\text{p}$ originate from the second (off-diagonal) term of (8.80) where

$$D_\text{d}, D_\text{p} \propto (p-n) \tag{8.84}$$

for $(\omega_c \tau)^2 \gg 1$. As expected from previous work on single crystal graphite, the observed anomaly in Fig. 8.38 should be qualitatively explained if

$$p - n \simeq 10^{23} \mathrm{m}^{-3} \qquad (8.85)$$

for these carbon fibers.

Quantitative interpretation of the TEP is sometimes difficult, since the phonon drag contribution plays an important role. However, the TEP provides useful information on the sign of the carriers, scattering processes for electrons or holes, the band structure, and the phonon mean free path.

## 8.7 Piezoresistance and the Effect of Hydrostatic Pressure

Piezoresistance is simply the change in resistance which occurs when a fiber is strained. Early measurements [Conor and Owston 1969; Owston 1970; Berg et al. 1972] showed that the piezoresistance of ex-PAN fibers was complex. Most fibers showed a linear increase in resistance above about 0.1% of the strain, but had variable behavior in most cases at low strain. Typical curves for ex-PAN fibers are shown in Fig. 8.41 where both positive and negative piezoresistance behavior is observed. Beetz [1981] showed that, under application of cyclic strain, the piezoresistance in ex-pitch fibers was reversible and reproducible until strains near the breaking point were reached; he attributed the non-recoverable increase in resistance at high strains to microcracking just prior to mechanical failure. For the remaining discussion, we will concentrate on the reproducible, reversible part of the piezoresistance curves.

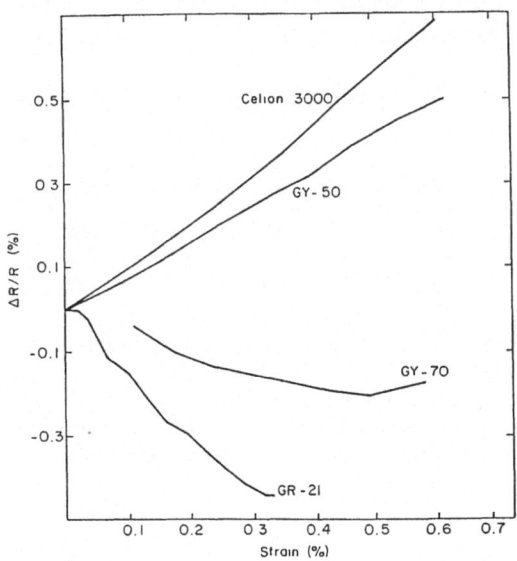

Fig. 8.41. Typical curves showing the change of resistance $\Delta R/R$ with increasing longitudinal strain for ex-PAN fibers [from Goldberg 1985]

221

There are several causes for piezoresistance phenomena. One obvious source of piezoresistance is the shape change which occurs whenever a fiber is strained. When a fiber is put under uniaxial tension, it will increase in length, and usually decrease in diameter. Both of these effects will increase the resistance of the fiber, assuming that the resistivity does not change. It is therefore useful to separate these geometric effects from changes in resistivity, which can produce either a negative or positive contribution to the piezoresistance. One therefore writes the following expression for the total piezoresistance:

$$\Delta R/R = \Delta L/L - \Delta A/A + \Delta \varrho/\varrho \quad , \tag{8.86}$$

where the first two terms are due to geometrical effects and the third term is due to changes in the resistivity itself. The various quantities in (8.86) are defined as follows: $R$ is the measured resistance, $\Delta R$ is the change in resistance per unit strain, $L$ is the sample length, $\Delta L$ is the change in length per unit strain (so that $\Delta L/L$ is the strain $\varepsilon$), $A$ is the cross-sectional area per unit strain, $\Delta A$ is the change in the cross-sectional area, $\varrho$ is the resistivity, and $\Delta \varrho$ is the change in resistivity per unit strain. Using the notation of Sect. 4.3.2, we can write

$$\varepsilon_{2'2'} = \Delta L/L , \quad \Delta A/A = -2\nu = \varepsilon_{1'1'} + \varepsilon_{3'3'} \quad . \tag{8.87}$$

The Poisson ratio ($\nu$) for the fiber is difficult to measure directly, but can be estimated using the uniform stress model, discussed in Sect. 4.3.2. Since ($\Delta A/A$) is negative, the Poisson ratio is defined as a positive quantity.

The qualitative behavior of the piezoresistance in carbon and graphite fibers can be summarized as follow:

– Low modulus (35-45 Msi, 240-310 GPa) fibers have a linear, positive piezoresistance. The linear, positive piezoresistance can be explained by the changes in shape discussed above (see Table 8.8 and Fig. 8.41)

Table 8.8. Piezoresistance[a] for various ex-PAN, ex-pitch, and vapor-grown carbon fibers[b] [Goldberg 1985]

| Fiber: | Celion 3000 | VGCF | GY-70 | P-100 | VG/B | VG/A |
|---|---|---|---|---|---|---|
| Total: $(\Delta R/R)/\varepsilon_{2'2'}$ | 1.5–2 | 1.3–2 | −1.5 | −2 | −3.4 | 13.4 |
| Geometrical: $(1 + 2\nu)$ | 1.5 | 1.6 | 2 | 2.7 | 1.6 | 1.6 |
| Resistivity: $(\Delta \varrho/\varrho)/\varepsilon_{2'2'}$ | 0 | 0 | −3.5 | −4.7 | −5 | 11.8 |

[a] Due to substantial nonlinearities in the piezoresistance and large uncertainties in estimates of the Poisson ratio, all values in this table should be viewed as rough approximations. The uncertainty is typically about ±0.5 for all the dimensionless quantities shown here.
[b] Data for ex-PAN Celion 3000 and GY-70 fibers, and for ex-pitch P-100 fibers are from Goldberg [1985], while data for the as-grown VGCF vapor-grown fibers and the type $A$ and type $B$ vapor-grown (VG) fibers heat treated to 3000°C are from Endo and Koyama [1980].

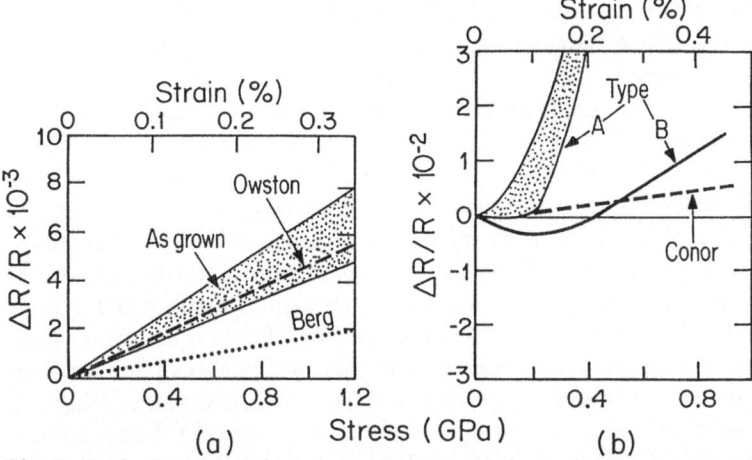

**Fig. 8.42a,b.** Fractional change in resistance ($\Delta R/R$) of (a) as-grown and (b) heat treated (to 3000°C) type A and type B vapor-grown fibers as functions of tensile stress and strain [Endo and Koyama 1980]. Also shown in the figure are comparisons to polymer-based fibers measured by Conor and Owston [1969], Owston [1970] and Berg et al. [1972] (see text)

[Goldberg 1985]. A similar behavior is found for the as-grown CCVD filaments shown in Fig. 8.42 [Endo and Koyama 1980]. Because of uncertainty in the estimate of the Poisson ratio, one cannot rule out some small contribution to the piezoresistance from a strain-induced change in the resistivity of these low modulus materials. Returning to Fig. 8.41, we note that the GY-50 fiber is more highly ordered than Celion 3000.
– High modulus ex-PAN and ex-pitch fibers ($> 70$ Msi, $> 500$ GPa) have a negative piezoresistance. As can be seen from Figs. 8.41 and 8.43, the piezoresistance for these ex-polymer fibers becomes increasingly negative as the modulus increases (as discussed below). We note however that the geometric effects contribute an increasing positive term as the modulus increases (see Fig. 8.43). All of the fibers in Fig. 8.41 have a room temperature resistivity which is dominated by defect and crystal

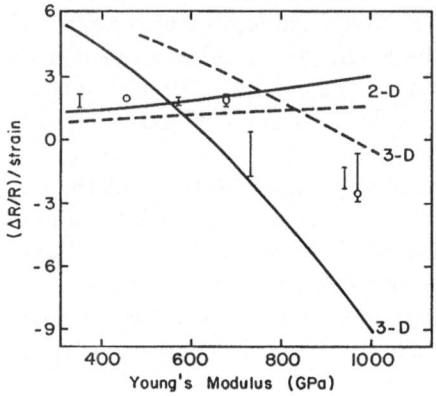

**Fig. 8.43.** Piezoresistance ($\Delta R/R$ per unit strain) vs the fiber modulus. The measured moduli were corrected for testing instrument compliance and for fiber porosity, and are therefore higher than are typically quoted. Solid lines are the result of calculations using the uniform stress model discussed in the text with $S_{44} = 14 \times 10^{-12}$ Pa$^{-1}$, $S_{13} = -1.8 \times 10^{-12}$ Pa$^{-1}$. The dashed lines correspond to $S_{44} = 20 \times 10^{-12}$ Pa$^{-1}$, $S_{13} = -0.3 \times 10^{-12}$ Pa$^{-1}$ [Goldberg 1985]

223

boundary scattering. The piezoresistance in these materials is not linear, and the largest negative slopes in the $\Delta R/R$ vs strain curves are found at low strains.

- Vapor-grown fibers heat treated to 3000°C (see Fig. 8.42) can have a piezoresistance similar to high modulus ex-PAN and ex-pitch fibers (type B fibers) or can show a very large positive piezoresistance (type A fibers) depending on the degree of order achieved in the fiber [Endo and Koyama 1980]. Type A fibers exhibit a positive slope for their resistance versus temperature behavior even at low temperatures [Chieu et al. 1983] and are clearly more graphitic than type B fibers which have a negative $(\partial \varrho/\partial T)$ slope. [Ex-PAN and ex-pitch fibers both show negative slopes $\partial \varrho/\partial T$ at low temperature in their $\varrho(T)$ curves.] Some values for the parameters describing the structural, electrical and mechanical properties of the vapor-grown fibers in comparison to PAN-based fibers are given in Table 8.9 for fiber types discussed in this section with regard to their piezoresistance properties.

**Table 8.9.** Parameters describing the structural, electrical and mechanical properties for various fibers[a,b]

|  | $\langle \phi \rangle$ [°] | Crystallite size [Å] | $\bar{c}$ Interlayer distance [Å] | Resistivity (at 300 K) [$\Omega$m] | Tensile strength [GPa] | Young's modulus [GPa] |
|---|---|---|---|---|---|---|
| VGCF | 30 | 10 | 3.48 | $1 \times 10^{-5}$ | 1–3 | 200–300[a] |
| GF-3000 | < 1 | > 1000 | 3.35 | $8 \times 10^{-7}$ | 1–5 | 200–350[b] |
| Celion G-30 | 32.5 | 28 | 3.54 | $1.5 \times 10^{-5}$ | 3.8 | 240 |
| Celion G-50 | 17.6 | 58 | 3.42 | $1.05 \times 10^{-5}$ | 2.5 | 350 |
| Celion GY-70 | 10.8 | 113 | 3.42 | $6.5 \times 10^{-6}$ | 2.2 | 500–600 |

[a] Fiber diameter is 5–6 $\mu$m. VGCF refers to as-grown CCVD fibers while GF-3000 refers to CCVD fibers after heat treatment to 3000°C [Endo et al. 1980]. The data for the Celion fibers are due to Goldberg [1985].
[b] The average degree of misorientation $\langle \phi \rangle$ is given in terms of the FWHM x-ray linewidths.

In addition to the effects discussed above, several explanations have been given in the literature for changes in the resistivity induced by uniaxial stress in carbon and graphite fibers. They include: changes in orientation [Curtis et al. 1968], changes in inter-particle contact [Komaki 1980; Fischbach et al. 1980], and changes in carrier density and other electronic effects [Goldberg 1985].

Any theory of the piezoresistance needs to explain the trend of the observed effect from a weakly positive piezoresistance for low modulus fibers, to negative values for higher modulus (discussed above), and finally to large positive values for the most highly ordered type A vapor-grown fibers.

If one views the observed changes in orientation in terms of Ruland's unwrinkling ribbon model (see Sect. 4.3.2), then it is clear that part of the effect of the strain is to improve the alignment of the fibrils along the fiber axis. This effect would reduce the increase in resistance expected from the geometric effects since one expects the electrons to follow a path along the wrinkled ribbons. As the ribbons unwrinkle, the electron would still be traveling along the length of the ribbons, whether wrinkled, or straightened out, so that the increased length of the actual electron path would be smaller than the $\Delta L$ assumed in the first term of (8.86). Thus the unwrinkling of ribbons would to a first approximation decrease the positive contribution to the piezoresistance expected from the geometrical effects, but would not lead to a net decrease in resistance with increasing strain. In addition, the amount of increase in the orientational alignment of the fibrils is largest in the low modulus fibers. Thus, it is difficult on the basis of this mechanism to explain the observed trend in terms of changes in orientation with strain.

Changes in inter-particle contacts could also lead to a decrease in resistance with increasing strain. One could imagine that, as the fibrils become better oriented, the scattering at crystallite boundaries is decreased, thus lowering the resistance. Alternatively, if one views the resistance as being dominated by scattering in the boundary between crystallites, then one might also expect that this process could be changed by straining the fiber. However, there is again no reason to expect that these effects would be more important in the higher modulus fibers than in low modulus fibers. In fact, boundary scattering would if anything be expected to be more important in low modulus fibers.

For these reasons Goldberg [1985] proposed that the resistivity changes in the piezoresistance could be understood in terms of electron density changes. The electron density changes occur because, as is well known, straining the fibers pushes the graphite basal planes closer together, so that the carrier density increases. For this effect to be applicable, it is necessary for the fiber to have three-dimensionally ordered graphite and not be completely turbostratic. The increase in 3D ordering has been shown by Stamatoff et al. [1983] to occur gradually with increasing modulus. Thus the trend towards more negative piezoresistance with increasing modulus has a natural qualitative explanation when this model is used. In order to quantitatively understand the data, one can write an expression for the change in resistivity with strain as follows:

$$\frac{\Delta\varrho}{\varrho}\frac{1}{\varepsilon_{2'2'}} = 7q\frac{\varepsilon_{33}}{\varepsilon_{2'2'}} \quad , \tag{8.88}$$

where the notation is the same as in Sect. 4.3.2 (i.e., the unprimed index 3 refers to the $c$-axis direction of a crystallite in the fiber while the index 2' refers to the fiber axis). The factor 7 comes from the following considerations. The band overlap is about $2|\gamma_2|$ and the logarithmic derivative of $\gamma_2$ with respect to the $c$-axis strain has been measured in single crystal graphite to be $\sim 7$. From this we get an estimate for the change in carrier density with strain [see Kelly (1981) for a review]. In (8.88), $q$ is a dimensionless parameter which depends on the degree of 3D order in the fiber. It is expected that in turbostratic carbon, the electron density will not depend on the $c$-axis strain, and thus $q$ goes from 0 (no 3D order) to 1 (complete 3D order). It should be noted that (8.88) does not include the effect of changes in orientation and interparticle contact discussed above. This equation further neglects any strain-induced changes in carrier mobility. This assumption should be reasonable for ex-PAN and ex-pitch fibers at room temperature and below, where the electron mobility is dominated by defect scattering. However, for the type A vapor-grown fibers used by Endo and Koyama [1980], there is a large phonon contribution to the scattering at room temperature so that increasing the strain is expected to decrease the carrier mobility, based on measurements showing a decrease in magnetoresistance with pressure [Hambourger 1986]. This effect could explain the positive piezoresistance for the type A vapor-grown fibers and the positive contribution to the piezoresistance at higher stress levels for the more highly ordered ex-PAN and ex-pitch fibers.

Referring to (8.88), the average compression of the basal planes of the crystallites can be evaluated using the uniform stress model, following the same techniques as described in Sect. 4.3.2 to yield

$$\frac{\varepsilon_{33}}{\varepsilon_{2'2'}} = E_Y[S_{13} + (S_{33} - S_{13})\langle \sin^2 \phi \rangle] \quad . \tag{8.89}$$

The Young's modulus $E_Y$ and elastic compliance constants $S_{ij}$ which are appropriate for these fibers are discussed in Sect. 4.3.2. The angle $\phi$ is the misorientation angle defined in Fig. 3.15. Following this type of analysis, a plot of the piezoresistance per unit strain is shown vs Young's modulus $E_Y$ in Fig. 8.43 along with the two theoretical limits $q = 0$ (indicated by 2D) and $q = 1$ (indicated by 3D). As in the analysis of Sect. 6.1, the moduli here were also corrected for the compliance of the tensile test equipment and for the porosity of the fibers. The experimental results lie between the 2D and 3D curves when $S_{13}$ is chosen to be $-1.8 \times 10^{-12}$/Pa (i.e., the value obtained in graphite from thermal expansion data instead of the value obtained from ultrasonic measurements, see Sect. 6.1.2 and B.T. Kelly [1981]).

### 8.7.1 Effect of Pressure on Resistance

Goldberg (1985) also computed the changes in resistance that would occur with hydrostatic stress (piezoresistance) using the uniform stress model (this is in contrast to the above calculation done for uniaxial stress). For the geometric part of the resistance change, he obtained (using the techniques described in Sect. 4.3.2)

$$\frac{\Delta R/R}{P} = \frac{(1 - 2\langle \sin^2 \phi \rangle)(2S_{13} + S_{33}) + 2\langle \sin^4 \phi \rangle(S_{11} + S_{12} + S_{13})}{3} \quad . \tag{8.90}$$

The compression of the basal planes due to hydrostatic pressure $P$ is given by

$$\varepsilon_{33} = -(2S_{13} + S_{33})P/3 \quad . \tag{8.91}$$

Thus, using the above described model one finds

$$\frac{\Delta R/R}{P} = \frac{(2S_{13}+S_{33})(1-7q-2\langle\sin^2\phi\rangle)+2\langle\sin^4\phi\rangle(S_{11}+S_{12}+S_{13})}{3} \quad , \tag{8.92}$$

where $q$ is related to the degree of three-dimensional order as previously discussed.

Fischbach [1985] reported a decrease in resistance with hydrostatic pressure for low modulus fibers, and an increase for high modulus fibers. This is opposite to what would be expected from the above equations. However, it has been observed [Spain 1987] that the decrease in resistance with increasing pressure occurs only when oil pressure is used. When helium was used as the pressure transferring medium, no decrease in resistance was observed. This is probably because the helium penetrates the pores of the fiber, and thus affects the fiber very differently from oil. One might suppose that the results of Fischbach are due to significant changes in the porosity of the fiber due to the oil pressure. There was no detailed characterization of the structure and 3D order in the fibers used by Fischbach [1985]. More detailed work on the hydrostatic pressure dependence of the resistance of fibers should provide additional insights into the validity of the analysis of the piezoresistance made by Goldberg [1985], as well as valuable information about fiber structure and its relationship to the electronic and mechanical properties.

### 8.7.2 Effect of Pressure on the Magnetoresistance

The effect of pressure on the magnetoresistance of high modulus (690 GPa) ex-pitch P-100 fibers has been measured by Hambourger [1986a, 1986b] at 4.2 K. Although the data for these partially graphitic fibers [see Hughes (1987) for data on these fibers] showed considerably scatter, those below 0.5 T could be fitted with an expression $d\ln(|\Delta\varrho/\varrho_0|)/dP = -6.1 \times 10^{-11}/\mathrm{Pa}$. Hambourger discussed these results in terms of Bright's model (see Sect. 8.4.2), using (8.52) to obtain

$$\frac{\Delta\varrho}{\varrho_0} \simeq -\ln 2 \, \frac{e^2\gamma_0^2 a_0^2 \tau^2}{\pi N_0 \hbar^4 \tilde{c}} B^2 \quad , \tag{8.93}$$

which is valid for $N_{\mathrm{I}} \ll N_0$ and $K \gg 1$, see (8.67) and (8.68). Since the interplanar separation, $\tilde{c}$, is expected to vary in a similar way to that of graphite, this term would increase the magnetoresistance effect by $2.8 \times 10^{-11}/\mathrm{Pa}$, see Kelly [1981]. The only other pressure-dependent parameter in (8.93) is the scattering time, $\tau$, which must decrease at the rate of $(1/\tau)(d\tau/dP) = -4.5 \times 10^{-11}/\mathrm{Pa}$, according to this equation. Further work needs to be carried out to relate this result to microscopic models and to the variation of the resistivity with pressure. Hambourger [1984, 1985] also studied the effect of pressure on the transport properties of CCVD filaments.

## 8.8 Non-ohmic Behavior

Since fibers are inhomogeneous materials, it is evident that non-ohmic effects should be observable. At lower $T_{HT}$ it is clear that conduction via localized hopping through boundaries between more highly conducting regions dominates the resistivity. At heat treatment temperatures below 1000°C, the fibers are not entirely composed of carbon, and the other elements can contribute significantly to the heterogeneity and thus to the observation of non-ohmic behavior. Figure 8.44 shows the type of non-ohmic behavior which has been observed in ex-PAN fibers heat treated between 600°C and 800°C [Kalnin et al. 1986; Goldberg et al. 1987]. This is the heat treatment range in which the fibers change from good insulators to conductors, and thus the heterogeneity is probably most significant in determining the measured properties. The data shown in Fig. 8.44 were taken with a very short voltage ramp (total time for the experiment about 1 ms), but not so short that heating effects are unimportant. One could conclude that there are intrinsic non-ohmic effects in these materials in addition to thermal effects because in many cases the more highly conducting state which is achieved at high currents persists for long times (days to weeks) after the current is removed.

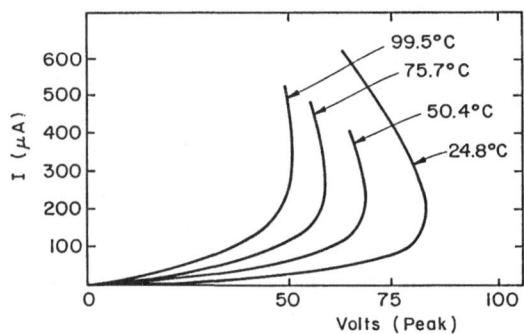

Fig. 8.44. Typical current vs voltage curves for low $T_{HT}$ ex-PAN fibers ($T_{HT}$ in the range 600°–800°C) in the non-ohmic region [from Kalnin et al. 1986]

Intrinsic non-ohmic behavior has not been observed in fibers heat treated above 1000°C. The thermal effects which are expected at high current densities have been discussed by Spain et al. [1983a] in terms of sample heating.

## 8.9 Electrical Noise

Measurement of the electrical resistance of carbon fibers immediately shows that the electrical noise is orders of magnitude above the thermal noise, $V_{th}$.

228

The Nyquist theorem [Kittel 1958] yields the expression for the voltage

$$V_{th} = 2(k_B T R \delta f)^{1/2} \quad , \tag{8.94}$$

where $R$ is the fiber resistance, $k_B$ is Boltzmann's constant, $\delta f$ the frequency bandwidth of the measurement, and $T$ the absolute temperature. [For a review of electrical noise in conductors, see Hooge et al. (1981).] The fibrillar structure of the fibers suggests that the noise would originate from fibril-fibril contacts. This implies that the noise voltage $V_n$, which would be proportional to the current $I$ in the fiber, would depend on frequency according to

$$V_n \propto \sqrt{\ln[1 + (\delta f / f)]} \sim (\delta f / f)^{1/2} \quad \text{for} \quad \delta f \ll f \quad , \tag{8.95}$$

where $f$ is the frequency of measurement [Conor and Owston 1969]. Similar behavior is observed in granular carbon resistors [Campbell and Chipman 1949] but the current noise varies as $1/f^n$ where $1 < n < 2$. The actual behavior observed for ex-PAN fibers [Owston 1970] was close to that predicted by (8.95) with $V_n \propto I$. The noise was proportionally higher for higher resistance fibers (i.e., lower $T_{HT}$ and low modulus). As discussed in Sect. 8.3.2 this is just what one would expect of a phenomenon which was dominated by the disordered regions between fibrils or crystallites. If these contacts are made more conducting when a fiber is put under tension (one of the possible contributions to negative piezoresistance discussed in Sect. 8.7) then the current-induced noise should decrease. One also expects the noise to be significantly less for the better-ordered CCVD fibers. However, no measurements of either of these effects have been reported.

# 9. High Temperature Properties

Very few measurements have been reported in the open literature on the structure and properties of carbon fibers above room temperature [Rowe and Lowe 1977; Sheehan 1987]. However, there is a somewhat larger literature on the properties of carbons at elevated temperatures [Lutcov et al. 1970; Tanaka and Suzuki 1972; Null et al. 1973; Leider et al. 1973]. Some high temperature properties of fibers can be estimated by analogy to bulk graphite results, and this approach will be used in this section where results on fibers are lacking.

## 9.1 High Temperature Thermal Properties

### 9.1.1 Thermal Conductivity

The property that has been most extensively investigated is the high temperature thermal conductivity of carbons. A comprehensive study of the thermal conductivity has been carried out up to 2500 K and the data have been classified according to the density of the carbon [Lutcov et al. 1970]. The carbon density can be roughly related to crystalline perfection, with the most highly ordered carbons approaching the mass density of single crystal graphite ($\varrho_m = 2.26$ g/cm$^3$). Below room temperature, the temperature dependence of the thermal conductivity exhibits a maximum (see Fig. 5.11) which increases in magnitude and moves to lower temperature as the crystalline perfection increases (see Fig. 9.1). Above room temperature, $\kappa(T)$ exhibits an exponential decrease $\kappa(T) = \exp[\theta/(bT)]$ with three characteristic linear segments in the plot of $\log\kappa(T)$ vs $1/T$ (see Fig. 9.2), corresponding to $\theta$ values of $\theta_1 = 179$ K, $\theta_2 = 900$ K, and $\theta_3 = 2480$ K for $b = 2$ [Krumhansl and Brooks 1953; Lutcov et al. 1970]. A similar dependence is expected for carbon fibers in this temperature range.

The phonon mean free path $l_\phi$ is the dominant factor in the temperature dependence of the thermal conductivity $\kappa(T) = C_v v l_\phi/3$, which is entirely due to the lattice contribution. At high temperature, $l_\phi$ is most sensitive to umklapp processes, which are expected to exhibit an exponential temperature dependence [Peierls 1929]. This dependence is especially strong in graphite because of its very high in-plane Debye temperature ($\theta_D = 2480$ K). The high temperature thermal conductivity data for carbons are well fit by this

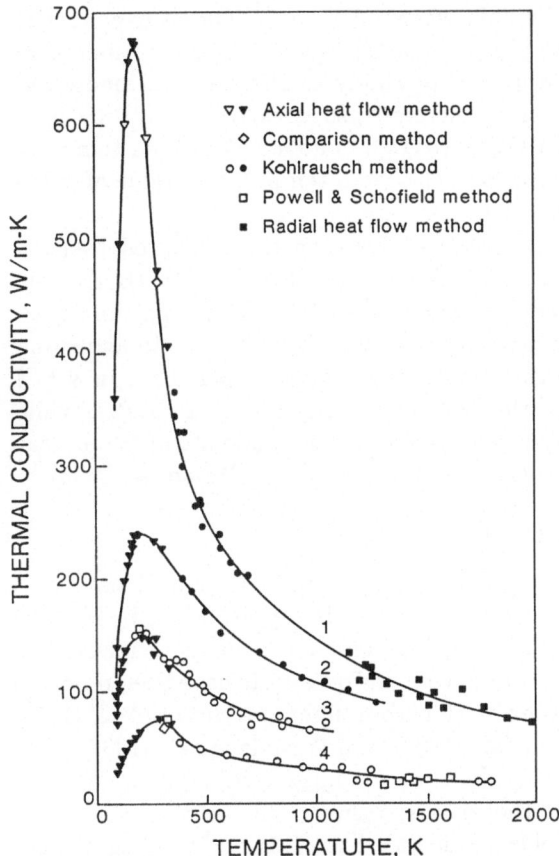

**Fig. 9.1.** The temperature dependence of the thermal conductivity ($\kappa_a$ and $\kappa_c$) of graphites with mass densities of $\varrho_m = 2.0$ g/cm$^3$ and 2.1 g/cm$^3$ (for $T_{HT} = 2400°C$) using various high temperature measurement techniques described by Lutcov et al. [1970]: curve 1 gives $\kappa_a(T)$ for $\varrho_m = 2.1$ g/cm$^3$; curve 2 gives $\kappa_a(T)$ for $\varrho_m = 2.0$ g/cm$^3$; curve 3 gives $\kappa_c(T)$ for $\varrho_m = 2.1$ g/cm$^3$; curve 4 gives $\kappa_c(T)$ for $\varrho_m = 2.0$ g/cm$^3$

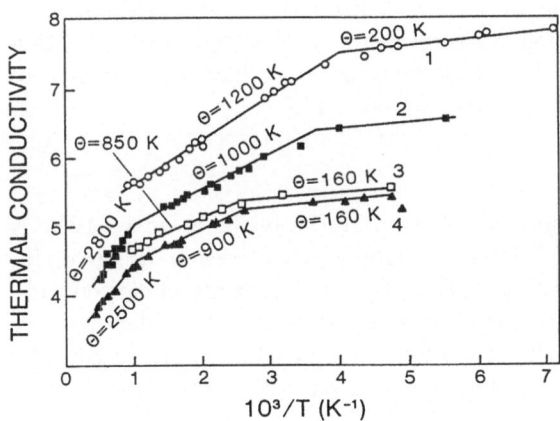

**Fig. 9.2.** The thermal conductivity of graphites (natural log scale) vs $1/T$ [Lutcov et al. 1970]: curve 1 gives $\kappa_a(1/T)$ for annealed pyrolytic carbon; curve 2 gives $\kappa_c(1/T)$ for $\varrho_m = 2.1$ g/cm$^3$; curve 3 gives $\kappa_c(1/T)$ for $\varrho_m = 2.0$ g/cm$^3$; curve 4 gives $\kappa_c(1/T)$ for $\varrho_m = 1.9$ g/cm$^3$

functional form, as shown in Fig. 9.2, where the three different exponential segments are clearly seen in the plot of $\log\kappa(T)$ vs $1/T$. As the temperature increases, the stiff modes contribute increasingly to the thermal conductivity. Also, as the crystalline ordering increases, so does the anisotropy in $\kappa$ [Tanaka and Suzuki 1972]. For HOPG, $(\kappa_a/\kappa_c)$ is about 65 at high temperatures (2700 K), somewhat reduced from the room temperature value of $\approx$200 (see Sect. 5.3) [Null et al. 1973].

Generally, good agreement is obtained between the measured thermal conductivity components $\kappa_a(T)$ and $\kappa_c(T)$ and the Kelly model [B.T. Kelly 1969b] for thermally annealed and stress annealed HOPG, though the agreement is not satisfactory for the as-deposited pyrographite prior to annealing [Null et al. 1973]. An anomalously small temperature dependence in $\kappa_c(T)$ is found above ~1500 K for the thermal- and stress-annealed HOPG, while $\kappa_c(T)$ for the as-deposited pyrographite shows an anomalous minimum near 2000 K, which is attributed to thermal expansion effects [Null et al. 1973].

### 9.1.2 Thermal Expansion Coefficient

Extensive measurements of the thermal expansion coefficient $\alpha_a(T)$ and $\alpha_c(T)$ have been reported in bulk graphite for temperatures up to 3000°C by Morgan [1972], showing large anisotropy and highly anomalous behavior, as described in Sect. 5.2. In the case of fibers, thermal expansion results have been reported up to ~1000°C for ex-PAN and ex-pitch carbon fibers by Butler et al. [1971] and more recently by Tanabe et al. [1987] and Yasuda et al. [1987]. The thermal expansion results for the fibers were analyzed in terms of suitable averaging over the values of the thermal expansion coefficients $\alpha_a$ and $\alpha_c$ for crystalline graphite using Reynolds' model (see Sect. 5.2), and good agreement between the measurements and model were obtained up to 700°C (see Figs. 5.9 and 5.10). For the highly ordered fibers, the thermal expansion along the fiber length is small, as would be expected from the anomalously low $\alpha_a$ for graphite. Thermal expansion measurements in the transverse direction have also been made on fibers, but these measurements have been limited to temperatures near room temperature (see Sect. 5.2).

### 9.1.3 Specific Heats

The high temperature heat capacity of graphite is available up to 1200 K from the JANAF tables [JANAF 1969] and empirical expressions are given for higher temperatures by Leider et al. [1973]. The heat capacity of carbons and graphite is unusual in exhibiting significant temperature dependence above room temperature because of the exceptionally high stiffness of the graphite lattice. Thus, new degrees of freedom are excited with increasing $T$ up to ~2000 K and beyond. Saturation of the heat capacity at the DuLong-Petit value $3R$/mole, where $R$ is the universal gas constant, is achieved well above

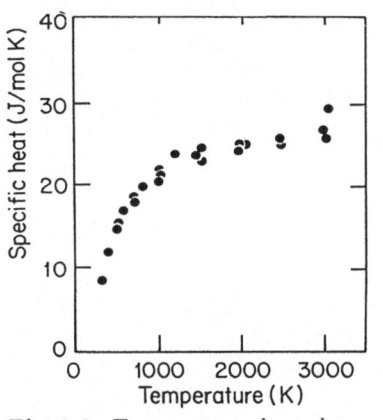

Fig. 9.3. Temperature dependence of the specific heat for graphite [adapted from Kelly (1981)]

Fig. 9.4. The temperature dependence ▶ of the electrical resistivity $\varrho$ of graphites with various mass densities $\varrho_m$ [Lutcov et al. 1970]: curve 1 gives $\varrho(T)$ for $\varrho_m = 1.0$ g/cm$^3$; curve 2 gives $\varrho_a(T)$ for $\varrho_m = 1.9$ g/cm$^3$; curve 3 gives $\varrho_c(T)$ for $\varrho_m = 1.9$ g/cm$^3$; curve 4 gives $\varrho_a(T)$ for $\varrho_m = 2.0$ g/cm$^3$; curve 5 gives $\varrho_c(T)$ for $\varrho_m = 2.0$ g/cm$^3$; curve 6 gives $\varrho_c(T)$ for $\varrho_m = 2.1$ g/cm$^3$; curve 7 gives $\varrho_a(T)$ for annealed pyrolytic carbon

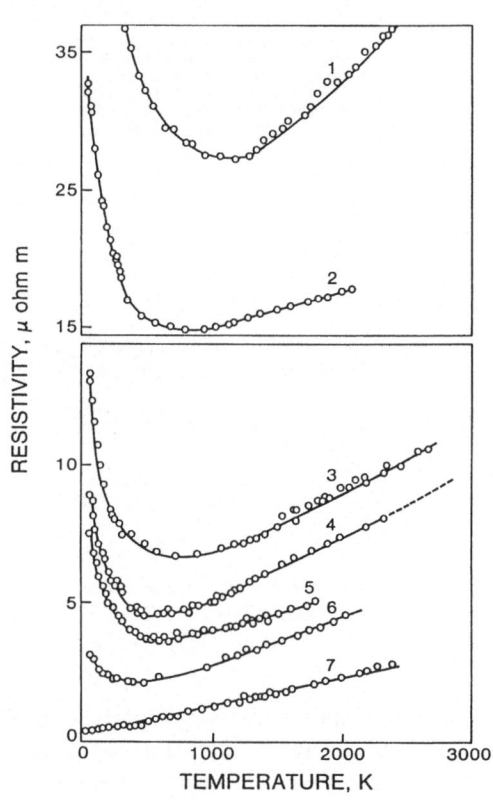

2000 K (see Fig. 9.3). In the lower temperature range (below 300 K), data for the temperature dependence of the heat capacity are given by Lutcov et al. [1970] as a function of the carbon density (corresponding to various degrees of structural perfection). No high temperature heat capacity data for carbon fibers are to our knowledge presently available. Thermodynamic functions (entropy, enthalpy, thermodynamic potential) are given by Lutcov et al. [1970] for various carbons (classified according to their densities), and a more complete analysis of the entropy, enthalpy and thermodynamic function for graphite up to the melting point is given by Leider et al. [1973]. It is anticipated that specific heats of fibers at high temperature would be similar to those of bulk carbons with the same mass density.

## 9.2 High Temperature Resistivity

The electrical resistivity $\varrho$ of graphites of various mass densities $\varrho_m$ (corresponding to various degrees of structural perfection) has been measured as a function of temperature above 1000 K (see Fig. 9.4). These data show that the high temperature resistivity increases approximately linearly in $T$ over

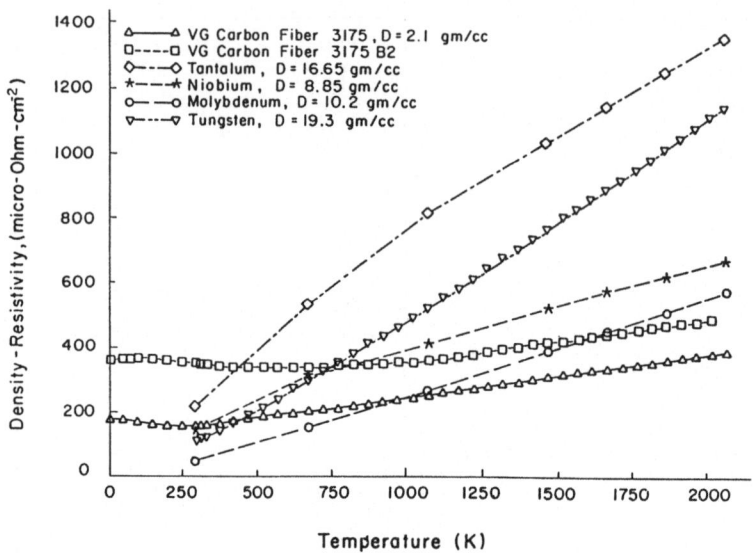

**Fig. 9.5.** The specific high temperature resistivity (density × resistivity) for two heat treated CCVD fibers ($T_{\text{HT}} = 3175$ K) in comparison with that of the refractory metals Ta, Mo and W [Ahmed et al. 1987; Woollam 1987]

a wide temperature range, consistent with the resistivity measurements on CCVD fibers above room temperature [Endo et al. 1983a] shown in Fig. 8.21. Measurements of the high temperature resistivity of as-deposited and thermally annealed vapor-grown fibers show an almost temperature-independent behavior over the temperature range $100 < T < 400$ K [Endo et al. 1983a], with a linear increase in $\varrho(T)$ at higher temperatures. The linear increase in $\varrho(T)$ above ∼400 K has been confirmed by three groups [Endo et al. 1983a; Ahmed et al. 1987; Heremans et al. 1988], and detailed measurements up to 2100 K are shown in Fig. 9.5. The results in Figs. 8.21 and 9.5 are consistent over the temperature range they have in common. Of particular interest is the low specific resistivity of carbon fibers at high temperatures, showing a lower specific resistivity above 1000 K than any of the refractory metals [Ahmed et al. 1987], as shown in Fig. 9.5. This observation could be of some practical importance.

In an effort to measure the resistivity change at the solid-liquid phase transition ($T_{\text{M}} \approx 4450$ K), approximate data on the high temperature resistance of CCVD carbon fibers have recently been obtained up to the melting point $T_{\text{M}}$, and the results are shown in Fig. 9.6 [Heremans et al. 1988]. The fibers were heated resistively by applying a single 28 μs current pulse with currents up to 20 A. The temperature of the fiber as a function of time was determined from the energy supplied in the pulse and the measured heat capacity for bulk graphite [Bundy et al. 1973] over this temperature range,

234

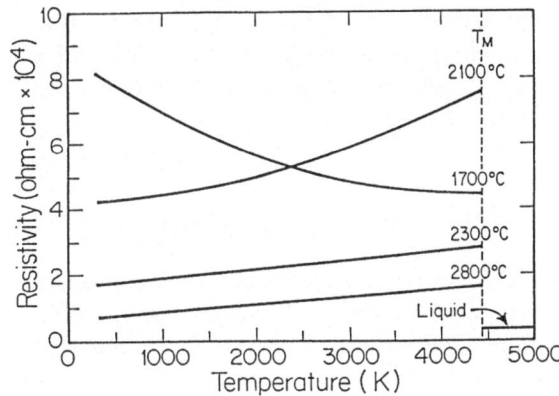

Fig. 9.6. The electrical resistivity vs temperature for CCVD carbon fibers with various heat treatment temperatures ($T_{\mathrm{HT}}$ = 1700, 2100, 2300, 2800°C). The sharp decrease in $\varrho(T)$ above ~4000 K is identified with the melting of the carbon fiber. The measured electrical resistivity for liquid carbon is shown [Heremans et al. 1988]

assuming that all the power dissipated in the current pulse was converted into thermal energy in the fiber. The results for well-graphitized fibers show an approximately linear temperature dependence for $\varrho(T)$ up to the melting temperature, where $\varrho(T)$ drops by nearly one order of magnitude, consistent with metallic conduction in liquid carbon (see Fig. 9.6). The results in Fig. 9.6 show that $\varrho(T)$ depends on $T_{\mathrm{HT}}$ in a complicated way. Since carbon is the highest melting point solid, the electrical resistivity of carbon fibers at very high temperatures could be of practical interest.

Also of interest is the observation that the product $\kappa\varrho$ for carbons at high temperature ($T > 1500$ K) is about the same for many carbons [Lutcov et al. 1970] and is independent of temperature and carbon density (for $T >$ 1500 K). It should be noted that this empirical relation is unusual and is not analogous to the Wiedemann-Franz law since for pristine carbon fibers, $\kappa(T)$ is dominated by phonons, while electrons play the dominant role in $\varrho(T)$.

Some justification for this empirical relation can however be found from the following argument. From the Wiedemann-Franz law, $\kappa_e\varrho = L_0 T$ where $\kappa_e$ is the electronic component of the thermal conductivity and $L_0$ is the Lorenz number. Since phonons dominate the high temperature thermal conductivity, we can write $\kappa \approx \kappa_{\mathrm{ph}}$ so that

$$\kappa\varrho \simeq \frac{\kappa_{\mathrm{ph}}}{\kappa_e} L_0 T = \frac{C_{\mathrm{ph}} v_{\mathrm{ph}}^2 \tau_{\mathrm{ph}}}{C_e v_e^2 \tau_e} L_0 T \quad , \tag{9.1}$$

where $v_{\mathrm{ph}}$ and $v_e$ are the phonon and electron velocities, respectively. To find the temperature dependence of the product $\kappa\varrho$ we note that at high $T$ the lattice heat capacity $C_{\mathrm{ph}}$ approaches a constant value ($\frac{3}{2} N k_{\mathrm{B}}$) according to the Dulong-Petit law, while the electronic heat capacity $C_e$ is approximated by $\gamma T$. Since umklapp processes dominate $\tau_{\mathrm{ph}}$ at high $T$, and since the temperature dependence of $\tau_{\mathrm{ph}}$ has the form $\tau_{\mathrm{ph}} \propto \exp(\theta/bT)$, then $\tau_{\mathrm{ph}}$ approaches a constant value at very high $T$. The usual temperature dependence $\tau_e \propto 1/T$ is assumed for the electron relaxation time $\tau_e$, while the kinetic energy of the electrons at high temperature is approximated by $\frac{1}{2}m_e v_e^2 \sim k_B T$ and $m_e$ is assumed to be independent of $T$. Combining the temperature dependence of all factors, it then follows that $\kappa\varrho \propto v_{ph}^2$. Thus $\kappa\varrho$ is expected to be independent of temperature at high T, as is observed experimentally.

## 9.3  High Temperature Mechanical Properties

Carbon fibers and carbon-carbon (C–C) composites exhibit superior mechanical properties at room temperature, as discussed in Chaps. 6 and 12. What is even more remarkable is the superior strength of carbon fibers and C-C composites at high temperatures.

Though few publications on the high temperature mechanical properties of carbons are available in the open literature [Kotlensky and Martens 1963], it is clear from what has been published that the tensile strength increases modestly until a temperature of $\approx 2000°C$ but falls at higher temperatures, while the Young's modulus decreases monotonically. In this connection, see Fig. 9.7 for the temperature dependence of $\sigma_T(T)$ and $E_Y(T)$ for Thornel P-50 fibers up to 2800°C [Rowe and Lowe 1977]. The decrease in $E_Y$ with rising temperature can be explained by the increasing anharmonicity of the lattice vibrations, thereby promoting plasticity and reducing the elastic response. With regard to the tensile strength, raising the temperature increases the plasticity and reduces the deleterious effect of point defects and local flaws along the fiber length. Furthermore, the high temperature may result in additional annealing when $T$ becomes comparable to $T_{HT}$, resulting in an increase in the graphene ribbon width, thereby leading to an increase in fiber strength. The first round of thermal cycling tends to reduce the fiber strength but subsequent cycling was found to have only a minor effect on $\sigma_T$. In con-

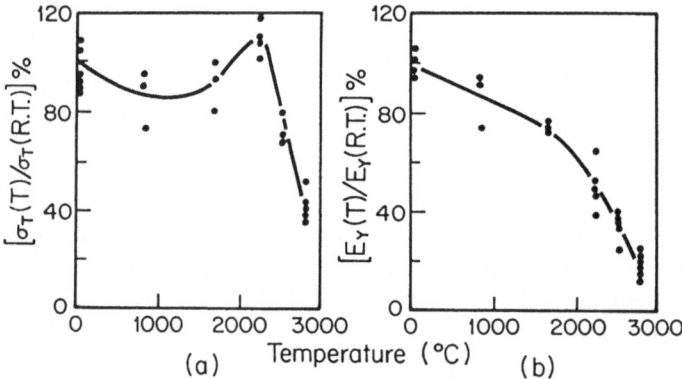

**Fig. 9.7.** Temperature dependence of (a) $\sigma_T$ and (b) $E_Y$ for a Thornel P-50 ex-pitch fiber plotted in terms of the percentage change in $\sigma_T(T)$ and $E_Y(T)$ relative to room temperature [Rowe and Lowe 1977]

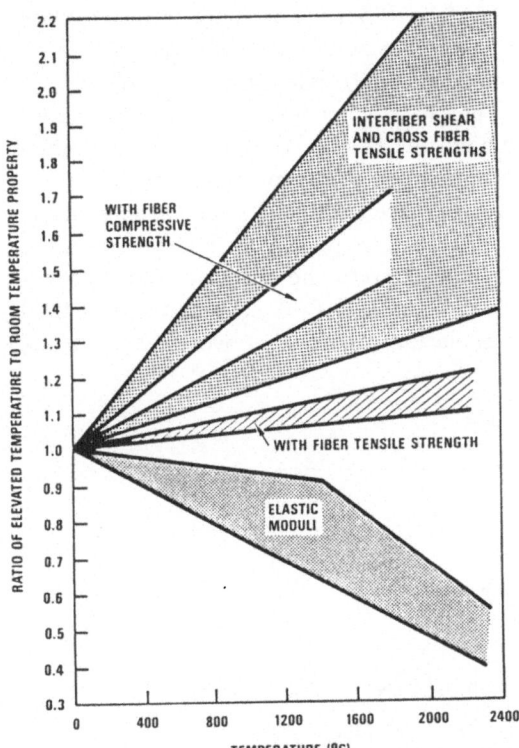

Fig. 9.8. Trends in the temperature dependence of various mechanical properties for C–C composites. It is to be noted that in general the mechanical properties improve with increasing temperature up to 2000°C. The shading indicates a range of values [Sheehan 1987]

trast, thermal cycling to 2800°C has almost no effect on $E_Y$ [Rowe and Lowe 1977].

Somewhat more data are available for the high temperature mechanical properties of carbon-carbon composites. This topic has recently been reviewed by Sheehan [1987]. The results on the carbon-carbon composites are consistent with the results on individual fibers (see Fig. 9.8). Specifically, this figure shows improvement in the mechanical properties (tensile strength, compressive strength and shear strength) as the temperature rises. Once again, the elastic moduli of the C–C composites decrease monotonically with rising $T$.

The physical basis for the improved mechanical performance of the C–C composites is explained as follows: As the C–C composite is cooled down from its processing temperature, cracks develop due to the differential thermal contraction of the fiber and the matrix. Increasing the temperature results in a healing of the cracks, and consequently a more homogeneous and coherent C–C composite is formed, with significantly increased strength [Sheehan 1987]. In designing C–C composites, the degree of bonding between the fibers and the matrix controls the mechanical properties of the composite. Strong bonding between the matrix and the fibers results in high shear strength while weak bonding increases the toughness, so that crack propagation in the matrix material can be arrested at the fiber surface. Fiber coatings are used to control the degree of matrix-fiber bonding to emphasize toughness or strength.

Carbon materials have an unusually low diffusivity [Kanter 1957] at high temperatures (at 2000°C, $D \sim 10^{-15}$ cm$^2$/s), two orders of magnitude smaller than Si in SiC and four orders of magnitude smaller than C in SiC [Sheehan 1987]. The low diffusivity and high anisotropy are responsible for the exceptionally low creep rate for carbon fibers, more than four orders of magnitude less than typical ceramics materials, also used frequently for high temperature applications, as is illustrated in Fig. 9.9. Despite the excellent high temperature mechanical properties of carbon fibers and C–C composites, practical applications are limited because of oxidation problems above 1000°C, as discussed further in Sect. 9.4.

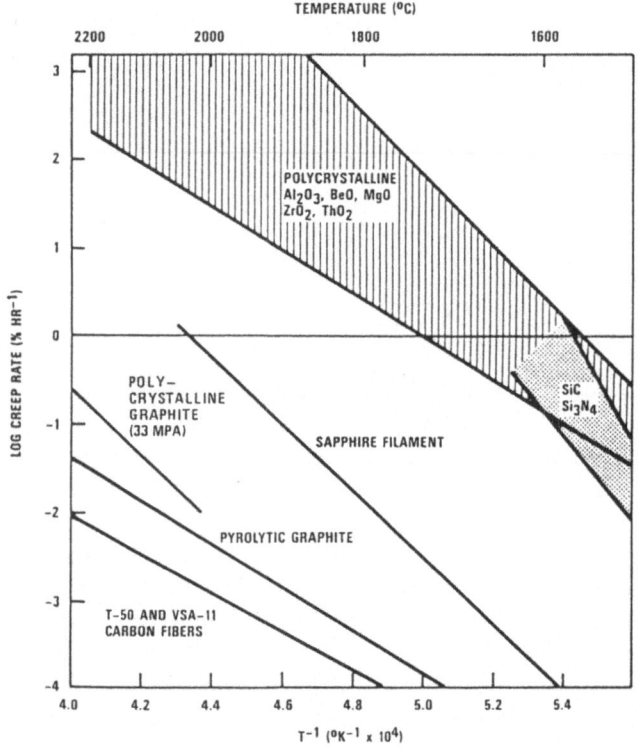

Fig. 9.9. Temperature dependence of the creep rate of C–C composites in comparison with several other materials [Sheehan 1987]

## 9.4 Oxidation Resistance

In an inert atmosphere or in vacuum, carbon fibers retain their excellent mechanical properties at elevated temperatures better than any known material.

Nevertheless, they do react with oxygen, especially in the presence of $H_2O$ vapor. In the graphite structure, edge sites are more easily oxidized than basal plane sites and many impurities are known to catalyze the oxidation reaction in carbons. The oxidation reaction evolves CO and $CO_2$ gas and leads to a loss of weight, and thus a loss in strength of the fibers or any composite made from them. Of course, if one protects the fibers from oxygen attack, they can be useful structural materials up to very high temperatures (see Sect. 9.3). For this reason there is serious interest in developing near-surface modifications or coatings to inhibit high temperature oxidation [Sheehan 1987].

Though little work has been reported in the literature on oxidation resistance for carbon fibers, it is well known that disordered carbons oxidize more readily than highly ordered graphitic carbons [McKee and Mimeault 1973; Donnet and Voet 1976]. A recent review on oxidation resistance in carbon fibers has been given by Eckstein [1986], and Sheehan [1987] has reviewed the use of coatings to achieve oxidation resistance. With regard to coatings for high temperature oxidation resistance, many are ceramics. But ceramic coatings on carbon fibers suffer from differences in the thermal expansion coefficient between the ceramic and graphite and have poor adhesion (see Sect. 9.4.2)

### 9.4.1 Experimental Results

The oxidation resistance of a number of commercially available carbon and graphite fibers has been reviewed by McMahon [1973, 1978]. At 400°C, many high strength carbon fibers are almost completely consumed, while others with similar Young's modulus have only a 3–5% weight loss (see Table 9.1). The different behavior of these fibers has been shown to be due to small (0.1–0.3%) amounts of alkali metal impurities uniformly distributed in fibers with poor oxidation resistance. The alkali metal is known to catalyze the oxidation of carbon. These impurities are present because of the detailed process used to spin the polymer precursor. Eckstein [1981, 1986] has shown that by increasing the carbon content of a carbon fiber (i.e., by decreasing the impurity content), the oxidation weight loss of PAN, rayon, pitch and mesophase pitch fibers (in the temperature range $230° < T < 375°C$) can be decreased. An empirical relation for the weight loss ($W_0 - W_t$) of ex-pitch fibers due to oxidation is

$$[(W_0 - W_t)/W_0]^n = kSt \quad , \tag{9.2}$$

where $W_0$ is the initial weight, $n$ is a constant (with values in the range 0.5–1.0), $k$ is a rate constant for the oxidation process and follows an Arrhenius relation for the temperature dependence, i.e., $k = k_0 \exp(E_a/k_B T)$, $S$ is the specific surface area, $E_a$ is an activation energy, $T$ is the temperature and $t$ is the time [Barr and Eckstein 1987]. Equation (9.2) has been verified to ~400°C.

**Table 9.1.** Weight loss for various carbon fibers after 3 hours at 400°C in air[a]

| Fiber | Carbon [%] | Modulus [GPa] | Alkali metal [ppm] | Weight loss [%] |
|---|---|---|---|---|
| High-modulus Celion | 99.7 | 413 | ND[b] | 0.05 |
| GY-70 | 99.3 | 517 | ND | 0.1 |
| High Modulus Celion | 99.6 | 345 | ND | 0.2 |
| UCC VS All pitch | | 345 | [c] | 0.6 |
| VM-0032(UCC non-woven mat) | | 220 | [c] | 0.7 |
| HMS | | 345 | | 3.3 |
| Celion 3000/6000 | 93.0 | 241 | ND | 3–5 |
| UCC pitch | 98.0 | 276 | <10 | 4.6 |
| Thornel 50 | | 345 | | 9.6 |
| Hercules AS | 93.9 | 241 | | 98.7 |
| Hercules AS 54-3-1 | 96.6 | 276 | 2880 | –[d] |
| Thornel 300 | 91.3 | 241 | >2100 | 99.3 |

[a] From McMahon [1978].

[b] ND indicates that no alkali metal could be detected in these fibers.

[c] A blank in the alkali metal column indicates that the measurement was not performed on that sample.

[d] No value is given in McMahon [1978], but the inference is that the weight loss might be similar to that for Hercules AS.

With regard to the effect of structural order on the oxidation resistance of carbon fibers, several studies have been carried out. For example, Fig. 9.10 clearly shows that graphitization of a PAN fiber retards the oxidation process [McMahon 1973, 1978]. Studies by Endo and Koyama [1980], and by G.W.Smith [1984] show that the weight loss due to oxidation is negligible up to some critical temperature, above which oxidation is rapid, as seen in Fig. 9.11. Their work also demonstrates that the fibers with higher structural perfection resist oxidation to higher temperatures; this is most clearly seen in Fig. 9.11 by comparing the oxidation resistance of a fiber prior to heat treatment (curve 5) with a similar fiber heat treated to 2800°C (curve 10). The

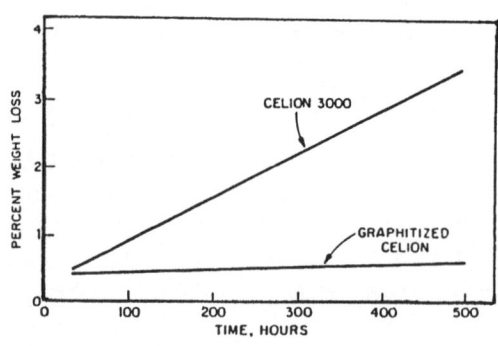

**Fig. 9.10.** The time dependence of the weight loss of annealed and non-annealed PAN fibers heated in air at 588 K [McMahon 1978]

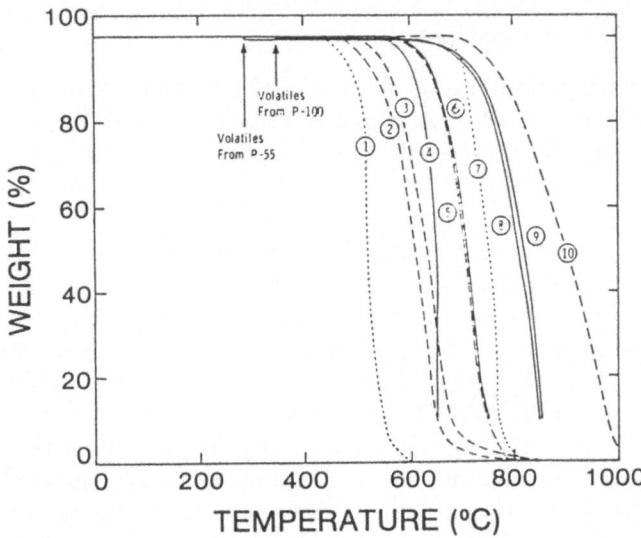

**Fig. 9.11.** Thermogravimetric analysis of weight loss curves for ten varieties of carbon fibers at a temperature scan rate of 5°C/min. The data of McKee and Mimeault [1973] were taken using a flowing oxygen atmosphere, whereas those of Endo and Koyama [1980] and Smith [1984] use air. The samples for the various curves are as follows. (1) low temperature treated ex-PAN (Modmor II); (2) low temperature ex-PAN; (3) low temperature ex-rayon; (4) ex-PAN (Union Carbide Thornel-300); (5) as-grown CCVD; (6) as-grown PYROGRAF; (7) high temperature ex-PAN (Modmor I); (8) high modulus mesophase ex-pitch (Union Carbide P-100); (9) intermediate modulus mesophase ex-pitch (Union Carbide P-55); (10) CCVD heat treated to 2800°C. From G.W. Smith [1984]

data of McKee and Mimeault [1973] were taken using a flowing oxygen atmosphere, whereas those of Endo and Koyama [1980], and Smith [1984] used air. Because of the close connection between structural perfection and the modulus of carbon fibers, their oxidation resistance can be correlated with the fiber modulus. Since the highly graphitic vapor-grown fibers ($T_{HT} \sim 2800°C$) expose a minimum number of edge sites and chemical impurities, it would appear that surface coatings and surface treatments of various types are needed to achieve oxidation resistance well beyond 800°C (see Sect. 9.4.2).

A novel concept for producing oxidation resistant coatings is based on the intercalation of the carbon fibers with a species such as $AlCl_3$ which is oxidized after intercalation [McQuillan and Reynolds 1986]. The aim of this procedure is to desorb the oxidized intercalant, leaving behind an oxidized ceramic coating which will inhibit further oxidation. What was actually observed upon heating such intercalated fibers in air to $\sim 800°C$ was the complete oxidation of the carbon fiber, leaving behind a porous $Al_2O_3$ fiber of approximately the same size and shape as the original carbon fiber.

## 9.4.2 Oxidation-Resistant Coatings

As reentry vehicle technology developed in the 1960s, the need for oxidation-resistant coatings on nuclear grade carbon nose cones was recognized [Zeitsch 1967; Goldstein et al. 1966]. From these efforts $B_2O_3$ glass coatings were developed, combining thermal stability with appropriate viscosity and wetting to provide oxidation protection to carbon surfaces over a wide temperature range [Chown et al. 1963]. In this connection, McKee [1987] has shown that molten borates are more successful in wetting a carbon substrate than other glass formers such as silicates, titanates, germanates, phosphates. In fact, the addition of boron as an impurity in the preparation of a carbon fiber also results in oxidation resistance. The physics behind this effect is the lowering of the Fermi level (the oxidation process also corresponds to a lowering of the Fermi level) by the boron addition, thereby reducing the probability that the carbon reacts with oxygen or some other oxidizing agent. In a parallel approach, non-oxide ceramic constituents can be incorporated into the carbon body, so that oxidation results in the formation of a borate glass coating which provides subsequent protection up to $\sim 1700°C$ for several hours, which is sufficient for many applications. The limitation of this coating is the volatility of the $B_2O_3$ above 1500°C [Soulen et al. 1955; McKee 1986].

Another common oxidation-resistant coating is SiC, which is applied to carbon fibers by a CVD process, giving oxidation resistance up to 1700°C, provided that care is taken to avoid flaws in the coating. Other Si-based ceramics can also be used for oxidation protective coatings, up to $\approx 1700°C$. Short term protection up to 2200°C has been achieved with sintered ZrC and $ZrB_2$ coatings. The high oxygen diffusion rates in the Zr and Hf oxides is responsible for the short term protection offered by these ceramic oxides [Sheehan 1987].

Since no known coating provides long term high temperature oxidation protection to a carbon surface, a multiple coating approach is used when a higher degree of oxidation protection is needed [Saito and Kogo 1980]. An effective oxidation protection coating should isolate the carbon from the oxidizing environment and should act as an efficient barrier to oxidation. To be effective, flaws in the coating due to fabrication processing, differences in the thermal expansion of the coating and the substrate, or stress-induced flaws must be eliminated. Thus, it is common to use a sealant glass-former in contact with the carbon surface to heal any flaws or cracks that are produced in the outer coating during thermal cycling. Borate glasses provide good sealants, and can be used in conjunction with CVD-deposited SiC or $Si_3N_4$ coatings.

For higher temperature use (above 1500°C), silicate glasses offer promise for replacing the borate glass, with $SiO_2$ forming the outer protective coating. One limitation of using $SiO_2$ glasses is the high viscosity of $SiO_2$ at low temperatures, limiting the flow necessary for sealing cracks and spreading over internal surfaces. For this reason, the $SiO_2$ glass is fluxed with some lower viscosity additives to provide effective operation at lower temperatures. The most oxidation resistant ceramic materials are $HfB_2$ and mixtures of $HfB_2$ with SiC. Such high temperature ceramics may form outer coatings above 1800°C, with an inner $SiO_2$ glass coating to serve as an erosion barrier and as a sealant for cracks [Sheehan 1987].

Though the concept of preparing oxidation resistant coatings for carbon fibers or for C–C composites is simple to formulate, complicated technology is needed for its effective implementation at increasing operating temperatures.

# 10. Intercalation of Graphite Fibers and Filaments

Intercalation compounds are formed by the insertion of atomic or molecular layers of a guest chemical species between layers in a host material such as graphite. Numerous reviews and conference proceedings on graphite intercalation compounds (GICs) are available [Vogel and Hérold 1977; Hérold 1979; Vogel 1980; Dresselhaus and Dresselhaus 1981; Pietronero and Tosatti 1981; Nishina et al. 1981; Solin 1982; Hérold and Guérard 1983; Dresselhaus et al. 1983, 1986; Eklund et al. 1984; Dresselhaus 1987]. The intercalation process occurs in highly anisotropic layered structures where the intraplanar binding forces are large in comparison with the interplanar binding forces. The guest species in an intercalation compound exhibits order, in contrast to doping where the guest species tends to occupy random locations. Intercalation provides the host material with a means for controlled variation of many physical properties over wide ranges. Intercalation can proceed with either donor intercalants which transfer electrons to the graphite host material or with acceptors which receive electrons from the graphite.

Intercalation can be achieved starting from a solid, liquid or gaseous reagent, though preparation from the vapor is the most common when samples with the highest degree of structural ordering are desired. Electrochemical techniques are also commonly used. A given intercalation compound can often be prepared by alternative growth techniques. Generally, atoms and simple molecules can be intercalated from the vapor, but large organic molecules are intercalated from solution. The intercalation rate and resulting intercalate concentration are strongly dependent on the intercalation conditions, such as pressure, the temperature difference between the host material and the intercalant, the physical sample dimensions, the degree of crystalline order and the defect density in the host material. When intercalation proceeds from the vapor, as for example in the case of the intercalation of bromine into graphite, the initiation of the intercalation process requires the intercalant partial pressure to exceed a threshold value that depends on temperature, the host material and the intercalant species. Intercalation proceeds with the insertion of intercalant layers sequentially from the surface layers inward, but starting from the edges through a nucleation process and advancing in the layer planes by diffusion.

Techniques have been developed for the intercalation of large numbers ($\gg 10^2$) of chemical reagents into graphite host materials, ranging from simple ionic species such as alkali metals, through diatomic molecules such as the halogens, to large organic molecules. The simpler binary and ternary compounds are usually prepared by direct synthesis, and the more complicated materials by a variety of stepwise intercalation procedures. A simple example of ternary intercalation is the synthesis of graphite compounds with the alloy $Na_xK_{1-x}$. Such an intercalation compound is of particular interest since sodium by itself does not readily intercalate into graphite but potassium does. Other interesting ternary compounds which have been synthesized include KH-GICs, KHg-GICs, and $KNH_3$-GICs. The KH-GICs allow packing densities of hydrogen atoms comparable to that in solid hydrogen, and the KHg-GICs are of interest for their unusual superconducting properties. Stepwise intercalation is often used for the introduction of organic intercalants such as benzene and tetrahydrofuran (THF) into a graphite host into which potassium has previously been intercalated.

Graphite intercalation compounds (GICs) are often very reactive and must be stored in an inert atmosphere. Some GICs (e.g., $CsBi_x$-GICs, $FeCl_3$-GICs and $SbCl_5$-GICs) are however quite stable in air. The reactivity of the intercalant does not necessarily determine the reactivity of the resulting GIC; for example pristine $FeCl_3$ is highly reactive while $FeCl_3$-GICs are relatively stable at room temperature. The reactivity of most GICs is greatly reduced by lowering the temperature to 77 K. Because of this reactivity, graphite intercalation compounds are normally encapsulated after preparation (e.g., in glass or quartz ampoules) to prevent intercalant desorption.

However, if the intercalation compound is removed from the encapsulating ampoule and exposed to the normal room-temperature environment, desorption proceeds for a period of time, after which the desorption is negligible. The resulting material, called a residue compound, is chemically stable and contains an intercalant concentration (usually small) dependent on the desorption temperature, on the intercalant concentration of the original lamellar compound and on the microstructure of the graphite host material.

Graphite intercalation compounds (GICs) are of particular physical interest because of their high degree of ordering. The most striking type of ordering is the staging phenomenon, which is defined by a periodic arrangement of $n$ graphite layers between sequential intercalate layers, where $n$ is the stage index (see Fig. 10.1). In this figure the potassium layers are indicated by dashed lines and the graphite layers by solid lines connecting circles, indicating schematically a projection of the carbon atom positions. The $ABAB$ graphite layer stacking for stages $n \geq 2$ is maintained between intercalate layers, although a "rhombohedral" stacking arrangement appears across intercalate layers. The stacking order is well-confirmed by x-ray diffraction ($hkl$) patterns [Dresselhaus and Dresselhaus 1981].

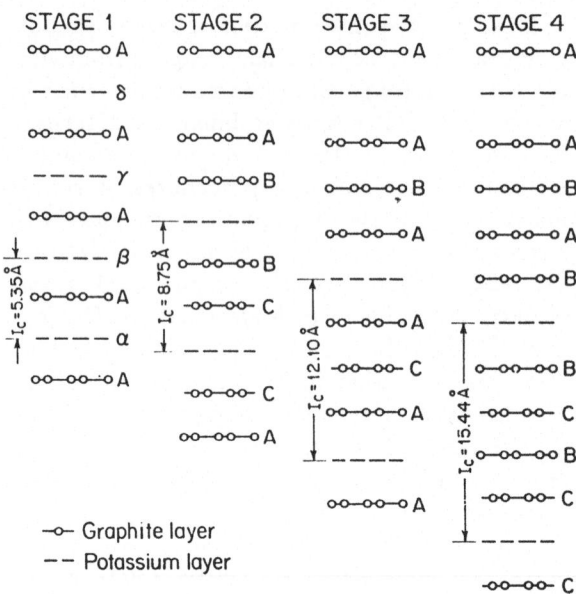

STAGE 1    STAGE 2    STAGE 3    STAGE 4

—○— Graphite layer
– – Potassium layer

Fig. 10.1. Schematic diagram illustrating the staging phenomenon in graphite-potassium intercalation compounds for stage index $1 \leq n \leq 4$ (adapted from Rüdorff and Schulze [1954])

It is possible to prepare single-staged materials with small (1–5%) admixtures of secondary-staged regions. X-ray diffractograms show that well-staged materials can be made up to stage ~10, indicating a c-axis repeat distance of $I_c \sim 40$ Å. A physical basis for the staging phenomenon is associated with the strong interatomic intercalant-intercalant binding relative to the intercalant-graphite binding, thereby favoring a close-packed in-plane arrangement, while the introduction of each intercalant layer adds a substantial strain energy, thereby favoring the minimum number of intercalant layers consistent with a given average intercalant concentration.

The intercalation process has a negligible effect on the in-plane ordering of the graphene layers and on the c-axis graphite interplanar spacing. The interlayer stacking of adjacent graphene planes remains $ABAB$ as in pristine graphite, although departures from this stacking can occur across an intervening intercalant layer. The most stable stacking arrangement of the graphene layers is the one which minimizes the volume needed to accommodate the intercalant in the interlamellar space.

The intercalant layers also exhibit ordering which may be similar to that of the parent material and may be commensurate or incommensurate with respect to the adjacent graphene layers. An example of a commensurate superlattice is a first-stage alkali-metal compound with potassium or rubidium intercalated into the graphite host (nominal chemical formulae $C_8K$ and

$C_8Rb$). The lattice constants of the intercalant layer are typically much larger than those of the graphene layers because the intercalant metallic radii or intramolecular bond lengths tend to be large compared with the carbon radius. Thus $C_8K$ and $C_8Rb$ exhibit a $(2 \times 2)$ in-plane superlattice with the in-plane unit cell area being four times greater than that for graphite.

The novel geometry of graphite fibers facilitates the study of various intercalation phenomena and is of particular importance for practical applications of graphite intercalation compounds. The synthesis, structure and properties of intercalated graphite fibers have much in common with intercalated bulk graphite host materials [Dresselhaus and Dresselhaus 1981], such as kish graphite [Austerman 1968] and highly oriented pyrolytic graphite (HOPG) [Moore 1973]. We therefore focus here on the unique structure and properties of intercalated graphite fibers and the differences between GICs prepared from fiber and bulk graphite host materials. Attention is also given to new physics that can be pursued with intercalated fibrous materials. The special applications of intercalated fibers are reviewed in Sects. 12.7 and 12.8.

## 10.1  Structural Order and Intercalation

Of particular significance to the properties of intercalated graphite fibers is the structure and structural order of the pristine fibers, which in turn is strongly dependent on the precursor material and on the heat treatment process of the pristine fibers [Dresselhaus 1984]. Graphite fibers based on polyacrylonitrile (ex-PAN), mesophase pitch (ex-pitch) and CCVD precursor (ex-CCVD) materials have all been successfully intercalated with many different intercalants to produce both donor- and acceptor-type GICs.

Intercalation is facilitated by a high degree of structural order of the graphite host material and is retarded by defects, structural imperfections and disorder. Good staging is only achieved in hosts with a high degree of structural perfection. Thus, basic scientific studies on intercalated graphite fibers have focused on highly ordered graphite fiber host materials (especially CCVD fibers). In contrast, the requirements of specific commercial applications of intercalated graphite fibers may involve rather different issues, such as the amount of intercalate uptake, quality control of the intercalation process, long term stability of the intercalated fibers under a variety of environmental conditions, and the availability of continuous intercalated fiber lengths.

Graphite fibers are normally intercalated by techniques similar to those used for the corresponding HOPG-based GICs, though the specific intercalation conditions will be different in detail. For example, the intercalation of a typical intercalant such as $CuCl_2$ into fiber hosts is successful over a smaller temperature range than for HOPG [Meschi et al. 1986]. In the case of CCVD fibers, intercalation is initiated at the free ends of the fibers and then proceeds along the fiber length. This is not the case for fibers made from polymeric precursors as will be discussed below. Heat treatment of the fibers to high temperatures ($\approx 3000°$C) prior to intercalation increases the structural ordering of the graphite planes, leading to enhanced intercalation and more uniform staging. In this context, CCVD fibers, heat treated to $\sim 3000°$C, exhibit crystalline coherence lengths of $L_a \sim 1000$ Å from which donor and acceptor compounds with well established staging order can be synthesized [Dresselhaus 1984].

In most cases the higher density of structural defects in the fibers tends to inhibit intercalation relative to bulk HOPG. Spain et al. [1964] and Ubbelohde [1969] have discussed the importance of defects in the graphite host material in connection with the threshold for bromine and nitrate intercalation. Hooley and Deitz [1978] demonstrated that the threshold pressure for halogen intercalation of fibers depended on their degree of graphitization; i.e., their structural perfection. The threshold pressure also was found to depend upon whether the fiber had been previously intercalated and with which intercalant species. If a fiber was de-intercalated and then subsequently re-intercalated, the threshold pressure was significantly reduced, as shown in Fig. 10.2, by the intercalation isotherms of $Br_2$ on ex-pitch (F54 II, $T_{HT}=3000°$C) and ex-PAN ($T_{HT}=2800°$C) fibers [Hooley and Deitz 1978]. The Celanese 2800 fibers do not intercalate bromine in the pressure range shown in the figure. The F54 II fibers do however intercalate bromine with a threshold pressure of $p/p_0 = 0.3$ on the first intercalation and a threshold pressure of $p/p_0 = 0.1$ on subsequent intercalations. The threshold pressure for HOPG is $p/p_0 \simeq 0.1$. Fibers which could not be intercalated with bromine because of an excessively high thresh-

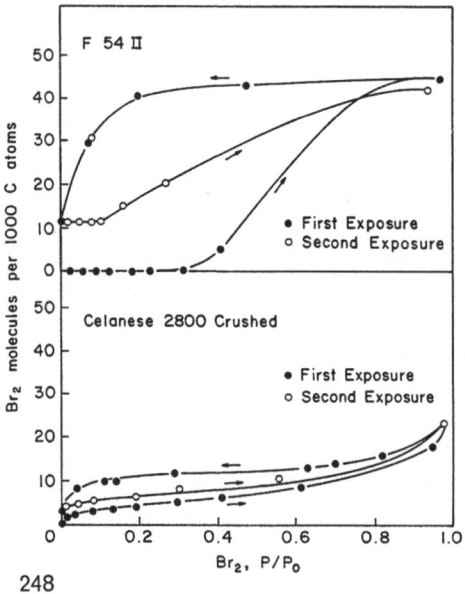

Fig. 10.2. Intercalation isotherms of $Br_2$ on ex-pitch (F54 II, $T_{HT}=3000°$C) and ex-PAN ($T_{HT}=2800°$C) fibers [Hooley and Deitz 1978]

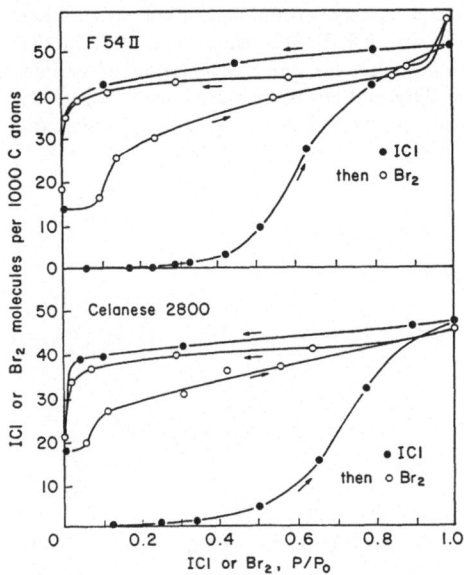

**Fig. 10.3.** Intercalation isotherms of ICl followed by $Br_2$ intercalation on the same fibers as used in Fig. 10.2 [Hooley and Deitz 1978]

old pressure could be first intercalated with ICl (see Fig. 10.3). This initial ICl intercalation presumably broke the bonds holding the graphite planes together and facilitated subsequent intercalation of bromine with a threshold pressure of $(p/p_0) \sim 0.1$.

These threshold phenomena were investigated electrochemically on a series of fibers made from both rayon and PAN by Besenhard et al. [1982a, 1982b]. By intercalating electrochemically, they were able to show a strong correlation with modulus (and thus presumably with structural perfection) of the threshold potential and the amount of charge which could be recovered upon de-intercalation (see Fig. 10.4). A higher threshold potential is related to greater forces pinning the layers together, while lower values of the charge recovery are related to higher densities of the intercalant remaining in

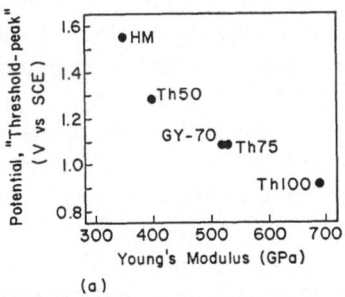

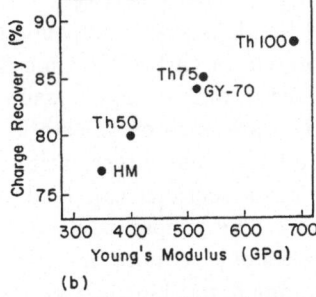

**Fig. 10.4a,b.** Correlation of the (a) potential threshold peak measured with respect to a standard calomel electrode (SCE) and (b) charge recovery with the Young's modulus for the fibers measured during the initial electrochemical intercalation of the fibers in 96% $H_2SO_4$ [Besenhard et al. 1982a]

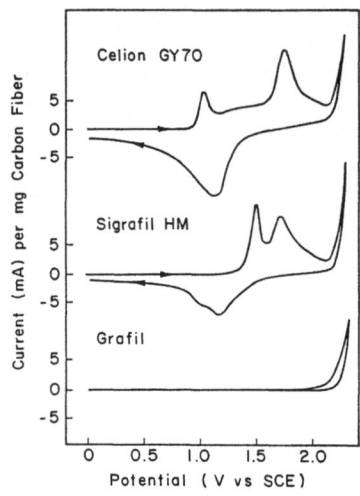

**Fig. 10.5.** Cyclic voltammograms of carbon fiber bundles in 96% $H_2SO_4$ showing a shift in the threshold peak for the intercalation of various fibers. The voltage is measured with respect to a standard calomel electrode (SCE) [Besenhard et al. 1982a]

a charged state in the fibers. The authors correlated the potential threshold peak and charge recovery with the Young's modulus of the fibers measured during the initial electrochemical intercalation of the fibers in 96% $H_2SO_4$ [Besenhard et al. 1982a, 1982b].

The electrochemical intercalation of sulfuric acid into three different graphite fibers is shown in Fig. 10.5. The advantage of the electrochemical technique is that it allows one to quantify the amount of energy (the integral of the voltage peak) associated with the irreversible threshold observed during the initial intercalation of a fiber. Besenhard et al. [1982a, 1982b] were thus able to estimate the density of $sp^3$-bonded carbons which, if responsible for inhibiting the initial intercalation, would explain the voltage vs current data. For GY-70, an ex-PAN fiber with $T_{HT} \sim 2200°C$, they estimated that approximately 0.25% of its carbon atoms were $sp^3$ bonded. Phenomena other than $sp^3$ bonding between the planes can be invoked to explain the additional resistance to opening the planes to the intercalant, such as the compressive stresses applied by neighboring ribbons (see Fig. 6.16) or the bending of the ribbons. The density of $sp^3$ bonds is just a convenient way to express the data. Electrochemical phenomena in carbon fluorine fiber compounds have also found practical applications in high energy density batteries (see Sect. 12.8).

Even though the presence of defects in fibers implies that one needs a larger driving force to initiate intercalation [Hooley 1977], the small dimensions of the fiber cross sections can lead to striking enhancements of the rate of intercalation. The preparation of stage 1 or 2 $AsF_5$ compounds in HOPG typically takes one or more days. Fibers, however, can be fully intercalated in less than one hour. This is because the fibers are small and because most commercially produced fibers have a sufficient number of exposed edges of basal planes so that intercalation occurs fairly uniformly along the fiber length. Thus, intercalation of a 180 cm long yarn of ex-PAN fibers with $AsF_5$ [Goldberg and Kalnin 1981] occurred at roughly the same rate as 2 cm single filaments [Kalnin and Goldberg 1981]. As already mentioned (and further discussed in Sect. 10.4.1), this is not what happens when one intercalates very well ordered CCVD fibers. In addition to enhanced kinetics in systems with large chemical driving forces for intercalation (such as $AsF_5$), it has been found that the small fiber cross section leads to enhanced intercalant uptake at least near the fiber surface, so that, for example, stage 1 $NiCl_2$ and $FeCl_3$ graphite intercalation compounds are more easily prepared in graphite fiber hosts than in bulk graphite [Endo et al. 1983b].

A comparison of some intercalation conditions for HOPG and two fiber hosts by Meschi et al. [1986] is given in Table 10.1, where it is seen that the more graphitic CCVD fibers (ex-benzene) intercalate more similarly to HOPG than to ex-mesophase pitch fibers annealed at comparable heat treatment temperatures $T_{HT}$ [Gaier and Slabe 1986]. When heat treated to high temperature (e.g., 2900°C), the CCVD fibers show better staging and a larger intercalant uptake than the ex-mesophase pitch fibers.

Table 10.1. Weight uptake and staging for $CuCl_2$ intercalation[a]

| Graphite host | T[°C] | $t$[days] | Composition[b] | Stage |
|---|---|---|---|---|
| Pyrographite (HOPG) | 535 | 7 | $C_{10}CuCl_2$ | 2 |
| Ex-mesophase pitch fiber $T_{HT} = 3200°C$ | 520 | 7 | $C_{12}CuCl_2$ | 2-3-4 |
| Ex-benzene (BDF) $T_{HT} = 2900°C$ | 500 | 2 | $C_{11}CuCl_2$ | 2 |

[a] From Meschi et al. [1986]. Here $T$ and $t$ are respectively the temperature and time for intercalation.
[b] From measurements of susceptibility, x-ray diffraction and weight uptake.

Of particular significance for both scientific studies and practical applications is the higher stability of intercalated CCVD fibers with regard to de-intercalation under ambient conditions [Gaier and Jaworske 1985; Gaier 1985; Shioya et al. 1986]. In fact, after the initial intercalant desorption, Gaier and Jaworske [1985], and Gaier and Slabe [1986] have shown that $Br_2$ intercalated pitch fibers (P-100) remain essentially stable for an indefinite time period as residue compounds under ambient conditions despite the large surface area of exposed edge planes. Perhaps defect sites serve to pin the intercalant in this case. However, finely chopped CCVD fibers tend to de-intercalate rapidly, especially through the open ends [Gaier and Slabe 1986].

## 10.2  Structure and Staging

As for bulk GICs, the structure of intercalated graphite fibers is described by ordering along the $c$-axis, which is dominated by the staging phenomenon and by in-plane ordering which can be either commensurate or incommensurate with respect to the adjacent graphite layers [Dresselhaus and Dresselhaus 1981]. For a given intercalant, the ideal $c$-axis and in-plane ordering of the intercalation compound is the same in fibrous graphite hosts as for HOPG, though the degree of ordering that is achieved in the actual samples tends to be significantly lower in the fibrous host materials. For example, the density of the fibers tends to be lower than that for HOPG, both before and after intercalation [Meschi et al. 1986], consistent with the presence of voids in the pristine and intercalated fibers. The larger variation in density after intercalation of the fibers is consistent with lower homogeneity of the intercalation process in fiber hosts (see Table 10.2). High resolution TEM studies [Endo et al. 1983a] show that the degree of ordering in the intercalated fibers increases with increasing order of the pristine fiber host.

Table 10.2. Fiber densities before and after intercalation with $CuCl_2$[a]

| Fibers | Density before intercalation [g/cm$^3$] | Density after intercalation [g/cm$^3$] |
|---|---|---|
| CCVD ($T_{HT} = 2900°C$) | $2.08 \pm 0.05$ | $2.40 \pm 0.1$ |
| Ex-mesophase pitch[b] ($T_{HT} = 2500°C$) | $2.12 \pm 0.05$ | $2.36 \pm 0.1$ |
| Ex-mesophase pitch[b] ($T_{HT} = 3200°C$) | $2.12 \pm 0.05$ | $2.45 \pm 0.1$ |
| Graphite[c] (calculated values) | 2.26 | 2.56 (2nd stage) |

[a] From Meschi et al. [1986].  [b] From Union Carbide Corp.  [c] Ideal limit.

Since the fiber geometry offers no special advantages for the study of in-plane ordering, the intercalated graphite fibers have not been used to study the interesting structural phase transitions that are observed in GICs [Dresselhaus and Dresselhaus 1981]. For these reasons, structural studies of intercalated graphite fibers tend to focus on staging coherence and staging fidelity, rather than the broader structural issues commonly studied in bulk GICs.

In this context, fibers provide a convenient host material for studying $c$-axis structural phenomena, because the lattice imaging method can be applied generally to thin fibers (diameter $\leq 1\,\mu m$), without further thinning. In contrast, c-axis lattice images are more difficult to observe in bulk graphite hosts since their observation requires the sample edges to be turned up to achieve the proper geometry for c-axis lattice imaging. Thus graphite fibers provide a convenient host material for studying lattice damage introduced by various means, as for example, by ion implantation (see Sect. 11.3) [Endo et al. 1984; Salamanca-Riba 1985; Salamanca-Riba et al. 1985].

To compare the properties of GICs based on fibers and on bulk graphite hosts, knowledge of the stage of the GIC is important. Therefore considerable effort has been directed to the characterization of intercalated graphite fibers with regard to both stage index and staging fidelity (longitudinal and transverse to the c-axis). The intercalated fibers can be characterized for stage by a number of techniques. For example, the staging of a bunch of intercalated graphite fibers can be conveniently measured using the Debye-Scherrer x-ray diffraction technique. This technique has been applied to characterize the staging in intercalated ex-PAN, ex-mesophase pitch and CCVD graphite fibers. An example illustrating the degree to which staging can be achieved in a vapor-grown graphite fiber is shown in Fig. 10.6, where a Debye-Scherrer

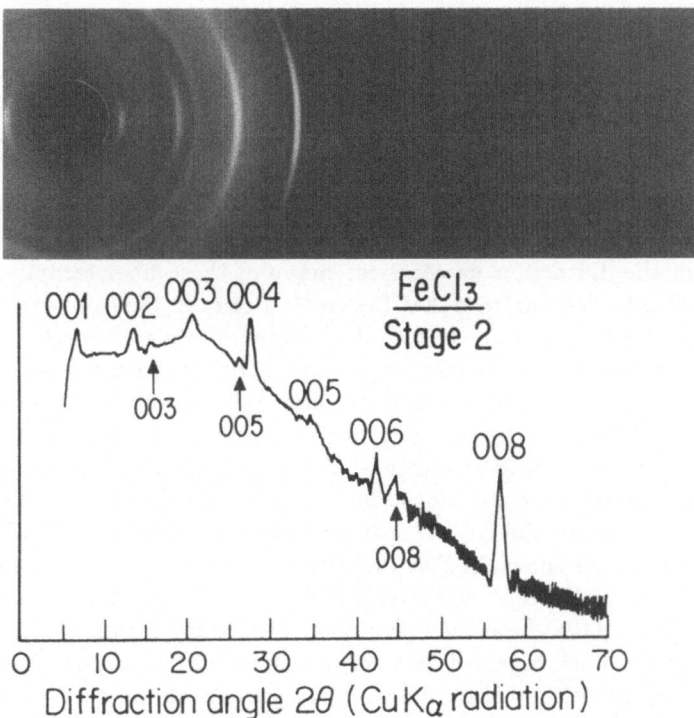

**Fig. 10.6.** Debye-Scherrer x-ray diffraction pattern taken with Cu $K\alpha$ radiation and the corresponding microdensitometer trace for a stage 2 $FeCl_3$-intercalated CCVD fiber. The $(00l)$ reflections are indexed, with labels above the trace for the stage 2 reflections, and below the trace for the small admixture of stage 3 regions [Chieu et al. 1983]

pattern for stage 2 FeCl$_3$-GIC fiber sample is shown [Chieu et al. 1983]. The (00$l$) reflections are indexed and the densitometer trace for this diffraction pattern shows only a small admixture of stage 3 regions. Debye-Scherrer patterns for intercalated ex-PAN and ex-mesophase pitch fibers show a much lower degree of staging, with the ex-mesophase pitch fibers showing more evidence for staging than the ex-PAN fibers [Kwizera et al. 1982; Meschi et al. 1986]. This is consistent with the relative degree of structural perfection found in these three kinds of fibers (see Chap. 3). The degree of staging fidelity that is achieved in a given fiber host material also depends on the intercalant that is used. For example, PAN fibers intercalated with AsF$_5$ were found to exhibit an unusual amount of staging relative to other intercalants [Goldberg 1985] though, even under the most favorable conditions, single stage fibers have not been realized in PAN host materials.

Measurement of the x-ray absorption as a function of energy near and above an absorption edge can provide information on: the chemical state, orientation, and local environment of an atom even in disordered materials (see Sect. 3.8). Bartlett et al. [1978] demonstrated that AsF$_5$ intercalation in HOPG is believed to proceed according to the reaction

$$2e + 3AsF_5 \rightarrow AsF_3 + 2AsF_6^-$$

based on the shape and position of the As $K$ shell absorption edge. The electrons in the above reaction come from the graphite, and the degree of charge transfer per intercalated AsF$_5$ unit depends on how far to the right the reaction goes.

Heald et al. [1983] examined the absorption edge of arsenic in low concentration AsF$_5$ intercalated fibers. Even though no evidence for intercalation was found through the diffraction patterns for many of these fiber samples, Heald et al. were able to demonstrate that the As$^{3+}$ species in the compound was oriented along the basal planes. In fact, the degree of orientation of the As-F bonds was higher than expected for AsF$_3$, implying that either the AsF$_3$ molecule is distorted upon intercalation or that other As$^{3+}$ species are present in these compounds. However, the shape of the absorption edge was consistent with the reaction proceeding as shown above (see Table 10.3). In these fibers with low intercalant concentrations, the reaction was found to go almost to completion, while for the HOPG host a considerable amount of AsF$_5$ was found in the sample, as shown in Fig. 10.7 [Heald et al. 1983]. This result differs from prior work of Bartlett et al. [1978] where it was concluded that the reaction in stage 2 HOPG-based samples went all the way to completion.

Since fibers have a much higher defect density than HOPG or powdered graphite, one would expect the higher defect density to affect the properties of an intercalation compound at low intercalant concentrations. Low concentration intercalation compounds in fiber hosts have been examined in two EXAFS studies: the one described above on AsF$_5$ intercalation and another

**Table 10.3.** The relative concentrations of intercalated species obtained from x-ray absorption data (Fig. 10.7) by Heald et al. [1983][a]

| Host: | $AsF_3$ | $AsF_5$ | $AsF_6^-$ |
|---|---|---|---|
| Fibers – perpendicular orientation | 0.30 | 0.08 | 0.62 |
| Fibers – parallel orientation | 0.36 | 0.09 | 0.55 |
| HOPG | 0.19 | 0.45 | 0.36 |

[a] Note the approximate 2:1 ratio of $AsF_6^-$ and $AsF_3$ in both fibers and HOPG in agreement with the reaction shown in the text and originally suggested by Bartlett et al. [1978]

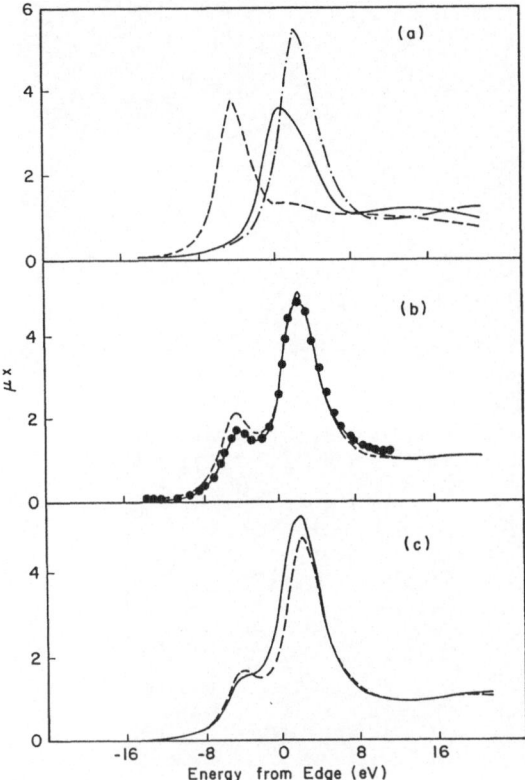

Fig. 10.7.
X-ray absorption spectra of: (a) $AsF_3$ (*dashed line*), $AsF_5$ (*solid line*) and $NaAsF_6$ (*dotted-dashed line*) standards; (b) $AsF_5$-intercalated Celion GY-70 fibers with the x-ray polarization parallel (*dashed line*) and perpendicular (*solid line*) to the fiber axis. (A fit to the data using a sum of the above standard traces is shown by the dots); (c) $AsF_5$-intercalated HOPG (*solid line*) and fibers (*dashed line*) with polarization perpendicular to the fiber axis [Heald et al. 1983]

by Feldman et al. [1983, 1984] based on bromine intercalation. By measuring the absorption spectrum of bromine residue compounds of both HOPG and ex-pitch fibers, they were able to show that, as in HOPG and grafoil [Heald and Stern 1980], bromine intercalates into the fibers and lies between the basal planes in ex-pitch fiber bromine residue compounds. Thus, the *intercalation* of bromine and $AsF_5$ (rather than a simple trapping of molecules at prismatic edge sites) was demonstrated in residue fiber compounds.

While Debye-Scherrer patterns provide average structural information of a statistical nature, they provide little information about the variation of stag-

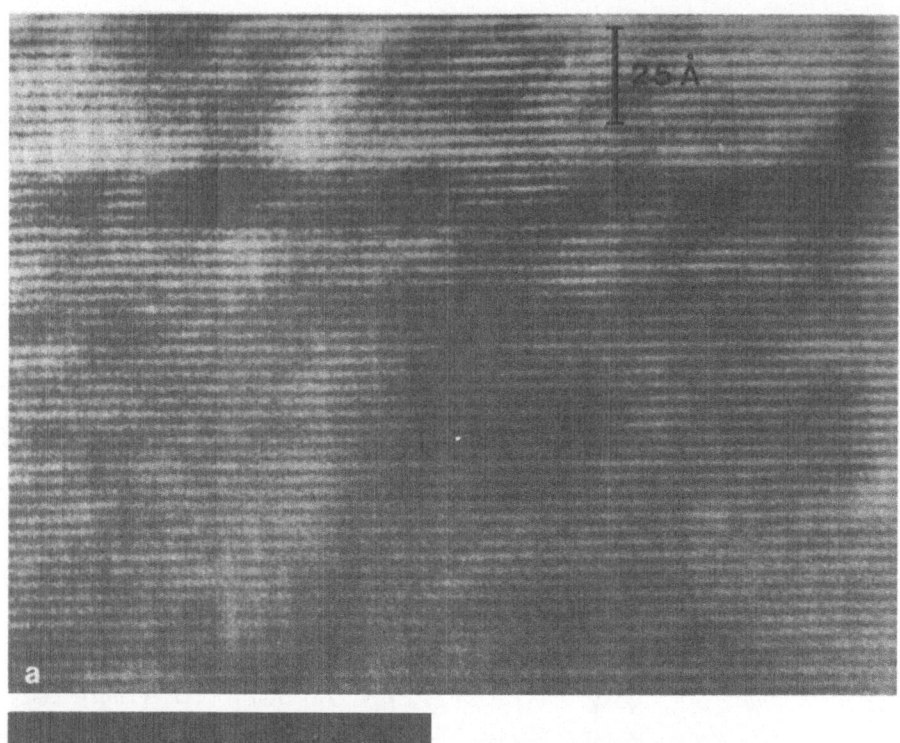

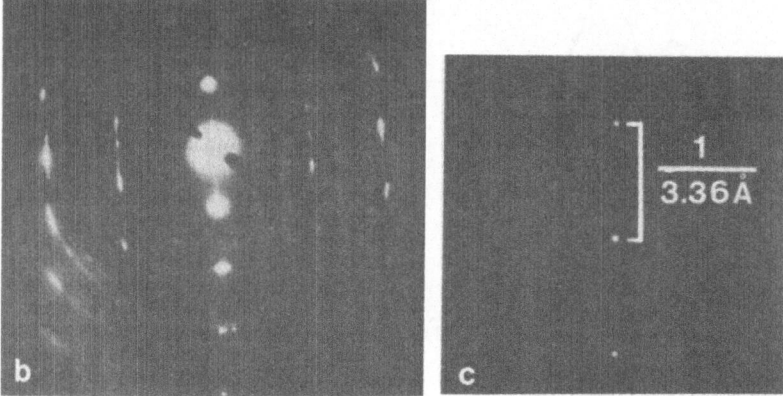

Fig. 10.8a-c. Caption see opposite page

ing from one fiber to another or along the length of a single fiber. To identify the staging in a single fiber and to determine the staging variation along the length of the fiber, characterization has more generally been carried out using either lattice imaging techniques with a high resolution transmission electron microscope (TEM) [Endo et al. 1983a, 1983b] or with a Raman microprobe [McNeil et al. 1985, 1986]. The field emission scanning transmission electron microscope (FESTEM) has also been used to determine the compositional

256

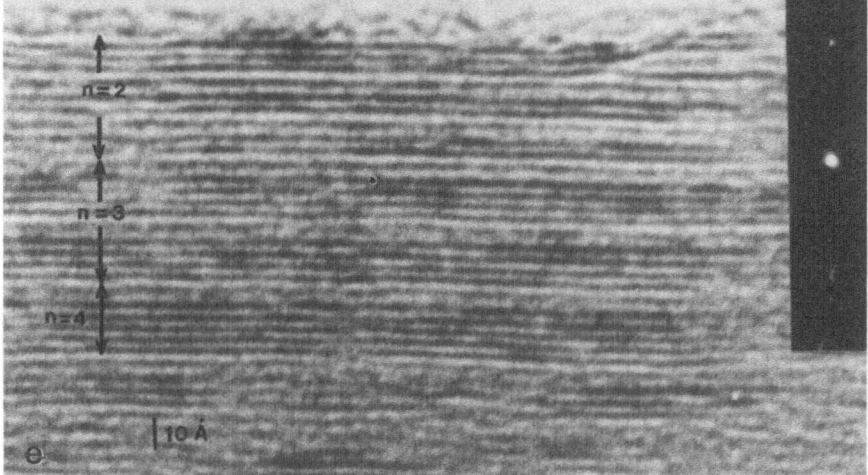

Fig. 10.8a-e. Lattice fringe image of a CCVD fiber heat treated to 2900°C: (a) lattice fringe pattern of the pristine fiber; (b) selected area diffraction pattern corresponding to (a); (c) optical diffractogram corresponding to (a); (d) lattice fringe image of a fiber similar to (a) after intercalation with $CuCl_2$ to a stage 2 compound in a benzene-derived fiber ($T_{HT}$ = 2900°C) and the optical diffractogram (*upper right*) corresponding to (d) [Endo et al. 1983a, 1983b; Dresselhaus 1984]. (e) Lattice fringe image from a mixed stage 2 and 3 $SbCl_5$ intercalated fiber [Salamanca-Riba et al. 1985]

variation along the fiber length, based on the x-ray fluorescence technique [Qian et al. 1987].

Lattice fringe images for CCVD graphite fibers heat treated to temperatures $T_{HT} \geq 2800°C$ show defect-free parallel lattice planes for distances extending at least 1000 Å (see Fig. 10.8a). The effect of intercalation is illustrated in Fig. 10.8d where layer sequences with a dominantly stage $n = 2$ $CuCl_2$-intercalated CCVD fiber ($T_{HT} = 2900°C$) are readily identified [Endo et al. 1983a, 1983b; Dresselhaus 1984], though some $n = 1$ and $n = 3$ staging infidelities can also be found in the photograph. For contrast, Fig. 10.8e shows lattice fringes that are not so straight for a similar fiber intercalated with $SbCl_5$ to yield a mixed stage 2 and stage 3 compound. The optical diffractograms were taken using the photographic film of the lattice fringes as a diffraction grating and give information on the staging homogeneity and staging fidelity (see Fig. 10.8d and e). This identification of staging by the lattice fringes is made possible by the different atomic numbers and hence different in-plane electron densities in the intercalant layers relative to the graphite layers, thereby showing contrast between electrons scattered by the intercalant and by the graphite layers. These contrasts vary from one intercalant to another, thereby giving rise to characteristic visual patterns for each intercalant. It is significant that the extent of the long, defect-free regions of graphite intercalation compounds is strongly dependent on the perfection of the host crystal, the characteristics of the intercalated species and the conditions of intercalation. Commensurate GICs using intercalants such as KHg and $Br_2$, which exhibit long coherence distances in bulk GICs, also tend to exhibit the largest defect-free regions in the corresponding intercalated fibers [Timp and Dresselhaus 1984].

The field emission scanning transmission electron microscope (FESTEM) has also been used to monitor the homogeneity of the intercalant concentration along the fiber length. Specific studies have been made for a partially desorbed $Br_2$ intercalated pitch-based fiber by mapping the intensity of the Br $K\alpha_1$ x-ray line while scanning the fiber length [Qian et al. 1987]. The FESTEM technique has also been applied to study the lateral homogeneity of the bromine distribution across the fiber cross section, showing that the bromine concentration is approximately uniform along the cross section for bromine-rich regions and follows a Gaussian distribution over the bromine-poor regions. This microscopic information was then used to model the macroscopic residual resistivity of the fiber [Qian et al. 1987].

It should be emphasized that the high resolution lattice fringe imaging technique provides microscopic information on the nature and density of defects found in intercalated graphite fibers. Statistical information is found by scanning the length of many similarly intercalated fibers. The most common types of defects in relatively well-staged fibers are staging infidelities whereby a stage $n - 1$ and/or $n + 1$ sandwich is observed directly on the micrograph of a nominally stage $n$ compound (see Fig. 10.8d). For less well-staged intercalated fibers, edge dislocations associated with the termination of an intercalant layer are commonly found (see Fig. 10.9). Note the wavy fringe pattern in Fig. 10.9 for the $KH_x$ intercalant [Salamanca-Riba et al. 1986] in comparison to Fig. 10.8d. In general, acceptor compounds tend to

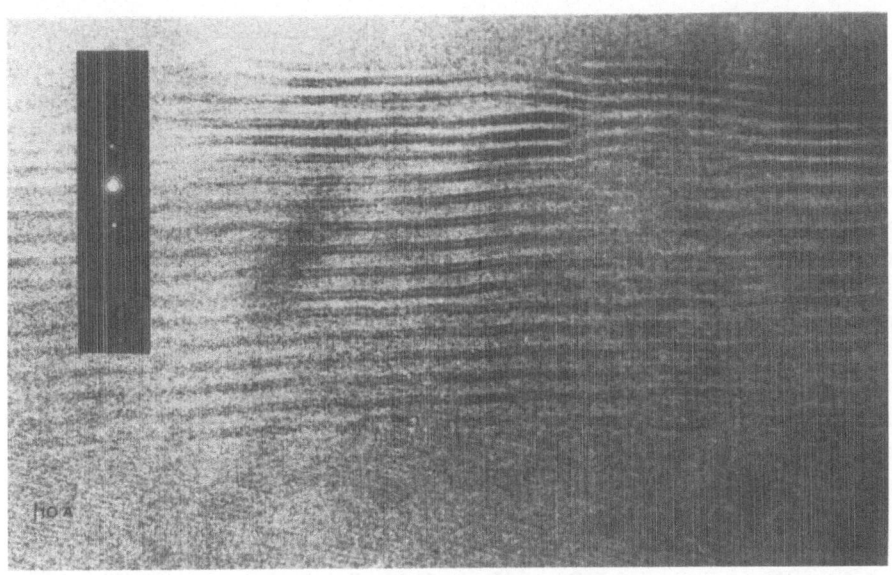

**Fig. 10.9.** Bright field $c$-axis lattice fringes of a stage 2 intercalated fiber (CCVD) prepared by chemical absorption of hydrogen into a stage 1 $C_8K$ intercalated fiber. The figure shows a $C_8KH_x$ region and a $C_{24}K$ (desorbed) region. The inset is an optical diffractogram taken from the negative of the lattice fringe photograph [Salamanca-Riba et al. 1986]

yield long straight fringes while alkali metal donor compounds show short wavy fringes, more similar to Fig. 10.9.

## 10.3 Lattice Properties

The high anisotropy of the structure of graphite and its intercalation compounds is also reflected in the lattice properties. The lattice modes show large in-plane dispersion, but weak coupling and small dispersion along the $c$-axis. Study of the lattice mode structure (for example, by the Raman effect) is of particular significance because the modes on the intercalant layer, the graphite bounding layers (adjacent to the intercalant) and the graphite interior layers (not adjacent to the intercalant) occur at different frequencies and therefore can be investigated independently as a function of intercalated species and concentration. The results show that the lattice mode frequencies of intercalated graphite fibers are the same as for similarly intercalated HOPG [Chieu et al. 1983]. Evidence that the interior graphite layers are essentially unaffected by intercalation is provided by Raman spectra observed for a variety of donor and acceptor compounds, in which the principal in-plane Raman-active optical mode appears essentially unshifted from its frequency in pristine graphite, but with an intensity that decreases with increasing intercalant concentration, corresponding to the decrease in the relative number of interior graphite layers.

Modifications to the graphite lattice modes are, however, observed on the bounding graphite layers adjacent to the intercalant layer planes [Dresselhaus and Dresselhaus 1982]. The characteristics of the Raman spectra show that the perturbation due to the intercalant layer is largely confined to the graphite bounding layers, which effectively screen the graphite interior layers from the charged intercalant layer. The modes associated with the bounding graphite layers $\hat{E}_{2g_2}$ are upshifted in frequency relative to the interior layer modes $E_{2g_2}^0$ by $\sim 20$ cm$^{-1}$ almost independent of intercalant species and concentration. The intensity of the upshifted modes grows with increasing intercalant concentration as the relative number of graphite bounding layers increases. The lattice mode structure of the intercalated layer depends in detail on the intercalant species and is related to the corresponding modes in the parent crystalline solid.

The use of Raman spectroscopy for the stage characterization of intercalated graphite fibers is essentially the same as that for intercalated graphite generally [Dresselhaus and Dresselhaus 1982; Dresselhaus 1984]. Since the stage dependence of Raman-active frequencies for both the graphite interior layers $\omega(E_{2g_2}^0)$ and graphite bounding layers $\omega(\hat{E}_{2g_2})$ is the same for HOPG and graphite fiber-based GICs, the measured Raman frequencies can be used to yield the stage of a given donor or acceptor fiber [Dresselhaus and Dresselhaus 1982; Dresselhaus 1984]. The relative intensities of the Raman lines associated with the graphite bounding layers and the graphite interior layers can also be used to provide information on the stage of the graphite fibers. As the intercalant uptake increases, and the number of graphite bounding layers relative to the number of graphite interior layers increases, so does the intensity of the bounding layer mode $\hat{E}_{2g_2}$ relative to that for the interior layers $E_{2g_2}^0$ mode. We note that for stage 1 and 2 compounds, all the graphite layers are adjacent to intercalate layers, so that the intensity for the $E_{2g_2}^0$ mode due to the graphite interior layers vanishes [Dresselhaus and Dresselhaus 1982].

For acceptor compounds, the frequencies are upshifted as a function of reciprocal stage (see Fig. 10.10), reflecting the decrease in the intraplanar bond lengths as electron charge is transferred from the graphite to the intercalant layers. Of significance is the similar upshift in frequency observed with a large number of acceptor intercalants. Similar Raman frequencies are found for graphite fibers and HOPG host materials intercalated to the same stage [Chieu et al. 1983; Chieu 1983]. Thus, using the results of Fig. 10.10, the stage of acceptor-intercalated graphite fibers can be found from the measured Raman frequencies $\omega(\hat{E}_{2g_2})$ and $\omega(E_{2g_2}^0)$. In contrast to the acceptor GICs, the donor GICs exhibit a characteristic downshift in frequency, since the bond lengths increase as electrons are transferred to the graphite from the donor-intercalant. Figure 10.11 shows the characteristic downshift in fre-

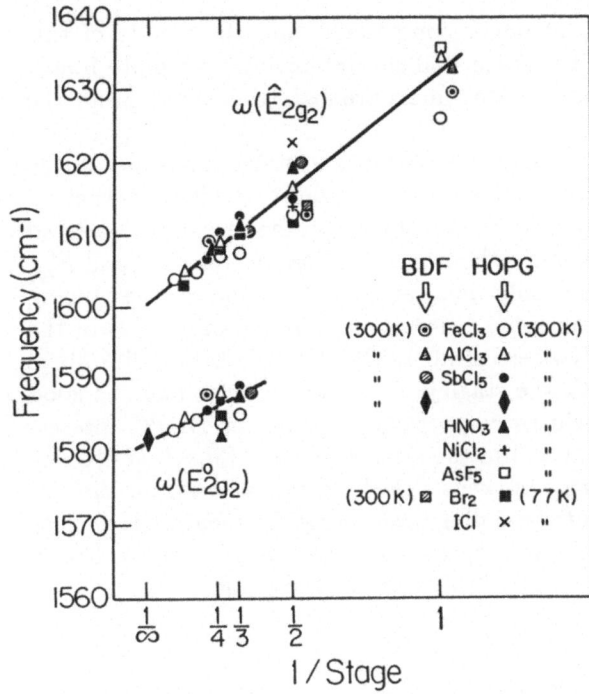

Fig. 10.10. Raman frequencies $\omega(\hat{E}_{2g_2})$ and $\omega(E^0_{2g_2})$ vs reciprocal stage for intercalated CCVD fibers (labeled BDF) and intercalated bulk HOPG. The stage of acceptor-intercalated graphite fibers can be found by comparison with the measured Raman frequencies [Chieu et al. 1983]

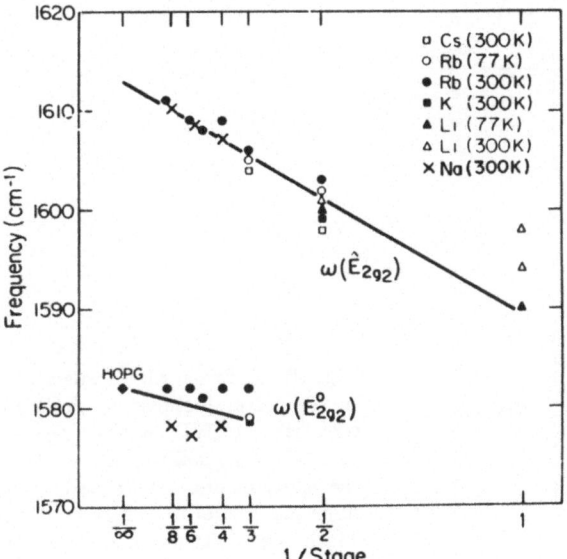

Fig. 10.11. Summary of the Raman frequencies obtained with a number of donor intercalants in graphite [Dresselhaus and Dresselhaus 1982]

quency $\omega(\hat{E}_{2g_2})$ for a variety of donor compounds, and the results of this figure are used to determine the stage of donor-intercalated graphite fibers. Raman spectra from donor intercalated fibers are less extensive than for the acceptors.

Narrow Raman lines occur only for well-staged samples. Thus, use of the small shifts in the Raman mode frequency to provide stage identification for graphite fibers requires the fibers to be well-staged. In the case of intercalated ex-PAN and mesophase pitch fibers, the Raman spectra for the $\hat{E}_{2g_2}$ and $E_{2g_2}^0$ lines are often very broad, and only qualitative information on the staging can be obtained from the peak frequencies or the relative intensities of the $\hat{E}_{2g_2}$ and $E_{2g_2}^0$ Raman lines [Kwizera et al. 1983]. For CCVD graphite fibers heat treated to $T_{HT} \geq 2900°C$, the staging of the intercalated fibers is good enough to show Raman linewidths comparable to those for HOPG intercalated with the same intercalant and to the same stage (see Fig. 10.12) [Chieu et al. 1983], thereby allowing definitive stage characterization from the Raman frequencies for the graphite bounding layer mode. This feature is utilized in the application of the Raman microprobe for the stage characterization of single graphite fibers [McNeil et al. 1986], as described below.

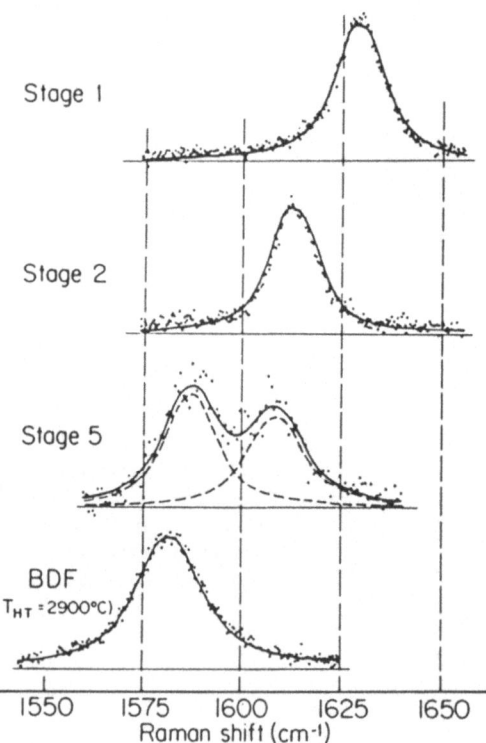

Fig. 10.12. First-order Raman spectra for several stages of $FeCl_3$-intercalated CCVD fibers ($T_{HT} = 2900°$ C). The solid lines are fits to the experimental points and determine the central frequency and linewidth of each Raman line. The spectrum for the $E_{2g_2}$ mode of pristine graphite CCVD fibers (labeled BDF) is also shown for comparison [Chieu et al. 1983]

A Raman microprobe (beam size of $\sim 2\,\mu$m) can be used to determine the stage index of single fibers and the staging fidelity along the length of a single fiber. Using the Raman microprobe, stage determination is achieved by measurement of both the mode frequencies and the mode intensities for the graphite bounding layers relative to those for the graphite interior layers. The stage determination of *single* intercalated graphite fibers by a non-destructive technique is important because it allows quantitative and reproducible measurements to be made of the intercalation-induced modifications to the properties of graphite fibers. To illustrate typical variations in the staging of a fiber, we show in Fig. 10.13 Raman microprobe spectra taken along the length of a nominally stage 3 FeCl$_3$-intercalated CCVD graphite fiber ($T_{HT} = 2900^\circ$C).

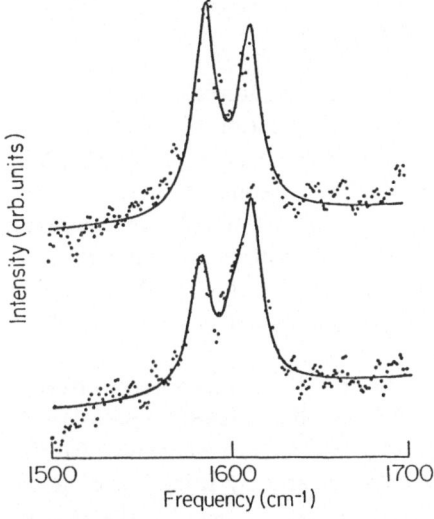

**Fig. 10.13.** Raman microprobe spectra of a nominal stage 3 FeCl$_3$-intercalated CCVD graphite fiber, showing two adjacent regions separated by 7.5 $\mu$m with different mixtures of stages 3 and 4 [McNeil et al. 1986]

The spectra show variations from one region to another, indicative of some admixture of stage 4 regions [McNeil et al. 1986]. The variation in the spectra along the length of an intercalated fiber (as demonstrated by the microprobe scans) tends to be significantly larger than for similar scans taken along an HOPG-based sample with the same intercalant and nominal stage [McNeil et al. 1986]. The same conclusions are reached by lattice fringing studies, which show individual staging defects (e.g., one sequence of stage 3 followed by several stage 2 sequences as shown in Fig. 10.8d). In contrast, the Raman microprobe provides staging information averaged over a small (2 $\mu$m) distance, which nevertheless contains many unit cells. While the Raman microprobe provides a convenient tool for rapid, non-destructive characterization of single fibers, the lattice fringing technique provides complementary microscopic information on individual defects. The lattice fringing method however cannot be used for rapid, non-destructive characterization of fibers which are to be subsequently used for other experiments [Endo et al. 1983a].

The Raman microprobe has also been applied to a comparative characterization of staging in ex-benzene CCVD and ex-mesophase pitch intercalated fibers, taking Raman microprobe spectra with light incident along the fiber axis and perpendicular to it [Meschi et al. 1986]. The results show that whereas the ex-benzene fibers are approximately homogeneously staged up to the fiber surface, the ex-mesophase pitch fibers show essentially no intercalant close to the surface, but only deeper into the fiber. This intercalant distribution has been used to account for the air stability of the ex-mesophase pitch fibers [Meschi et al. 1986]. The relative absence of edge sites in the CCVD fiber structure is likely responsible for their air stability.

## 10.4  Electrical Transport Properties

In the past few years, attention has been focused on the intercalation-enhanced electrical conductivity of graphite fibers for the fabrication of practical conductors [Vogel et al. 1977; Goldberg and Kalnin 1981; Natarajan et al. 1983; Natarajan and Woollam 1983; Jaworske et al. 1987; Manini et al. 1983; Murday et al. 1984; Manini et al. 1985; Meschi et al. 1986]. Typically the increase in conductivity upon the intercalation of highly crystalline graphite fibers is about one order of magnitude. Furthermore, the fiber geometry offers advantages relative to HOPG for increasing the compositional stability of GICs under ambient conditions [Endo et al. 1981, 1983c].

The increase of the electrical conductivity of bulk graphite upon intercalation has been reviewed by Dresselhaus and Dresselhaus [1981], Spain [1981] and others. The density of carriers increases by injection of electrons into the graphite planes in the case of donor intercalants, and by injection of holes from acceptors. The decrease in the mobility is outweighed by the carrier density increase, so that conductivity enhancement ensues. The carriers are located in the graphene planes, and for high stage compounds ($n \geq 2$) the carrier density on sequential graphene planes varies as a result of screening.

Transport measurements on bulk graphite and its intercalation compounds have in the past presented a host of difficulties due to the large anisotropy of the conductivity [Spain et al. 1967; Issi 1987]. The large magnitude of the ratio of the fiber length to cross sectional area ($L/A$) is highly favorable for reliable and accurate measurements of many of the important transport properties. Hence graphite fibers in their pristine and intercalated forms have recently been exploited for high resolution $a$-axis electrical conductivity, thermal conductivity, thermopower and magnetoresistance measurements [Issi 1987]. Especially interesting have been the measurements on single fibers and the new physics that has thus been learned about electrical transport phenomena, as discussed below. Also of interest is the potential of intercalated fibers for special-purpose applications (see Sect. 12.7).

## 10.4.1 Electrical Conductivity

Though both graphite fibers and HOPG experience an order of magnitude enhancement in the in-plane electrical conductivity $\sigma_a$, the actual conductivities achieved by the intercalation of graphite fibers are generally lower than for the same intercalate species and stage in a HOPG host (see Table 10.4). This is due to the greater structural disorder in graphite fibers, consistent with the lower conductivities of the fiber hosts prior to intercalation. However, in some cases, intercalation-enhanced fiber conductivities comparable to the corresponding enhancements in HOPG have been reported [Shioya et al. 1986].

**Table 10.4.** Enhancement of room temperature conductivity of carbon fibers by intercalation[a]

| Pristine fiber[b] | Intercalant | Stage | $\varrho_0$ [$\mu\Omega$m] | $\hat{\varrho}$ [$\mu\Omega$m] | Enhancement factor [$\hat{\varrho}_0/\hat{\varrho}$] |
|---|---|---|---|---|---|
| Thornel 50[c] | $HNO_3$ | – | $7.5 \times 10^4$ | $7.0 \times 10^3$ | 11 |
| CCVD[d] (2900°C) | $HNO_3$ | – | $8.0 \times 10^3$ | $6.0 \times 10^2$ | 13 |
| CCVD[d] (2900°C) | $Br_2$ | – | $8.0 \times 10^3$ | $1.5 \times 10^3$ | 5.3 |
| CCVD[e] (3300°C) | $AsF_5$ | – | $5.5 \times 10^3$ | $1.1 \times 10^2$ | 50 |
| P-100[f] (3000°C) | $CuCl_2$ | 2 | $1.9 \times 10^4$ | $(1.3-2.0) \times 10^3$ | 11 |
| P-100[f] (3200°C) | $CuCl_2$ | 2 | $2.0 \times 10^4$ | $(2.0-2.2) \times 10^3$ | ~10 |
| P-100[f] (2500°C) | $CuCl_2$ | – | $(3.1-3.4) \times 10^4$ | $(3.5-4.0) \times 10^3$ | ~9 |
| CCVD[f] (2900°C) | $CuCl_2$ | 2,3 | $1.0 \times 10^4$ | $2.0 \times 10^3$ | 5 |
| CCVD[f] (2900°C) | $CuCl_2$ | 1 | $1.0 \times 10^4$ | $8.0 \times 10^2$ | 12 |
| HMPVA3[g] (2800°C) | $C_xF$ | – | $8.3 \times 10^4$ | $1.4 \times 10^4$ | 5.9 |
| VSB-32[g] (2800°C) | $C_xF$ | – | $4.0 \times 10^4$ | $5.0 \times 10^3$ | 8.0 |
| CCVD[h] (2900°C) | $C_xF$ | – | $7.7 \times 10^3$ | $9.1 \times 10^2$ | 8.5 |
| CCVD[h] (2900°C) | $CuCl_2$ | 2 | $7.0 \times 10^3$ | $9.0 \times 10^2$ | 7.7 |
| CCVD[h] (2900°C) | $FeCl_3$ | 1 | $7.0 \times 10^3$ | $7.0 \times 10^2$ | 10 |
| CCVD[h] (2900°C) | $CoCl_2$ | – | $7.0 \times 10^3$ | $8.5 \times 10^2$ | 8.2 |
| P-100[i] | $NiCl_2$ | – | $2.5 \times 10^4$ | $1.9 \times 10^3$ | 13 |
| P-100[i] | $CuCl_2$ | – | $2.5 \times 10^4$ | $2.5 \times 10^3$ | 10 |
| GY70[i] | $NiCl_2$ | – | $3.6 \times 10^4$ | $2.05 \times 10^3$ | 18 |
| GY70[i] | $CuCl_2$ | 2,3 | $3.6 \times 10^4$ | $2.5 \times 10^3$ | 14 |
| TP41043[j] | $CdCl_2$ | – | $2.0 \times 10^4$ | $(1.5-2.0) \times 10^3$ | ~11 |

[a] The resistivity of pristine fibers is denoted by $\varrho_0$, and after intercalation by $\hat{\varrho}$ so that the intercalation-induced enhancement of the conductivity is given by $(\hat{\varrho}_0/\hat{\varrho})$.
[b] Estimated $T_{HT}$ values are given in parenthesis. All CCVD fibers in the table are benzene derived. The Hercules HMPVA3 and Celanese GY70 are PAN fibers while the Union Carbide VSB-32, P-100 and TP41043 are mesophase pitch fibers.
[c] Vogel [1976]
[d] Endo et al. [1981]
[e] Shioya et al. [1986]
[f] Meschi et al. [1986]
[g] Nakajima et al. [1986]
[h] Chieu et al. [1983]
[i] Bagga et al. [1984]
[j] Dominguez and Murday [1983]

An extensive effort to produce a high conductivity graphite fiber was carried out with $AsF_5$ as the intercalant [Shioya et al. 1986]. The choice of $AsF_5$ for the intercalant was motivated by previous measurements on HOPG-based GICs [Vogel et al. 1977] showing that the intercalant $AsF_5$ yields the highest room temperature electrical conductivity. The reason why $AsF_5$ intercalation yields higher conductivities than other intercalants is not yet well understood from a fundamental standpoint. The choice of CCVD fibers heat treated to high temperatures (e.g., 2900°C) was motivated by the high conductivities achieved in these fibrous host materials with high $T_{HT}$.

Several important conclusions were reached in the conductivity studies of $AsF_5$ intercalated CCVD fibers [Shioya et al. 1986]. In these experiments the conductivity was monitored as the intercalation proceeded and the intercalated fibers were subsequently characterized by SEM and TEM for structural order and by electron microprobe analysis with regard to the distribution of $AsF_5$. These measurements confirmed that intercalation is enhanced by increased structural order of the host fiber material. The best structural order was achieved by benzene-derived fibers heat treated to $T_{HT} = 3300°C$. The experiments conclusively confirm that for these fibers, intercalation starts from the fiber ends and proceeds along the fiber length. A maximum room temperature conductivity ($9 \times 10^7 \Omega^{-1} m^{-1}$) for the $AsF_5$ intercalated fibers was apparently obtained (see Fig. 10.14) significantly above that for copper ($5.9 \times 10^7 \Omega^{-1} m^{-1}$) and previous measurements on $AsF_5$-intercalated fibers [Davis et al. 1981]. These high values for $\sigma_a$ were obtained while the fiber was surrounded by $AsF_5$ gas in the intercalation ampoule. This high value of the conductivity must at this time be regarded as open to further confirmation, since contact potentials and thermal effects may have contributed to a lowering of the measured potential.

Steps observed in the conductivity vs time curve (Fig. 10.14) may be identified with staging transitions. It was further found that the structural organization of the CCVD fibers inhibited $AsF_5$ desorption, both in vacuum

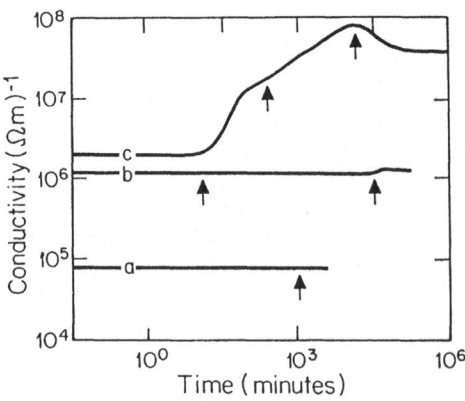

Fig. 10.14. In situ electrical conductivity vs time during $AsF_5$ intercalation of CCVD fibers (↑ indicates $AsF_5$ recharge) for the following conditions: (a) before heat treatment with both fiber ends open; (b) after heat treatment to 3300°C but with both fiber ends closed; (c) as in (b) but with fiber ends open [Shioya et al. 1986]

and under ambient laboratory conditions. The physical basis for this effect is the great reduction of surface area where exposed intercalate edge planes could lead to intercalate desorption. Long term stability (hundreds of days) was achieved for residue compounds of $AsF_5$-intercalated fibers. These residue fibers have conductivities in excess of $10^7 \Omega^{-1} m^{-1}$, high enough for useful applications as a lightweight conductor. Comparative long term (i.e., 4 months) stability studies on the desorption of $C_x F$ [Nakajima et al. 1986] show that desorption occurs more slowly going from PAN to CCVD fibers, indicating that structural disorder and defects in the fiber host material retard desorption.

For the highly ordered fibers, the correlation between the conductivity before and after intercalation is reproducible, as are the absolute values of the conductivity for specified conditions of intercalation [Shioya et al. 1986]. Also shown in Fig. 10.14 is information on the kinetics. Before heat treatment of the fibers there is negligible enhanced conductivity of the fibers by intercalation. When the ends of the heat treated fibers are sealed, again the increase in $\sigma$ is very small, consistent with a small intercalant uptake. This indicates that for the CCVD fibers intercalation starts at the fiber ends, consistent with the visual appearance of the movement of the color front during the intercalation process. The conductivity enhancements achieved with the desorbed $AsF_5$ intercalated CCVD fibers are somewhat higher than those achieved with air-stable metal chloride intercalants ($FeCl_3$, $PdCl_2$, $MnCl_2$, $CuCl_2$, $CoCl_2$, and $NiCl_2$) [Chieu et al. 1983; Murday et al. 1984], though the differences are not great.

Intercalation-induced increases (by factors of 5–12) in the electrical conductivity relative to that of the pristine fibers are common for many kinds of fibers and intercalants [Meschi et al. 1986; Jaworske and Miller 1985], though the intercalation-induced maximum conductivity is lower and the sample-to-sample variation is much larger than for the corresponding intercalation into HOPG.

Studies of the temperature dependence of the conductivity are of particular interest with regard to identification of the conduction mechanism. In contrast to the behavior in many pristine graphite fibers, low-stage intercalated graphite fibers typically show a metallic temperature dependence of the conductivity $\sigma$ (see Fig. 10.15), as is found for intercalated bulk host graphite materials [Chieu et al. 1983; Issi 1987]. That is, for low-stage intercalated fibers, the resistivity decreases with decrease of temperature. Values of the residual resistance ratio (RRR = $R_{300K}/R_{4.2K}$) provide a figure of merit of the defect concentration of the fibers. Intercalation generally causes the RRR to increase, consistent with the large increase in carrier concentration and smaller decrease in carrier mobility. In this context, note that the RRRs for intercalated fibers in Fig. 10.15 are higher than for the corresponding pristine fibers. The larger RRR values for HOPG-based GICs relative to fibers inter-

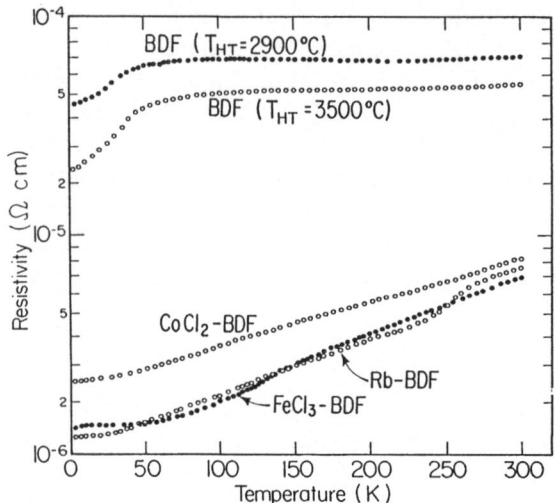

Fig. 10.15. Temperature dependence of the resistivity for pristine benzene-derived CCVD graphite fibers (BDF) heat treated to 2900°C and 3500°C and for fibers ($T_{HT} = 2900$°C) intercalated with a donor (desorbed Rb) and air-stable acceptors ($FeCl_3$, $CoCl_2$) [Chieu et al. 1983]

calated with the same intercalant to the same stage are consistent with the higher defect density of the fibers. Comparative studies on the intercalation of $C_xF$ in various fibers [Nakajima et al. 1986] show that the intercalation-induced increase in conductivity is greatest for CCVD fibers, less for ex-pitch fibers, and still less for ex-PAN fibers (see Table 10.4).

For practical utilization of intercalated fibers as lightweight conductors, thermal stability is important. Long term (3 months) stability has been demonstrated to temperatures above 100°C in $CuCl_2$-intercalated ex-pitch P-100 fibers [Ansart et al. 1987], and for shorter times to even higher temperatures for CCVD fibers [Endo et al. 1983a]. The failure mode of the intercalated fibers upon heating is also of interest. SEM photographs [Endo et al. 1981] show that heated fibers may rupture as a local exfoliation takes place, presumably accompanied by the release of the intercalant as a gas (see Sect. 10.8). Also of importance for practical applications is the high current density ($4 \times 10^4$ A/cm$^2$) that the P-100 fibers (intercalated with $CuCl_2$) could withstand without de-intercalation [Ansart et al. 1987].

Study of the low temperature electrical conductivity of intercalated graphite fibers has also revealed new physics. In general, 3D metals at low temperatures follow the Bloch-Grüneisen law for the temperature dependence of the resistivity:

$$\varrho_{BG}(T) = A_{BG} + B_{BG}T^n \quad , \tag{10.1}$$

where $n \sim 5$. The low temperature $T^5$ regime is applicable only when carrier scattering is dominated by low energy phonons through small angle scattering. At high temperatures where all phonons can scatter, a linear temperature dependence for $\varrho(T)$ is found [Ziman 1972]. Now for GICs, the functional form of the temperature dependence of the resistivity has been found by many authors to be

$$\varrho(T) = A + BT + CT^2 \quad , \tag{10.2}$$

where $A$, $B$ and $C$ are constants. This functional form has been verified experimentally for many intercalants for both intercalated graphite fibers and for intercalated HOPG [Manini et al. 1983]. The room temperature resistivity is predominantly due to carrier-phonon scattering, while defect scattering plays a significant role at low temperatures.

The functional form of (10.2) for the intercalated fibers is very different from that for the pristine fibers (see Fig. 10.15), which is also different from that of pristine HOPG [Kalnin and Goldberg 1983; Dresselhaus 1984]. Both the magnitude of the room temperature resistivity for fiber GICs and the temperature dependence of $\varrho(T)$ depend strongly on $T_{HT}$ for a given precursor material [Manini et al. 1983]. Size-dependent resistivity measurements of the pristine fibers show that the residual resistivity increases as the fiber diameter decreases [Tahar et al. 1986], consistent with a higher fraction of the sample being disordered as the diameter decreases.

Recently, high resolution resistivity measurements [Piraux et al. 1986b; Issi and Piraux 1986], made possible by the favorable geometry of intercalated graphite fibers, have been carried out at low temperatures to investigate whether a Bloch-Grüneisen effect exists for GICs at low temperatures. High resolution resistivity measurements on $CuCl_2$-intercalated CCVD fibers down to low temperatures (1.5 K), confirm the validity of (10.2), showing that the usual low temperature 3D Bloch-Grüneisen $T^n$ behavior with $n \sim 5$ is not applicable to GICs [Piraux et al. 1986b].

The dominant scattering process for carriers in pristine graphite above 50 K is due to the interaction of carriers with in-plane phonons (denoted by mode-1 in Sect. 4.2.1) [Ono and Sugihara 1966; Spain 1973]. Since the sound velocity of the in-plane carbon vibration $v_1$ is unchanged by intercalation, a typical energy for phonons interacting with carriers is

$$\hbar v_1 k_F / k_B \simeq 240\,\text{K} \quad , \tag{10.3}$$

where the values $v_1 = 2.1 \times 10^6$ cm/s, and $k_F = 1.5 \times 10^7$ cm$^{-1}$ for the Fermi wave vector are used. This implies that the scattering process associated with mode-1 phonons is ineffective at low temperatures. Actually, the temperature dependence of the calculated resistivity for this scattering process $\varrho_1$, is expected to obey a $T^4$ law below 40 K rather than $T^5$, but its magnitude is much smaller than the observed value for $\varrho$. This different temperature dependence results from the approximately 2D nature of the Fermi surfaces. On the other hand, the calculated resistivity due to the interaction of carriers with the out-of-plane carbon mode (mode-2) leads to the following temperature-dependence:

$$\varrho_2 \propto \begin{cases} T^{1.5} & \text{for} \quad 20\,\text{K} < T < 40\,\text{K} \\ T^{1.1} & \text{for} \quad 40\,\text{K} < T < 300\,\text{K} \end{cases} , \tag{10.4}$$

and the magnitude of the $\varrho_2$ contribution is close to the value observed experimentally for $\varrho$ [Sugihara 1984d]. In obtaining these results the following parameters were used: $D_1 = 16\,\text{eV}$, $D_2 = 3.5\,\text{eV}$, $v_1 = 2.1 \times 10^6$ cm/s, and $v_2 = 3.6 \times 10^5$ cm/s, where $D_i$ denotes the electron-phonon coupling constants [Sugihara 1984d]. Though the above theory explains the qualitative features of the resistivity above 40 K, it cannot account for the low temperature resistivity, which is represented by (10.2) even at liquid helium temperature. The temperature dependence of $\varrho(T)$ for $T \ll 40$K may possibly be explained by introducing an interaction with the low frequency intercalant modes, which are excited at helium temperature (see for example Al-Jishi and Dresselhaus [1982b] for a discussion of low frequency intercalant phonon modes). It should be noted that the out-of-plane phonons for GICs are almost dispersionless and have very low energies, thus giving rise to unusual low temperature transport phenomena, which are somewhat dependent on the specific intercalant species. Carrier-carrier interaction may provide an alternative mechanism of significance for the interpretation of low temperature transport phenomena in GICs.

The favorable fiber geometry has also made possible accurate measurements of charge transfer phenomena in GICs by study of the effect of ammoniation on the in-plane resistivity of the ternary $C_{24}K(NH_3)_x$ GIC system as a function of $NH_3$ concentration $x$ [Huang et al. 1987]. As $x$ was increased, the resistance of a fiber intercalated with $K(NH_3)_x$ was found to remain at the same value as for the stage 2 compound for $x = 0$, namely $C_{24}K$, until a value of $x = 1$ was reached. At the value $x = 1$ (where the concentration of $NH_3$ was approximately equal to that of the K), the resistance increased by more than a factor of 4, as the transition to a stage 1 compound took place. This dramatic increase in resistivity was explained by a back donation of electrons to the $(NH_3)$, thereby significantly reducing the carrier density and conductance in the graphite layers. As $x$ was further increased to $\sim 4.3$, a 10% decrease in resistance was observed [Huang et al. 1987] and was identified with charge transport in the intercalant layer as the intercalant molecules are squeezed together more tightly.

## 10.4.2 Magnetoresistance

Magnetoresistance measurements on intercalated graphite fibers are of particular interest because of the information they provide on the temperature dependence of the carrier mobility and because of the favorable geometry of the fibers for magnetoresistance measurements [Rahim et al. 1986] (see Sect. 8.4). To the extent that the magnetoresistance of graphites and graphite fibers yields values for an average carrier mobility $\mu$ through the simple relation $\Delta\varrho/\varrho(0) = (\mu B)^2$, measurements of $\Delta\varrho/\varrho(0) = [\varrho(B) - \varrho(0)]/\varrho(0)$ can be used in conjunction with zero-field resistivity $\varrho(0)$ measurements to estimate the temperature dependence of the carrier density and the carrier mobility for the intercalated fibers.

The simple two-band (STB) formula $\Delta\varrho/\varrho(0) = (\mu B)^2$ must be used with great caution for intercalation compounds, since the restrictive conditions (see Sect. 8.3.3) for its applicability are not satisfied as they are in the case of graphite where there is near equality of hole and electron average mobilities and densities. In fact, only one carrier type is often present in intercalation compounds – electrons in donor compounds, and holes in acceptors. A non-zero magnetoresistance arises from effects such as non-parabolicity of the dispersion relation, the presence of two or more bands of carriers with different effective masses, and from sample inhomogeneity. The appropriateness of the STB formula for intercalation compounds was considered by Spain et al. [1983a] for simple models of the dispersion relations and carrier scattering processes. They concluded that corrections to the STB formula could be sizable. A test of the formula could be made by fitting the magnetic field dependence of both the magnetoresistance and the Hall coefficient, but these data are not generally available for fibers, especially not for intercalated fibers. In addition, quantitative band models for the electron and phonon states are not generally available [Al-Jishi and Dresselhaus 1982b]. Therefore, the STB formula should only be considered as giving a "magnetoresistance mobility" which may differ by as much as a factor of ten from the "conductivity mobility" ($\mu = \sigma/ne$).

A variety of different detailed behavior is observed for the magnetoresistance of intercalated carbon fibers, depending on the ordering of the pristine fiber prior to intercalation and on the intercalation parameters, such as species and stage. For example, the magnetoresistance of intercalated CCVD fibers with $T_{\mathrm{HT}} = 2800°$–$3000°\mathrm{C}$ was found to be positive for $B < 1\,\mathrm{T}$ and $T < 300\,\mathrm{K}$ for residue compounds with $Br_2$ and $HNO_3$, and to show a quadratic field dependence at low fields [Endo et al. 1981]. A more complex field dependence is observed for intercalated fibers based on less-graphitic hosts. For example, P-100 pitch based fibers intercalated with $CuCl_2$ and $PdCl_2$ show a negative magnetoresistance at low fields ($< 1\,\mathrm{T}$), followed by a positive magnetoresistance at higher fields when the measurements are made at low temperature ($4\,\mathrm{K}$ and $78\,\mathrm{K}$) [Woollam et al. 1986b]. The reported ($\Delta\varrho/\varrho_0$) behavior for intercalated P-100 fibers bears some resemblance to that for the corresponding pristine fiber (see Fig. 8.30), though much smaller in magnitude and due to a different mechanism. The reason why the mechanism must be different is the much larger Fermi energy for GICs, so that the negative magnetoresistance is no longer dominated by a single $N = 0$ Landau level. A linear field dependence for $\Delta\varrho/\varrho_0$ vs $B$ is obtained at intermediate field values for intercalated CCVD filaments [Chieu et al. 1983], which is also observed in pristine fibers.

An angular $\cos^2 \theta$ dependence of $\Delta\varrho/\varrho_0$ was observed out to about 70°C, as is also observed approximately for unintercalated fibers (see Sect. 8.4.1); here $(\theta + \frac{\pi}{2})$ denotes the angle of the magnetic field with the fiber axis and $\varrho_0$ denotes the resistivity in zero magnetic field. The significant decrease in carrier mobility associated with intercalation is reflected in the order of magnitude decrease in the magnetoresistance of the intercalated CCVD graphite fibers relative to the pristine fibers, as shown in Fig. 10.16 [Chieu et al. 1983]. A similar decrease in $(\Delta\varrho/\varrho_0)$ was also observed by Endo et al. [1981] at low magnetic fields upon intercalation of CCVD fibers with $HNO_3$ and $Br_2$. Intercalation reduces the carrier mobility by increasing the dimensions of the Fermi surface and hence increasing the carrier effective masses, while at the same time introducing additional intercalation-induced scattering centers.

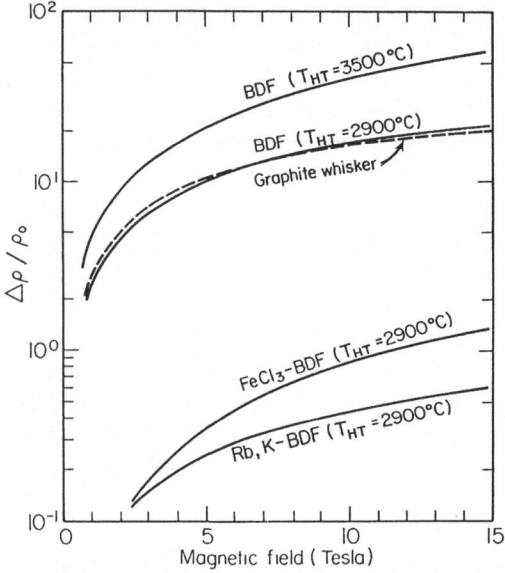

**Fig. 10.16.** Transverse magnetoresistance (log scale) vs magnetic field for pristine CCVD fibers (BDF) heat treated to 2900°C and 3500°C, and for acceptor $FeCl_3$ and donor (Rb and K) intercalated graphite fibers ($T_{HT} = 2900$°C). The measurements were made at 4 K [Chieu et al. 1983]

Intercalation of well-ordered graphite fibers yields sufficiently high carrier mobilities to satisfy the condition for the observation of quantum oscillatory effects. Several reports have already been made on the observation of the Shubnikov-de Haas effect in intercalated graphite fibers. It is found that intercalation enhances observation of quantum oscillatory effects [Chieu et al. 1983; Natarajan et al. 1983]. Shubnikov-de Haas quantum oscillations have even been observed on intercalated fibers exhibiting a negative magnetoresistance at low temperatures, specifically in P-100 ex-pitch fibers intercalated with $CuCl_2$ and $PdCl_2$, with the magnitude of the Shubnikov-de Haas oscillations decreasing as the magnitude of the negative magnetoresistance effect increases, consistent with identification of the negative magnetoresistance effect in intercalated carbon fibers with disorder [Woollam et al. 1986b].

Because of the fiber geometry, there is no easy way to apply a magnetic field along a single crystallographic direction, except along the fiber axis (the $a$-axis). Because of the cylindrical nature of the $\pi$-electron Fermi surfaces, the most interesting field direction is along the $c$-axis, which unfortunately is not along a single spatial direction in a fiber. Therefore, the fiber geometry is unfavorable for quantitative interpretation of the observed Shubnikov-de Haas periodicities in terms of band structure models.

### 10.4.3 Weak Localization

Intercalation with acceptors increases the resistivity anisotropy ratio $\varrho_c/\varrho_a$ by 2 or 3 orders of magnitude, making the transport behavior in GICs more two-dimensional than in graphite itself and in alkali metal donor GICs. Accordingly, the graphite fibers intercalated with acceptors are good candidates for the study of 2D weak localization and electron-electron interaction effects on the transport properties. Both localization and interaction effects appear as a logarithmic increase in the resistivity with decreasing $T$ at very low temperatures [Lee and Ramakrishnan 1985].

Recently, high resolution electrical resistivity and magnetoresistance measurements were performed on low-stage acceptor graphite fiber intercalation compounds in partly disordered fiber host materials. A logarithmic increase of the resistivity with decreasing temperature and a negative magnetoresistance were observed in the low temperature range [Piraux et al. 1985, 1986a; Piraux 1987]. Figure 10.17 illustrates the low temperature variation of the electrical resistivity of three different samples and Fig. 10.18 shows the transverse magnetoresistance as a function of magnetic field for a set of five samples at 4.2 K [Piraux et al. 1986a]. The magnitude of the localization effect increases with the residual resistivity of the intercalated fiber and is directly correlated with the magnitude of the negative magnetoresistance of the acceptor GIC fibers [Issi 1987; Piraux et al. 1987]. Magnetoresistance measurements have been used to distinguish between weak localization and carrier-carrier interaction effects [Piraux et al. 1987]. Negative magnetoresistance phenomena are also observed in disordered non-intercalated graphite fibers (see Sect. 8.4.2), where the phenomena are ascribed to a field-induced (linear) increase in carrier density for a 2D system in the low quantum number limit [Bright 1979]. These phenomena are discussed below in terms of weak localization and electron correlation effects.

It was proposed by Abrahams et al. [1979] that all electronic states in a disordered two-dimensional system are localized, regardless of how weak the disorder is. This conclusion is based on a scaling argument which leads to a conductivity at zero temperature

$$\sigma = \sigma_{\mathrm{B}} + \sigma_{\mathrm{loc}} , \quad \text{where} \tag{10.5}$$

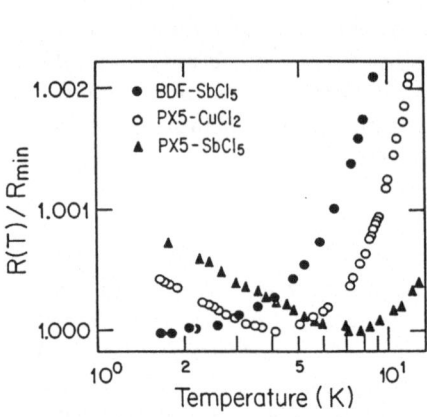

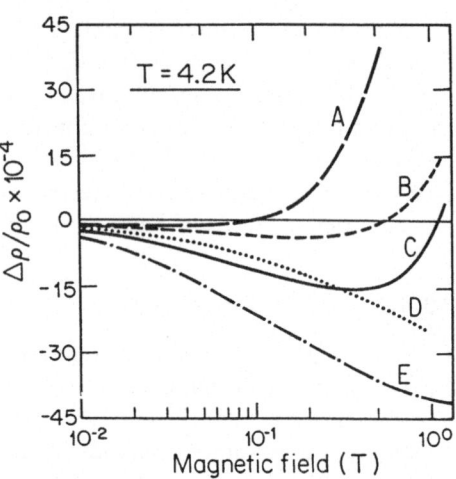

Fig. 10.17. Temperature dependence of the normalized resistance for three intercalated fiber samples: BDF-SbCl$_5$ ($\bullet$), PX5-CuCl$_2$ ($\circ$), and PX5-SbCl$_5$ ($\triangle$) [Piraux et al. 1986a]

Fig. 10.18. Transverse magnetoresistance as a function of magnetic field for a set of five intercalated fiber samples at 4.2 K: (A) BDF-SbCl$_5$, (B) PX5-CuCl$_2$, (C) PX5-SbCl$_5$, (D) VSC25-CuCl$_2$, and (E) P-100-4-CuCl$_2$ [Piraux et al. 1987]

$$\sigma_{\text{loc}} = -\frac{e^2}{\pi^2 \hbar} \ln L/l \qquad (10.6)$$

and $\sigma_{\text{B}} = ne^2\tau/m^*$ is the usual metallic conductivity for bulk samples, $L$ is the dimension of the sample and $l$ is the elastic mean free path defined by $l = v_{\text{F}}\tau$. Equation (10.5) is valid for $\sigma_{\text{B}} \gg \sigma_{\text{loc}}$. The reason why the conductivity depends on the sample size $L$ is as follows. The amplitude of the electron wave function $\psi_{\text{e}}$, even if $\psi_{\text{e}}$ is not localized, fluctuates spatially due to the effect of random potentials (e.g., arising from disorder) and it is unfavorable for electrons to pass through the region of the sample where the amplitude of $\psi_{\text{e}}$ is small. However, if $L$ is small enough, electrons can go through from one end of the sample to the other end by choosing the most favorable path with the largest amplitude for $\psi_{\text{e}}$. For this most favorable path, the conductivity is largest.

At finite temperatures, $L$ in (10.6) is replaced [Anderson et al. 1979] by the inelastic scattering length $L_{\text{in}}$ given by

$$L_{\text{in}} = \sqrt{\mathcal{D}\tau_{\text{in}}} \quad , \qquad (10.7)$$

where $\tau_{\text{in}}$ is the inelastic lifetime and the diffusion coefficient is given by $\mathcal{D} = v_{\text{F}}^2\tau/d$ (where $d = 2$ in 2D). Then we have

$$\sigma_{\text{loc}} = -\frac{e^2}{2\pi^2 \hbar} \ln \tau_{\text{in}}/\tau \quad , \qquad (10.8)$$

where it is assumed that $\tau_{\text{in}} \gg \tau$, and a negative sign for $\sigma_{\text{loc}}$ implies a decrease in electrical conductivity due to weak localization effects.

The ac conductivity for non-interacting electrons was calculated by Gorkov et al. [1979]. They found for a 2D system

$$\sigma_{\text{loc}} = -\frac{e^2}{2\pi^2\hbar}\ln(1/\omega\tau) \quad , \tag{10.9}$$

where $\omega$ is the frequency of the field. For finite temperatures such that $\hbar\tau_{\text{in}}^{-1} \gg \hbar\omega$, the factor $1/\omega$ should be replaced by $\tau_{\text{in}}$ and the Abrahams et al. [1979] result (10.8) is obtained for weak localization.

A totally different theory based on electron-electron correlation was first proposed by Altshuler and Aronov [1979] for disordered three-dimensional systems and later extended by Altshuler et al. [1980a] to two-dimensional systems. The theory of Altschuler and co-workers does not assume localization. This theory is called the interaction approach and it leads to a decrease in conductivity due to electron-electron correlations given by

$$\sigma_{\text{int}} = -\frac{e^2}{2\pi^2\hbar}\gamma\ln(k_{\text{B}}T\tau/\hbar) \quad , \tag{10.10}$$

where $\gamma$ is a factor which depends on the screening by charge carriers and $\gamma$ is close to unity in the limit of weak screening and to 0.35 in the strong screening case. These results indicate that both the interaction effect [Altshuler et al. 1980a] and a diffusion-type effect [Abrahams et al. 1979; Gorkov et al. 1979] must be taken into account for the case of the intercalated fibers. If the conductivity is written as

$$\sigma = e^2 g(E_{\text{F}})\mathcal{D} \quad , \tag{10.11}$$

where $g(E)$ is the density of states, then

$$\delta\sigma/\sigma = \delta g/g + \delta\mathcal{D}/\mathcal{D} \quad . \tag{10.12}$$

Here $\delta g/g$ is given by the interaction approach, and $\delta\mathcal{D}/\mathcal{D}$ by the theory of Abrahams et al. [1979] or Gorkov et al. [1979].

These localization and interaction effects both predict an increase of $\delta R_\square(T) = R_\square(T) - R_\square$ in the resistance as the temperature is lowered from $T_0$ to $T$, where $T_0$ is an arbitrary temperature below the minimum resistance of the fiber (see Fig. 10.17). This increase in resistance is given by

$$\delta R_\square(T)/R_\square = -\frac{e^2}{2\pi^2\hbar}R_\square(T_0)(p+\gamma)\ln(T/T_0) \quad , \tag{10.13}$$

where $R_\square$ is the resistance per square, which is related to the measured residual resistivity $\varrho_0$ at temperature $T_0$ and to the repeat distance $I_c$ by $R_\square = \varrho_0/I_c$ and $T^p$ is the temperature dependence of $1/\tau_{\text{in}}$. Since the exponent $p$ has the approximate value $p \simeq 1$ for temperatures $T > 1.5\,\text{K}$ [Piraux et al. 1986a], the weak localization and electron-electron interaction effects both contribute in a similar way to the increase in resistance with decreasing temperature.

**Table 10.5.** Typical resistivity parameters[a] for various types of fibers after intercalation. P55, P-100-4, VSC25 and PX5 are pitch-based carbon fibers. BDF is a benzene-derived CCVD carbon fiber [Piraux et al. 1986a]

| Sample | Intercalant | $\varrho_0$ [$\mu\Omega$m] | Stage | $T_c$ [K] | RRR | $p + \gamma$ |
|--------|-------------|------|-------|------|------|--------|
| P55 | CuCl$_2$ | 24 | 3 | 15 | 1.15 | 0.45 |
| P-100-4#1 | CuCl$_2$ | 35 | 1 & 2 | 22 | 1.14 | 0.89 |
| P-100-4#2 | CuCl$_2$ | 31 | 1 & 2 | 20 | 1.16 | 0.71 |
| VSC25#1 | CuCl$_2$ | 22 | 1 & 2 | 15 | 1.23 | 0.62 |
| VSC25#2 | CuCl$_2$ | 19 | 1 & 2 | 15 | 1.20 | 0.66 |
| VSC25 | CoCl$_2$ | 50 | 2 | 27 | 1.12 | 0.61 |
| PX5#1 | CuCl$_2$ | 7.5 | 4 | 4.2 | 1.37 | 0.45 |
| PX5#2 | CuCl$_2$ | 5 | 4 | 1.5 | 2.12 | 0.45 |
| PX5 | SbCl$_5$ | 20 | 2 | 9 | 1.30 | 0.45 |
| BDF | SbCl$_5$ | 7 | 2 | 1.6 | 2.01 | 0.45 |
| HOPG | CuCl$_2$ | 8.6 | – | 3.2 | 1.74 | 0.54 |

[a] Here $\varrho_0$ denotes the residual resistance, $T_c$ denotes the critical temperature below which the resistance increases with decreasing temperature, RRR denotes the residual resistance ratio $R(300\,\mathrm{K})/R(4.2\,\mathrm{K})$, and $p$ is the exponent in $T^p$ which gives the temperature dependence of the inverse inelastic lifetime $(1/\tau_{\mathrm{in}})$ and $\gamma$ is a dimensionless factor which depends on the screening by charge carriers

For each sample in Fig. 10.17, the resistance first decreases with decreasing temperature and exhibits a minimum at a given critical temperature $T_c$, below which it starts to increase. Below $T_c$, the resistance of all the fibers except the CCVD fiber intercalated with SbCl$_5$ (BDF-SbCl$_5$) shows a logarithmic increase with decreasing temperature and the observed results are fitted by using the parameters in Table 10.5.

Since a magnetic field breaks down the time reversal symmetry, the localization effect is partly destroyed by a magnetic field and a negative magnetoresistance effect is induced. In two-dimensional systems the following magnetoresistance is obtained [Altshuler et al. 1980b; Hikami et al. 1980]:

$$\sigma_{\mathrm{loc}}(B,T) - \sigma_{\mathrm{loc}}(0,T) = \frac{e^2}{2\pi^2\hbar}[\psi(\tfrac{1}{2} + \tfrac{1}{x}) + \ln x] > 0 \quad , \tag{10.14}$$

where $\psi$ is the digamma function and

$$x = \frac{4L_{\mathrm{in}}^2 eB}{\hbar c} = \frac{2v_F^2 \tau \tau_{\mathrm{in}} eB}{\hbar c} \quad . \tag{10.15}$$

For fields such that $x \gg 1$, Eq. (10.14) predicts a $\ln B$ behavior. On the other hand, the resistivity change due to the interaction effect is positive in contrast with the localization effect, see (10.14), and its magnitude is much smaller than that of (10.14) [Lee and Ramakrishnan 1985]. Accordingly, the resistance change corresponding to (10.14) is

$$\delta R_\square\,(B)/R_\square\,(0) = -R_\square\,(0)\frac{e^2}{2\pi^2\hbar}[\psi(\tfrac{1}{2}+\tfrac{1}{x})+\ln x] < 0 \quad, \qquad (10.16)$$

where $\delta R_\square\,(B) = R_\square\,(B) - R_\square\,(0)$.

Equations (10.13) and (10.16) should be compared with Figs. 10.17 and 10.18 given above. It should be noted that (10.13) and (10.16) are proportional to $R_\square = \varrho_0/I_c$, where $\varrho_0$ is the residual resistivity. This relation was satisfied in most of the samples employed by Piraux et al. [1986a]. From Fig. 10.18, it follows that the magnetoresistance is negative so that weak localization effects seem to play a role in the low $T$ anomalous behavior in $R(T)$. It can be ascertained that the term $\ln x$ is dominant in (10.16). Those samples showing larger $\delta R_\square\,(B)/R_\square\,(0)$ tend to show larger magnetoresistance effects as would be expected if the weak localization mechanism plays a significant role.

Recent detailed studies of the effect of a magnetic field on the low temperature resistivity of a large variety of acceptor GIC fibers [Piraux et al. 1987] show that the magnetoresistance curves are well described by a weak localization theory for two-dimensional disordered systems. From analysis of the temperature and magnetic field dependence of the low temperature magnetoresistance data, the temperature dependence of the inelastic scattering time is obtained. It is found that the inelastic lifetime of the conduction carriers is determined by both hole-hole ($T^2$ dependence) and hole-phonon (linear $T$ dependence) scattering. The contributions from the weak localization and hole-hole scattering effects have been separated through magnetoresistance measurements [Piraux et al. 1987], and since the magnetic field induces only a small change in the coefficient of the logarithmic temperature dependence, it is concluded that the logarithmic temperature dependence is largely due to hole-hole scattering.

## 10.5 Thermal Transport Properties

Because of the greatly enhanced carrier density in GICs relative to graphite, both phonons and electrons contribute to the low temperature thermal conductivity

$$\kappa = \kappa_e + \kappa_l \quad, \qquad (10.17)$$

where the lattice contribution $\kappa_l$ is well approximated by the Debye relation

$$\cdot\ \kappa_l = \tfrac{1}{3}C_v v L_\phi \qquad (10.18)$$

and the electronic contribution $\kappa_e$ by the Wiedemann-Franz law

$$\kappa_e = L_o T \sigma \quad, \qquad (10.19)$$

as discussed in Sect. 5.3 where the Lorenz number $L_o$ is given. The ability to measure the electrical conductivity $\sigma$ accurately in fibrous GICs thus allows

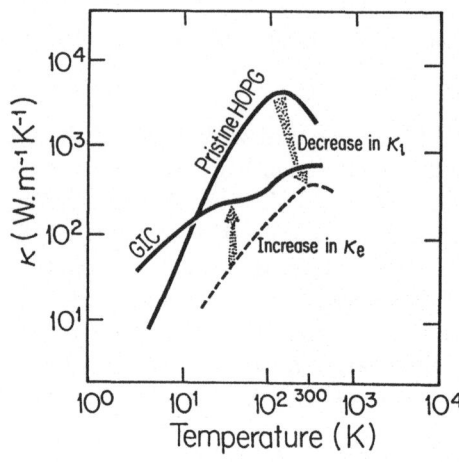

**Fig. 10.19.** Schematic representation of the effect of intercalation (GIC) on the temperature dependence of the lattice and electronic contributions $\kappa_l$ and $\kappa_e$ to the thermal conductivity of pristine graphite (HOPG) [Issi 1987]

determination of $\kappa_e$ and separation of the electronic and lattice contributions [Piraux et al. 1985; Piraux et al. 1986c] without the need for application of high magnetic fields [Heremans et al. 1982].

For typical metals (such as copper), $\kappa$ exhibits a linear temperature dependence at low $T$, where $\kappa_e$ dominates. In this low temperature regime, $\sigma$ is constant and is dominated by electron scattering from static lattice defects. At higher temperatures, $\kappa_e$ for a pure metal increases to a maximum value, then decreases and eventually saturates to a constant value, reflecting the linear $1/T$ dependence of $\sigma$ in the regime where electron-phonon scattering dominates (see Fig. 10.19). Very pure and well-ordered metals show a sharp maximum in $\kappa(T)$ while metals with a high defect concentration exhibit a broad maximum or no maximum at all. Though the Wiedemann-Franz law only applies when the same relaxation time can be used for both electrons and phonons, the Wiedemann-Franz law is a good approximation for $\kappa_e$ in the case of GICs at low temperatures and above the maximum in $\kappa(T)$ (see Fig. 10.19). Even for intermediate temperatures, the Wiedemann-Franz law provides a rough approximation for $\kappa_e$ [Issi 1987].

Intercalation causes a large increase in carrier density so that $\kappa_e$ becomes dominant at low temperatures and the in-plane thermal conductivity $\kappa_a$ increases relative to that in pristine graphite. At higher temperatures near the maximum in $\kappa_a$, the phonon contribution dominates. Since intercalation introduces a large density of static lattice defects, $\kappa_l$ is reduced, as is shown schematically in Fig. 10.19. A contribution to $\kappa_a$ is also made by phonons associated with intercalant vibrational modes $\kappa_{l_I}$, but this contribution is expected to be small because the intercalant modes are soft relative to those of the graphitic phonons.

The various contributions to the thermal conductivity of intercalated fibers can be described with the aid of Fig. 10.20. First we note that the maximum total thermal conductivity is of the order of a few hundred $Wm^{-1}K^{-1}$.

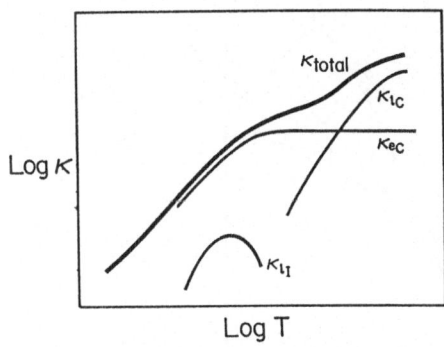

**Fig. 10.20.** Log-log schematic representation of the various contributions to the thermal conductivity of a GIC below 300 K [Issi 1987]

At low temperatures the thermal conductivity of the intercalated graphite increases linearly with temperature due to the dominant contribution from the carriers in the carbon layers $\kappa_{e_C}$ for samples with low defect and impurity concentrations. An additional lattice contribution $\kappa_{l_I}$ associated with the intercalant may also appear at low temperatures. Around room temperature the thermal conductivity of the intercalated graphite is dominated by the lattice contribution of the carbon layers $\kappa_{l_C}$. The resulting total conductivity $\kappa_{total}$ is thus characterized by a linear $T$ dependence at low temperatures, a hump at intermediate temperature, followed by a broad maximum at higher temperatures (see Fig. 10.20).

A more detailed analysis of the thermal conductivity of GICs can be made invoking the various scattering mechanisms, so that the total scattering probability per unit time is then given by summing over the dominant scattering processes:

$$\frac{1}{\tau} = \frac{1}{\tau_N} + \frac{1}{\tau_U} + \frac{1}{\tau_{BS}} + \frac{1}{\tau_D} \quad , \tag{10.20}$$

where $\tau_N$ and $\tau_U$ refer to relaxation times associated with normal electron-phonon and umklapp processes, respectively. Boundary scattering gives rise to a relaxation time

$$\tau_{BS} = L_{BS}/v \quad , \tag{10.21}$$

which is independent of phonon frequency $\omega$, and $L_{BS}$ is the boundary scattering length; scattering from Daumas-Hérold domain walls [Daumas and Hérold 1969] could also contribute to this term. The relaxation time $\tau_D$ associated with point defects is highly dependent on the phonon frequency ($\tau_D \sim \omega^r$ where $r \sim 4$ for 3D solids and $r \sim 3$ for the 2D case).

Typical results for the thermal conductivity of donor and acceptor GICs are shown in Fig. 10.21 for bulk graphite host materials and in Fig. 10.22 for intercalated CCVD fibers. The linear temperature dependence of $\kappa_a$ at low temperatures for all the GIC samples is consistent with the dominance of an electronic contribution to $\kappa_a$ at low $T$. Further support for the electronic contribution to $\kappa_a$ comes from the electrical conductivity measurements made

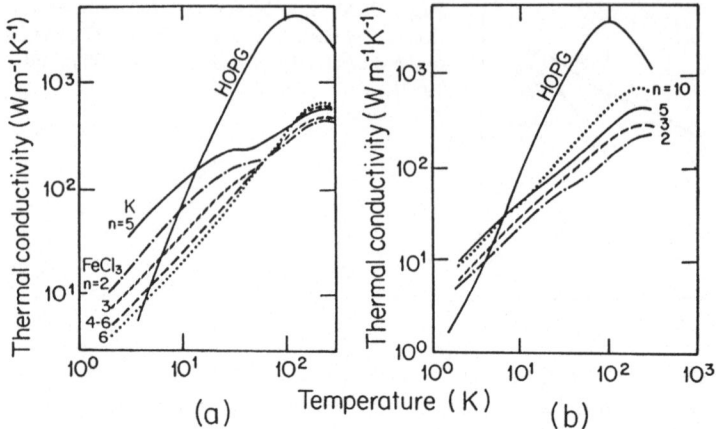

**Fig. 10.21a,b.** Temperature variation of the in-plane thermal conductivity of various bulk graphite intercalation compounds compared to that of pristine graphite (HOPG) **(a)** for pure stage 2, 3 and 6 samples, and a mixed stage (4–6) FeCl$_3$-GIC sample, and for a stage 5 potassium donor bulk GIC [Issi et al. 1983], **(b)** for various stages of SbCl$_5$-GICs [Elzinga et al. 1982]

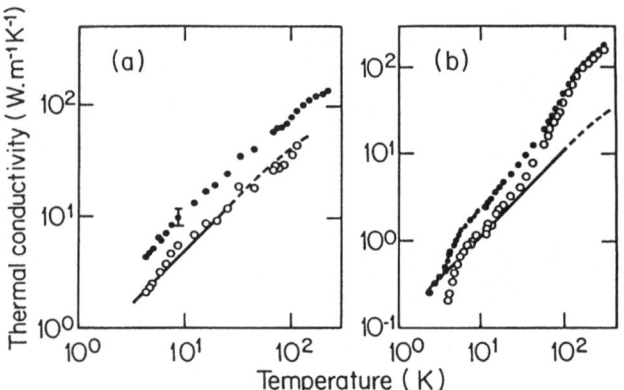

**Fig. 10.22a,b.** The temperature variation of the thermal conductivity of intercalated CCVD fibers: **(a)** for a stage 2 CuCl$_2$-intercalated fiber and **(b)** for a mixed stage (1 and 2) CuCl$_2$-intercalated fiber. In both cases the curves represent the electronic thermal conductivities $\kappa_e$ computed from the electrical resistivities measured on the same sample. The dark circles represent the values of the measured total thermal conductivity $\kappa$ and the open circles represent the lattice thermal conductivity, where $\kappa_l = \kappa - \kappa_e$ [Piraux et al. 1985; Piraux et al. 1986c]

on the same fibers [Piraux et al. 1985, 1986c], allowing direct computation of $\kappa_e$ by (10.19). With the CCVD fibers, it is possible to achieve sufficiently large $L/A$ (fiber length/area) values so that four-probe measurements can be carried out quantitatively on a single fiber. Thus a direct determination of $\kappa_l$ can be made by subtraction $\kappa_l = \kappa - \kappa_e$, thus allowing separation of $\kappa$ into $\kappa_e$ and $\kappa_l$ without the need for high magnetic fields [Heremans et al. 1982].

## 10.6 Thermopower

Measurement of the thermoelectric power (TEP) or Seebeck coefficient $S$ provides important information about the sign of the carrier type dominating the transport properties. Thus, as expected, donor GICs exhibit negative thermopower and positive values are found for acceptors, as shown in Fig. 10.23 for bulk GICs. Of particular interest in the $S_a(T)$ curves for GICs are the following features [Heremans et al. 1981; Elzinga et al. 1982; Issi et al. 1982; Sugihara 1983a, 1983b]:

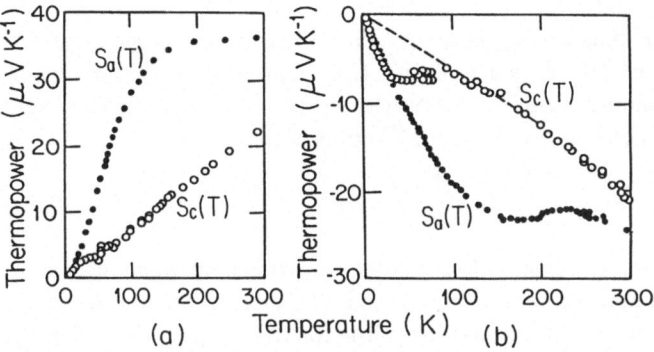

Fig. 10.23a,b. Temperature variation of the in-plane ($\bullet$) and $c$-axis ($\circ$) thermopower [denoted by $S_a(T)$ and $S_c(T)$, respectively] of graphite intercalation compounds [Issi et al. 1982]: (a) stage 2 graphite FeCl$_3$ *acceptor* compound with a positive thermopower in both directions, (b) stage 5 graphite-potassium *donor* compound with a negative thermopower in both directions

1. The $S_a(T)$ curves for GICs are qualitatively different in functional form from those of pristine graphite (see Sect. 8.6). Anomalies in $S(T)$ of GICs in Fig. 10.23 are attributed to phase transitions, which are seen more sensitively in $S_c(T)$ than in $S_a(T)$.
2. At low temperature ($\leq 100$ K), $S(T)$ follows a linear temperature dependence both $\parallel$ and $\perp$ to the layer planes [$S_a(T)$ and $S_c(T)$, respectively].
3. The anisotropy in $S(T)$ is relatively small compared with other transport properties.
4. The in-plane component $S_a(T)$ saturates above $\sim$150 K for both donor and acceptor GICs, but the $c$-axis component $S_c(T)$ remains linear in $T$ up to room temperature and beyond.

The thermopower of a single intercalated CCVD fiber has also been measured [Endo et al. 1981; Piraux et al. 1986c]. For the case of fibers, the thermopower is measured along the fiber axis, which corresponds approximately to the $a$-axis, $S_a(T)$. The temperature dependence of the thermopower for a single fiber is very similar to that of $S_a(T)$ for the corresponding bulk GICs [Endo et al. 1981; Piraux et al. 1986c], both in functional form and in the magnitude of $S_a$ at saturation for the case of the $Br_2$, $HNO_3$ and $CuCl_2$ intercalants.

The plateau in the $S_a(T)$ curves in the high temperature regime (see Fig. 10.23) is not sensitive to stage index except for very dilute compounds, as required for compatibility with the $S_a(T)$ curve for pristine graphite (see Sect. 8.6 and Fig. 8.38). Sugihara [1983a, 1983b] has presented a model based on the phonon drag effect to explain the saturation in $S_a(T)$. From intuitive considerations, an expression for the phonon drag thermoelectric power is obtained as

$$S_{ph} = \langle C_{ph} R \rangle / (3eN) \quad , \tag{10.22}$$

where $N$ is the carrier density, $C_{ph}$ is the specific heat of the phonon system interacting with carriers and $R$ is a quantity that represents the momentum transfer from the phonon system to the carrier system. In treating the thermopower problem, $\langle C_{ph} R \rangle$ takes the form [Sugihara 1983b]

$$\langle C_{ph} R \rangle = \frac{2\pi}{I_c} \frac{2\pi}{(2\pi)^2} \sum_i \int_0^{2k_{Fi}} dq \, q \frac{(\hbar \omega_q)^2}{k_B T^2} N_q (N_q + 1) R_{ii}(q) \quad , \tag{10.23}$$

where we consider two-dimensional in-plane vibrations of wave vector $q$, frequency $\omega_q$ and occupation number $N_q$, and $R_{ii}(q)$ is given by $R_{ii}(q) = \tau(q)/\tau_{ii}(q)$. Here $\tau(q)$ is the total relaxation time for the $q^{th}$ phonon and $\tau_{ii}(q)$ denotes the phonon relaxation time associated with the electron-phonon intraband scattering. The Fermi momentum for carriers in the $i^{th}$ band is $k_{Fi}$ and $I_c$ denotes the separation between sequential intercalant layers. In (10.23) interband transitions are neglected. From the definition of $R_{ii}(q)$ we see that it is less than unity. An explicit expression for $1/\tau_{ii}(q)$ is given by

$$\frac{1}{\tau_{ii}(q)} = \frac{D^2 \hbar q k_{Fi}}{\varrho_m I_c p_0^2} = a_i q \quad , \tag{10.24}$$

where $D$ is the electron-phonon coupling constant ($\sim 16\,eV$), $\varrho_m$ is the mass density of the sample and

$$p_0 = \sqrt{3} \gamma_0 a_0 / 2 \quad . \tag{10.25}$$

The phonon relaxation rate $1/\tau(q)$ is represented by

$$\frac{1}{\tau(q)} = \frac{1}{\tau_{BS}} + \frac{1}{\tau_c} + \frac{1}{\tau_I} + \frac{1}{\tau_{ph}} \quad , \tag{10.26}$$

where (i) $1/\tau_{BS}$ is due to boundary scattering which is independent of $q$ and $T$, where ($1/\tau_{BS} \simeq v_s/L$, where $L$ is the sample dimension and $v_s$ is the velocity of sound). (ii) $1/\tau_c$ is due to the carrier-phonon scattering process given by (10.24), so that $1/\tau_c$ is proportional to $q$. (iii) $1/\tau_I$ is the scattering rate due to the strain field around point defects or impurities. For two-dimensional phonons [Carruthers 1961; B.T. Kelly 1967, 1969a], this scattering rate is proportional to $q^3$:

$$1/\tau_{\rm I}(q) = f q^3 \quad . \tag{10.27}$$

In a previous theory for the thermopower in GICs [Sugihara 1983b], $1/\tau_{\rm I}$ was attributed to Rayleigh scattering. However, scattering due to a strain field is generally larger than Rayleigh scattering. While Rayleigh scattering occurs through a local potential, the strain field is non-local. The fact that the functional form for $1/\tau_{\rm I}$ is the same for both types of scattering is purely coincidental. Hence (10.27) should be viewed as expressing also the contribution of strain-field scattering. (iv) $1/\tau_{\rm ph}$ is due to the normal phonon-phonon scattering, which for two-dimensional phonons has the functional form $1/\tau_{\rm ph} = C q T^3$, where $C$ is a constant.

At very low temperatures where $(2\hbar v_s k_{\rm F}/k_{\rm B}T) \gg 1$ is satisfied, boundary scattering is the predominant process (where we note that the velocity of sound $v_s = 2.1 \times 10^4$ m/s). Then $R \simeq \tau_{\rm BS}/\tau_{\rm c} \propto q$ so that from (10.22) and (10.23) it follows that $S_{\rm ph} \propto T^3$. This $T^3$ temperature dependence was confirmed in thermopower measurements on SbCl$_5$-GICs up to 20 K [Elzinga et al. 1982].

At intermediate temperature, where $\exp(2\hbar v_s k_{\rm F}/k_{\rm B}T) \gg 1$ is realized, boundary scattering can be neglected and $1/\tau$ is represented by

$$1/\tau \simeq 1/\tau_{\rm I} + 1/\tau_{\rm c} + 1/\tau_{\rm ph} \quad . \tag{10.28}$$

For GICs there is a range of temperature where strain-field scattering can be more important than carrier-phonon and phonon-phonon scattering so that $1/\tau_{\rm I} \gg 1/\tau_{\rm c}$, $1/\tau_{\rm ph}$ is satisfied. By introducing intercalant layers into pristine graphite, it is expected that many point defects are formed. Owing to the functional dependence $1/\tau_{\rm I}(q) \propto q^3$, strain-field scattering becomes the predominant process at high temperatures in GICs which have a large cross-sectional diameter of the Fermi surface ($k_{\rm F} \sim 10^9$ m$^{-1}$), while in pristine graphite this process is unimportant since $k_{\rm F} \sim 10^8$ m$^{-1}$.

In this case the $T$ dependence of the phonon drag TEP is given by

$$S_{\rm ph} \propto T^{-2} \int_{q_0}^{q_{\rm max}} dq\, q N_q (N_q + 1) \quad , \tag{10.29}$$

where $q_{\rm max} = 2k_{\rm F}$ and the lower limit $q_0$ is estimated by the condition

$$\frac{1}{\tau_{\rm I}(q_0)} = \frac{1}{\tau_{\rm c}(q_0)} + \frac{1}{\tau_{\rm ph}(q_0)} \quad . \tag{10.30}$$

Equation (10.29) can be approximated by

$$S_{\rm ph} \propto \int_{x_0}^{x_{\rm max}} dx\, \frac{e^x}{(e^x - 1)^2} = -\ln x_0 + 0.46 \quad , \tag{10.31}$$

where $x_{\rm max} = \hbar v_s q_{\rm max}/k_{\rm B}T > 1$ and $x_0 = \hbar v_s q_0/k_{\rm B}T < 1$. Equation (10.31) is only weakly temperature dependent, thereby accounting for the saturation behavior in the thermopower.

If the energy dependence of the carrier relaxation time is assumed to be $\tau(E) \propto E^p$, the diffusion thermoelectric power $S_d$ is given by

$$S_d \simeq \frac{\pi^2}{3e} \frac{k_B^2 T}{E_F}(1+p) \quad . \tag{10.32}$$

If $E_F = 0.7\,\text{eV}$ is inserted into (10.32), we obtain

$$S_d \simeq 3.5(1+p) \times 10^{-2} T \quad \mu V/K \quad . \tag{10.33}$$

This value is too small (a few $\mu V/K$ at $T > 100\,\text{K}$) to explain the observed TEP at high temperature. However, the diffusion term $S_d$ should not be neglected and should be added to $S_{ph}$ so that the resulting $S = S_d + S_{ph}$ provides a weak temperature dependence, which is in qualitative agreement with the observed results in Fig. 10.23 [Issi et al. 1982].

The thermopower of intercalated fibers is most sensitive to $S_a$ for graphite and only weakly dependent on $S_c$ because of the fiber geometry. Since intercalation occurs predominantly in highly ordered graphite host materials, the average misalignment angle (Fig. 3.15) will be small. Thus, approximating the TEP measured in the fiber by $S_a$ should be a good approximation. The lower thermopower for donors relative to acceptors can be explained by the higher carrier density in the donor GICs.

## 10.7  Mechanical Properties

Strong fibers with high electrical conductivity can be achieved by intercalation, resulting in an order of magnitude increase in electrical conductivity with a lesser decrease in the tensile strength $\sigma_T$ and in Young's modulus $E_Y$. This is shown in Fig. 10.24 for intercalated P-100 mesophase pitch precursor fibers, and for methane-derived and benzene-derived CCVD fibers [Meschi et al. 1986]. For each precursor, results for $\sigma_T$ and $E_Y$ are shown after intercalation with several intercalants, including some staged compounds (the stage index being given in parenthesis). The results show that residue compounds exhibit the smallest degradation (by a factor of between 3 and 9) in mechanical properties and the largest degradation is found in the stage 1 alkali metal compounds [Murday et al. 1984]. This behavior suggests that defects introduced during intercalation pin the basal planes together and inhibit shear (i.e., defects lower $S_{44}$ as discussed in Sect. 4.3.1). No systematic study of the stage dependence of the degradation of the tensile strength $\sigma_T$ in intercalated fibers has been carried out to date. Nevertheless, it has been established through studies on CCVD fibers intercalated with $HNO_3$ and $Br_2$ that the stress-strain curves for these intercalated fibers are similar to those for the fibers prior to intercalation [Endo et al. 1981]. In this work it was also found that the tensile strength was approximately inversely proportional to

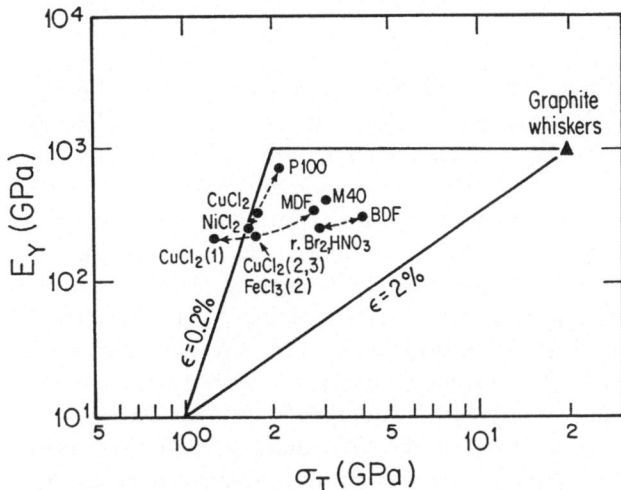

**Fig. 10.24.** Young's modulus ($E_Y$) vs tensile strength ($\sigma_T$) at room temperature for a series of graphite fibers before and after intercalation (connected by dashed lines). BDF and MDF refer to benzene- and methane-derived fibers and r refers to a residue compound. Lines for $(\Delta L/L) = 0.2\%$ and 2% are indicated. [Meschi et al. 1986]

the fiber diameter. The triangle in Fig. 10.24 gives reference values for fiber elongations to the break point of $\varepsilon = \Delta L/L$ of 0.2% and 2%. The figure shows that intercalation tends to decrease the fiber strength more than the modulus for each type of fiber precursor that has been investigated [Endo et al. 1981; Dominguez et al. 1983; Oshima et al. 1983; Murday et al. 1984]. Specifically for CCVD fibers, Endo et al. [1981] found that the Young's moduli decreased by up to 10% and the tensile strength by up to 35%. Intercalation makes the fibers more fragile (more likely to fracture upon bending), though this effect is less pronounced with smaller diameter fibers [Murday et al. 1984].

The charge transfer to the $\pi$-electrons is not expected to affect the mechanical properties significantly. However, intercalation also results in a large sample expansion in the $c$-direction, which is expected to have a larger effect on $\sigma_T$ than on $E_Y$, due to the increase in local stress resulting from an expansion of the well-graphitized regions of the fiber which intercalate, and no expansion of the disordered regions that do not intercalate. The lower tensile strength of carbon fibers relative to single crystal graphite is likely due to the introduction of defects or flaws along the length of the fibers. Thus short lengths of fiber are less likely to have flaws and are more likely to exhibit better mechanical properties (as discussed in Sect. 6.2.1). In this context, intercalation is expected to lower the fiber strength through the introduction of additional flaws and defects. Because of the unusually high values of the tensile strength and Young's modulus prior to intercalation, the intercalated residue fibers still show excellent mechanical properties.

# 10.8    Exfoliation

When a graphite intercalation compound (GIC) is heated, a thermal expansion, large ($\sim 3.5 \times 10^{-5}/°C$) compared with most materials, occurs along the c-axis, while the in-plane lattice constant remains almost unchanged. This anisotropic thermal expansion behavior is reversible and can be modeled in terms of a superposition of the c-axis thermal expansion coefficients of the constituent layers [Salamanca-Riba and Dresselhaus 1986]. However, when a GIC is heated above a critical temperature $T_B$, a gigantic c-axis expansion can occur, with sample elongations $\Delta L_S/L_S$ of as much as a factor of 300 [Inagaki et al. 1983]. This gigantic elongation, called *exfoliation*, is generally not reversible and gives rise to spongy, foamy, low density, high surface area material (e.g., 85m$^2$/g [Thomy and Duval 1969]). The exfoliation effect compromises the structural integrity of the GIC material and therefore is undesirable for applications of GICs where structural integrity is necessary.

On the other hand, the spongy, foam-like material is desirable for other applications, such as gas adsorption. The commercial version of the exfoliated "wormy" material is called vermicular graphite and is used for high surface area applications. When pressed into sheets, it is called grafoil and is widely used as a high temperature gasket or packing material because of its flexibility, chemical inertness, low transverse thermal conductivity and ability to withstand high temperatures. Grafoil-type products can be used to contain molten corrosive liquid metals at high temperatures and also to extinguish metal fires [S.H. Anderson and Chung 1984].

To date, there have only been a few reports on the exfoliation of intercalated graphite fibers [S.H. Anderson and Chung 1983, 1984; Jiménez-González et al. 1986]. Because of the different structure and texture of the exfoliated fibers, and the integrity of an exfoliated fiber along its length, special applications for exfoliated graphite fibers (in contrast to exfoliated HOPG) may be found.

The general exfoliation phenomenon involves an explosive sample elongation along the c-axis when the intercalated sample is heated above a temperature $T_B$, as shown in Fig. 10.25 [Martin and Brocklehurst 1964] for a bromine residue GIC where only modest elongations ($\Delta L_S/L_S \approx 2$) are found. By increasing the structural perfection, the intercalant concentration, and the temperature of exfoliation, and by proper selection of intercalant, the elongation can be greatly increased [Dowell 1975]. For example, values as high as ($\Delta L_S/L_S$) $\sim 300$ have been reported for single crystal graphite flakes intercalated with K and tetrahydrofuran (THF) $\simeq 800°C$ [Inagaki et al. 1983]. If repeated heating and cooling cycles are employed, lower total expansion ($\Delta L_S/L_S$) is obtained on subsequent exfoliation, though quasi-reproducible behavior is observed after the initial heating and cooling cycle [Martin and Brocklehurst 1964; S.H. Anderson and Chung 1984].

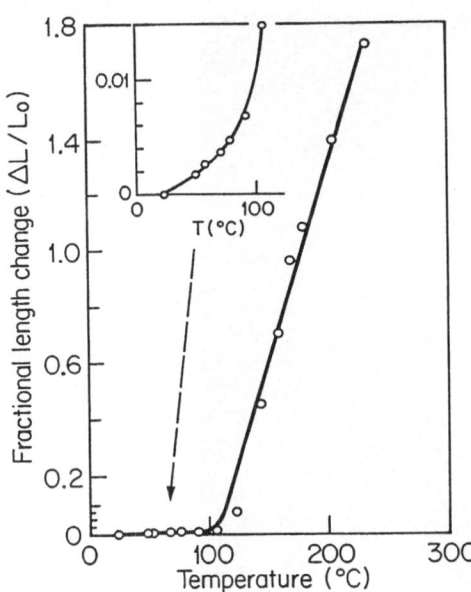

Fig. 10.25. The thermal expansion of a pyrolytic graphite-bromine residue compound (residual bromine content 2.2 at.% Br/C). The inset shows the fractional length changes on an expanded scale for temperatures below 100°C [Martin and Brocklehurst 1964]

The microstructure of the exfoliated material has been studied by various workers [Inagaki et al. 1983; S.H. Anderson and Chung 1984; Jiménez-González et al. 1986], and all report the type of spongy structure seen in Fig. 10.26 for exfoliated graphite fibers when viewed by the scanning electron microscope (SEM) under low and medium magnification. The microstructure in the figure suggests that the web material in the exfoliated fiber consists of many graphite layers which remain essentially intact. Because of the violent gas release associated with the exfoliation process described below, the residual webs of graphene layers are aligned at arbitrary angles with respect to each other, but with some preferred orientations. This is confirmed by the electron diffraction pattern of Fig. 10.27, showing sharp spots due to the preservation of graphitically ordered material. The spots are arranged in a ring due to the quasi-random orientation of the graphitic material. The background rings are likely due to the highly disordered graphitic regions in the sponge-like material [Jiménez-González et al. 1986]. The disorder of the graphitic material was verified by Raman spectra and lattice fringe images. The preservation of the intercalant spot pattern in the exfoliated material [S.H. Anderson and Chung 1984] indicates that some intercalant remains trapped between the graphite layers with a structural arrangement as is found in the intercalation compound prior to exfoliation. X-ray fluorescence measurements on the exfoliated CCVD fibers confirmed the presence of small amounts of intercalant in the exfoliated material, but no evidence for staging was found. Inagaki et al. [1983] found that by heat treatment of the exfoliated material to elevated temperatures (e.g., 1200°C), the intercalant (K and THF) could be expelled.

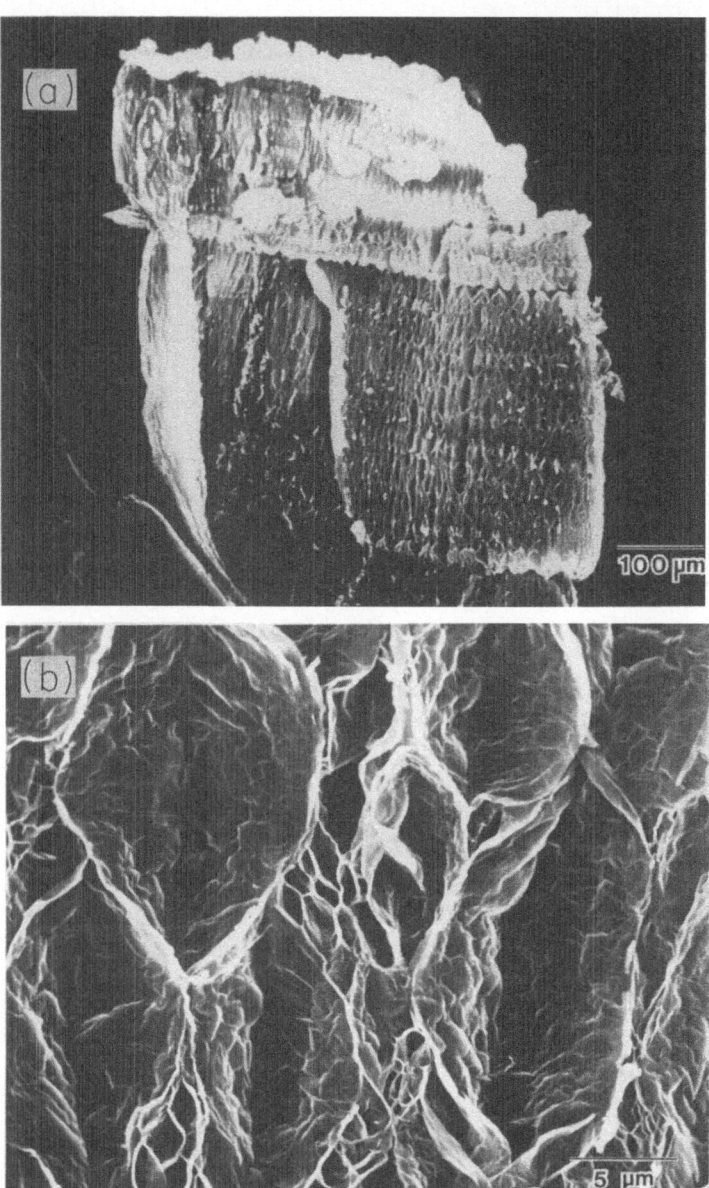

Fig. 10.26a,b. SEM micrograph of a benzene-derived carbon fiber after exfoliation. For this fiber, $T_{HT} \sim 3000°C$ and $SbCl_5$ was used as the intercalant. (a) The general structure (low magnification) and (b) a more detailed view (higher magnification) [Jiménez-González et al. 1986]

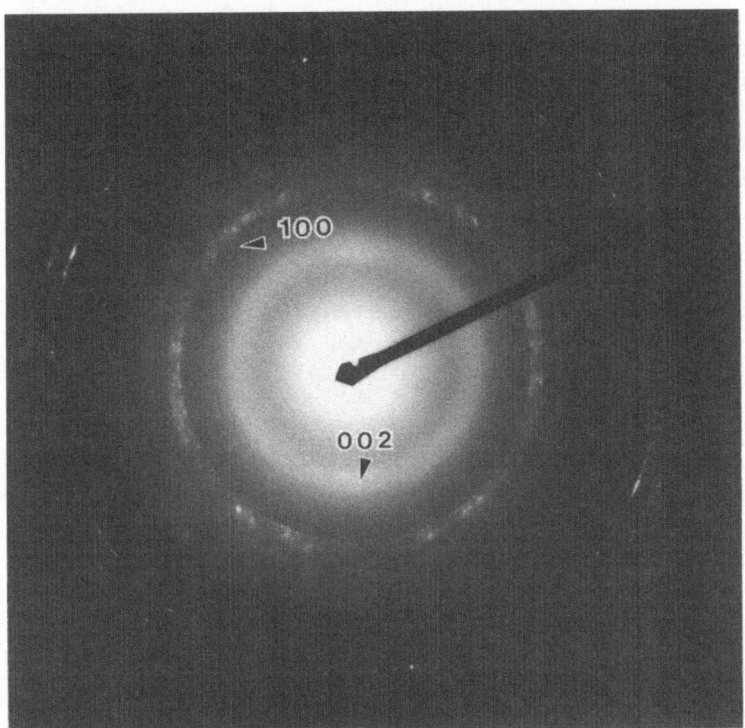

**Fig. 10.27.** Typical electron diffraction pattern from an exfoliated fiber such as that shown in Fig. 10.26 [Jiménez-González et al. 1986]

In their work, Inagaki and coworkers were interested in developing an exfoliation product which would have an extremely low vapor pressure of corrosive gas species, and for this reason they worked with K + THF to avoid use of acids such as $HNO_3$ and $H_2SO_4$.

Since exfoliation is accompanied by structural changes as discussed above, there are also significant changes in properties, such as in-plane and $c$-axis electrical conductivity, thermal conductivity and thermopower [Olsen et al. 1970; Uher and Sander 1983; Chung and Wong 1985]. For example, in the bulk samples, an increase in the resistivity of ~200 in the grafoil rolling direction of a bulk exfoliated sample and of ~20 normal to the grafoil rolling direction was reported [Chung and Wong 1985], so that the exfoliated material is much less anisotropic, probably due to the nearly random orientations of the graphitic planes. For the vapor-grown fiber samples, the increase in resistance along the fiber length was less than a factor of 10, indicative of substantial integrity of the fiber upon exfoliation [Jiménez-González et al. 1986].

The basic explanation for the exfoliation process was provided by Martin and Brocklehurst [1964] based on detailed studies of the effect of restraining loads on suppressing the onset of exfoliation. According to their model for exfoliation, the intercalant undergoes a phase change to the vapor phase, forming disc-shaped bubbles of radius $r$ and height $I_c$ in the interlayer region between graphite planes, with a gas pocket accumulating in certain regions where diffusion is blocked by defects. Exfoliation then occurs when the gas pressure exceeds the internal stress parallel to the $c$-axis. Their model is developed by writing the gas pressure in terms of the ideal gas law

$$P = nk_{\mathrm{B}}T/\pi r^2 I_c \tag{10.34}$$

where $I_c$ is the stage repeat distance and the tensile stress $F$ parallel to the $c$-axis as

$$F = \left[\frac{\pi \gamma G}{2(1-\nu^2)r}\right]^{1/2} = P - \sigma_{\mathrm{appl}} \quad , \tag{10.35}$$

in which $\gamma$ is the effective surface energy per unit area of the bubble, $\nu$ is Poisson's ratio for graphite, $G$ is the shear modulus of graphite and $\sigma_{\mathrm{appl}}$ is the externally applied stress. On this basis, Martin and Brocklehurst [1964] were able to explain the strong stress dependence observed for $T_B$ and the very

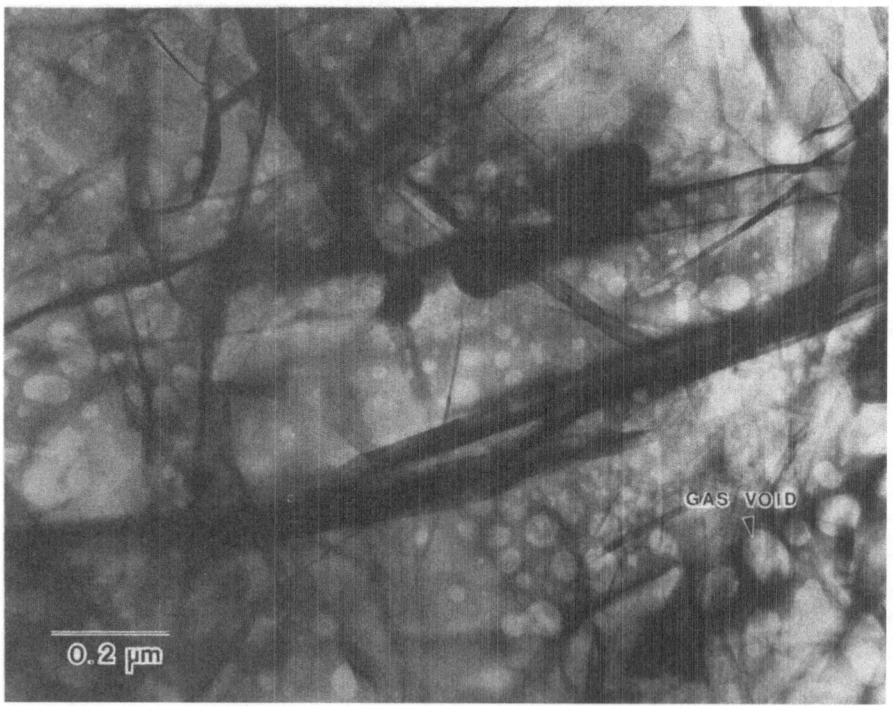

**Fig. 10.28.** Overall structural appearance of an exfoliated fiber (such as that shown in Fig. 10.26) as seen by the TEM. Note the high concentration of gas voids [Jiménez-González et al. 1986]

weak stress dependence observed for the $c$-axis thermal coefficient of expansion $\alpha_c = (d\tilde{c}/dT)/\tilde{c}$. Trapped bubbles in the interlamellar region have been directly identified by high resolution TEM studies in the exfoliated fibers, as shown in Fig. 10.28 [Jiménez-González et al. 1986]. Also, evidence for a high stress level in the intercalation compound during exfoliation is given by the enthalpy changes measured for the exfoliation process of a bulk $Br_2$-GIC sample which were far higher (by a factor of 2) than the enthalpy change associated with vaporization of the same amount of $Br_2$ intercalant [Mazieres et al. 1975].

Because of the high degree of structural order and the favorable geometry of the faceted CCVD fibers heat treated to $\sim$3000°C, a much greater volume increase upon exfoliation was observed for CCVD fibers than for fibers based on other precursor materials. PAN-based fibers are not expected to favor exfoliation because of their relatively poor crystallinity. Pitch-based fibers have much more structural order, but the radial geometry of the layer planes provide a strong stress field, inhibiting large $c$-axis expansion. An increase in fiber diameter by about a factor of 3 has been reported upon exfoliation of an ICl-intercalated Thornel P-100-4 fiber [S.H. Anderson and Chung 1983]. The exfoliated pitch fibers were found to exhibit a reentrant cross section and a roughened surface, with little effect of the exfoliation process on the fiber length. In contrast, heat treated CCVD fibers offer a polygonal cross section with the $c$-axis directed approximately normal to the fiber surface. This is a highly favorable geometry for large $c$-axis expansion. These concepts have been verified by direct measurements on exfoliated fibers [Jiménez-González et al. 1986], which show a typical lateral fiber elongation by a factor of 25–45 (or more) in a single direction, yielding thin ribbon-shaped porous filaments.

Carbon fiber composites containing exfoliated graphite flakes have been shown to exhibit increased ductility in both unidirectional and bidirectional composites [Li et al. 1987], though decreases in composite density and tensile strength were also reported. These composites were also found to arrest crack propagation. Such composite materials may have applications for seals and gaskets.

Study of the structure/property relations and possible applications of exfoliated graphite fibers is still at an early stage.

# 11. Ion Implantation of Graphite Fibers and Filaments

Ion implantation is an important technique for modifying material properties through the introduction of impurity atoms or the creation of lattice defects in a controlled way. The technique is important in the semiconductor industry for making $p$-$n$ junctions by, for example, implanting $n$-type impurities into $p$-type host materials. From a materials science point of view, ion implantation allows essentially any element of the periodic table to be introduced into the near-surface region of essentially any host material, with quantitative control over the depth and composition profile of the impurity by proper choice of ion energy and fluence (i.e., the total number of implanted ions per unit area of sample). Furthermore an important application of ion implantation is in the synthesis of metastable alloys which could not be produced by other means.

In the implantation process, ions of energy $E$ and beam current $i_b$ are incident on a sample surface and come to rest at some characteristic depth $R_p$ with a Gaussian distribution of half width at half maximum $\Delta R_p$ (see Fig. 11.1) [Lindhard et al. 1963; Mayer et al. 1970]. Typical values of the implantation parameters (for carbon in graphite) are: ion energies $E \sim 100\,\mathrm{keV}$,

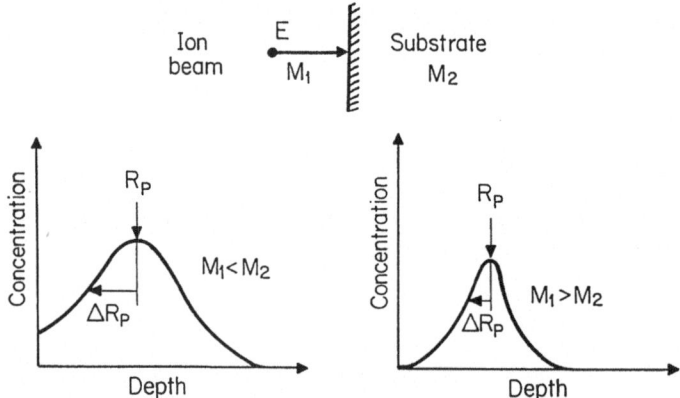

**Fig. 11.1.** The depth distribution of implanted atoms in an amorphous target for the case in which the ion mass is less than or greater than the mass of the substrate atoms. To a first approximation, the mean depth $R_p$ depends on ion mass $M_1$ and incident ion energy $E$, whereas the relative width $\Delta R_p / R_p$ of the distribution depends primarily on the ratio between the masses of the ions and the substrate carbon atoms $M_2$. In almost all practical cases for the implantation of carbon fibers $M_1 \gg M_2$ [Mayer et al. 1970]

beam currents $i_b \sim 50\,\mu A$, penetration depths $R_p \sim 100$ nm with half widths $\Delta R_p \sim 30$ nm [Gibbons et al. 1975]. In the implanted near-surface regions, implant concentrations of $10^{-3}$ to $10^{-5}$ relative to the host materials are typical. In special cases, local concentrations as high as 20 at.% (atomic percent) of implants have been achieved.

## 11.1   The Ion Implantation Process

Some characteristic features of ion implantation are the following [Mayer et al. 1970; Brice 1975; Carter and Grant 1976; Hirvonen 1980; Picraux 1984; Picraux and Peercy 1985]:

1. In general, the ions penetrate the host material to a small depth ($R_p \ll 1\,\mu m$). Thus ion implantation is a *near-surface* phenomenon. To achieve a large percentage of impurity ions in the host, the host material must be thin (comparable to $R_p$), and high fluences of implants must be used ($\phi > 10^{16}/cm^2$), where the fluence is defined as the number of incident ions per unit surface area. Ion implantation can therefore affect the structure and properties of thin fibers significantly.

2. The depth profile of the implanted ions ($R_p$) is controlled by the ion energy and the ion mass. The doping level (or concentration) is controlled by the ion fluence $\phi$.

3. Implantation is a non-equilibrium process. Therefore there are no solubility limits on the introduction of dopants. With ion implantation one can thus introduce high concentrations of dopants, exceeding the normal solubility limits. For this reason, ion implantation permits the synthesis of metastable materials.

4. The implantation process is highly directional with little lateral spread. Thus it is possible to implant materials according to prescribed patterns using masks. Implantation proceeds in the regions where the masks are not present.

5. The diffusion process is commonly used for the introduction of impurities into semiconductors and for intercalation. Efficient diffusion occurs at high temperatures. With ion implantation, impurities can be introduced at much lower temperatures, as for example room temperature. This is a major advantage of ion implantation.

6. The versatility of ion implantation is another important characteristic. With the same ion implanter, a large number of different implants can be introduced by merely changing the ion source. The implanted atoms are introduced in an atomically dispersed fashion, another desirable feature. Furthermore, no oxide or interfacial barriers are formed in the implantation process.

7. The maximum concentration of ions that can be introduced is limited. As the implantation process proceeds, the incident ions participate in

both implantation into the bulk and the sputtering of atoms off the surface. Sputtering occurs because the surface atoms receive sufficient energy to escape from the surface during the collision process. The dynamic equilibrium between the sputtering and implantation processes limits the maximum concentration of the implanted species that can be achieved. Sputtering causes the surface to recede slowly during implantation.

8. Implantation causes radiation damage. For many applications, this radiation damage is undesirable. To reduce the radiation damage, the implantation can be carried out at elevated temperatures and/or the materials can be annealed after implantation. In practice, the elevated temperatures used for implantation or for post-implantation annealing are much lower than typical temperatures used for the diffusion of impurities.

A variety of techniques are used to characterize the implanted materials. Rutherford ion backscattering of light ions at higher energies (e.g., 2 MeV He$^+$) is used to determine the composition versus depth [Chu et al. 1978]. Resonant nuclear reaction techniques are also used to enhance the sensitivity of the backscattering yield for a particular atomic species. Depth profiling by sputtering in combination with Auger or secondary ion mass spectroscopy is also used. Electron microscopy, glancing angle x-ray analysis and ion channeling in single crystals [Feldman et al. 1982] provide detailed information on the local atomic structure of the alloys formed by ion implantation. Because ion implantation primarily causes modification of the near-surface region, Raman scattering provides an especially sensitive technique for the characterization of the lattice damage induced by ion implantation and the subsequent annealing of this damage.

Energetic ions traveling through solids lose energy by collisions with the atoms of the material. There are two dominant mechanisms for this energy loss:

1. The interaction between the incident ion and the *electrons* of the host material. This inelastic scattering process gives rise to *electronic energy loss*. This process dominates at high ion energies.
2. The interaction between the incident ions and the *nuclei* of the host material. This is an elastic scattering process which gives rise to *nuclear energy loss*. This process dominates at low ion energies. For the case of implantation of carbon ions into carbon fibers, the nuclear and electronic energy loss processes are of about equal magnitude at 15 keV. At this ion energy, ions lighter than carbon would favor electronic energy loss while heavier ions would exhibit more nuclear energy loss.

## 11.2 Application to Carbon Fibers

Since almost any element of the periodic table can be implanted in significant concentrations (e.g., up to 10–20 at.% at the maximum of the implant distribution), ion implantation provides a general technique for surface modification of graphite fibers, and the technique is applicable over a wide temperature range. Because of their small diameters, carbon fibers are more sensitive than bulk graphite for studying implantation-induced changes in the structure and properties of graphite and significant work in this area has been carried out [Thrower 1969; Hall et al. 1982; McNeil et al. 1984; Elman et al. 1982, 1983, 1984, 1985; Venkatesan et al. 1984; Endo et al. 1984; Salamanca-Riba et al. 1985]. By appropriate selection of the fiber diameter, implant mass and energy, it is possible to control the fraction of the fiber that contains a significant concentration of the implanted species. For typical choices of the implantation parameters, the dominant energy loss mechanism in implanted carbon fibers is due to nuclear energy loss processes. Vapor-grown fibers have predominantly been used for ion implantation studies in graphite fibers because the vapor-grown fibers have the highest structural perfection when heat treated to $T_{HT} \geq 2900°C$, thereby allowing a quantitative study of the implantation-induced modifications to the near-surface structure [Salamanca-Riba et al. 1985]. In contrast to chemical intercalation, the ion implantation technique is expected to provide doped materials which are stable to higher temperature.

The special fiber geometry is also particularly useful for convenient imaging of the $c$-axis lattice planes by transmission electron microscopy (TEM), allowing direct measurement of the structural coherence distances $L_a$ and $L_c$ parallel and perpendicular to the lattice planes in studies of the regrowth of crystalline graphite. These studies have been complemented with TEM studies on ion implanted HOPG to provide the corresponding information on the microstructure in the basal plane of graphite [Salamanca-Riba et al. 1985]. Observations have been made on fibers directly after implantation and after subsequent annealing at various temperatures $T_a$ and for a range of annealing times $t_a$. In this way, information has been obtained on the recrystallization and regrowth kinetics of highly damaged graphite. The annealing studies on the fibers were carried out in the temperature range 1500°C $< T_a < 2800°C$ under a constant argon gas flow for annealing times in the range $15 < t_a < 90$ min. Analysis of these results in both implanted vapor-grown fibers and HOPG have provided the time dependence for the activation energies for both in-plane and $c$-axis recrystallization processes as discussed in Sect. 11.3 [Salamanca-Riba et al. 1985].

Ion implantation has been used to modify the properties of the graphite host material, especially the introduction of Fe ions to produce magnetic behavior in the near-surface region of the host material [Koon et al. 1984]. Though this approach would be expected to have even larger relative effects in implanted carbon fibers (because of their small diameters), no reports on magnetic implants in fibers have yet been published.

From the applications point of view, the implantation of guest species in the near-surface region should enhance the adhesion of other species to fiber surfaces, by greatly increasing the density of active surface sites.

## 11.3    Implantation-Induced Structural Modifications

In comparing the structure of ion-implanted and pristine fibers, bright field, (002) dark field and selected area electron diffraction patterns and lattice images have been examined using high resolution transmission electron microscopy. Lattice images were taken from areas within a few hundred angstroms of the fiber edge, corresponding to thicknesses $\leq 500$ Å. The graphite planes were oriented with the graphite $c$-axis normal to the electron beam direction. Dark field and lattice image measurements were made on both the implanted and non-implanted faces (see Fig. 11.2). The lattice images were obtained by placing a circular aperture that encompasses at minimum the direct beam and the (002) reflections observed in the back focal plane of the objective lens of the TEM. The (002) dark field images were obtained using an aperture encompassing only the (002) reflection and tilting the beam to align the (002) reflection along the axial direction. The implantation-induced structural modifications were studied as a function of ion mass (Fig. 11.3), ion fluence (Fig. 11.4) and annealing temperature (Fig. 11.5) [Endo et al. 1984; Salamanca-Riba et al. 1985].

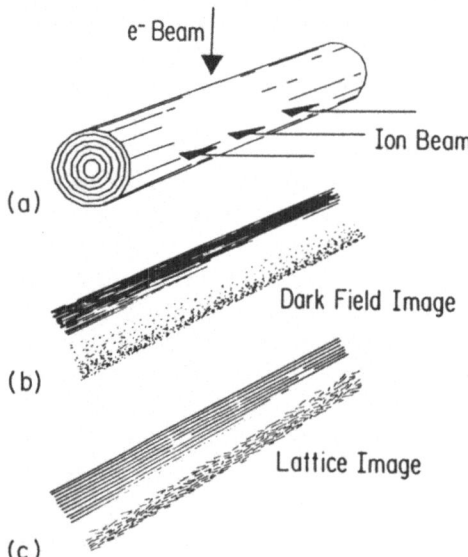

(a)

Dark Field Image

(b)

Lattice Image

(c)

Fig. 11.2. (a) Schematic representation of the directions of the ion beam for implantation and the electron beam for electron microscopy. (b) Schematic representation of the (002) dark field image and (c) Schematic of the lattice fringe images of the implanted and unimplanted sides of the graphite fiber observed in the transmission electron microscope [Endo et al. 1984]

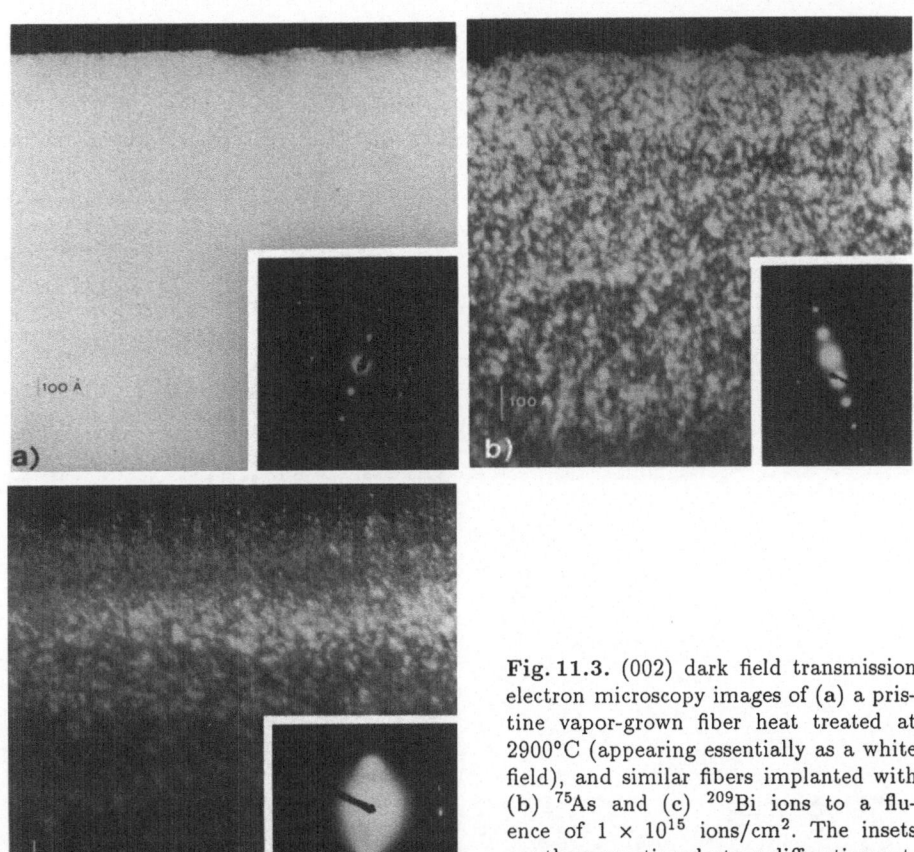

**Fig. 11.3.** (002) dark field transmission electron microscopy images of (a) a pristine vapor-grown fiber heat treated at 2900°C (appearing essentially as a white field), and similar fibers implanted with (b) $^{75}$As and (c) $^{209}$Bi ions to a fluence of $1 \times 10^{15}$ ions/cm$^2$. The insets are the respective electron diffraction patterns showing predominately (00$l$) reflections [Endo et al. 1984]

More specifically, Fig. 11.3 shows (002) dark field images of an unimplanted highly ordered fiber (Fig. 11.3a) in comparison with the same fiber implanted with 30 keV $^{75}$As ions (Fig. 11.3b) and with 30 keV $^{209}$Bi ions (Fig. 11.3c), both to a fluence of $1 \times 10^{15}$ ions/cm$^2$. The low ion energy was selected to concentrate the implanted ions near the surface. The corresponding selected area diffraction patterns are shown as insets to each of the dark field images. Before implantation, the fibers exhibit sharp (002) and (112) diffraction spots, consistent with three-dimensional ordering, as well as a bright Bragg band for the dark field image, indicative of large coherent crystallites. Ion implantation reduces the crystallites size to 20 Å and 50 Å respectively for the $^{209}$Bi and $^{75}$As implants. The lattice damage of the $^{209}$Bi is far greater than for the $^{75}$As implants, as seen in the diffraction pattern. The smaller thickness of the damaged region in the case of the $^{209}$Bi implant

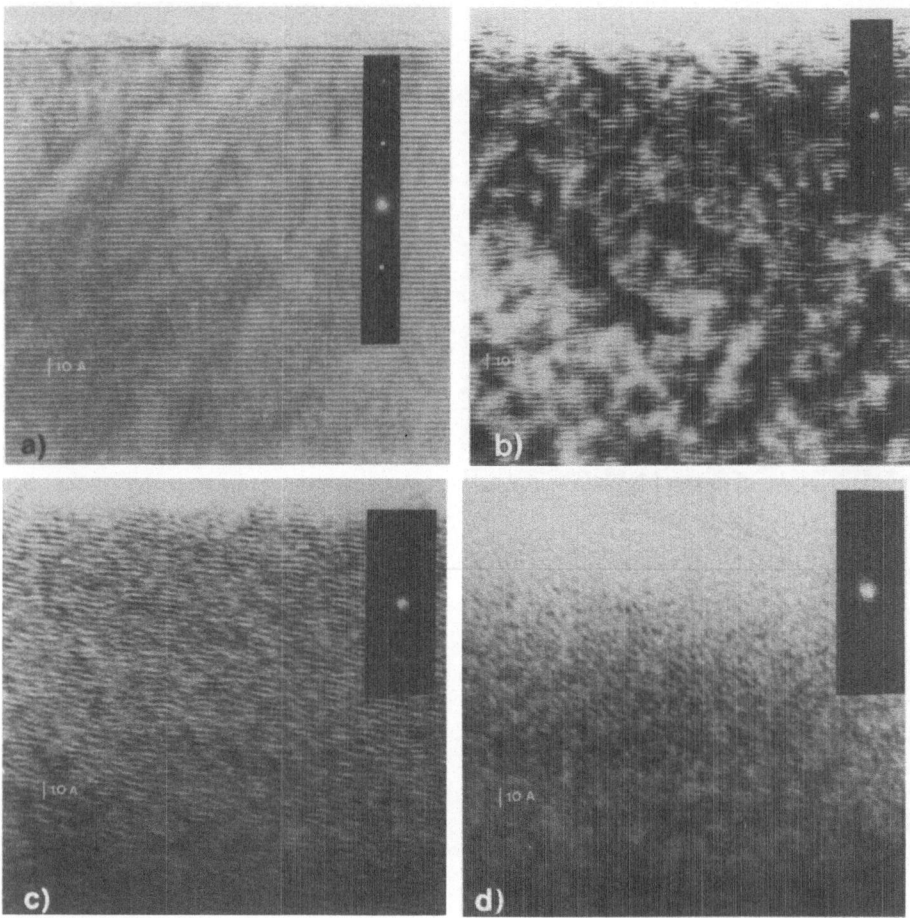

**Fig. 11.4.** (002) lattice images of (**a**) a pristine fiber and (**b**)-(**d**) fibers ion implanted with $^{122}$Sb ions at 30 keV to fluences of $5 \times 10^{12}$, $1 \times 10^{14}$ and $1 \times 10^{15}$ ions/cm$^2$, respectively. Strictly speaking, antimony has two isotopes with atomic masses of 121 and 123, occurring in almost equal abundances. The insets are optical diffractograms taken from the negatives of the corresponding lattice images [Endo et al. 1984]

is due to the smaller penetration depth of the heavy mass ion. The results indicate that the heavier the implanted ion, the greater the lattice damage that is induced, since a larger amount of energy must then be dissipated in a smaller sample volume, keeping other parameters fixed. The extent of the lattice damage was measured from the lattice fringe patterns in terms of the coherence length $L_{a\parallel}$ of the lattice fringes along the fiber length. A log-log plot of in-plane coherence lengths $L_a$ vs the ion mass $M_i$ shows a functional dependence $L_a \approx M_i^{-1/2}$ (see Fig. 11.6). Similar results were obtained with dark field and lattice fringe measurements of $L_a$ [Salamanca-Riba et al. 1985].

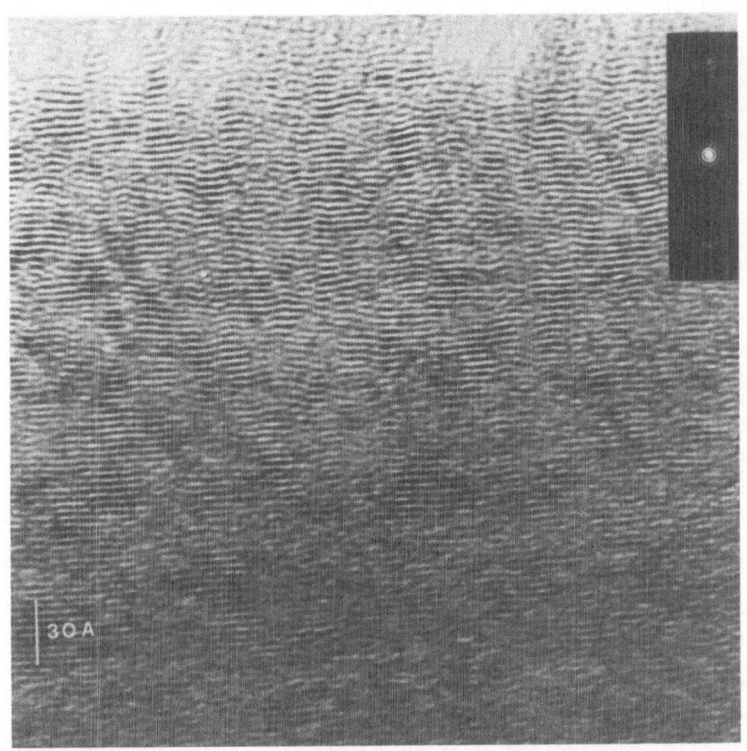

**Fig. 11.5.** (002) lattice image of fibers implanted with $^{209}$Bi ions at 30 keV to a fluence of $1 \times 10^{15}$ ions/cm$^2$ and post-implantation annealed at $T_a = 1500°C$. The inset is an optical diffractogram taken from the negative of the lattice image [Endo et al. 1984]

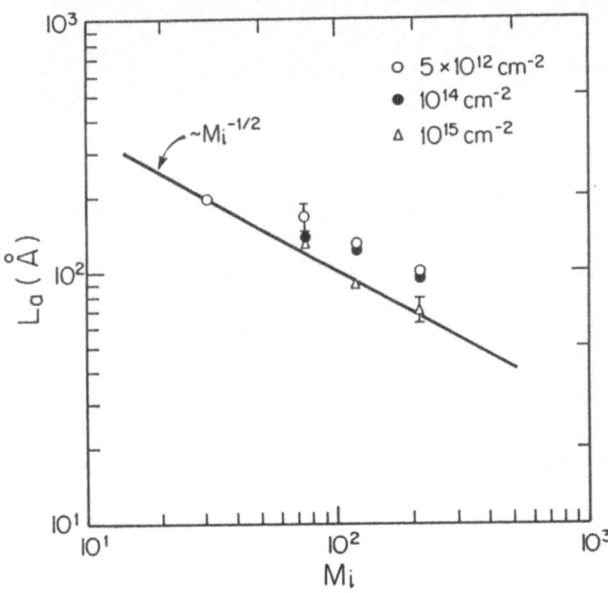

**Fig. 11.6.** Dependence of in-plane coherence length $L_a$ on ion mass $M_i$ for various fluences. The solid line represents a $M_i^{-1/2}$ dependence [Endo et al. 1984]

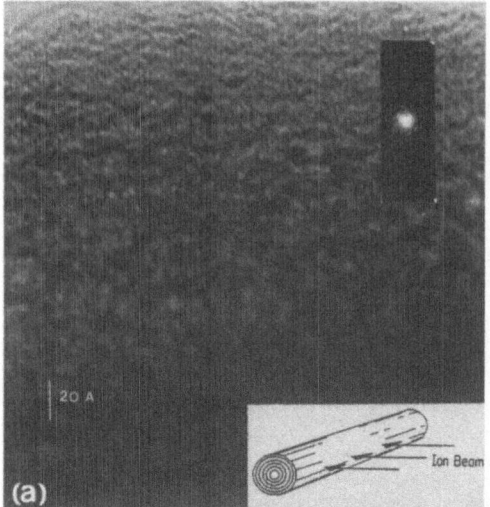

Fig. 11.7. (a) (002) lattice image of fibers implanted with $^{209}$Bi ions at 30 keV to a fluence of $1 \times 10^{15}$ ions/cm$^2$. (b) Selected area electron diffraction pattern corresponding to (a), and (c) electron diffraction of the pristine CCVD fiber ($T_{HT} = 2900°C$) prior to implantation [Endo et al. 1984]

The lattice fringe image pattern in Fig. 11.7 shows the amorphization of the near-surface region of the Bi-implanted fiber. The effect of the implantation on the selected area ($hk0$) electron diffraction pattern is to give a weak broad ring with a diameter equal to the (100) diffraction spots, the spots on the diffraction pattern are due to the graphite layers beyond the ion penetration depth.

The dependence of the ion damage on ion dose is illustrated in Fig. 11.4 where (002) lattice fringe images are shown for 30 keV $^{122}$Sb ions implanted to fluences of $5 \times 10^{12}$cm$^{-2}$, $1 \times 10^{14}$cm$^{-2}$ and $1 \times 10^{15}$cm$^{-2}$. The results show that implantation to a fluence of $1 \times 10^{14}$cm$^{-2}$ yields significant lattice damage, and at the highest fluence, the fringes corresponding to the graphite layers

300

have essentially disappeared and the texture is similar to that of amorphous carbon. The extent of the lattice damage can also be evaluated by examination of the optical diffractograms taken from the lattice image patterns; the optical diffractograms corresponding to each lattice image pattern are shown as insets (see discussion below).

Finally in Fig. 11.5 we see that high temperature annealing is required to reduce the lattice damage. For $^{209}$Bi ions implanted to a fluence of $1 \times 10^{15}$ cm$^{-2}$ at 30 keV (see the fringe pattern in Fig. 11.7), a highly disordered near-surface region is observed, and this disorder is significantly annealed out by 1500°C. The onset of significant ordering in the (002) lattice fringes (Fig. 11.5) of fibers annealed at $T_a = 1500$°C for 1 h indicates that some recrystallization occurs for $T_a \leq 1500$°C. The multiple spot pattern found in the optical diffractogram (inset to Fig. 11.5) confirms that the parallelism with respect to the fiber axis has not been completely restored at $T_a = 1500$°C. Interplanar spacings $\tilde{c}$ in the range $3.36 \pm 0.08$ Å $\leq \tilde{c} \leq 3.85 \pm 0.08$ Å were measured from the optical diffractogram of Fig. 11.5 assuming the opposite (unimplanted) side of the fiber to have $\tilde{c} = 3.36$ Å. Annealing temperatures of $T_a \sim 2800$°C (not shown) fully restored the ordered lattice fringe pattern which was lost by ion implantation, but the interplanar spacing obtained from the optical diffractogram indicates $\tilde{c}$ values as large as $\sim 3.80 \pm 0.08$ Å in the annealed material. The in-plane ($L_a$) and c-axis ($L_c$) correlation lengths were obtained from the (002) lattice images by directly measuring the length of the fringes and the distance over which the fringes were parallel, respectively. $L_a$ and $L_c$ values were also obtained from the (002) dark field images by measuring the longitudinal and transverse widths of the bright 'spots', and from the full width at half maximum of the (002) spots of the electron diffraction patterns.

Figure 11.8a shows Arrhenius plots for $L_a$ and $L_c$ as a function of $1/T_a \times 10^4$ K$^{-1}$ for post-implantation annealed fibers. For these data, activation energies $E_a \sim 0.66 \pm 0.08$ eV/atom and $E_c \sim 0.78 \pm 0.08$ eV/atom are obtained from the $L_a$ and $L_c$ plots for the in-plane and c-axis grain growth processes, respectively [Salamanca-Riba et al. 1985]. In this work a broad ring was found in the electron diffraction pattern of post implantation annealed HOPG ($T_a \sim 1500$°C) with a wave vector $1.78 \pm 0.03$ Å$^{-1}$ corresponding to the (002) reciprocal lattice vector of graphite. This observation was interpreted to indicate that random recrystallization takes place together with the 2D regrowth. This process gives rise to the formation of small crystallites that have their c-axes randomly oriented. The random recrystallization process takes place because the implanted region of the sample was amorphized to the extent that some of the memory of its former crystallinity was lost. Since in-plane (2D) regrowth also takes place at this annealing temperature, it is concluded that not all the memory is lost during ion implantation and many of the crystallites grow from tiny seeds with their c-axes parallel to the c-axis of the substrate.

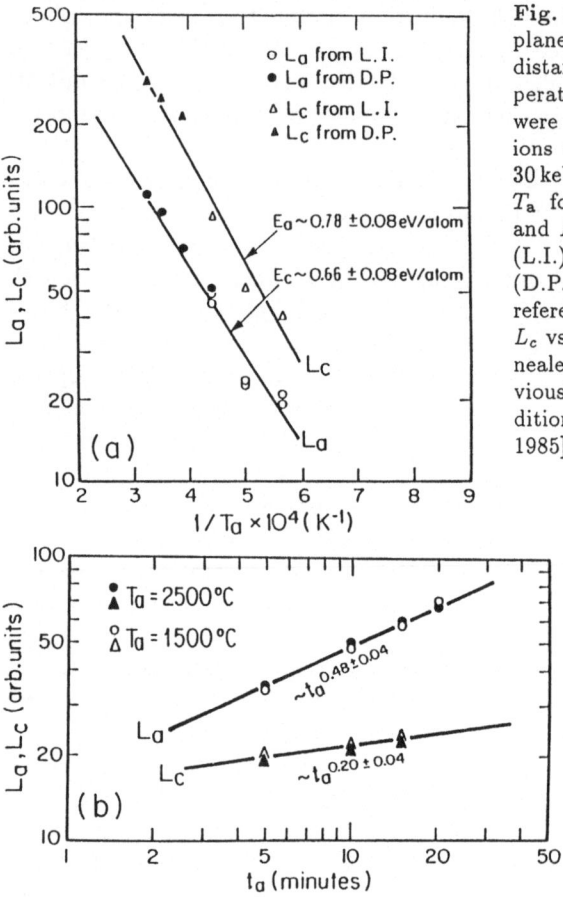

**Fig. 11.8.** (a) Arrhenius plot of in-plane ($L_a$) and c-axis ($L_c$) coherence distances vs reciprocal annealing temperature $1/T_a$. The vapor-grown fibers were previously implanted with $^{209}$Bi ions to a fluence of $1 \times 10^{15}$/cm$^2$ at 30 keV and subsequently annealed at $T_a$ for 1 h. The measurements of $L_a$ and $L_c$ were made from lattice images (L.I.) and from the diffraction patterns (D.P.), but without use of a calibrated reference. (b) Log-log plots of $L_a$ and $L_c$ vs annealing time $t_a$ for samples annealed at 1500°C and 2500°C and previously implanted under the same conditions as in (a) [Salamanca-Riba et al. 1985]

## 11.3.1 Characterization of Samples

Rutherford backscattering spectroscopy (RBS) measurements from the near-surface region of the implanted HOPG samples after annealing to $T_a > 1500$°C showed no evidence of $^{209}$Bi ions present in the samples. Using the RBS technique Bi was detected in the as-implanted samples, while trace amounts were also present in those annealed at 1500°C. This is in agreement with previous RBS results for high temperature annealing of HOPG implanted with a $1 \times 10^{15}$ fluence of $^{75}$As ions/cm$^2$ [Elman et al. 1985], showing that the $^{75}$As ions diffuse between the graphite planes and out of the implanted region of the sample for $T_a \geq 2300$°C. It is therefore inferred that the implants diffuse out of the sample at higher temperatures, consistent with the usual steps of the graphitization process (see Sect. 3.7.2).

## 11.3.2 Regrowth

Studies of the kinetics of the recrystallization process in fibers for annealing times $t_a$ in the range $15 \leq t_a \leq 90$ min showed different time dependences for the in-plane ($\sim t_a^{1/2}$) and $c$-axis ($\sim t_a^{1/4}$) regrowth processes. Assuming that diffusion limits the grain growth, the time dependence of grain growth is expected to be proportional to $t_a^{1/2}$ and a diffusion activation energy of either $2E_a$ or $2E_c \sim 1.5$ eV was deduced. This is an intermediate value between reported activation energies for the diffusion of single interstitials in the basal plane ($\sim 0.4 \pm 0.1$ eV) and along the $c$-axis ($\sim 2.9 \pm 0.3$ eV) [Thrower 1969] (see Sect. 3.1.4). It is clear that the in-plane grain growth requires diffusion of atoms. The regrowth along the $c$-axis, on the other hand, is achieved by annealing of dislocations, which is required to get the proper stacking of the graphite layers. This is a diffusion-activated process in which three-dimensional diffusion producing climb of dislocations takes place [Thrower 1969]. The slower time dependence for the 3D grain growth process supports graphitization studies on carbons [Oberlin 1984] which imply that substantial 2D ordering is necessary before the alignment of the graphite planes takes place. Thus the $t_a^{1/4}$ dependence of annealing corresponds to the random regrowth process [Salamanca-Riba et al. 1985].

Since intercalation depends on a nucleation/diffusion growth process, it has been thought that ion implantation might stimulate the nucleation phase and thereby enhance the intercalation process. This effect has been demonstrated to a limited extent for the enhanced intercalation of sodium in HOPG [Menjo et al. 1984]. Implantation-enhanced intercalation has not yet been studied in fibers.

# 12. Applications of Graphite Fibers and Filaments

This chapter considers some applications of carbon fibers. This topic is not treated exhaustively. Instead, an attempt is made to give an overview, with emphasis given to those applications which involve physical principles.

## 12.1  Economic Considerations

The price of commercially available carbon and graphite fibers ranges from about $20/kg for the least expensive, low modulus ex-PAN fibers to above $2000/kg for ultra-high modulus ex-pitch fibers (Union Carbide P-100). Most of the current applications of carbon fibers utilize high strength, low modulus ex-PAN fibers which cost $20–$60/kg. The consumption of these materials has been growing rapidly in recent years (see Fig. 12.1), although the price has not dropped significantly. This high price is due to the fact that the PAN precursor fiber is relatively expensive, and the yield is less than 50% (as discussed in Sect. 2.1). When ex-pitch precursor fibers were introduced in the 1970s, it was expected that they would ultimately be much cheaper than those made from PAN because of lower raw material costs and higher yields. This has not happened, presumably because of difficulties in preparing and spinning pitch which lead to significantly higher costs [Fitzer 1985].

The primary justification for using carbon and graphite fibers despite their relatively high cost is most often their superior mechanical properties

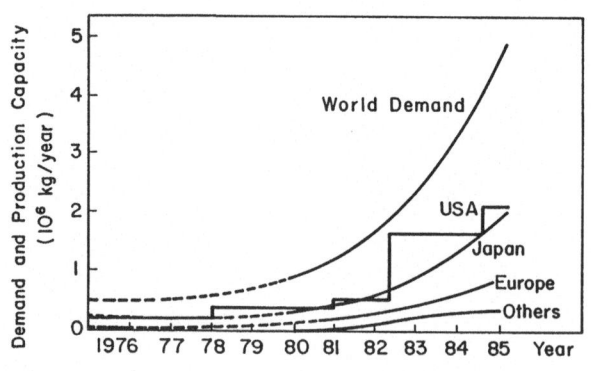

Fig. 12.1. Worldwide carbon fiber market and production capacity [Fitzer 1985]

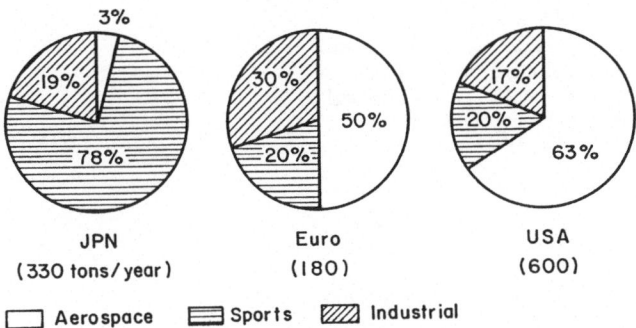

**Fig. 12.2.** End use distribution of carbon fibers in Japan, Europe and the United States (basis is for 1981 with total consumption of 1110 tons/year) [Fitzer 1985]

and low density; i.e., their high specific strength and specific modulus (see Fig. 6.1). When one can justify increased cost in order to make a structure lighter, carbon fibers can economically replace traditional materials. Weight is obviously of extreme importance in aircraft and space applications, and it was for these applications that carbon fibers were first used [see B.W. Anderson (1987), Krieger (1986)]. In addition, many high-performance sporting goods have been developed which utilize carbon fiber composites. As shown in Fig. 12.2, these two applications dominate today's demand for carbon fibers.

Automotive applications have always been viewed as the largest potential market for carbon fibers because of their high specific strength. One pound of carbon fiber in every car would still require more than the total current world production. However, the high price has prevented such a large penetration into the automotive market. Some significant applications have been found recently such as in the drive shaft on small General Motors trucks and for the chassis in racing cars [Clarke 1986]. Another large volume application area is in reinforced cement for the construction of large buildings. Such an application of isotropic ex-pitch carbon fibers has recently been made in the successful construction of the "Ark Hills" twin tower complex in Tokyo.

A much lower priced fiber would have significant advantages in presently marginal applications such as transportation. The best hope for such a fiber lies in CCVD filaments. This is because CCVD fibers can be made from extremely cheap raw materials and can be produced in a single step with properties reasonably close to high strength chopped ex-PAN fibers. The new catalytic floating method for fabricating a continuous flow of thin CCVD filaments [Endo et al. 1987a] has considerable potential for relatively high throughput and low cost. One disadvantage is that CCVD fibers cannot be produced as a continuous filament, but rather consist of chopped and milled filaments. Although most structural applications use continuous fibers so that one gets the maximum benefit from the properties of the fibers in the final

305

**Table 12.1.** Economic considerations

| Fiber type | Modulus [GPa] | Elongation [%] | Price [U.S.$/kg] |
|---|---|---|---|
| ex-PAN | 240 | 1.6 | 26–40 |
| ex-PAN | 300 | 1.7 | 100 |
| ex-PAN | 360 | 0.7 | 120 |
| ex-PAN | 500 | 0.4 | 1430 |
| ex-mesophase-pitch | 380 | 0.5 | 90–140 |
| ex-mesophase-pitch | 520 | 0.4 | 700–1320 |
| ex-mesophase-pitch | 700 | 0.3 | 2000–3300 |
| ex-mesophase-pitch | 830 | 0.3 | 2640 |
| ex-isotropic-pitch | 42 | 0.2 | 16 |

composite, it may be advantageous to sacrifice some of the properties if lower cost chopped and milled filaments can be developed.

The price of both ex-pitch and ex-PAN fibers increases rapidly with increasing modulus (see Table 12.1). This is partly due to the small market for high modulus fibers and partly due to the cost of heat treatment of any material near 3000°C. Substantial cost reduction may be possible for large-scale fiber production plants. From an economic standpoint, applications requiring very high modulus fibers necessitate even more performance advantages than those which use low modulus fibers. It is important to remember that only these high modulus fibers can be intercalated easily, and thus intercalated graphite fibers will also be expensive. Here again CCVD fibers prepared by the catalytic floating method and heat treated to ~2000°C may provide a cost effective host material for intercalation. The filaments thus prepared could have specific advantages over carbon blacks for electromagnetic shielding applications because of their enhanced electrical conductivity and higher aspect ratios.

Whenever one tries to optimize the cost/performance of a composite material, it is important to look at all the important mechanical, electrical, thermal, and chemical properties. One can often mix a variety of materials, such as glass or polymer fibers, in order to keep the costs down and still achieve the desired performance. In other applications the unique properties of carbon fiber composites, such as low thermal expansion, high thermal conductivity or high corrosion resistance, will dictate that carbon fibers be used (see Fig. 12.3). For example, a thermally conducting polymer made as a CCVD fiber/polymer composite could perhaps find applications which would justify the higher costs of the constituents. Coatings can also enable one to combine the mechanical properties of carbon fibers with other desired attributes such as high electrical conductivity, superconductivity, or improved oxidation resistance.

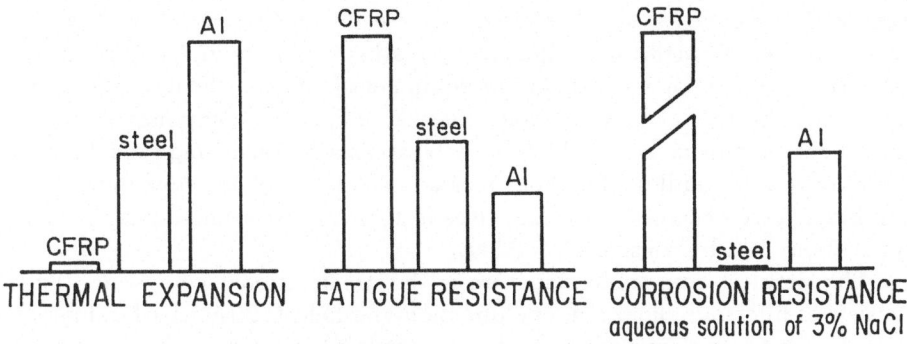

Fig. 12.3. Further advantages of the properties of advanced composites (CFRP = carbon fiber reinforced plastics) in contrast to conventional materials such as steel and aluminum. Thermal expansion is important for such applications as parabolic antennas, cryogenic applications, measuring instruments; fatigue resistance is important for machinery, sporting goods, musical instruments; corrosion resistance is important for biocompatibility, chemical apparatus and pipeline applications [Fitzer 1985]

## 12.2 Structural Applications of Composites

### 12.2.1 Carbon Fiber-Polymer Composites

We have extensively discussed the fact that carbon fibers can be produced with high specific strength and stiffness compared to most conventional construction materials. In order to take advantage of the mechanical properties of the fibers in structural applications, they must be put in composites comprising fibers and matrix. The matrix must have the ability to transfer stress between fibers so that all the fibers used are effective in bearing the load. In addition, the matrix should allow for economic processing of the composite into useful shapes. Other properties of the matrix such as chemical inertness, thermal stability, and environmental stability are also important for many applications.

A useful summary of applications of composites has been given by Sittig [1980]. An idea of the scope of these applications can be gleaned from a listing of those considered in this text: Automotive brake-linings, drive shafts, rotary engine seals (with a metal matrix), springs, tire cords, wheels, structural components for racing cars; electrical brushes, conductive papers and plastics; x-ray shields (fibers containing transition-metal compounds to absorb the x-ray radiation); radar absorbers; superconducting cables; thermocouples; linear-positioning motor components; aircraft – structural components, high-stress bearings; helicopters – rotor blades, fuselage components, shafts; centrifuge rotors; flywheels; gas-turbine components; machine-tool supports; masts for radar and radio beacons; satellite reentry shields and space satellite structural components; speaker cones; lightweight components for manufacturing machinery; medical equipment – prosthetic devices and blood filters, support structures for x-ray diagnosis equipment (utilizing the low scattering and absorption properties of carbon); musical instrument components; sports equipment – tennis rackets, golf club shafts, archery bows, fishing rods, hockey sticks, oars and paddles, skis, boat masts ... This list shows how much carbon fibers have become part of our lives in the last few years. Some of these applications will be considered further.

The simplest composite structure is the unidirectional composite. This consists of fibers all aligned in one direction and held together with a matrix resin. Carbon fibers are usually elastic (i.e., have a linear relationship between their stress and strain) right up to their breaking strain (see Sect. 6.2). Organic matrix materials are usually elastic up to strains which are much higher than the breaking strain of carbon filaments. Thus, the composite will behave elastically as long as it is not about to fail. Since the strain on both the reinforcing fibers and the matrix will be the same (i.e., the fibers are not pulling out of the matrix), the force per unit area exerted on the unidirectional composite, $\sigma_T$, will be supported by both the fibers and matrix according to the following equation:

$$\sigma_T = \text{force/unit area} = [\beta E_{Yf} + (1 - \beta)E_{Ym}]\varepsilon \quad , \tag{12.1}$$

where $\beta$ is the volume fraction of the fibers, $\varepsilon$ is the strain, $E_{Yf}$ is the Young's modulus of the fibers, and $E_{Ym}$ is the Young's modulus of the matrix. The theoretical limit for the volume fraction of cylindrical fibers in a unidirectional composite is 91%. However, since it is critical that every fiber be wetted with matrix, the practical limit for useful materials is about 70% [Parratt 1972].

The modulus of the carbon fibers in carbon fiber-polymer composites is more than 20 times greater than the modulus of the matrix, and unidirectional composites have fiber volume fractions of 50%–70%. Therefore, using (12.1), one finds that the composite modulus $E_{Yc}$ parallel to the fiber axis ($E_{Yc} = \sigma_T/\varepsilon$) is given by

$$E_{Yc} = \beta E_{Yf} + (1 - \beta)E_{Ym} \tag{12.2}$$

and is dominated by the carbon fibers.

When the unidirectional composite is put under tension along the fiber direction, the fibers fail at much lower strain than the matrix. Thus the strength will be limited by the strength of the fibers times the volume fraction of fibers. One must remember that carbon fiber strengths are limited by a statistical distribution of defects (see Sect. 6.2). The effective strength of a composite will therefore be reduced from the average fiber strength times the volume fraction by an amount which depends on the width of the distribution of fiber strengths. When the weaker fibers start to fail, the load they carried must be supported by the remaining fibers.

308

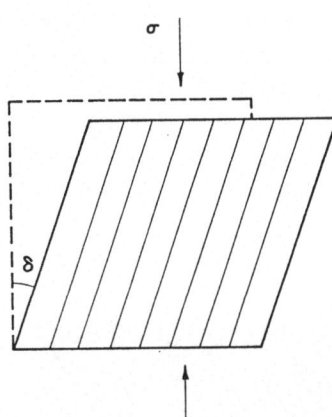

**Fig. 12.4.** Shear microbuckling within a composite. Under compressive loads, a composite can deform via a shear mode more easily than a longitudinal compressive mode [Parratt 1972]

Although one can obtain impressive properties for unidirectional composites under uniaxial tension, the properties are not as good when placed under compressional or shear forces. The resistance to shear is determined entirely by the matrix and the strength of the matrix-fiber bond. As shown in Fig. 12.4, the shear properties of the composite are important in determining the mechanical properties under compressive loads because of the possibility of microbuckling within the composite. Some typical properties (tensile strength, bulk modulus, etc.) of unidirectional composites under both tension and compression are given in Table 12.2.

Unidirectional composites can be made simply by aligning fiber bundles in a mold, impregnating the fibers with a resin, compacting them while squeezing

Table 12.2. Typical properties of various graphite epoxy composites[a]

|  | Tensile | | Compressive | |
|  | Strength [GPa] | Modulus [GPa] | Strength [MPa] | Modulus [GPa] |
|---|---|---|---|---|
| Chopped fiber | | | | |
| molding compound | 0.35 | 108 | 0.47 | – |
| Unidirectional[b] | | | | |
| high strength fiber(0°) | 1.8 | 138 | 0.99 | 113 |
| high strength fiber(45°) | 0.50 | 57.3 | 0.50 | 50.4 |
| medium strength fiber(0°) | 1.5 | 147 | 1.4 | 131 |
| medium strength fiber(45°) | 0.50 | 64.9 | 0.50 | 64.2 |
| high modulus fiber(0°) | 1.2 | 215 | 76 | 177 |
| Fabric (woven) | | | | |
| medium strength fiber | 0.51 | 70.4 | 0.51 | 63.5 |

[a] From Watts [1985].
[b] The number $(n°)$ refers to the angle at which the stress is applied. (0°) means that the stress is applied along the fiber direction, while (45°) means the stress is applied at 45° to the fiber axis.

out excess resin, and then curing the resin. Such simple molding techniques are rarely used in large-scale production because of difficulties in maintaining uniformity of the resin and alignment of the fibers [Gill 1972]. The most common way of handling carbon fibers in a commercial environment is in pre-preg form. Pre-preg is a sheet of carbon fiber impregnated with uncured resin which has well-controlled thickness, fiber loading, and fiber alignment. A typical production of pre-preg is outlined in Fig. 12.5.

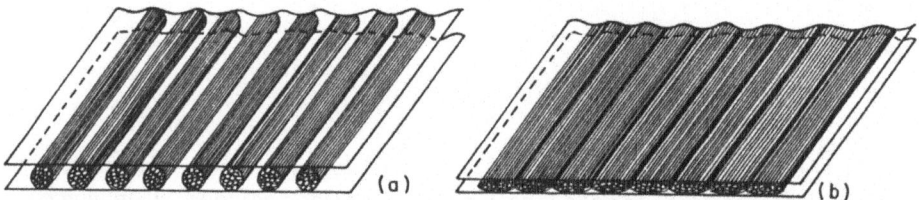

**Fig. 12.5a,b.** Production of pre-preg sheet: (a) fiber tows are placed between two sheets of plastic film or siliconized paper (tows are first coated with resin and are carefully spaced and aligned parallel to each other; the plastic film may be peeled off at a later stage); (b) fiber tows after being rolled between the sheets of plastic (tows have spread outward and adjacent tows now meet; fiber alignment is preserved and the action of rolling removes trapped air and ensures an even thickness; heat is usually applied to assist the rolling operation) [Gill 1972]

Pre-pregs are usually used to form laminated structures whose mechanical properties have been optimized (see Fig. 12.6). The orientation of the fibers in each layer can be chosen in order to provide the desired mechanical characteristics of the final sheet (see Fig. 12.7). This technique also provides the opportunity for optimizing the cost/performance by laminating pre-pregs made from different reinforcing fibers which are less expensive and/or have different properties than carbon (such as glass or Kevlar).

There are numerous other techniques for forming composite parts in addition to laminating and molding techniques using pre-pregs. They have been

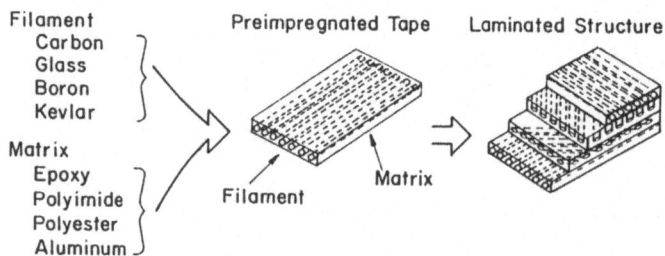

**Fig. 12.6.** Schematic of the production of a laminated structure from pre-preg sheets [Riggs 1985]

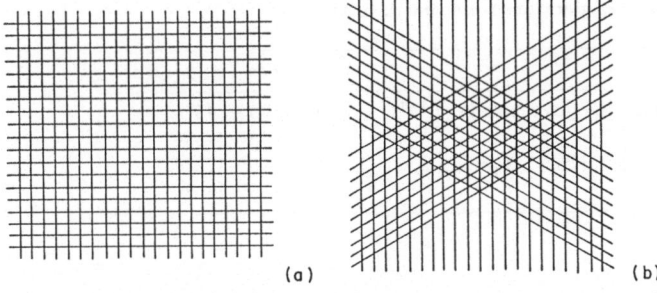

**Fig. 12.7.** Examples of different designs of composites showing layers of pre-pregs with fibers at different relative angles [Gill 1972]

reviewed recently by Hayes [1985a] and Riggs [1985] and include filament winding and pultrusion (the continuous pulling of resin impregnated fiber through a hot dye which continuously cures the resin). Carbon fabrics are also used in both woven and non-woven form in the fabrication of sheet materials.

The choice of resin is critical, and the factors which govern that choice have been reviewed by Hayes [1985b]. A comparison of composite properties (laminate tensile strength vs strand tensile strength) made with two different resins is shown in Fig. 12.8. The most commonly used resins are epoxies which are thermally cured. High temperature (170°–180°C) cured systems provide excellent thermal and environmental stability, including maintenance of properties under wet conditions. Lower temperature cured systems (120°–130°C) are less expensive to fabricate, and can be made with excellent impact

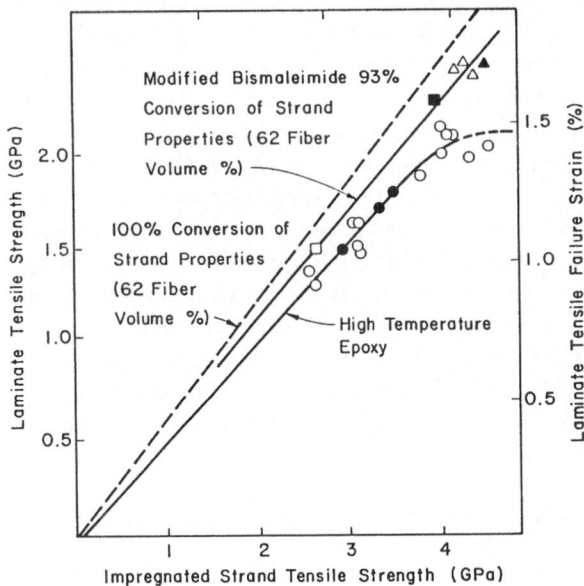

**Fig. 12.8.** The strength of carbon fiber laminates in two different matrices (see Table 12.3) [Riggs 1985]

311

**Table 12.3.** Potential high temperature resin matrix materials[a]

| Material | Use Temperature |
|---|---|
| Bismaleimides and maleimide-epoxy blends (BMI) | 205°–245°C |
| Polyimides (PI) | 260°–315°C |
| Polyquinoxalines (PQ) and polyphenylquinoxalines (PPQ) | 315°–370°C |
| Polybenzimidazoles (PBI) | 315°–370°C |

[a] From Riggs [1985].

strength, toughness, adhesion to underlying honeycombs, and fatigue resistance.

New resins are also being developed for improved toughness and high temperature performance. Table 12.3 lists some of the high temperature resins currently being considered. BMIs are already used commercially, and can be processed using pre-preg technology just like epoxies, while PIs are also finding practical applications in aerospace systems [Riggs 1985]. The other systems still need considerable development in either properties (PQ and PPQ) or processing techniques (PBI).

Another class of matrix materials includes thermoplastics. These materials offer the advantage that they do not need to be cured, but only need to be melted during the processing of the part. Thus fabrication time (and cost) is significantly reduced. Most of the present applications use polymers such as nylon, polypropylene, and polyethylene loaded with 10%–25% chopped fiber by volume. These materials can then be injection-molded into complex shapes. Table 12.4 shows an example of the properties obtained at low loading in these materials. It is clear that there is a more significant enhancement of the modulus than the strength. This is because the strength is critically dependent on the bonding of the matrix to the fiber. In fact, the shear strength of all composites can be significantly improved by careful attention to the matrix fiber interface [Riggs 1985 and references therein].

Commercial uses of molding compounds made with low (less than 25%) loadings of chopped carbon fibers are most often in applications which require both the increased stiffness and increased electrical conductivity which results from using this expensive filler material. One such application is in housings for computers and video terminals where the electrical conductivity is important for electromagnetic interference (EMI) shielding reasons, discussed in Sect. 12.3.1. CCVD fibers (especially the ultra-thin fibers) have been seriously considered for such applications. Coating the carbon fibers with metals or intercalating them provides even higher conductivities and thus improved shielding; this is further discussed in Sect. 12.3.1.

An important drawback of most thermoplastics for aerospace applications is their low use temperatures. Table 12.5 lists some higher temperature ther-

312

**Table 12.4.** Properties of thermoplastics reinforced with chopped carbon fibers[a]

| Thermoplastic | Fiber content [%vol.] | Tensile strength [MPa] | Flexure strength [MPa] | Flexure modulus [GPa] |
|---|---|---|---|---|
| Polypropylene | 0 | 30 | 1.4 | – |
| | 4.4 | 34 | 52 | 3.2 |
| | 8.9 | 35 | 56 | 4.6 |
| ABS[b] | 0 | 43 | – | 2.3 |
| | 5.0 | 55 | 95 | 4.7 |
| | 9.6 | 42 | 73 | 5.8 |
| Styrene-acrylonitrile | 0 | 69 | 106 | 3.3 |
| | 5.1 | 66 | 101 | 6.4 |
| | 9.9 | 66 | 104 | 10.4 |
| Nylon 6[c] | 0 | 69 | – | 2.3 |
| | 5.4 | 78 | 132 | 4.9 |
| | 10.2 | 92 | 158 | 7.7 |

[a] As presented in Gill [1972]; data from Hollingsworth and Sims [1969]. The flexure modulus and flexure strength are determined in a three-point bend test such as described in ASTM 790-49T.
[b] ABS is a copolymer of acrylonitrile, butadiene and styrene monomers.
[c] Nylon is a general name for polyamides with nylon 6 having the specific formula $H[HN(CH_2)5CO]_n OH$, with 6 carbon atoms in each repeat unit.

**Table 12.5.** New higher temperature thermoplastics[a]

| Material | Use temperature [°C] |
|---|---|
| Polyamide-imide (PAI) | 260 |
| Polyphenylene sulfide (PPS) | 200–230 |
| Polysulfone (PS) | 150 |
| Polyethersulfone (PES) | 175 |
| Polyetherimide (PEI) | 200 |
| Polyetheretherketone (PEEK) | 150 |

[a] From Riggs [1985].

moplastics which are being considered as matrix materials. One which has shown promise as a matrix for continuous carbon fibers is PEEK. Uniaxial tapes are made by coating each fiber with PEEK. These tapes can then be molded in a laminate construction similar to epoxy pre-pregs, but not requiring any curing time. These composites offer a number of potential advantages over epoxy materials. PEEK has low moisture absorption, and thus the properties are not affected by water. In addition, PEEK is not as brittle as epoxies, and thus leads to a tougher composite [Dickson et al. 1985]. This is done without sacrificing the bonding strength of the matrix to the fiber. Thus PEEK carbon fiber composites have greater strengths when tested at an angle to the orientation of the fibers than epoxy composites (see Table 12.6).

In addition to excellent mechanical properties (strength, modulus, fatigue, etc.), carbon and graphite composites can have added value because of their other physical properties such as thermal conductivity, low thermal expansion, and reasonably high electrical conductivity. Composite thermal expansion properties (see Sect. 5.2) are given in Table 12.7, and thermal conductivity properties (see Sect. 5.3) are in Table 12.8. These properties can also be controlled by design of the composite, in particular, the orientation of the different laminates. The ability to make composites with almost zero thermal expansion is important in a number of aerospace applications such as

**Table 12.6.** Comparison of mechanical properties of carbon fiber-PEEK and carbon fiber-epoxy composites: 0°, 45°, 90° laminates[a] (as defined in Table 12.7)

| Material | Orientation of test | Tensile strength [MPA] | Tensile modulus [GPA] | Fracture strain [%] |
|----------|--------------------|------------------------|----------------------|---------------------|
| CF-PEEK  | 0° or 90°          | 740                    | 61                   | 1.1                 |
| CF-epoxy | 0° or 90°          | 932                    | 83                   | 1.1                 |
| CF-PEEK  | 45°                | 194                    | 14                   | 4.3                 |
| CF-epoxy | 45°                | 126                    | –                    | 1.3                 |

[a] From Dickson et al. [1985]. In the table CF denotes carbon fiber.

**Table 12.7.** Coefficient of thermal expansion of composites[a,b]

| Material type | Laminate type[c] | |
|---------------|:---:|:---:|
| | 0° | 0°,45°,90° |
| High strength carbon fiber | −0.11 | 0.56 |
| Medium strength carbon fiber | −0.14 | 0.44 |
| High modulus graphite fiber | −0.17 | 0.25 |
| Fiberglass | 1.9 | 3.3 |
| Aluminum | 174 | |
| Steel | 3.9 | |
| Nylon (6/6) | 25 | |

[a] From Watts [1985].
[b] Units of the coefficient of thermal expansion are $10^{-6}/°C$.
[c] A laminate type 0° is unidirectional, a 0°,45°,90° laminate is one in which there are three layers of fibers at 0°, 45° and 90°.

**Table 12.8.** Thermal conductivity of composites[a,b]

| Material type | Laminate type[c] | |
|---|---|---|
| | 0° | 0°,45°,90° |
| High strength carbon fiber | 2–3 | 1.5–2.5 |
| Medium strength carbon fiber | 7–12 | 2–4 |
| High modulus carbon fiber | 17–19 | 9–10 |
| Fiberglass | 1 | 0.1 |
| Aluminum | 46–72 | |
| Steel | 5–16 | |

[a] From Watts [1985].
[b] Units of thermal conductivity are W/m°C.
[c] Laminate types are defined in Table 12.7.

antennae and mirrors which must maintain their dimensions to within a very small tolerance. The thermal conductivity of carbon and graphite fibers (as well as the electrical conductivity) increases with increasing heat treatment temperature and modulus (see Sect. 5.3 for example). If oriented composites of CCVD fibers were made, their thermal conductivity could be raised by at least one more order of magnitude beyond what is shown in Table 12.8 (see Fig. 5.11). The practical methods of controlling the electrical conductivity of carbon and graphite fibers are discussed in Sects. 8.3, 10.4.1 and 12.3. It is important to note, however, that in unidirectional composites the resistivity at 90° to the fiber orientation is controlled to a large extent by the matrix, and is typically 0.01 to 0.1 $\Omega$m [Watts 1985]. One possible method for reducing the resistivity would be the addition of ultra-thin vapor-grown fibers [Endo et al. 1987a].

Equation (12.1) gives an example of a property that can be computed using the rule of mixtures. Other properties, using the general symbol $X$, can be computed using the same equation with $E_Y$ replaced by $X$. An example is the mass density, $X = \varrho_m$. The electrical conductivity can sometimes be computed using this equation. If the current flow is along the continuous fibers, then they make up a system of conductors in parallel with the matrix, so that the parallel conductivity, $\sigma_\parallel$ is

$$\sigma_\parallel = \beta\sigma_f + (1 - \beta)\sigma_m \sim \beta\sigma_f \quad , \tag{12.3}$$

where $\beta$ is the fractional concentration of fibers and the contribution from the matrix can usually be neglected since $\sigma_m \ll \sigma_f$.

However, the transverse conductivity $\sigma_\perp$ will not be given by this formula in general. Two distinct regimes can be envisaged above or below a critical fraction $\beta_c$ [Carmona and Mouney 1987; Carmona and El Amarti 1987]. Below this fraction there will not be a continuous path through the fibers, as illustrated in Fig. 12.9a, while above it, continuous paths can be

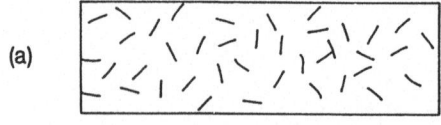

(a)

(b)

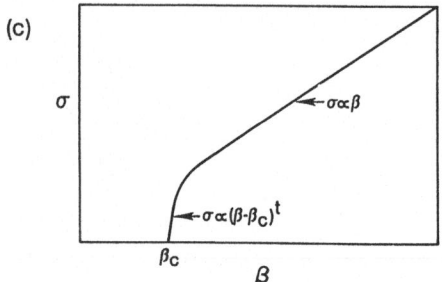

(c)

**Fig. 12.9a-c.** Conductivity of chopped filaments in a resin matrix: (a) Schematic of the composite for $\beta < \beta_c$ and in (b) for $\beta > \beta_c$. The conductivity as a function of filling factor $\beta$ is given schematically in (c)

found (Fig. 12.9b). $\beta_c$ is the fractional fiber concentration at the percolation threshold, which can be estimated from computer simulation studies to be $\beta_c = 0.5\%–2.0\%$. The conductivity of the fiber across the cross section has not been measured. Both inter-fiber contacts and fiber-electrode contacts will be important in determining the effective conductance; both of these types of contacts are difficult to characterize. Percolation is considered further in Sect. 12.3.3.

If the fibers are at an angle $\psi$ to the current then the effective conductivity can be computed as $\sigma_\parallel \cos^2 \psi + \sigma_\perp \sin^2 \psi$ to a first approximation, where $\sigma_\parallel$ is obtained from (12.3) and $\sigma_\perp$ from Fig. 12.9c. If the mats of fibers in the composite are overlayed, as in Fig. 12.7, then the total effective conductivity is the sum of that from each layer.

Polymer composites of carbon and graphite fibers have enormous commercial and military importance, and thus have been studied in great detail. The properties can be controlled both by modifying the fibers used and by choice of the matrix material. In addition, surface modification of the carbon fibers so as to control the bonding between the resin and the fibers plays an important role in determining the final properties. Finally, the orientation and fabrication techniques can be used to further optimize the properties of the finished article. In this review we have only touched on some of these important issues, and the reader is referred to other reviews such as Watts [1985], Riggs [1985], and Fitzer [1985] and the references therein for more information.

## 12.2.2    Carbon-Carbon Composites

As mentioned frequently in this review, the carbon-carbon bonds in the graphite planes ($\sigma$-bonds) are the strongest known. Carbon should therefore have the potential of extremely high strength. The major problem that has to be overcome is the control of microstructure, so that the strong in-plane bonds, rather than the much weaker interlayer bonds, control the fracture. These considerations have already been mentioned in Chap. 6. Carbon-carbon (C-C) composites represent the ultimate attempt to utilize these strong bonds in ultra-high strength materials.

Much of the literature on these materials is proprietary or classified, and most of the available literature is in the form of brief conference reports. Recent reviews include Fritz et al. [1978], Fitzer and Heym [1978], Fitzer and Hüttner [1981], Pacault and Marchand [1983], Lamicq [1984] and Fitzer [1987].

The basic idea of C-C composites is to use carbon filaments in a carbon matrix. Preparation procedures for the matrix are usually via impregnation [Fitzer and Terwisch 1972], but other routes such as chemical vapor deposition [Spear 1982] and plasma-enhanced CVD [Lachter et al. 1985] are possible. The arrangement of the filaments in the matrix can be uni-, bi-, or tri-directional. The use of a Raman microprobe (see Sect. 4.5) allows characterization of the structural order in the fiber and matrix of the composite independently [Doll et al. 1987]. The novel feature of this work is the ability to synthesize a 2D C-C composite with a high degree of ordering in both the mesophase pitch matrix and fiber components.

From a mechanical point of view, there are similarities between the C-C composites and steel-reinforced concrete. The fibers have very high elastic moduli (e.g., 250–500 GPa, see Table 6.1) compared to the glassy carbon, or pyrolytic carbon (30 GPa) matrix. The elastic modulus of the bulk composite can often be calculated by simple volume-fraction proportionality. The flexural strength depends on whether the matrix is ductile or brittle. If ductile, then the strength of the composite is controlled by the fibers, whereas failure is initiated in a brittle matrix. Other factors which are important are the pores and cracks in the matrix, which need to be controlled by suitable processing.

Figure 12.10 summarizes the flexural strength as a function of Young's modulus for various forms of C-C composites. Fitzer and Hüttner [1981] note that many carbons are two-phase in nature, and can be regarded as composites without fiber reinforcement. The new composites utilizing carbon fibers as the skeleton clearly exceed the mechanical properties of bulk carbons. Uni-directional composites are comparable to steels, but retain their strength to very high temperatures (2500°C) (see Sect. 9.3). These properties, coupled with their relatively low density, make them important structural components, such as for use in rocket nozzles [Herud 1973].

317

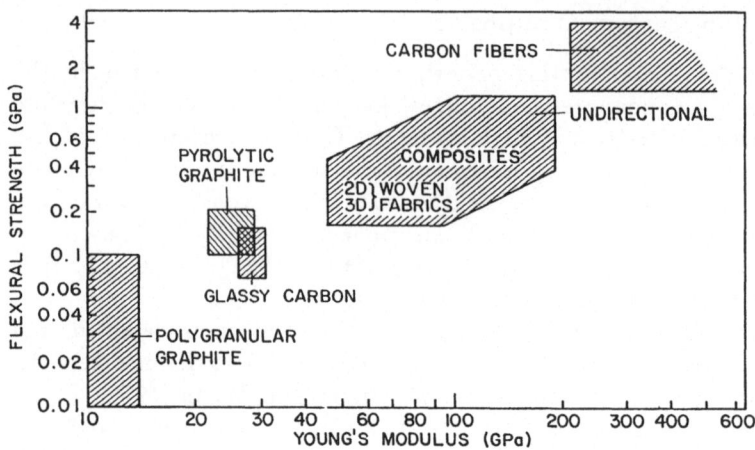

**Fig. 12.10.** Flexural strength as a function of Young's modulus for various forms of carbon-carbon composites [Fitzer and Hüttner 1981]

It is important that highly graphitized carbon-carbon composites be used for nuclear applications because of the resistance of highly graphitic fibers to shrinkage under intense neutron irradiation damage [Price et al. 1985] as discussed in Sect. 12.5. It has recently been found that highly graphitized C–C composites can be intercalated with both donor (e.g., alkali metal) and acceptor (e.g., $CuCl_2$) intercalants [Doll et al. 1987]. Such intercalated C–C composites may find applications where high electrical conductivity is important.

## 12.3  Electrical Applications for Carbon Fibers

### 12.3.1  Electromagnetic Interference Shielding

Most modern electronic systems (computers, televisions, etc.) use molded thermoplastic enclosures. Unfortunately, they are transparent to low (e.g., radio) frequency electromagnetic radiation, so that any signals generated in the apparatus are able to pass without attenuation into the environment. In addition, it is known that thermo-nuclear explosions produce large electromagnetic disturbances which can interact with electronic systems if they are not adequately shielded. For these reasons it has been important to develop materials which act as effective electromagnetic interference (EMI) shields, yet are compatible with the inexpensive molding techniques used today.

The performance of shielding is measured simply as the ratio of the intensity of the incident radiation ($I_i$) to that transmitted ($I_t$) in decibels:

$$\text{Attenuation} \quad (\text{dB}) = 10 \log I_i/I_t \quad . \tag{12.4}$$

Laws laid down by regulating agencies in Europe and the USA usually require attenuation levels of 40 dB or so, which implies that the transmitted intensity must be reduced to $10^{-4}$ of the incident intensity.

Shielding materials are usually in the form of sheets, and their resistance $R$ is measured in ohms per square ($\Omega/\Box$). A formula relating the sheet resistance to the attenuation is

$$\text{Attenuation} \quad (\text{dB}) = 20 \log Z_0/4R \quad , \tag{12.5}$$

where $Z_0$ is the electromagnetic impedance of empty space ($377\,\Omega$). If an attenuation level of 40 dB is required, then substituting in (12.5) gives $R = 0.92\,\Omega/\Box$ as the upper limit for the sheet resistance. This formula is only valid over a restricted frequency range, since skin effects can become important at higher frequencies, but it is useful for most applications and gives a rough guide to the performance needed for the sheet.

The sheet resistance attainable in composites is restricted by the resistance per unit length of the fibers and the packing fraction. As a result, it is impracticable to utilize untreated fibers for this application. The possibilities which are open are to use either intercalated, or metal-coated fibers. The latter is usually preferred, since the cost of heat treating the fibers to achieve the degree of intercalation needed to obtain a sufficiently high electrical conductivity has been prohibitive. Recent developments in the intercalation of CCVD fibers [Endo et al. 1987a] may make this approach economically viable.

Nickel coated graphite (NCG) fibers have been developed by the American Cyanamid Company [Beetz and Brinen 1983; Luxon 1986]. A continuous electroplating technique typically deposits a layer of Ni ($0.4\,\mu$m thick) onto fibers of diameter $d_1$. If the outer diameter of the Ni is $d_2 = 7.8\,\mu$m, its resistivity is $\varrho_2$, and that of the fiber is $\varrho_1$, then the average, or effective, dc resistivity $\langle \varrho \rangle$ along the fiber axis is

$$\langle \varrho \rangle = \left[ \frac{d_1^2}{\varrho_1 d_2^2} + \frac{(d_2^2 - d_1^2)}{\varrho_2 d_2^2} \right]^{-1} \quad . \tag{12.6}$$

Using the values $\varrho_1 = 10^{-4}\,\Omega$m and $\varrho_2 = 3.6 \times 10^{-7}\,\Omega$m, then $\langle \varrho \rangle = 7 \times 10^{-6}\,\Omega$m. Table 12.9 gives the properties of these coated fibers. It is noted that the Ni coating dominates the conductivity. If the frequency of the electromagnetic radiation is sufficiently high so that the skin depth in the Ni is less than its thickness, then the effective resistance of the coated fiber is that of the metal skin depth. This occurs above a frequency of $\sim$100 GHz for the Ni coating discussed, which is at the high frequency end of today's electronics technology.

Tables 12.9 and 12.10 give the effective attenuation levels achieved with various fiber-loading fractions using chopped filaments. It can be seen that 50 dB attenuation can be achieved with a weight fraction of fibers of 20%, and up to 95 dB with fully loaded composites. The molding behavior of the com-

**Table 12.9.** Properties of Ni-coated fibers [Luxon 1986]

| Property | Value (typical) |
|---|---|
| Diameter | 7.8 $\mu$m |
| Metal coating thickness | 0.4 $\mu$m |
| Specific gravity | 3.0 |
| Tensile strength ($\sigma_t$) | 3.1 GPa |
| Tensile modulus ($E_Y$) | 220 GPa |
| Elongation at fracture, $E_t$ | 1.4% |
| Electrical resistivity | $7 \times 10^{-8}$ $\Omega$m[a] |
| Thermal conductivity | 640 W/m K |
| Coeff. thermal expansion, $\alpha_{\parallel}$ | $-0.8 \times 10^{-6}$/K |

[a] Resistivity of nickel coating.

**Table 12.10.** Values for the attenuation coefficient of Ni-coated, chopped fiber composites[a] at various frequencies [Luxon and Murthy 1986]

| Fiber loading | Attenuation [dB] | | | | |
|---|---|---|---|---|---|
| [wt.%] | 30 | 100 | 300 | 1000 | (MHz) |
| 5 | 19 | 16 | 37 | 11 | |
| 10 | 33 | 34 | 42 | 25 | |
| 15 | 40 | 32 | 59 | 69 | |
| 20 | 50 | 38 | 59 | 72 | |

[a] The composites in this table are dry-blend (dB-100) nickel-coated carbon fibers of the American Cyanamid Company.

**Table 12.11.** The mechanical properties of nickel-coated fibers in polycarbonate resin[a]

| Fiber wt.% | 0 | 10 | 15 | 20 |
|---|---|---|---|---|
| Tensile strength [MPa] | 39 | 72 | 83 | 86 |
| Elongation to fracture [%] | – | 4 | 3 | 3 |
| Flexural strength [MPa] | 91 | 103 | 110 | 124 |
| Flexural modulus [GPa] | 2.2 | 4.4 | 6.7 | 8.3 |

[a] CYCOM, NCG Composite of American Cyanamid Company [Luxon 1986].

posite is claimed to be identical with the resin itself so that manufacture of the composite is practical. The mechanical properties of the sheets are excellent, also, as may be seen from Table 12.11. The drawback of these metal-coated fibers is their increased density ($\sim$3 gm/cm$^3$). It is possible that intercalated fibers could give the same shielding performance with lower density.

Equation (12.5) is only valid in the far-field approximation for the electromagnetic radiation. This is the situation where the magnetic and electric components of the radiation are related through the impedance of vacuum

(377 $\Omega$). In the region close to a wire carrying current, the magnetic component is significantly higher than this, and the magnetic properties of the composite determine the effectiveness of the shielding at low frequency. Fibers with high magnetic permeability coatings would be very useful for this application, since the effective permeability of spherical magnetic particles of permeability $\mu$ is reduced to a value of about 3 by their shape factor even if $\mu = 1000$ [e.g., see discussion of shape factors in Kittel (1985)]. However, the shape factor does not reduce the effective permeability in the case of highly elongated particles with the field aligned along the axis. If the magnetically coated fibers are randomly oriented in the composite, then the permeability would be reduced to approximately $\mu/3$. Such magnetic particles could be of interest for practical applications.

Another interesting application of chopped fibers for EMI shielding is in building structures. A low density of chopped fibers mixed with the construction concrete can provide effective shielding for televisions [Endo 1987b]. A practical test of this concept has been carried out successfully in the construction of the Ark Hills twin towers in Tokyo, as mentioned in Sect. 12.1.

### 12.3.2 Antistatic Coatings

Antistatic coatings are important in many industrial situations where the build-up of charge can produce unwanted effects. Examples are in microelectronic circuitry, where static discharges can damage components, and in the chemical industry where discharges can ignite explosive gases. Values of the surface resistivity less than about $10^5$ $\Omega/\square$ are usually sufficient to protect components, and only low densities of carbon fibers are needed to achieve this value. This can be seen by considering a coating with commercial high-strength carbon fibers, which have a resistance of about $10\,000$ $\Omega/cm$. If effective contact is established with only one fiber per centimeter across a square sample, the surface resistance would be $10\,000$ $\Omega/\square$. Of course, the fibers are not arranged in a regular, continuous, fashion, so that a higher density is required in practice to reach this value, but the example illustrates the point. Much lower resistances could be achieved with metal-coated fibers, as outlined in Sect. 12.3.1.

A simple way of preparing antistatic coatings is in lightweight papers. A review of this topic has been given by Walker [1986]. These antistatic papers are prepared by producing a slurry containing fibers, removing the slurry with meshes, and pressing and drying the papers. The carbon fibers can be mixed with other fibers such as glass, cellulose, or aramid, and lie in the plane of the paper, with random orientation within the plane. The resulting papers can be handled like normal ones, and are capable of being cut and rolled. Thinner papers (below 100 g/m$^2$) can be molded to conform to irregular shapes.

Preliminary studies on the applicability of pristine and intercalated fibers for electromagnetic interference (EMI) shielding and for protection of aircraft skins against lightning strikes have been carried out using continuous P-100 pitch-based fibers impregnated with polystyrene to form a carbon/epoxy composite [Centanni and Clark 1986]. The difference in the power transmission parallel and perpendicular to the fiber axis was measured for microwave radiation in the $(3.95$–$5.85)\times10^9$ Hz range. The measurements for the anisotropy in the power transmission $\parallel$ and $\perp$ to the fiber directions was $\sim 1 : 2$ for the unintercalated fibers, but $\sim 1 : 6$ for the bromine intercalated (residue) fibers. This observation can be explained by the increased anisotropy in the conductivity $\sigma$ for the intercalated fibers.

### 12.3.3 Electrical Devices Based on Carbon-Resin Composites Near the Percolation Threshold

It was mentioned briefly in Sect. 12.3.1 that there is a critical volume fraction of chopped fibers required in a resin before the composite becomes conducting. The critical value, $\beta_c$, at which the conductivity increases rapidly is called the percolation threshold. There are several interesting devices which utilize the unusual properties in this region.

Percolation theory is a geometrical theory describing the structure of random particles such as filaments in a resin matrix as a function of their volume fraction, $\beta$. It describes the way in which the particles come into contact and form clusters. As such, it is different from other theories of heterogeneous materials, such as mean-field theory, which ignore the clustering (aggregation) of the particles. The clusters are small in size below $\beta_c$, and disconnected. At the percolation threshold, $\beta_c$, the clusters reach a very large size, and the connectivity of the clusters increases rapidly as $\beta$ approaches $\beta_c$. Electrical conduction occurs through the fibers and connections between the fibers, so that the geometrical threshold for connectivity at $\beta_c$ is also the threshold for conduction.

The transition from local to essentially infinite connectivity is similar to a second-order thermodynamic phase transition, and can be treated using scaling concepts. Thus, there are relationships between physical properties with power-law exponents, as there are for many systems such as fluids, magnetic materials, etc., near their critical points [see Stanley (1971, 1984)]. Much information on the percolation threshold has been obtained with computer simulations, as well as physical experiments, which are compared to the predications of first-principles theories. A review of concepts and experiments relevant to percolation has been given by Kirkpatrick [1973].

Chopped filaments of length $L$, diameter $d$, and conductivity $\sigma$ in a resin matrix offer an excellent system with which to study the percolation threshold. The fibers typically have conductivities $10^5$–$10^6$ S/m, and the resin $10^{-16}$ S/m. If the fibers are randomly oriented in the resin, then both computer simulations [Boissonade et al. 1983] and fundamental theories [Balberg

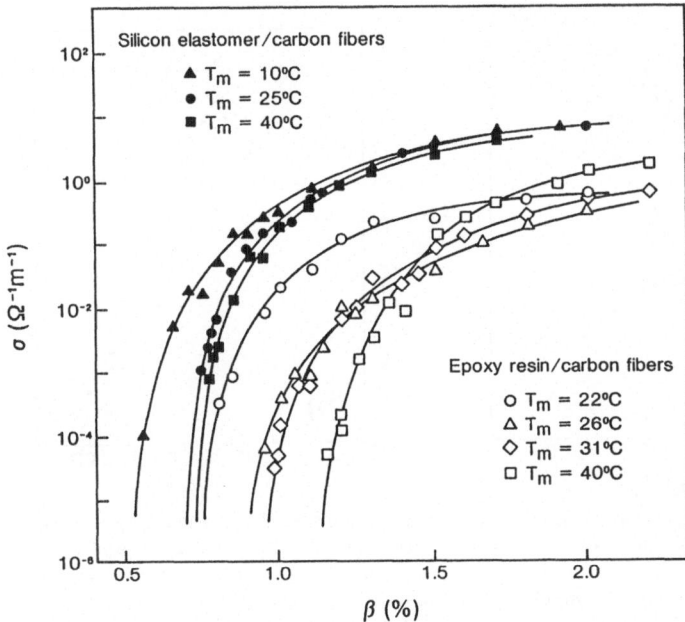

**Fig. 12.11.** Semi-log plot of the conductivity $\sigma$ versus the fiber volume fraction $\beta$ in an insulating matrix for different mixing temperatures $T_m$ and for epoxy resin and silicone elastomer matrix materials [Carmona and Mouney 1987]

et al. 1984] predict that $\sigma \propto d/L$ above $\beta_c$. However, Carmona et al. [1981] showed experimentally a dependence of $\sigma \propto (d/L)^2$. Carmona and Mouney [1987] then showed that parameters other than $1/d$ were of importance in determining $\beta_c$, namely the polymer surface tension and viscosity, which can be varied during the preparation of the composite by controlling the setting temperature. Figure 12.11 illustrates several experimental curves of $\sigma$ versus $\beta$ for different processing temperatures $T_m$. It may be seen that $\beta_c$ varies significantly for these different cases, lying between about 0.5% and 2.0%. The critical exponent $t$, where $\sigma \propto (\beta - \beta_c)^t$ was found to be 3 in these experiments, in contrast to the expected value of 2 for isotropic composites.

Carmona and El Amarti [1987] and Carmona and Mouney [1987] have investigated the conductivity of short, aligned fibers in a resin matrix. The conductivities in two principal directions perpendicular to the alignment axis, $\sigma_\perp$, were found to agree reasonably well, but Fig. 12.12 shows that the parallel conductivity, $\sigma_\parallel$, can be significantly different from $\sigma_\perp$, particularly near $\beta_c$. The anisotropy $\sigma_\parallel/\sigma_\perp$ falls from about $10^4$ to $\sim$20 in the narrow region $\beta$ = 1.6–1.75%. Shklovskii [1978] had predicted that the conductivity would be isotropic far from $\beta_c$. The behavior in Fig. 12.12 is surprising at first sight. However, conduction occurs by tortuous paths in the directions both parallel and perpendicular to the alignment axis. The dashed line in Fig. 12.12b is

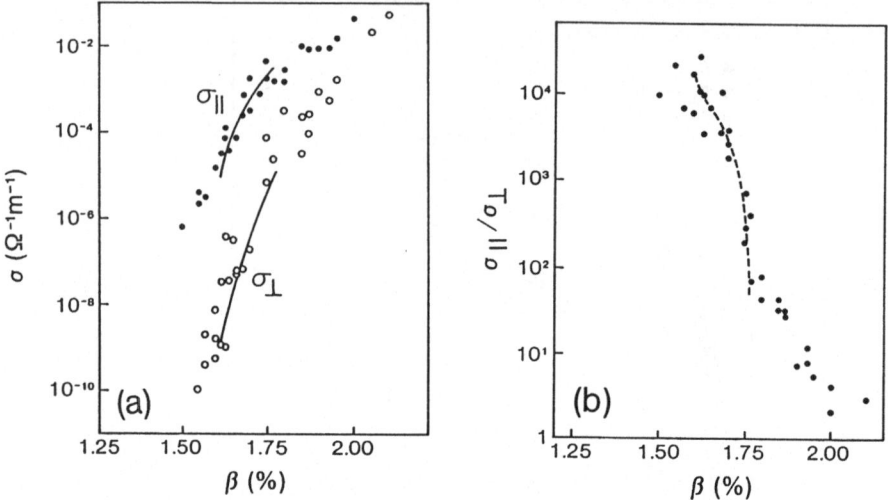

Fig. 12.12a,b. The anisotropy of the conductivities, $\sigma_\parallel$, and $\sigma_\perp$ for aligned chopped fibers as a function of fiber volume fraction in a resin matrix. (a) Conductivity $\sigma_\parallel(\beta)$ and $\sigma_\perp(\beta)$, (b) anisotropy ratio $\sigma_\parallel(\beta)/\sigma_\perp(\beta)$ vs $\beta$. The points are experimental and the lines are calculated fits to the percolation model [Carmona and El Amarti 1987]

the prediction of theory [Shklovskii 1978]:

$$\sigma_\parallel/\sigma_\perp = (L/d)^2 \mid 1 + A(\beta - \beta_c)^\lambda \mid \quad , \tag{12.7}$$

where $A$ is a constant and $\lambda$ the scaling exponent. The value $\beta_c = 1.65\%$ was found from fitting experimental data to this expression.

The rapid rise in the conductivity with $\beta$ near $\beta_c$ can be used in several devices. For instance, the volume fraction can be varied by applying pressure, since the resin is much more compressible than the fibers. As a result, the predictable, reversible, increase in the conductivity with pressure is much higher than in conventional pressure transducers based on the electrical resistivity. The range of pressure over which the transducer is useful can be varied by selecting the compressibility of the resin.

### 12.3.4 Fibers with Superconducting Coatings

Superconducting magnets are widely used in special applications, such as experimental apparatus, particle accelerators, and NMR imaging devices. These magnets utilize multifilament windings to obtain favorable properties. Ideally, such filaments have a combination of properties, including: high current-carrying capacity, particularly in high magnetic fields; the ability to withstand high stresses produced by the currents and fields; ease of preparation and handling, so that coils can be wound easily and cheaply; the ability to withstand thermal and field cycling; and sometimes, the ability to withstand irradiation. It was realized in 1976 [Pike et al. 1976; W.D. Smith et al. 1976] that coated carbon fibers could be of interest for this application. These authors coated carbon fibers with NbN. The superconductor NbN can, with suitable processing, achieve an upper critical magnetic field $H_{c2}(0)$, sometimes referred to as $B_{c2}(0)$, of 45 T [Ashkin et al. 1984], which makes it of technological interest when considered with its superconducting transition temperature of $T_c = 16$ K. These early workers only reached $H_{c2}(0)$ values of 13 T with their coated fibers, so that other superconducting materials were of greater interest.

Further work has been stimulated by the observation that useful superconducting properties of NbN depend critically on the deposition procedures, giving a fine, homogeneous microstructure that pins the flux vortices, thereby resulting in a raising of $H_{c2}$. Efforts have therefore been made to deposit NbN (or, more accurately, $NbN_{1-x}C_x$) in the presence of controlled plasmas [Dietrich et al. 1983; Dietrich and Dustmann 1984; Ohshima et al. 1984; Dietrich 1985]. This has led to useful improvements in properties. In particular, critical current densities to nearly $10^{10}$ A/m$^2$, and $H_{c2}(0)$ values up to 25 T have been reported [Dietrich and Dustmann 1984].

It was found that the residual stress between the coating and the fiber substrate could be controlled by processing conditions. However, thermal expansion was best controlled by using a layer of SiC between the coating and the substrate. Best results were obtained using high-strength ex-PAN fibers, with promise of enhanced performance for diameters of 5 $\mu$m or below. Copper cladding could be applied with electrodeless chemical deposition, and the filaments could be bonded to a wire.

It is possible that these superconducting filaments can be of technological interest. With the discovery of high $T_c$ superconducting materials [Bednorz and Müller 1986; Wu et al. 1987; Cava et al. 1987; Uchida et al. 1987; Takagi et al. 1987], such as $(La_{1-x}Ba_x)_2CuO_{4-\delta}$ or $YBa_2Cu_3O_7$, critical fields of over 100 T are expected. If suitable deposition and heat treatment techniques can be developed, superconducting coated carbon fiber technology may head in new directions and find a much wider range of applications, since refrigeration with liquid nitrogen will now be sufficient.

### 12.3.5  CCVD Carbon on Ex-polymer Fibers

The practical exploitation of the high conductivity of intercalated fibers has been limited by a number of factors, insofar as the highest conductivity fibers cannot be made in continuous lengths and the fibers that can be made in continuous lengths do not have high enough conductivities to be of commercial interest in most cases, even though their conductivities can be increased by an order of magnitude by intercalation.

Recently Matsumura et al. [1985], Tsukamoto et al. [1986] and Fukuda et al. [1987] have developed a continuous fiber which combines the advantages of the continuous ex-pitch fiber with the advantages of the high degree of structural perfection that can be achieved with CCVD filaments. In their process, highly conductive fibers ($\varrho = 55\,\mu\Omega\text{cm}$) are prepared starting with a continuous ex-polymer carbon fiber substrate and preparing a CCVD coating on this fibrous substrate by pyrolysis of cyanoacetylene and subsequent heat treatment to 3000°C. The CCVD coating that is thus prepared is similar to the coatings added during the thickening process of the very thin fibrils prepared by CCVD (see Sect. 2.3). Since the mass of the original fiber can be increased by up to about 10 times by the deposition process, which is not inherently expensive, the specific cost of as-thickened fibers may be lower than that of the original ex-polymer fiber. The subsequent heat-treatment to 3000°C is inherently expensive. It is probable that these fibers will only be of commercial use if useful properties can be obtained with lower $T_{\text{HT}}$, thereby lowering costs.

Intercalation of this fiber with $HNO_3$ was reported to decrease the resistivity by an order of magnitude ($\varrho \simeq 7.1\,\mu\Omega\text{cm}$). Upon exposure to air, the resistivity was found to increase slightly, but after this initial increase, the resistivity remained stable (over 90 days). During this time the intercalant transformed from an $\alpha$ phase to a more stable $\beta$ phase. Though at an early stage of development, this general approach to the exploitation of the high structural perfection of the CCVD coating has many attractive features for practical applications.

## 12.4 Thermal Applications

As discussed in Sect. 5.3, the thermal conductivity of well-graphitized carbon fibers can reach high values, suggesting potential applications. As an example, carbon fibers have been discussed as a means to improve the thermal conductivity of substrates for integrated circuits. PAN fibers are already finding a commercial use by one computer manufacturer for this application. The thermal conductivity of composites has been discussed by James et al. [1987].

Carbon fibers are attractive for high temperature conduction applications because of their ability to withstand very high temperatures (melting point is ~4450 K [Heremans et al. 1988]). For high temperature applications, surface oxidation must be avoided, and in this connection SiC coatings are of great value as an inhibitor of surface oxidation insofar as they provide protection against oxidation up to 1700°C (see Sect. 9.4.2).

Another example is the use of fiber-reinforced sheets of composites or papers for heating elements. Laminar heating elements can be constructed easily using fiber-reinforced papers. Potential applications are protection heaters for pipes and tanks, temperature-controlled vessels, panel heaters in electronic equipment, etc. The electrical current must be calculated on the basis that a current $I$ is applied uniformly to the edges of the panel (of dimensions $a \times b$).

The voltage needed to drive the current through the edges of dimension $a$ is

$$V = IRb/a \ , \tag{12.8}$$

where $R$ is the resistance per square. The power dissipated per unit area, $P'$, is

$$P' = I^2 R/a^2 \tag{12.9}$$

and the total power, $P$, is

$$P = \frac{I^2 Rb}{a} = \frac{V^2 a}{Rb} \ . \tag{12.10}$$

The actual operating performance will depend on details such as the inter-fiber contact resistances, and the effective thermal resistances of the composite and the composite-surrounding medium interfaces, which cannot be predicted.

## 12.5  Nuclear Reactor Applications

Because of the excellent high temperature strength, high toughness, low impurity concentration, low atomic number and low neutron absorption cross section, carbon fibers and carbon-carbon (C–C) composites are attractive materials for nuclear reactor core applications, either for fission or fusion reactors. However, commercial PAN and rayon-based fibers are highly prone to neutron radiation damage and can shrink by 30% or more [Gray 1978; Bullock 1979] upon fast neutron irradiation to a fluence of $\sim 5 \times 10^{25}$ n/m². In contrast, crystalline graphite is relatively radiation stable [Price 1974], so that by further heat treatment of commercial PAN- and pitch-based carbon fibers to 3100°C, the shrinkage resulting from neutron irradiation can be reduced by roughly a factor of 4 [Price et al. 1985] (see Fig. 12.13 and Table 12.12). For the more graphitic mesophase pitch fibers, the shrinkage can be reduced to $\sim$1% after heat treatment to 3100°C, which is perhaps

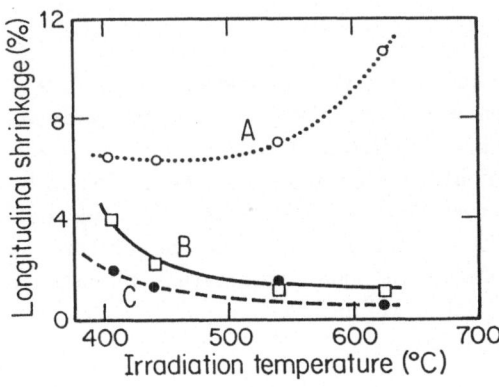

**Fig. 12.13.** Dependence of longitudinal shrinkage of carbon fibers on irradiation temperature for carbon materials of different crystallinity. Data are normalized to a neutron fluence of $5 \times 10^{25}$ m⁻² ($E > 50\,\mathrm{keV}$) for the following fibers: ($A$) ex-PAN, high strength, as-received fiber; ($B$) ex-pitch fiber annealed at 3100°C; ($C$) highly oriented pyrolytic graphite (HOPG) [Price et al. 1985]

**Table 12.12.** Neutron irradiation-induced changes[a] in dimensions and tensile properties of carbon fibers, SiC fibers and HOPG[b]

| Material | $T_{HT}$ [°C] | $L_c$ [Å] | Condition | $\Delta L/L$ [%] | $\Delta d/d$ [%] | $\overline{\sigma}_T$ [GPa] | $\overline{E}_Y$ [GPa] |
|---|---|---|---|---|---|---|---|
| Ex-PAN fiber 1 | – | 17 | Unirradiated | – | – | 3.05 | 220 |
| | | – | Irradiated[a] | –33.8 | +17.5 | – | – |
| | 2650 | 61 | Unirradiated | – | – | 1.78 | 310 |
| | | – | Irradiated[a] | –8.4 | +3.0 | – | – |
| | 3100 | 77 | Unirradiated | – | – | 1.63 | 320 |
| | | – | Irradiated[a] | –6.7 | +0.7 | – | – |
| Ex-PAN fiber 2 | – | 54 | Unirradiated | – | – | 2.12 | 300 |
| | | – | Irradiated[a] | –10.7 | +3.6 | – | – |
| | 2650 | 67 | Unirradiated | – | – | 1.88 | 330 |
| | | – | Irradiated[a] | –4.6 | +1.6 | – | – |
| | 3100 | 101 | Unirradiated | – | – | 1.50 | 320 |
| | | – | Irradiated[a] | –3.8 | +0.1 | – | – |
| Ex-pitch fiber 1 | – | 112 | Unirradiated | – | – | 1.97 | 370 |
| | | – | Irradiated[a] | –13.2 | +7.3 | – | – |
| | 2650 | 150 | Unirradiated | – | – | 1.57 | 430 |
| | | – | Irradiated[a] | –5.8 | +4.2 | 1.79 | 310 |
| | 3100 | 161 | Unirradiated | – | – | 1.22 | 420 |
| | | – | Irradiated[a] | –4.5 | +3.4 | 1.79 | 310 |
| Ex-pitch fiber 2 | – | 168 | Unirradiated | – | – | 2.17 | 930 |
| | | – | Irradiated[a] | –4.1 | +3.5 | – | – |
| | 2650 | 193 | Unirradiated | – | – | 2.77 | 770 |
| | | – | Irradiated[a] | –2.6 | +1.9 | – | – |
| | 3100 | 336 | Unirradiated | – | – | 1.97 | 590 |
| | | – | Irradiated[a] | 1.1 | +1.5 | 2.40 | 570 |
| CCVD fiber | 3100 | 1600 | Irradiated[a] | –0.2 | – | – | – |
| SiC fiber | – | – | Unirradiated | – | – | 3.39 | 210 |
| | – | – | Irradiated[a] | –1.8 | –0.3 | 4.56 | 330 |
| HOPG | – | – | Irradiated[a] | –0.5 | +1.9 | – | – |

[a] Neutron fluence is $5.0 \times 10^{25} \mathrm{n/m^2}$ ($E > 50\,\mathrm{KeV}$) at an irradiation temperature of 620°C. Here, the dimensional length is $L$ and the diameter is $d$. Average tensile data (strength $\sigma_T$ and modulus $E_Y$) are given for a neutron fluence of $4.2 \times 10^{25} \mathrm{n/m^2}$ ($E > 50\,\mathrm{keV}$) at an irradiation temperature of 440°C.
[b] From Price et al. [1985].

an acceptable shrinkage level for nuclear reactor applications. Preliminary measurements on vapor-grown fibers were particularly encouraging, indicating negligible shrinkage ($\sim$0.2%) when irradiated with fast ($>$50 keV) neutrons to a fluence of $\sim 5 \times 10^{25} \mathrm{n/m^3}$ at a fiber temperature estimated to be between 410° and 620°C. As the crystalline order increases, the longitudinal shrinkage tends to decrease with increasing irradiation temperature, as shown in Fig. 12.13. This observation of the weak temperature dependence of the longitudinal fiber shrinkage makes well-ordered graphitic fibers even more attractive for these nuclear applications.

Neutron irradiation was not found to have serious deleterious effects on the mechanical properties of the high modulus pitch-based fibers: the tensile strength was actually found to increase (by between 15% and 45%) while the Young's modulus decreased slightly. The tensile strength of PAN-based fibers was also found to increase upon neutron irradiation. From these results we can conclude that either mesophase pitch or vapor-grown fibers, heat treated to 3000°C and above, are attractive materials for nuclear reactor core applications. The feasibility of preparing suitable carbon-carbon composite materials for nuclear applications is presently under investigation [Engle 1987]. A variety of ex-mesophase pitch C–C composite materials have recently been prepared with a high degree of crystallinity in both the matrix and fiber components, as established by both Raman and resistivity measurements [Doll et al. 1987].

There have been several studies of the effects of neutron irradiation on fiber/epoxy composites. As discussed in Sect. 6.2.1 the mechanical properties of fibers can be improved somewhat by irradiation, provided that the environment is oxygen-free. These irradiated fibers can then be used to fabricate composites with enhanced mechanical properties [McKague et al. 1973]. Bullock [1972c] however showed that the irradiation of fiber/epoxy composites at room or liquid nitrogen temperature to low fluences (e.g., $6 \times 10^{21}$ n/m$^2$ for $E > 1$ MeV) resulted in a reduction of flexural strengths and moduli.

Fornes et al. [1981] carried out tests on several types of fiber/epoxy composites irradiated with 0.5 MeV electrons to doses of 8000 rads, and showed that there was a slight increase in the flexural strength and modulus. No differences could be observed in the x-ray diffraction patterns after irradiation [Kent et al. 1984].

From these studies it may be concluded that fiber/epoxy composites are not radiation tolerant for neutrons. However, radiation-tolerant carbon-carbon composites can be prepared and used in practical applications.

## 12.6 Optical Applications

### 12.6.1 Application of Carbon Filaments as Obscurants

The application of carbon filaments to the extinction of electromagnetic radiation was first proposed in 1965 [Pedersen and Pedersen 1965]. Unfortunately, much of the work was classified [see Pedersen et al. (1969)]. At the present time it is well known that conducting filaments having radii less than, or of the same order as, the incident radiation can exhibit very high extinction coefficients (extinction is the sum of scattering and absorption) [see for example Bohren and Huffman (1983)]. The electrical and geometrical properties of such particles can be tailored to provide predictable extinction spectra. The possibility of using pristine and metal-coated carbon filaments for this application is discussed by Dresselhaus et al. [1985].

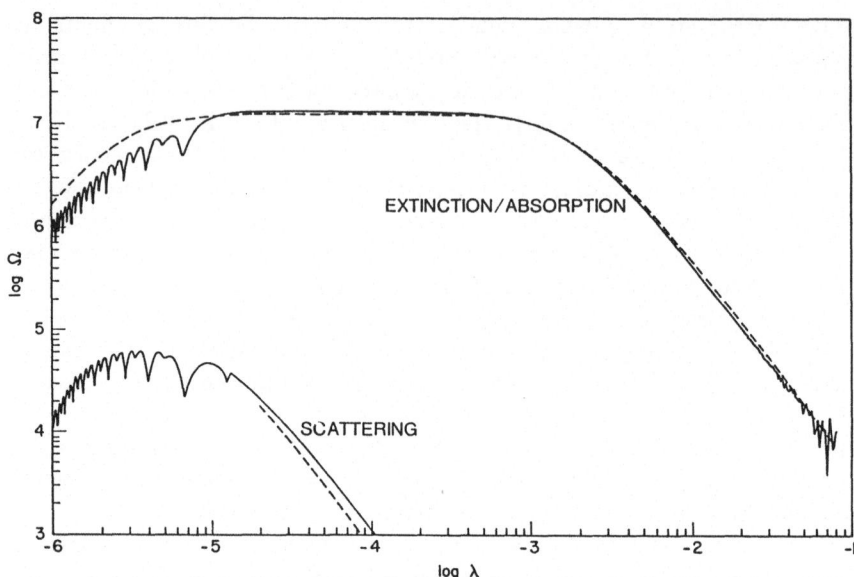

**Fig. 12.14.** Plot of the calculated scattering and extinction cross sections for electromagnetic radiation of wavelength $\lambda$ in a *pristine* carbon filament (MKS units). The full curve is the exact theory, and the dashed curve the quasi-static. The fiber radius is 200 Å, length is 10 $\mu$m, and the electrical conductivity is $10^{-6}$ $\Omega$m [Pedersen et al. 1987]

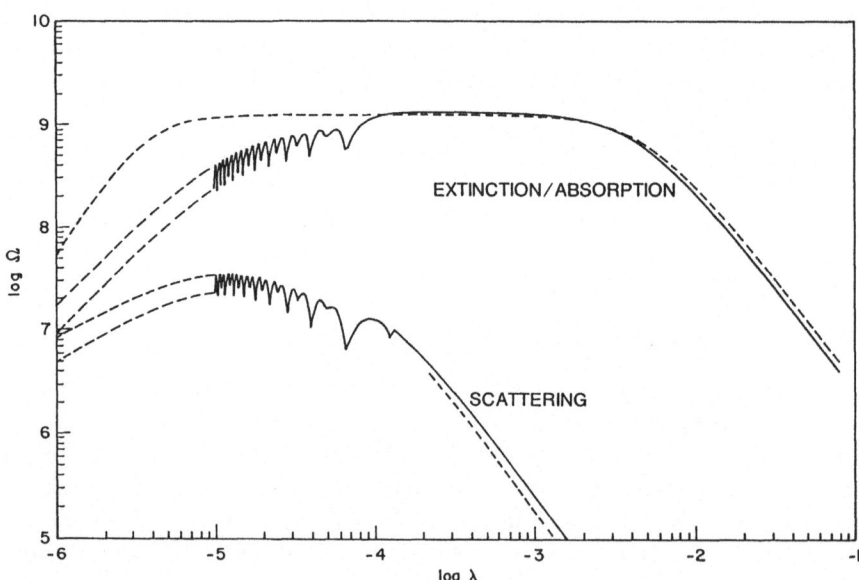

**Fig. 12.15.** Plot of the calculated scattering and extinction cross sections for electromagnetic radiation of wavelength $\lambda$ in a *metal-coated* carbon filament (MKS units). The fiber radius is 100 Å, length is 10 $\mu$m, the electrical conductivity of the coating is $10^{-7}$ $\Omega$m, the coating thickness is 50 Å and the outer diameter is 100 Å [Pedersen et al. 1987]

Figure 12.14 shows a curve of the calculated extinction cross section for a carbon filament of radius 20 nm, length 10 $\mu$m, and conductivity $10^5$ $\Omega^{-1}$m$^{-1}$. A similar curve is plotted in Fig. 12.15 for a filament of the same outer diameter coated with a 5 nm thick layer of metal of conductivity $10^7$ $\Omega^{-1}$m$^{-1}$ [Pedersen et al. 1987]. For both of these cases the absorption dominates the extinction so that only the scattering contribution is shown explicitly (the curves of extinction and absorption are almost identical). The dashed curves are the prediction of an approximate theory (the quasi-static theory discussed below) while the full lines are a more exact theory [Pedersen et al. 1985].

The general form of the plots can be understood in terms of quasi-static theory. Consider the case of electromagnetic radiation of wavelength $\lambda$, long compared to the length of the filament $L$ (i.e., $\lambda \gg L$). If the applied electric field, $E_0$, is parallel to the filament axis, then a charge builds up on the ends of the filament, producing an opposing field, $E_c$. The internal field, $E_i$, which couples the radiation to the electrons is then

$$E_i = E_0 - E_c = \frac{E_0}{\{1 + i[\ln(L/a) - 1](a/L)^2(\sigma\lambda/2\pi c\epsilon_0)\}} \quad , \tag{12.11}$$

where $\epsilon_0$ is the permittivity of free space, $c$ is the speed of light in vacuum, $\sigma$ is the electrical conductivity of the filament, $a$ and $L$ are the radius and length of the filament and $i$ is the imaginary unit. The term in brackets can be obtained from standard electromagnetic theory. The absorption cross section $\Omega$ is related to the dissipated power $P$ by

$$\Omega = P\sqrt{\frac{\mu_0}{\epsilon_0}} \Big/ E_0^2 \quad , \tag{12.12}$$

where $\mu_0$ is the magnetic permeability of free space and $P$ is obtained from

$$P = \Re\{\boldsymbol{J} \cdot \boldsymbol{E}\}V/2 \quad , \tag{12.13}$$

where $\Re$ denotes the real part, $\boldsymbol{J}$ is the electric current induced in the filament by the external field, and $V$ is the volume of the particle, so that $\Omega$, the absorption cross section, becomes

$$\Omega = \frac{V\sigma\sqrt{\mu_0/\epsilon_0}}{1 + \{[\ln(L/a) - 1](a/L)^2(\sigma\lambda)/(2\pi c\epsilon_0)\}^2} \quad . \tag{12.14}$$

Equation (12.14) predicts that $\Omega$ decreases as $\lambda^{-2}$ in the long wavelength limit, and assumes a constant value

$$\Omega = V\sigma\sqrt{\mu_0/\epsilon_0} \tag{12.15}$$

when

$$\lambda \leq \frac{2\pi c\epsilon_0}{\sigma}\left(\frac{L}{a}\right)^2 \Big/ [\ln(L/a) - 1] \quad . \tag{12.16}$$

In the short wavelength limit $\lambda \ll L$, the external field sets up a varying charge distribution along its length, so that the length $L$ in (12.16) is replaced by

$$L \to \frac{\lambda}{2\pi} \quad , \tag{12.17}$$

and the absorption cross section of (12.14) predicts a $\lambda^2$ dependence of $\Omega$ at short wavelengths.

The curves plotted in Figs. 12.14 and 12.15, which are averaged over all directions of electric field (this introduces a factor of 1/3 into the formulae for $\Omega$), follow the predictions of the quasi-static theory, at least in the overall forms of the curves. It is noted that higher extinction is obtained for the metal-coated filaments, and that strong extinction can be obtained over a wide wavelength range corresponding to about 1 $\mu$m to 10 cm. The oscillatory behavior at lower wavelengths is not predicted by the quasi-static theory, and is related to resonance phenomena where the wavelength of the field is an integral fraction of the length of the filament.

There have been several experiments carried out to test the theory, usually with thicker filaments, and the calculations appear to fit the data within experimental error. Further experimental work extending the range of filament sizes and coating parameters is being undertaken.

It is probable that practical obscurants will utilize these materials in a polymeric medium to avoid health problems associated with small diameter filaments.

### 12.6.2 Low-Reflectivity Surfaces

It has been shown that ion-bombarded carbon surfaces, such as that illustrated in Fig. 2.18, have low optical reflectivities out to about 10 $\mu$m [Culver et al. 1985]. This work has been extended to include surfaces on which CCVD filaments have been grown [Bowers et al. 1987]. Figure 12.16 illustrates reflectivity curves obtained out to 25 $\mu$m (hemispherical reflectance between 2 and 2.5 $\mu$m, and specular from 2.5 to 25 $\mu$m) on several types of surfaces. The most interesting result is the observation that CCVD filaments of diameters about 100 nm grown on Si surfaces have reflectivities below 1% from the near UV out to beyond 25 $\mu$m. This is possibly the lowest reflectivity found for any surface in this wavelength region, comparing favorably with commercial infrared black materials [Pompea et al. 1984].

Reflection occurs at any interface in which there is a change of refractive index. If the complex refractive indices of the two media at the interface are $n_1$ and $n_2$, then the normal incidence reflectivity, $\mathcal{R}$, is

$$\mathcal{R} = \left| \frac{n_1 - n_2}{n_1 + n_2} \right|^2 \quad . \tag{12.18}$$

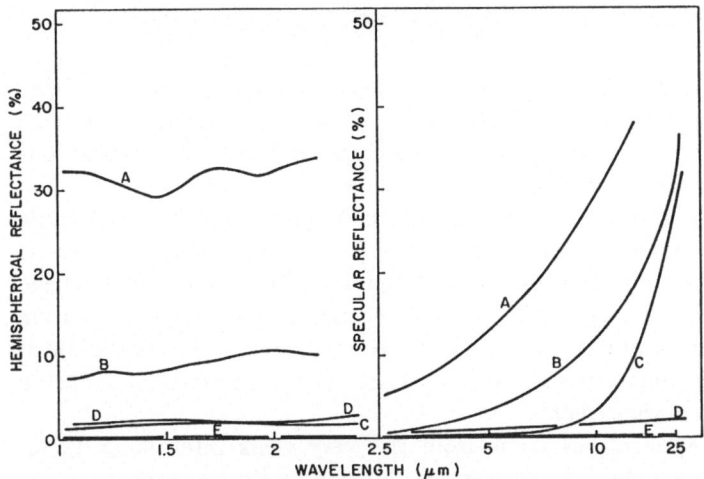

**Fig. 12.16.** Typical experimental curves of the hemispherical and specular reflectance as a function of wavelength for graphite surfaces. (*A*) Basal reflectivity from highly oriented graphite (HOPG), (*B*) fine-grain nuclear graphite, (*C*) ion-bombarded HOPG, (*D*) CCVD filaments on fine-grain nuclear graphite, and (*E*) CCVD filaments on pure silicon. [Data selected from Bowers et al. (1987)]

Since the absorption is represented by an imaginary component of $n$, an interface between a propagating medium (e.g., air) and an absorbing medium will have a discontinuity in $n$ and reflect a fraction of the radiation. It is essential, therefore, that the incoming radiation interact with a surface which has a gradation of properties, where the refractive index gradually changes from that of the propagating medium, to highly absorbing.

The physical mechanism of absorption by filamentary surfaces is probably related to the gradation of properties that arises near the growth surface. The incoming electromagnetic wave experiences an absorption coefficient of increasing value as the density of the filamentary materials increases towards the surface. As discussed in the preceding section, the diameter and length of the filaments is important in determining the wavelength region over which strong coupling can be achieved with the incident radiation. Further absorption can occur at the surface of the substrate, covered with a flocculent deposit. Finally, any reflected component from this surface will again be absorbed by the filaments. It is not known why the surfaces prepared by growing CCVD filaments on Si surfaces are more absorbing than those on carbon, but it may be related to the more graphitic (and therefore more highly conducting) nature of the filaments (see Sect. 2.2), or to the nature of the flocculent material, which may be predominantly non-stoichiometric SiC.

These surfaces may find applications in optical instrumentation, such as monochromators, detectors, etc., where applications are potentially space telescopes (IR), and fusion reactor Tokomaks.

## 12.7 Applications for Intercalated Carbon Fibers

Although battery applications (discussed separately in Sect. 12.8) have been most important for commercial exploitation, a number of other interesting applications of graphite intercalation compounds in the bulk or in fibrous form have been discussed and developed. In the area of graphite intercalation compounds (GICs) about 30 patents per year are issued, about 2/3 being filed in Japan. Most patents are filed by companies, indicating commercial interest in these materials [Setton 1987]. Though earlier patents emphasized bulk GICs, intercalated graphite fibers have become increasingly important in recent years. The most important GIC by far from a commercial standpoint is graphite-fluoride. Applications of GICs have been reviewed recently by Setton [1987] and are summarized here.

A number of applications come from the very weak interplanar bonding and high chemical stability of graphite-fluoride and other acceptor compounds, making these materials attractive for high temperature lubricants, low friction coatings, protective wear-resistant coatings, lubrication for ball bearings and self-lubricating watches, sound absorbing media, additives for abrasives, additives for polishing preparation, additives in combustion engine oil, among others. The chemical inertness of some acceptor intercalants accounts for applications under discussion as liners for auto exhaust tubes to prevent tar deposition. The exfoliation property which is exploited in grafoil and vermicular graphite (used as high surface area materials and high temperature seals) is under consideration in fibrous form for applications as extinguishers for metal fires and as a fire retardant [Setton 1987]. By enhancing the adhesion between fibers and the matrix materials, intercalation may lead to increased mechanical strength of carbon fiber-epoxy composites [Chung et al. 1987].

Many applications of intercalated fibers have been considered using the intercalated fibers as catalysts, including among others the synthesis of acetylene, ammonia, carbonyl sulfide and various polymerization reactions. For reactive GICs, such as alkali metal GICs, the easy intercalation and deintercalation has been used as a source and getter for metal vapors.

The order of magnitude increase in electrical conductivity of GICs has found a number of applications. A conducting paint containing chopped intercalated thin CCVD fibers has been discussed for printed circuit board (and related) applications. Such a paint could be made more conducting than materials now used for this application. A combination of electrical, optical and lubrication properties have found applications in toner inks, colored pencil leads, preparations for photosensitive imaging materials and electrochromic displays.

The very different gas absorption of K-GICs for hydrogen, deuterium and tritium makes this an attractive system for isotope separation and the

334

enrichment of tritium by passage of the $^1$H, $^2$H, and $^3$H mixtures through $C_{24}K$ powder [Terai and Takahashi 1984].

## 12.8  Batteries and Other Electrochemical Applications

Although many of the applications already mentioned utilize the unique chemical properties of carbon and graphite fibers, their use as electrodes in a variety of electrochemical applications seems particularly attractive because of the unique combination of favorable electrical conductivity, chemical stability, fiber geometry, and in many cases graphitic structure. Graphite powder is often used in electrodes in order to make the electrodes more conducting. Most primary batteries in use today have at least one graphite or carbon electrode or current collector.

The unique ability of graphite to intercalate both negative and positive ions electrochemically has long been recognized as an important source of applications. Armand and Touzain [1977] have reviewed the potential of acceptor-intercalated graphite for use as a cathode material. The electrochemical applications of graphite intercalation compounds has been reviewed by Besenhard et al. [1982a, 1982b], including primary battery applications of partially oxidized graphite and partially fluorinated graphite.

A primary battery based on fluorine and lithium is theoretically the best system [Nakajima 1983, 1986; Palchan and Talinaker 1987] from the standpoint of a high energy density power source. Such a battery has been realized using $(CF)_n$ as a cathode material, Li as an anode material and 1M $LiClO_4$-propylene carbonate (PC) as the electrolyte [Touhara et al. 1984b]. From a theoretical standpoint, an electromotive force (EMF) of 4.59 V is calculated from the free energies of formation of CF and LiF at 298 K. Yet the open circuit voltage (OCV) realized for this battery is only 3.2 V. The difference between the EMF and OCV is attributed to the reaction with the electrolyte forming an intermediate phase during the electrochemical reaction (C-F-Li-$z$PC) where $z$ is the solvation number for the propylene carbonate (PC). This intermediate phase is not a crystalline compound or an intercalation compound and eventually will decompose to graphite and LiF when all solvents are removed. The discharge reactions are summarized by

anode reaction:  $Li + zPC \longrightarrow Li^+ \cdot zPC + e^-$
cathode reaction:  $e^- + CF + Li^+ \cdot zPC \longrightarrow C\text{-}F\text{-}Li \cdot zPC$
total reaction:  $(CF)_n + mLi + mzPC \longrightarrow (CF)_{n-m} + m(C\text{-}F\text{-}Li \cdot zPC)$
$\longrightarrow mC + mLiF + mzPC.$

Physically, the $Li^+$ diffuses between the layered structures of the CF cathode, and the discharge reaction proceeds according to the disproportionation reaction forming the intermediate phase. As the discharge reaction proceeds (over

335

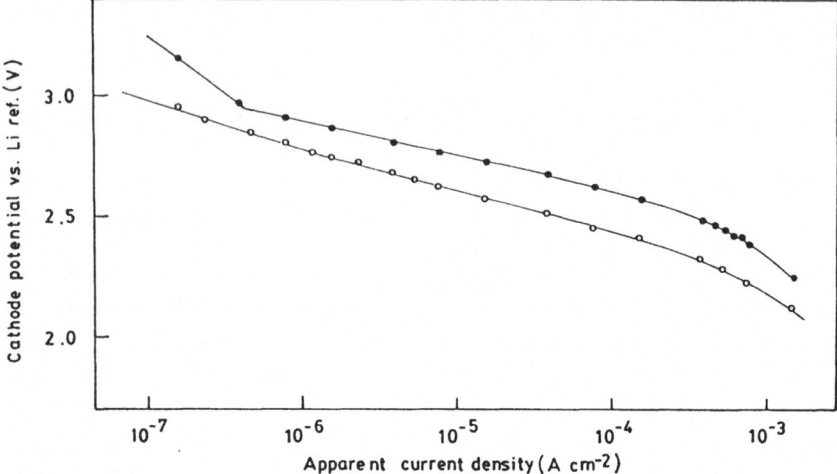

Fig. 12.17. Potential of the CF electrode relative to the Li reference electrode vs apparent current density of 25% discharge. Improved performance is found for the ● fluorine intercalated graphite fiber electrode ($CF_{0.68}$) relative to the ○ graphite fluoride electrode $CF_{0.64}$ prepared from natural graphite [Touhara et al. 1987a]

long times), the intermediate product eventually decomposes to graphite and LiF by decomposition and desolvation, each process contributing a free energy term unavailable to the electrochemical discharge.

The factors of interest for high performance battery operation include a high open circuit voltage (OCV), a low over-potential, high stability of the discharge potential, and a high percentage utilization of the cathode material in the energy production. It is found that the use of graphite fibers rather than natural graphite flakes results in improved performance, as shown in Fig. 12.17 [Touhara et al. 1987a].

With regard to the cathode material, it should be noted that there are two classes of graphite intercalation compounds based on fluorine: *ionic* (metallic) compounds, where charge transfer occurs between the fluorine and the planar graphene planes, and *covalent* (highly insulating) compounds, where there is no charge transfer and the carbon atoms are arranged in puckered layers (cyclohexane rings) where the A and B atoms are displaced in opposite directions relative to the graphene plane. It should be emphasized that the cathode material used for batteries is the covalent compound, and being insulating, carbon fibers or carbon powder is added to make the cathode electrode conducting. With regard to the covalent compounds, there are two main types: $(CF)_n$, which is normally prepared at $\sim 600°C$, forming a stage 1 compound, and $(C_2F)_n$, which is prepared at lower temperatures ($\sim 340$ to $\sim 400°C$) and forms a stage 2 compound.

Though the commercial batteries are based on $(CF)_n$, the $(C_2F)_n$ compounds have subsequently been shown to have a somewhat higher open cir-

336

cuit voltage (by $\sim0.2V$) and a lower over-potential (by $\sim0.1V$), in addition to other advantages cited below. Even better performance is achieved when the $(C_2F)_n$ is prepared from CCVD fibers [Touhara et al. 1984a] rather than bulk graphite, thereby achieving about a 0.5 V increase in the discharge potential at a current density of 0.5 mA/cm$^2$. This improved performance of $C_2F$ is in part explained by the higher open circuit voltage attributed to the higher activity of F$^-$ in $(C_2F)$ relative to CF in accordance with the relation to the open circuit voltage

$$OCV = E^\circ + \frac{RT}{F} \ln \frac{a_{F^-}^{C_xF}}{a_{F^-}^{int} a_{Li^+}^{int}} \qquad (12.19)$$

where $a_{F^-}^{int}$ and $a_{Li^+}^{int}$ are the activities of the F$^-$ and Li$^+$ ions in the intermediate phase, and $a_{F^-}^{C_xF}$ is the activity of the F$^-$ ion in the intercalated $C_xF$ compound, subject to the equilibrium conditions for the electrochemical potential

$$\overline{\mu}_{F^-}^{C_xF} = \overline{\mu}_{F^-}^{int} \qquad (12.20)$$

$$\overline{\mu}_{Li^+}^{int} = \overline{\mu}_{Li^+}^{bulk} \qquad (12.21)$$

The smaller over-potential (and consequently higher discharge voltage) for fiber-based cathodes is attributed to improved diffusion of the lithium ions into the inner cathode material because of the presence of defective microdomain structures in the fibers. [Touhara et al. 1987a]. For both $(C_2F)_n$ and $(CF)_n$ cathode materials, the over-potential is constant through the discharge, thereby providing a stable discharge voltage under load. The fiber host materials have additional advantages insofar as the intercalation can be done at lower temperatures and for shorter times [Touhara et al. 1984a], and also higher F/C ratios can be achieved in the stage 2 compound, thereby allowing more energy storage (discharge capacity) in the battery [Touhara et al. 1984a]. The use of $(C_2F)_n$ rather than the $(CF)_n$ also offers additional advantages with regard to providing a more constant battery discharge voltage and higher cathode utilization ($\sim80\%$) [Touhara et al. 1987a]. Further improvements in performance (discharge potential and cathode utilization) have been achieved with ultra-thin ($< 1\,\mu m$) CCVD fibers [Endo et al. 1987b].

Recently a fiber host material based on an activated rayon-derived precursor has been successfully used in $(CF)_n$/Li batteries. With these fibers the reaction temperature could be reduced to 200°C and below and the nature of the C-F bond could be controlled by the reaction temperature; the higher the temperature, the more covalent the C-F bond. For low temperature (below 150°C) fluorination, the OCV was about 0.3 V higher than for bulk $(CF)_n$ but the discharge characteristics were not as good. Discharge potentials comparable to the CCVD fibers were obtained with the activated carbon fibers with cathode utilization factors higher than 70% [Touhara et al. 1987b].

Although it is clear that one should be able to make rechargeable batteries based on intercalation, Besenhard et al. [1982a, 1982b] point out that one can typically store only one charge for every 24 carbon atoms in most electrochemical intercalation reactions. Use of the stage 1 compound $C_6Li$ allows one charge for every 6 carbon atoms, which significantly improves the practicality of these materials for battery applications. Nogami et al. [1982] reported on the use of an activated carbon fiber electrode which could be doped with either acceptors or donors. They compared this system with the much publicized polymer batteries [Nigrey et al. 1981; Shacklette et al. 1982], emphasizing the advantages of the high conductivity and stability of carbon and graphite over polyacetylene and poly ($p$-phenylene) used in the polymer battery systems.

Another application which would utilize the intercalation capability of graphite fibers is in concentration cells designed to convert thermal energy directly into electricity. Lalancette and Roussel [1976] described such a cell based on bromine intercalation of graphite whereby the amount of bromine intercalated at the two electrodes would be different due to the dependence of bromine activity on temperature. Endo et al. [1980] later showed that the energy conversion efficiency of such a cell could be improved if CCVD fibers were used at the electrodes. Maeda et al. [1983] demonstrated distinct electrochemical advantages to using CCVD fibers (instead of powder) in a similar concentration cell based on nitric acid.

Using bromine as an intercalant with CCVD fibers ($T_{HT} \sim 2900°C$) for the electrodes of the thermocell, an open circuit voltage of 200 mV and a short circuit current of $\sim$2–10 mA was obtained keeping the anode and cathode temperatures at 80°C and 15°C, respectively [Endo et al. 1983c]. Better performance and greater durability was achieved through better bromine circulation and a better cell configuration. Though the efficiency of such a thermal cell is low ($\sim$10%), it still may be of interest for energy conversion systems.

Carbon fiber electrodes have also been found to give improved stability in Na-S high temperature (325°–350°C) batteries [Breiter 1977; Joo 1978; Robinson 1982]. An application of carbon fiber electrodes which has been commercialized is the plating of trace metals from industrial waste water. The electrodes are typically made of carbon fiber fabrics [Fleet and Das Gupta 1977, 1978], and utilize the inherent strength, the small size and relative high surface area, the chemical inertness, and the relatively high electrical conductivity of carbon fibers. Low modulus, high strength fibers are adequate for this type of application, and thus the cost of the electrodes is not prohibitive.

Carbon fiber composites have been shown to have unique stability and erosion properties [Stafford 1980; Stafford et al. 1987] which make them well suited for electrochemical disinfection systems [Natishan et al. 1987]. In these systems the disinfectant is generated electrochemically. Thus chlorine would be produced at one electrode in a NaCl-containing system. The other electrode would be producing hydrogen. Reversing the driving voltage is important for keeping the electrode surfaces clean. Carbon is one of only a few materials which can be used for both reactions. The carbon surface is eroded slowly by oxidation, but carbon fiber composites erode without changing the electrochemical performance of the electrode.

To summarize, numerous electrochemical applications of carbon fibers have been discussed in the literature. Because of the relatively high cost of carbon fibers, only when carbon fibers offer significant advantages over more conventional electrode materials (such as graphite powder, bulk graphite, and porous carbons) will these applications be used commercially. We have here briefly summarized some of the battery applications where significant advantages of the use of fibers have been reported.

## 12.9  Medical Applications

It is well known that carbon is a useful material for implants because of its biocompatibility, and its relatively low specific gravity compared to most metals. Carbon fibers were tried as tendon replacements, but found to be unsuitable, because they tended to fragment after several months [Bokros 1979]. As a result, carbon fibers must normally be used in a composite, and coated with an isotropic carbon. There are also recent uses of carbon fibers in polymer composites [Ruyter and Ekstrand 1987] for dental applications.

An interesting application of carbon fibers in medical technology simultaneously exploits two favorable characteristics of composites - their stiffness and transparency to x-rays of energy $> 10\,\mathrm{KeV}$. Cooper and Leopold [1976] patented a patient support structure which allows great flexibility in the operation of x-ray diagnostic equipment, while permitting x-rays to be transmitted with low absorption loss and backscatter. This basic idea has since been used in several pieces of x-ray equipment.

# References

Abrahams, E., Anderson, P.W., Licciardello, D.C., Ramakrishnan, T.V. (1979): Phys. Rev. Lett. **42**, 673

Agrawal, A., Yinnon, H., Uhlmann, D.R., Pepper, R.T., Desper, C.R. (1986): J. Mater. Sci. **21**, 3455

Ahmed, A.U., Rost, M.C., Meyer, D., Woollam, J.A., Lake, M. (1987): Bull. Am. Phys. Soc. **32**, 489

Al-Jishi, R., Dresselhaus, G. (1982a): Phys. Rev. **B 26**, 4514

Al-Jishi, R., Dresselhaus, G. (1982b): Phys. Rev. **B 26**, 4523

Allen, S.A., Cooper, G.A., Mayer, R.M. (1969): Nature **224**, 684

Allen, S., Cooper, G.A., Johnson, D.J., Mayer, R.M. (1971): In Gregory, J.G., Rawlins, C.J. (eds.): 3rd Conference on Industrial Carbons and Graphite, London (Society of Chemical Industries, London) p. 456

Altshuler, B.L., Aronov, A.G. (1979): Solid State Commun. **30**, 115

Altshuler, B.L., Aronov, A.G., Lee, P.A. (1980a): Phys. Rev. Lett. **44**, 1288

Altshuler, B.L., Khmel'nitzkii, D., Larkin, A.I., Lee, P.A. (1980b): Phys. Rev. **B 22**, 5142

Amelinckx, S., Delavignette, P., Heerschap, M. (1965): In Walker, Jr., P.L. (ed.): *Chemistry and Physics of Carbon*, Vol. 1 (Marcel Dekker, New York) p. 1

Anderson, B.W. (1987): J. Phys. **D 20**, 311

Anderson, P.W., Abrahams, E., Ramakrishnan, T.V. (1979): Phys. Rev. Lett. **43**, 718

Anderson, S.H., Chung, D.D.L. (1983): Synth. Met. **8**, 343

Anderson, S.H., Chung, D.D.L. (1984): Carbon **22**, 253

Ansart, A., Meschi, C., Flandrois, S. (1988): In Selig, H., Davidov, D. (eds.): Proceedings of the 4th International Symposium on Graphite Intercalation Compounds, Jerusalem. Synth. Met. **23**, 455

Arai, S., Tsuzuku, T. (1972): Tanso **71**, 120 (in Japanese)

Armand, M., Touzain, P. (1977): J. Mater. Sci. Eng. **31**, 319

Ashkin, M. Gavaler, J.R., Greggi, J., Decroux, M. (1984): J. Appl. Phys. **55**, 1044

Audier, M., Coulon, M., Oberlin, A. (1980): Carbon **18**, 73

Audier, M., Oberlin, A., Oberlin, M., Coulon, M., Bonnetain, L. (1981): Carbon **19**, 217

Austerman, S.B. (1968): In Walker, Jr., P.L. (ed.): *Chemistry and Physics of Carbon*, Vol. 7 (Marcel Dekker, New York) p. 137

Ayasse, J.B., Ayache, C., Jager, B., Bonjour, E., Spain, I.L. (1979): Solid State Commun. **29**, 659

Azzeer, A.M., Silber, L.M., Spain, I.L., Patton, C.E., Goldberg, H.A. (1985): J. Appl. Phys. **57**, 2529

Bacon, D.J., Nicholson, A.P.P. (1975): Phys. Status Solidi **A 28**, 613

Bacon, D.J., Nicholson, A.P.P. (1977): J. Phys. **C 10**, 2295

Bacon, G.E. (1956a): J. Appl. Phys. **38**, 3585

Bacon, G.E. (1956b): J. Appl. Chem. **6**, 477

Bacon, R. (1958): In Doremus, R.H., Roberts, B.W., Turnbull, D. (eds.): *Growth and Perfection of Crystals* (Wiley, New York) p. 197

Bacon, R. (1960): J. Appl. Phys. **31**, 283

Bacon, R. (1973): In Walker, Jr., P.L., Thrower, P.A. (eds.): *Chemistry and Physics of Carbon*, Vol. 9 (Marcel Dekker, New York) p. 1

Bacon, R. (1980): Philos. Trans. R. Soc. London A **294**, 29

Bacon, R., Schalamon, W.A. (1969): Appl. Polym. Symp. **9**, 285

Bacon, R., Pallozza, A.A., Slosarik, S.E. (1966): Society of Plastics Industry, 21st. Tech. and Management Conf., Section 8 E

Bagga, G.R., Touzain, Ph., Bonnetain, L. (1984): In Pacault, A., Flandrois, S. (eds.): Extended Abstracts of the International Conference on Carbon, Bordeaux, France, p. 292

Bailey, A.C., Yates, B. (1970): J. Appl. Phys. **41**, 5088

Baird, T., Fryer, J.R., Grant, B. (1971): Nature (London) **233**, 329

Baird, T., Fryer, J.R., Grant, B. (1974): Carbon **12**, 591

Baker, C., Kelly, A. (1964): Philos. Mag. **9**, 927

Baker, R.T.K. (1987): Reported in the Pettinos Award Lecture, in Extended Abstracts of the 18th Biennial Conference on Carbon, Worcester, MA (American Carbon Society, University Park, PA)

Baker, R.T.K., Harris, P.S. (1973): J. Phys. E **5**, 793

Baker, R.T.K., Harris, P.S. (1978): In Walker, Jr., P.L., Thrower, P.A. (eds.): *Chemistry and Physics of Carbon*, Vol. 14 (Marcel Dekker, New York) p. 83

Baker, R.T.K., Waite, R.J. (1975): J. Catal. **37**, 101

Baker, R.T.K., Barber, M.A., Harris, P.S., Feates, F.S., Waite, R.J. (1972): J. Catal. **26**, 51

Baker, R.T.K., Harris, P.S., Thomas, R.B., Waite, R.J. (1973): J. Catal. **30**, 86

Baker, R.T.K., Chludzinski, J.J., Lund, C.R.F. (1987): In Extended Abstracts of the 18th Biennial Conference on Carbon, Worcester, MA (American Carbon Society, University Park, PA) p. 155

Balberg, I., Anderson, C.H., Alexander, S., Wagner, N. (1984): Phys. Rev. B **30**, 3933

Bale, H.D., Schmidt, P.W. (1984): Phys. Rev. Lett. **53**, 596

Barr, J.B., Eckstein, B.H. (1987): In Extended Abstracts of the 18th Biennial Conference on Carbon, Worcester, MA (American Carbon Society, University Park, PA) p. 9

Bartlett, N., Bianconi, R.N., McQuillan, B.W., Robertson, A.S., Thompson, A.C. (1978): J. Chem. Soc., Chem. Commun. 200

Bassani, F., Pastori Parravicini, G. (1967): Nuovo Cimento B **50**, 95

Batson, P.E. (1987): Private communication

Bednorz, J.G., Müller, K.A. (1986): Z. Phys. B **64**, 189

Beetz, Jr., C.P. (1981): In Extended Abstracts of the 15th Biennial Conference on Carbon, Philadelphia, PA (American Carbon Society, University Park, PA) p. 302

Beetz, Jr., C.P. (1982a): Fibre Sci. Technol. **16**, 45

Beetz, Jr., C.P. (1982b): Fibre Sci. Technol. **16**, 81

Beetz, Jr., C.P. (1982c): Fibre Sci. Technol. **16**, 219

Beetz, Jr., C.P. (1987): Private communication

Beetz, Jr., C.P., Brinen, J.S. (1983): In Extended Abstracts of the 16th Biennial Conference on Carbon, San Diego, CA (American Carbon Society, University Park, PA) p. 535

Beetz, Jr., C.P., Budd, G.W. (1983): Rev. Sci. Instrum. **54**, 1222

Bennett, S.C., Johnson, D.J. (1979): Carbon **17**, 25

Bennett, S.C., Johnson, D.J., Johnson, W. (1983): J. Mater. Sci. **18**, 3337

Berg, C.A., Cumptson, H., Rinsky, A. (1972): J. Text. Res. **42**, 486

Bergenlid, U., Hill, R.W., Wolf, E.J., Wilks, J. (1954): Philos. Mag. **45**, 851

Besenhard, J.O., Mohwald, H., Nickl, J.J. (1982a): Rev. Chim. Miner. **19**, 588

Besenhard, J.O., Theodoridou, E., Mohwald, H., Nickl, J.J. (1982b): Synth. Met. **4**, 211

Boellard, E., de Bokx, P.K., Kock, A.J.H.M., Geus, J.W. (1985): J. Catal. **96**, 481

Bohren, C.F., Huffman, D. (1983): *Absorption and Scattering of Light by Small Particles* (Wiley, New York)

Boissonade, J., Barreau, F., Carmona, F. (1983): J. Phys. A **16**, 2777

Bokros, J.C. (1979): U.S. Patent 4,149,277

Bowers, C.W., Culver, R.B., Solberg, W.A., Spain, I.L. (1987): J. Appl. Opt. **26**, 4625

Bowers, C.W., Zhao, Y.-Z., Spain, I.L. (1988): Carbon, in press

Breiter, M.W. (1977): U.S. Patent 4,053,689

Brice, D.K. (1975): *Ion Implantation Range and Energy Deposition Distributions* (Plenum, New York). A useful table for ion ranges and disorder depth distributions

Brice, D.K., Ashby, C.I.H. (1984): J. Chem. Phys. **81**, 6244

Bright, A.A. (1979): Phys. Rev. **B 20**, 5142

Bright, A.A., Singer, L.S. (1979): Carbon **17**, 59

Bullock, R.E. (1971): Radiat. Eff. **11**, 107

Bullock, R.E. (1972a): Radiat. Eff. **14**, 263

Bullock, R.E. (1972b): J. Mater. Sci. **7**, 964

Bullock, R.E. (1972c): J. Mater. Sci. Eng. **10**, 178

Bullock, R.E. (1974): Fibre Sci. Technol. **7**, 157

Bullock, R.E. (1979): Carbon **17**, 447

Bullock, R.E., Gordon, D.E., Deaton, B.C. (1973): Carbon **11**, 418

Bundy, F.P., Strong, H.M., Wentdorf, Jr., R.H. (1973): In Walker, Jr., P.L., Thrower, P.A. (eds.): *Chemistry and Physics of Carbon*, Vol. 10 (Marcel Dekker, New York) p. 213

Butler, B.L., Diefendorf, R.J. (1969): In Summary of Papers of the 9th Conference on Carbon, Boston, MA (American Carbon Society, University Park, PA) p. 161

Butler, B., Duliere, S., Tidmore, J. (1971): In Extended Abstracts of the 10th Biennial Conference on Carbon, Lehigh University, PA (American Carbon Society, University Park, PA) p. 45

Campbell, R.H., Chipman, R.A. (1949): Proc. IRE **37**, 938

Carmona, F. (1987): Private communication

Carmona, F., El Amarti, A. (1987): In Extended Abstracts of the 18th Biennial Conference on Carbon, Worcester, MA (American Carbon Society, University Park, PA) p. 442

Carmona, F., Mouney, C. (1987): In Extended Abstracts of the 18th Biennial Conference on Carbon, Worcester, MA (American Carbon Society, University Park, PA) p. 440

Carruthers, P. (1961): Rev. Mod. Phys. **33**, 92

Carter, G., Grant, W.A. (1976): *Ion Implantation of Semiconductors* (Edward Arnold, London)

Castle, J.G. (1953): Phys. Rev. **92**, 1063

Cava, R.J., Batlogg, B., van Dover, R.B., Murphy, D.W., Sunshine, S., Siegrist, T., Remeika, J.P., Rietman, E.A., Zahurak, S., Espinosa, G.P. (1987): Phys. Rev. Lett. **58**, 1676

Centanni, M.A., Clark, E. (1986): In Dresselhaus, M.S., Dresselhaus, G., Solin, S.A. (eds.): Extended Abstracts of the 1986 MRS Symposium on Graphite Intercalation Compounds, Boston (Materials Research Society, Pittsburgh, PA) p. 219

Chambers, R.G. (1950): Proc. R. Soc. London *A* 202, 378

Chau, C.K., Liu, S.Y. (1974): J. Low Temp. Phys. **15**, 447

Chausse, J.P., Hoarau, J. (1969): J. Chim. Phys. **66**, 1062

Cherville, J., Bothorel, P., Pacault, A. (1963): In Mrozowski, S., Studebaker, M.L., Walker, Jr., P.L. (eds.): Proceedings of the 5th Conference on Carbon, University Park, PA, 1961 (Pergamon, New York) p. 29

Chi, Z., Chou, T.-W., Shen, G. (1985): J. Mater. Sci. **20**, 3319

Chieu, T.C. (1983): PhD Thesis, Massachusetts Institute of Technology, Department of Electrical Engineering and Computer Science

Chieu, T.C., Dresselhaus, M.S., Endo, M. (1982): Phys. Rev. **B 26**, 5867

Chieu, T.C., Timp, G., Dresselhaus, M.S., Endo, M., Moore, A.W. (1983): Phys. Rev. **B 27**, 3686

Chieu, T.C., Elman, B.S., Salamanca-Riba, L., Endo, M., Dresselhaus, G. (1984): In Hubler, G.K., Holland, O.W., Clayton, C.R., White, C.W. (eds.): MRS Symposium on Ion Implantation and Ion Beam Processing of Materials, Boston 1983, Vol. 27 (North-Holland, New York) p. 487

Chown, J., Deacon, R.F., Singer, N., White, A.E.S. (1963): In Popper, P. (ed.): *Special Ceramics* (Academic, New York) p. 81

Chu, W.K., Mayer, J.W., Nicolet, M.A. (1978): *Backscattering Spectrometry* (Academic, New York)

Chung, D.D.L., Wong, L.W. (1985): Synth. Met. **12,** 533
Chung, D.D.L., Wong, L.W. (1986): Carbon **24,** 443
Chung, D.D.L., Li, P., Li, X. (1987): In Extended Abstracts of the 18th Biennial Conference on Carbon, Worcester, MA (American Carbon Society, University Park, PA) p. 446
Chwastiak, S., Barr, J.B., Didchenko, R. (1979): Carbon **17,** 49
Clarke, G.P. (1986): In *Carbon Fibres,* ed. The Plastics and Rubber Institute, London (Noyes, Park Ridge, NJ) p. 113
Collins, M.F., Haywood, B.C. (1960): Carbon **7,** 663
Conor, P.C., Owston, C.N. (1969): Nature (London) **223,** 1146
Cooper, A.A.G., Leopold, P.M. (1976): U.S. Patent 3,947,686
Cooper, G.A., Mayer, R.M. (1971): J. Mater. Sci. **6,** 60
Corbató, F.J. (1956): PhD Thesis, Massachusetts Institute of Technology, Department of Physics
Corbató, F.J. (1957): In Mrozowski, S., Studebaker, M.L., Walker, Jr., P.L. (eds.): Proceedings of the 3rd Conference on Carbon (Pergamon, New York) p. 173
Cottrell, A.H. (1964): *The Mechanical Properties of Matter* (Wiley, New York)
Coulson, C.A. (1952): *Valence* (Oxford University Press, Oxford)
Coulson, C.A., Poole, M.D. (1968): Carbon **2,** 275
Coulson, C.A., Schaad, L.J., Burnell, L. (1957): In Mrozowski, S., Studebaker, M.L., Walker, Jr., P.L. (eds.): Proceedings of the Third Conference on Carbon (Pergamon, New York) p. 27
Coulson, C.A., Herraez, M.A., Leal, M. Santos, E., Senent, S. (1963): Proc. R. Soc. London **A 247,** 461
Culver, R.B., Solberg, W.A., Robinson, R.S., Spain, I.L. (1985): J. Appl. Opt. **24,** 924
Cuomo, J.J., Harper, J.M.E. (1977): IBM Tech. Discl. Bull. **20,** 775
Curtis, G.J., Milne, J.M., Reynolds, W.N. (1968): Nature (London) **220,** 1024
Da Silva, J.L.G., Johnson, D.J. (1984): J. Mater. Sci. **19,** 3201
Daumas, N., Hérold, A. (1969): C. R. Acad. Sci., Series C **268,** 373
Davis, G.P., Endo, M., Vogel, F.L. (1981): In Extended Abstracts of the 15th Biennial Conference on Carbon, Philadelphia, PA (American Carbon Society, University Park, PA) p. 363
de Bokx, P.K., Kock, A.J.H.M., Boellard, E., Klop, W., Geus, J.W. (1985): J. Catal. **96,** 454
de Combarieu, A. (1968): J. de Phys. **66,** 1062
Delhaès, P. (1971): In Walker, Jr., P.L., Thrower, P.A. (eds.): *Chemistry and Physics of Carbon,* Vol. 7 (Marcel Dekker, New York) p. 193
Delhaès, P., Carmona, F. (1981): In Walker, Jr., P.L., Thrower, P.A. (eds.): *Chemistry and Physics of Carbon,* Vol. 17 (Marcel Dekker, New York) p. 89
Delhaès, P., Lemerle, M., Blondet-Gonte, G. (1971): C. R. Acad. Sci. **272,** 1285
Delhaès, P., deKepper, P., Uhlrich, M. (1974): Philos. Mag. **29,** 1301
Denison, P., Jones, F.R., Watts, J.F. (1985): J. Mater. Sci. **20,** 4647
Denison, P., Jones, F.R., Watts, J.F. (1987): J. Phys. **D 20,** 306
DeSorbo, W., Nichols, G.E. (1958): J. Phys. Chem. Solids **6,** 352
DeSorbo, W., Tyler, W.W. (1953): J. Chem. Phys. **21,** 1660
Dickson, R.F., Jones, C.J., Harris, B., Leach, D.C., Moore, D.R. (1985): J. Mater. Sci. **20,** 60
Diefendorf, R.J., Tokarsky, E.W. (1975): J. Polymer Eng. Sci. **15,** 150
Dietrich, M. (1985): IEEE Trans. MAG-**21,** 455
Dietrich, M., Dustmann, C.H. (1984): In *Advances in Cryogenic Engineering Materials* (Plenum, New York) p. 683
Dietrich, M., Dustmann, C.H., Schmaderer, F., Wahl, G. (1983): IEEE Trans. MAG-**19,** 406
Dillon, R.O., Woollam, J.A., Katkanant, V. (1984): Phys. Rev. **B 29,** 3482
Dischler, B., Bubenzer, A., Koidl, P. (1983): Solid State Commun. **48,** 105
Doezema, R.E., Datars, W.R., Schaber, H., Van Schyndel, A. (1979): Phys. Rev. **B 19,** 4224

Doll, G.L., Sakya, R.M., Nicholls, J.T., Speck, J.S., Dresselhaus, M.S., Engle, G.B. (1988): In Selig, H., Davidov, D. (eds.): Proceedings of the 4th International Symposium on Graphite Intercalation Compounds, Jerusalem. Synth. Met. **23**, 481

Dominguez, D.D., Murday, J.S. (1983): In Extended Abstracts of the 16th Biennial Conference on Carbon, San Diego, CA (American Carbon Society, University Park, PA) p. 276

Dominguez, D.D., Lakshmanan, J.L., Barbano, E.F., Murday, J.S. (1983): In Dresselhaus, M.S., Dresselhaus, G., Fischer, J.E., Moran, M.J. (eds.): Proceedings of the Symposium on Intercalated Graphite, Vol. 20 (North-Holland, New York) p. 63

Doni, E., Pastori-Parravicini, G. (1969): Nuovo Cimento **B 64**, 117

Donnet, J.B., Voet, A. (1976): *Carbon Black* (Marcel Dekker, New York) p. 147

Donnet, J.B., Fitzer, E., Köchling, K.H. (1982): Carbon **20**, 445

Donnet, J.B., Fitzer, E., Köchling, K.H. (1983): Carbon **21**, 517

Donnet, J.B., Fitzer, E., Köchling, K.H. (1985): Carbon **23**, 601

Donnet, J.B., Fitzer, E., Köchling, K.H. (1986): Carbon **24**, 246

Donnet, J.B., Fitzer, E., Köchling, K.H. (1987a): Carbon **25**, 317

Donnet, J.B., Fitzer, E., Köchling, K.H. (1987b): Carbon **25**, 449

Dowell, M.B. (1975): In Extended Abstracts of the 12th Biennial Conference on Carbon, Pittsburgh, PA (American Carbon Society, University Park, PA) p. 31

Dresselhaus, M.S. (1984): J. Chim. Phys. **81**, 739

Dresselhaus, M.S. (ed.) (1987): *Intercalation in Layered Materials* (Plenum, New York)

Dresselhaus, M.S., Dresselhaus, G. (1965): Phys. Rev. **140 A**, 401

Dresselhaus, M.S., Dresselhaus, G. (1979): In Lévy, F. (ed.): *Physics and Chemistry of Materials with Layered Structures*, Vol. 6 (Reidel, Dordrecht) p. 423

Dresselhaus, M.S., Dresselhaus, G. (1981): Adv. Phys. **30**, 139

Dresselhaus, M.S., Dresselhaus, G. (1982): "Light Scattering in Graphite Intercalation Compounds" in Cardona, M., Güntherodt, G. (eds.): *Light Scattering in Solids III*, Topics Appl. Phys., Vol. 51 (Springer, Berlin, Heidelberg) p. 3

Dresselhaus, M.S., Dresselhaus, G., Fischer, J.E. (1977): Phys. Rev. **B 15**, 3180

Dresselhaus, M.S., Dresselhaus, G., Fischer, J.E., Moran, M.J. (eds.) (1983): Proceedings of the Symposium on Intercalated Graphite, Vol. 20 (North-Holland, New York)

Dresselhaus, M.S., Goldberg, H.A., Spain, I.L. (1985): In Kohl, R.H. (ed.): Proceedings of the 1985 CRDC Scientific Conference on Obscuration and Aerosol Research, pp. 291–324

Dresselhaus, M.S., Dresselhaus, G., Solin, S.A. (eds.) (1986): Extended Abstracts of the 1986 MRS Symposium on Graphite Intercalation Compounds, Boston (Materials Research Society, Pittsburgh, PA)

Dreyfus, B., Maynard, R. (1967): J. de Phys. **28**, 955

Dyson, F.J. (1955) Phys. Rev. **98**, 349

Eckstein, B.H. (1981): Fibre Sci. Technol. **14**, 139

Eckstein, B.H. (1986): SAMPE J. 22–5, 18th International SAMPE Technical Conference, Seattle, WA

Edie, D.D., Fox, N.K., Barnett, B.C., Fain, C.C. (1986): Carbon **24**, 477

Eklund, P.C., Dresselhaus, M.S., Dresselhaus, G. (eds.) (1984): Extended Abstracts of the 1984 MRS Symposium on Graphite Intercalation Compounds (Materials Research Society, Pittsburgh, PA)

Elman, B.S., Shayegan, M., Dresselhaus, M.S., Mazurek, H., Dresselhaus, G. (1982): Phys. Rev. **B 25**, 4142

Elman, B.S., McNeil, L., Nicolini, C., Chieu, T.C., Dresselhaus, M.S., Dresselhaus, G. (1983): Phys. Rev. **B 28**, 7201

Elman, B.S., Braunstein, G., Dresselhaus, M.S., Dresselhaus, G., Venkatesan, T., Gibson, J.M. (1984): Phys. Rev. **B 29**, 4703

Elman, B.S., Braunstein, G., Dresselhaus, M.S., Venkatesan, T. (1985): Nucl. Instrum. Methods Phys. Res. **B 7/8**, 493 (IBMM Conference, July 16–20, 1984, Cornell University)

Elzinga, M., Morelli, D.T., Uher, C. (1982): Phys. Rev. **B 26**, 3321

Endo, M. (1975): PhD Thesis, University of Orleans, Orleans, France (in French)

Endo, M. (1987): Private communication

Endo, M. (1988a): J. Mater. Sci. **23**, 598

Endo, M. (1988b): Chem. Technol., in press

Endo, M., Komaki, K. (1983): In Extended Abstracts of the 16th Biennial Conference on Carbon, San Diego, CA (American Carbon Society, University Park, PA) p. 523

Endo, M., Koyama, T. (1977): Kotai Butsuri **12**, 1 (in Japanese)

Endo, M., Koyama, T. (1980): IEE of Japan **100 A**, 633 (in Japanese)

Endo, M., Shikata, M. (1985): Ohyo Butsuri **54**, 507 (in Japanese)

Endo, M., Ueno, H. (1984): In Eklund, P.C., Dresselhaus, M.S., Dresselhaus, G. (eds.): Extended Abstracts of the 1984 MRS Symposium on Graphite Intercalation Compounds (Materials Research Society, Pittsburgh, PA) p. 177

Endo, M., Koyama, T., Hishiyama, Y. (1976): Jpn. J. Appl. Phys. **15**, 2073

Endo, M., Oberlin, A., Koyama, T. (1977a): Jpn. J. Appl. Phys. **16**, 1519

Endo, M., Tamagawa, I., Koyama, T. (1977b): Jpn. J. Appl. Phys. **16**, 1771

Endo, M., Koyama, T., Inagaki, M. (1980): Ohyo Butsuri **49**, 563 (in Japanese)

Endo, M., Koyama, T., Inagaki, M. (1981): Synth. Met. **3**, 177

Endo, M., Hishiyama, Y., Koyama, T. (1982a): J. Phys. D **15**, 353

Endo, M., Komaki, K., Koyama, T. (1982b): International Symposium on Carbon (Toyohashi 1982) p. 535

Endo, M., Chieu, T.C., Timp, G., Dresselhaus, M.S., Elman, B.S. (1983a): Phys. Rev. **B 28**, 6982

Endo, M., Chieu, T.C., Timp, G., Dresselhaus, M.S., Elman, B.S. (1983b): Synth. Met. **8**, 251

Endo, M., Yamagishi, Y., Inagaki, M. (1983c): Synth. Met. **7**, 203

Endo, M., Salamanca-Riba, L., Dresselhaus, G., Gibson, J.M. (1984): J. Chim. Phys. **81**, 803

Endo, M., Katoh, A., Sugiura, T., Shiraishi, M. (1987a): In Extended Abstracts of the 18th Biennial Conference on Carbon, Worcester, MA (American Carbon Society, University Park, PA) p. 151

Endo, M., Momose, T., Touhara, H., Watanabe, N. (1987b): J. Power Sources, **20**, 99

Engle, G.B. (1987): Private communication

Ergun, S. (1970): Phys. Rev. **B 1**, 3371

Espelette, P., Marchand, A. (1987): Carbon **25**, 447

Ezekiel, H.M. (1969): In Summary of Papers of the 9th Conference on Carbon, Boston, MA (American Carbon Society, University Park, PA) p. 712

Ezekiel, H.M. (1970): J. Appl. Phys. **41**, 5351

Ezekiel, H.M. (1973): In Extended Abstracts of the 11th Biennial Conference on Carbon, Gatlinburg, TN (American Carbon Society, University Park, PA) p. 273

Feher, G., Kip, A.F. (1955): Phys. Rev. **98**, 337

Feldman, L.C., Mayer, J.W., Picraux, S.T. (1982): *Materials Analysis by Ion Channeling* (Academic, New York)

Feldman, L.J., Skelton, E.F., Ehrlich, A.C. Dominguez, D.D., Elam, W.T., Quardi, S.B., Lytle, F.W. (1983): In Extended Abstracts of the 16th Biennial Conference on Carbon, San Diego, CA (American Carbon Society, University Park, PA) p. 209

Feldman, L.J., Elam, W.T., Ehrlich, A.C., Skelton, E.F., Dominguez, D.D., Qaurdi, S.B., Chung, D.D.L., Lytle, F.W. (1984): In Hodgson, K.O., Hedman, B., Penner-Hahn, J.E. (eds.): *EXAFS and Near Edge Structure III*, Springer Proc. Phys., Vol. 2 (Springer, Berlin, Heidelberg) p. 464

Fink, J., Muller-Heinzerling, T., Pfluger, J., Bubenzer, A., Koidl, P., Crecelius, G. (1983): Solid State Commun. **47**, 687

Fink, J., Muller-Heinzerling, T., Pfluger, J., Scheerer, B., Dischler, B., Koidl, P., Bubenzer, A., Sah, R.E. (1984): Phys. Rev. **B 30**, 4713

Fischbach, D.B. (1985): In Extended Abstracts of the 17th Biennial Conference on Carbon Lexington, KY (American Carbon Society, University Park, PA) p. 229

Fischbach, D.B., Srinivasagopalan, S. (1978): 5th Conference on Industrial Carbons and Graphite, London (Society of Chemical Industries, London) p. 389

Fischbach, D.B., Komaki, K., Srinivasagopalan, S. (1980): Final Technical Report to U.S. Army Research Office (Grant No. DAA G29-76-G-0169) (unpublished)
Fitzer, E. (1985): In Fitzer, E. (ed.): *Carbon Fibres and Their Composites* (Springer, Berlin, Heidelberg) p. 3
Fitzer, E. (1987): Carbon **25**, 163
Fitzer, E. Heym, M. (1978): High Temp.-High Pressures **10**, 29
Fitzer, E., Hüttner, W. (1981): J. Phys. D **14**, 347
Fitzer, E., Terwisch, N. (1972): Carbon **10**, 383
Fleet, B., das Gupta, S. (1977): U.S. Patent 4,046,663
Fleet, B., das Gupta, S. (1978): U.S. Patents 4,108,745 and 4,108,757
Floro, J.A., Rossnagel, S.M., Robinson, R.S. (1983): J. Vac. Sci. Technol. **A 1**, 1398
Fornes, R.E., Memory, J.D., Naranong, N. (1981): J. Appl. Polym. Sci. **26**, 2061
Fourdeux, A., Herinckx, C., Perret, R., Ruland, W. (1969): C. R. Acad. Sci. **269 C**, 1597
Fourdeux, A., Perret, R., Ruland, W. (1971): In Proceedings of the International Conference on Carbon Fibers, Their Composites, and Applications, London. Plastics and Polymer Conference Supplement No. 5 (The Plastics Institute, London) p. 57
Freise, E.J., Kelly, A. (1963): Philos. Mag. **8**, 1519
Friedel, G. (1922): Ann. Phys. (Leipzig) **18**, 273
Fritz, W., Hüttner, W., Hartwig, G. (1978): In Clark, A.F., Reed, R.P. (eds.): *Nonmetallic Materials and Composites at Low Temperatures* (Plenum, New York) p. 245
Fuchs, K. (1938): Proc. Cambridge Philos. Soc. **34**, 100
Fujita, S. (1968): Carbon **6**, 746
Fujita, S., Izuki, K. (1961): J. Phys. Soc. Jpn. **16**, 1032
Fukuda, S., Tsukamoto, J., Takahashi, A., Matsumura, K. (1987): Synth. Met. **18**, 491
Fukuda, S., Takahashi, A., Tsukamoto, J. (1988): Synth. Met. **23**, 475
Gaier, J.R. (1985): NASA TM-86859 Report (unpublished)
Gaier, J.R., Jaworske, D.A. (1985): Synth. Met. **12**, 525
Gaier, J.R., Slabe, M.E. (1986): In Dresselhaus, M.S., Dresselhaus, G., Solin, S.A. (eds.): Extended Abstracts of the 1986 MRS Symposium on Graphite Intercalation Compounds, Boston (Materials Research Society, Pittsburgh, PA) p. 216
Ganguli, N., Krishnan, K.S. (1941): Proc. R. Soc. London **A 117**, 168
Gay, R., Gasparoux, H. (1965): *Les Carbones, Collection de Chimie Physique*, Tome I (Masson, Paris) p. 61
Gibbons, J.F., Johnson, W.S., Mylroie, S.W. (1975): *Projected Range Statistics* (Halstead, New York)
Gill, R.M. (1972): In *Carbon Fibres in Composite Materials* (Published for The Plastics Institute, London, by ILIFFE Books, London)
Goggin, P.R., Reynolds, W.N. (1967): Philos. Mag. **16**, 317
Goldberg, H.A. (1985): Final Report to US Army Research Office, Contract No. DAAE29-81-C-0016 (unpublished)
Goldberg, H.A. (1987): Unpublished work of the Celanese group
Goldberg, H.A., Kalnin, I.L. (1981): Synth. Met. **3**, 169
Goldberg, H.A., Kalnin, I.L., Spain, I.L., Volin, K.J. (1983): In Extended Abstracts of the 16th Biennial Conference on Carbon, San Diego, CA (American Carbon Society, University Park, PA) p. 281
Goldberg, H.A., Kalnin, I.L., Williams, C.W., Spain, I.L. (1987): US Patent 4,642,664
Goldstein, E.M., Carter, E.W., Klutz, S. (1966): Carbon **4**, 273
Goma, J., Oberlin, A. (1980): Thin Solid Films **65**, 221
Gorkov, L.P., Larkin, A.I., Khmel'nitzkii, D. (1979): JETP Lett. **30**, 229
Granato, A., Lücke, K. (1956): J. Appl. Phys. **27**, 583
Gray, W.J. (1978): Nucl. Technol. **40**, 194
Griffith, A.A. (1920): Philos. Trans. R. Soc. London **A 221**, 163
Griffith, Jr., O.K., Gayley, R.I. (1966): Carbon **3**, 541
Guigon, M., Oberlin, A., Desarmot, G. (1984a): Fibre Sci. Technol. **20**, 55
Guigon, M., Oberlin, A., Desarmot, G. (1984b): Fibre Sci. Technol. **20**, 177
Hall, T.M., Wagner, A., Thompson, L.F. (1982): J. Appl. Phys. **53**, 3997

347

Hamada, T., Nishida, Sajiki, Y., Furuyama, M., Tomioka, T. (1987a): In Extended Abstracts of the 18th Biennial Conference on Carbon, Worcester, MA (American Carbon Society, University Park, PA) p. 225

Hamada, T., Nishida, Sajiki, Y., Matsumoto, M., Endo, M. (1987b): J. Mater. Res. **2**, 850

Hambourger, P.D. (1984): In Eklund, P.C., Dresselhaus, M.S., Dresselhaus G. (eds.): Extended Abstracts of the 1984 MRS Symposium on Graphite Intercalation Compounds, Boston (Materials Research Society, Pittsburgh, PA) p. 199

Hambourger, P.D. (1985): Bull. Am. Phys. Soc. **30**, 239

Hambourger, P.D. (1986a): Appl. Phys. Commun. **5**, 223

Hambourger, P.D., Boyer, R.A., Du, Z.L., Hoz, P.H., Chu, C.W. (1986b): Bull. Am. Phys. Soc. **31**, 643

Hambourger, P.D. (1987): Private Communication

Hawthorne, H.M., Baker, C., Bentall, R.H., Linger, K.R. (1970): Nature (London) **227**, 946

Hayes, B.J. (1985a): In Fitzer, E. (ed.): *Carbon Fibres and Their Composites* (Springer, Berlin, Heidelberg) p. 143

Hayes, B.J. (1985b): In Fitzer, E. (ed.): *Carbon Fibres and Their Composites* (Springer, Berlin, Heidelberg) p. 81

Heald, S.M., Stern, E.A. (1980): Synth. Met. **2**, 87

Heald, S.M., Goldberg, H.A., Kalnin, I.L. (1983): In Bianconi, A., Incoccia, L., Stipcich, S. (eds.): *EXAFS and Near Edge Structure*, Springer Ser. Chem. Phys., Vol. 27 (Springer, Berlin, Heidelberg) p. 141

Heerschap, M.M., Delavignette, P. (1967): Carbon **5**, 383

Herrschap, M.M., Delavignette, P., Amelinckx, S. (1964): Carbon **1**, 235

Hennig, G. (1952): J. Chem. Phys. **20**, 1438

Hennig, G. (1965): Science **147**, 733

Hennig, G.R., Smaller, B., Yasaitis, E.L. (1954): Phys. Rev. **95**, 1088

Henrici-Olivé, G., Olivé, S. (1983): In *Industrial Developments*, Advances in Polymer Science, Vol. 51 (Springer, Berlin, Heidelberg) p. 1

Heremans, J. (1985): Carbon **23**, 431

Heremans, J., Beetz, Jr., C.P. (1985): Phys. Rev. **B 32**, 1981

Heremans, J., Issi, J.-P., Zabala-Martinez, I., Shayegan, M., Dresselhaus, M.S. (1981): Phys. Lett. **84 A**, 387

Heremans, J., Shayegan, M., Dresselhaus, M.S., Issi, J.-P. (1982): Phys. Rev. **B 26**, 3338

Heremans, J., Rahim, I., Dresselhaus, M.S. (1985): Phys. Rev. **B 32**, 6742

Heremans, J., Olk, C.H., Eesley, G.L., Steinbeck, J., Dresselhaus, G. (1988): Phys. Rev. Lett. **60**, 452

Hérold, A. (1979): In Lévy, F. (ed.): *Physics and Chemistry of Materials with Layered Structures*, Vol. 6 (Reidel, Dordrecht) p. 323

Hérold, A., Guérard, D. (eds.) (1983): Synth. Met. **8**

Herud, F. (1973): U.S. Patent 3,710,572

Hikami, S., Larkin, A.I., Nagaoka, Y. (1980): Prog. Theor. Phys. **63**, 707

Hirvonen, J.K. (1980): *Ion Implantation, Treatise on Materials Science and Technology*, Vol. 18 (Academic, New York)

Hishiyama, Y., Kaburagi, Y. (1986): In Boehm, H.P. (ed.): Proceedings of the 4th International Carbon Conference, Baden-Baden, FRG (German Ceramics Society)

Hishiyama, Y., Kaburagi, Y., Yoshida, A. (1984): In Inagaki, M. (ed.): *International Symposium on Carbon Fibers* (Toyohashi Inst. of Technology, Toyohashi, Japan) p. 21

Hitchon, J.W., Phillips, D.C. (1979): Fibre Sci. Technol. **12**, 217

Hoarau, J., Volpilhac, G. (1976): Phys. Rev. **B 14**, 4045

Holland, M.G., Klein, C.A., Straub, W.D. (1966): J. Phys. Chem. Solids **27**, 903

Holleyman, C.F. (1986): In *Carbon Fibres*, ed. The Plastics and Rubber Institute, London (Noyes, Park Ridge, NJ) p. 29

Hollingsworth, B.L., Sims, D. (1969): Composites **1**, 80

Hooge, F.N., Kleinpenning, T.G.M., Vandamme, L.K.J. (1981): Rep. Prog. Phys. **44**, 479

Hooker, C.N., Ubbelohde, A.R., Young, D.A. (1963): Proc. R. Soc. London **276**, 83

Hooley, G. (1977): In Leith, R.M.A. (ed.): *Preparation and Crystal Growth of Materials with Layered Structures* (Reidel, Dordrecht) p. 1

Hooley, J.G., Deitz, V.R. (1978): Carbon **16**, 251

Houska, C.R., Warren, B.E. (1954): J. Appl. Phys. **25**, 1503

Huang, Y.Y., Fan, Y.B., Solin, S.A., Zhang, J.M., Eklund, P.C., Heremans, J., Tibbetts, G.G. (1987): Solid State Commun. **64**, 443

Hughes, J.D.H. (1986a): Carbon **24**, 551

Hughes, J.D.H. (1986b): Met. Mater. **2**, 365

Hughes, J.D.H. (1987): J. Phys. D **20**, 276

Hughes, J.D.M., Morley, H., Jackson, E.E. (1980): J. Phys. D **13**, 921

Hutcheon, J.M. (1970): In Blackman, L.C.F. (ed.): *Modern Aspects of Graphite Technology* (Academic, New York) p. 1

Inagaki, M., Muramatsu, K., Maeda, Y., Maekawa, K. (1983): Synth. Met. **8**, 335

Inoue, M. (1962): J. Phys. Soc. Jpn. **17**, 808

Ishitani, A. (1981): Carbon **19**, 269

Issi, J.-P. (1987): In Dresselhaus, M.S. (ed.): *Intercalation in Layered Materials* (Plenum, New York) p. 347

Issi, J.-P., Piraux, L. (1986): Ann. Phys. (Paris), Colloque 2, Suppl. 2, **11**, 165

Issi, J.-P., Boxus, J., Poulaert, B., Heremans, J., Dresselhaus, M.S. (1982): Solid State Commun. **44**, 449

Issi, J.-P., Heremans, J., Dresselhaus, M.S. (1983): Phys. Rev. B **27**, 1333

Izuka, Y., Norita, T., Nishimura, T., Fujisawa, K. (1986): In *Carbon Fibres*, ed. The Plastics and Rubber Institute, London (Noyes, Park Ridge, NJ) p. 14

James, B.W., Wostenholm, G.H., Keen, G.S., McIvor, S.D. (1987): J. Phys. D **20**, 261

JANAF (1969): **168**, 370, 370–1, 370–2 (Dow Chemical Co, Midland, MI 1960–1966) Supplement 32, December 31, 1969

Jansen, H.J.F., Freeman, A.J. (1987): Phys. Rev. B **35**, 8207

Jaworske, D.A., Miller, J.D. (1985): NASA Technical Memorandum 87217 (unpublished)

Jaworske, D.A., Gaier, J.R., Maciag, C., Slabe, M.E. (1987): In Extended Abstracts of the 18th Biennial Conference on Carbon, Worcester, MA (American Carbon Society, University Park, PA) p. 235

Jay-Gerin, J.P., Maynard, R. (1970): J. Low Temp. Phys. **3**, 337

Jiménez-González, H.J., Speck, J.S., Roth, G., Dresselhaus, M.S., Endo, M. (1986): Carbon **24**, 627

Johnson, D.J. (1980): Philos. Trans. R. Soc. London A **294**, 443

Johnson, D.J. (1987a): J. Phys. D **20**, 286

Johnson, D.J. (1987b): Thrower, P.A. (ed.): *Chemistry and Physics of Carbon*, Vol. 20 (Marcel Dekker, New York) p. 1

Johnson, D.J., Tyson, C.N. (1969): J. Phys. D **2**, 787

Johnson, D.J., Tyson, C.N. (1970): J. Phys. D **3**, 526

Johnson, D.J., Tomizuka, I., Watanabe, O. (1975): Carbon **13**, 529

Johnson, J.W. (1969): Appl. Polym. Symp. **9**, 229

Johnson, J.W. (1979): Nature (London) **279**, 142

Johnson, J.W., Thorne, D.J. (1969): Carbon **7**, 659

Johnson, L.G., Dresselhaus, G. (1973): Phys. Rev. B **7**, 2275

Johnston, D.F. (1955): Proc. R. Soc. London A **227**, 349

Jones, B.F., Duncan, R.G. (1971): J. Mater. Sci. **6**, 289

Jones, B.F., Wilkins, B.J.S. (1972): Fibre Sci. Technol. **5**, 315

Jones, J.B., Singer, L.S. (1982): Carbon **20**, 379

Jones, J.B., Barr, J.B., Smith, R.E. (1980): J. Mater. Sci. **15**, 2455

Jones, W.R., Johnson, J.W. (1971): Carbon **9**, 645

Joo, L.A. (1978): U.S. Patent 4,127,634

Jouquet, G., Schill, R. (1971): In Proceedings of the International Conference on Carbon Fibers, Their Composites, and Applications, London. Plastics and Polymer Conference Supplement No. 5 (The Plastics Institute, London) p. 16

Kalnin, I.L., Goldberg, H.A. (1981): Synth. Met. **3**, 159

Kalnin, I.L., Goldberg, H.A. (1983): Synth. Met. **8**, 277

Kalnin, I.L., Goldberg, H.A., Williams, C.C. (1986): U.S. Patent 4,577,979
Kandani, N., Coulon, M., Bonnetain, L. (1984): In Pacault, A., Flandrois, S. (eds.): Extended Abstracts of the International Conference on Carbon, Bordeaux, France, p. 142
Kanter, M.A. (1957): Phys. Rev. **107**, 655
Kawamura, K., Kaneko, S., Tsuzuku, T. (1983): J. Phys. Soc. Jpn. **52**, 3936
Kazumata, Y. (1983): J. Phys. Chem. Solids **44**, 1025
Kazumata, Y., Nakano, Y., Yugo, S., Kimura, T. (1986): Phys. Status Solidi B **136**, 125
Keesom, P.H., Pearlman, N. (1955): Phys. Rev. **99**, 1119
Kelly, A. (1966): *Strong Solids* (Oxford University Press, Oxford) [3rd ed. 1986]
Kelly, B.T. (1967): Carbon **5**, 247
Kelly, B.T. (1969a): The Reactor Group Report: 1733(C) (unpublished)
Kelly, B.T. (1969b): In Walker, Jr., P.L. (ed.): *Chemistry and Physics of Carbon*, Vol. 5 (Marcel Dekker, New York) p. 119
Kelly, B.T. (1981): *Physics of Graphite* (Applied Science, London)
Kelly, B.T., Gilchrist, K.E. (1969): Carbon **7**, 355
Kelly, B.T., Taylor, R. (1973): In Walker, Jr., P.L., Thrower, P.A. (eds.): *Chemistry and Physics of Carbon*, Vol. 10 (Marcel Dekker, New York) p. 1
Kelly, B.T., Martin, W.H., Price, A.M., Bland, J.T. (1966): Philos. Mag. **14**, 343
Kennedy, A. (1960): Proc. Phys. Soc. London **75**, 607
Kent, M., Wolf, K., Memory, J.D., Fornes, R.E., Gilbert, R.D. (1984): Carbon **22**, 103
Kirkpatrick, S. (1973): Rev. Mod. Phys. **45**, 574
Kittel, C. (1958): *Elementary Statistical Physics* (Wiley, New York) p. 141
Kittel, C. (1985): *Introduction to Solid State Physics*, 6th ed. (Wiley, New York)
Klein, C.A. (1962): J. Appl. Phys. **33**, 3388
Klein, C.A. (1964): J. Appl. Phys. **35**, 2947
Klein, C.A. (1966): In Walker, Jr., P.L. (ed.): *Chemistry and Physics of Carbon*, Vol. 2 (Marcel Dekker, New York) p. 217
Knibbs, R.H. (1971): J. Microsc. **94**, 273
Kock, A.J.H.M., de Bokx, P.K., Boellard, E., Klop, W., Geus, J.W. (1985): J. Catal. **96**, 468
Komaki, K. (1980): Ph.D. Thesis, University of Washington, Seattle, WA (unpublished)
Komatsu, K. (1955): J. Phys. Soc. Jpn. **10**, 346
Komatsu, K., Nagamiya, T. (1951): J. Phys. Soc. Jpn. **6**, 438
Koon, N.C., Pehrsson, P., Weber, D., Schindler, A.I. (1984): J. Appl. Phys. **55**, 2497
Kotlensky, W.V., Martens, H.E. (1963): In Mrozowski, S., Studebaker, M.L., Walker, Jr., P.L. (eds.): Proceedings of the 5th Conference on Carbon, University Park, PA, 1961 (Pergamon, New York) p. 625
Koyama, T. (1972): Carbon **10**, 757
Koyama, T., Endo, M. (1982): Kogyo Zairyo **30**, 109
Koyama, T., Onuma, Y. (1963): Oyo Buturi **11**, 857 (in Japanese)
Koyama, T., Endo, M., Onuma, Y. (1972): Jpn. J. Appl. Phys. **11**, 445
Koyama, T., Endo, M., Hishiyama, Y. (1974): Jpn. J. Appl. Phys. **13**, 1933
Krieger, R.B. (1986): In *Carbon Fibres*, ed. The Plastics and Rubber Institute, London (Noyes, Park Ridge, NJ) p. 146
Krumhansl, J., Brooks, H. (1953): J. Chem. Phys. **21**, 1663
Kwizera, P., Dresselhaus, M.S., Uhlmann, D.R., Perkins, J.S., Desper, C.R. (1982): Carbon **20**, 387
Kwizera, P., Dresselhaus, M.S., Dresselhaus, G. (1983): Carbon **21**, 121
Lachter, A., Trinquecoste, M., Delhaes, P. (1985): Carbon **23**, 111
Lalancette, J.M., Roussel, R. (1976): Can. J. Chem. **54**, 3541
Lamicq, P.J. (1984): In Pacault, A., Flandrois, S. (eds.): Extended Abstracts of the International Conference on Carbon, Bordeaux, France, p. 735
Lee, P.A., Ramakrishnan, T.V. (1985): Rev. Mod. Phys. **57**, 287
Leider, H.R., Krikorian, O.H., Young, D.A. (1973): Carbon **11**, 555
Lespade, P., Al-Jishi, R., Dresselhaus, M.S. (1982): Carbon **20**, 427
Li, P., Li, X., Chung, D.D.L. (1987): In Extended Abstracts of the 18th Biennial Con-

ference on Carbon, Worcester, MA (American Carbon Society, University Park, PA) p. 239

Lindhard, J., Scharff, M., Schiøtt, H.E. (1963): Mat.-Fys. Medd. K. Dan. Vidensk. Selsk. **33,** 14

Lipson, H., Stokes, A.R. (1942): Proc. R. Soc. London **A 181,** 93

Lomer, W.M. (1958): Proc. R. Soc. London **A 227,** 300

Lovell, D.R. (1986): In *Carbon Fibres,* ed. The Plastics and Rubber Institute, London (Noyes, Park Ridge, NJ) p. 39

Lutcov, A.I., Volga, V.I., Dymov, B.K. (1970): Carbon **8,** 753

Luxon, B.A. (1986): In *Carbon Fibres,* ed. The Plastics and Rubber Institute, London (Noyes, Park Ridge, NJ) p. 64

Luxon, B.A., Murthy, M.V. (1986): Proc. Soc. of Plastics Eng., 44th Annual Tech. Conf., p. 869

MacFarlane, J.M., McLintock, I.S., Orr, J.C. (1970). Phys. Status Solidi A **3,** K239

Maeda, M., Kuramoto, Y., Horie, C. (1979): J. Phys. Soc. Jpn. **47,** 337

Maeda, Y., Kitamura, H., Itoh, E., Inagaki, M. (1983): Synth. Metals **7,** 211

Maire, J., Mering, J. (1958): Proc. First Conference of the Society of Chemical Ind. Conf. on Carbon and Graphite (London), 204

Manini, C., Marêché, J.-F., McRae, E. (1983): Synth. Met. **8,** 261

Manini, C., McRae, E., Marêché, J.-F., Hérold, A. (1985): Carbon **23,** 465

Manocha, L.M., Bahl, O.P. (1982): Fibre Sci. Technol. **17,** 221

Marshik, B., Apple, T., Meyer, D., Wagoner, G., Woollam, J.A. (1986): In Dresselhaus, M.S., Dresselhaus, G., Solin, S.A. (eds.): Extended Abstracts of the 1986 MRS Symposium on Graphite Intercalation Compounds, Boston (Materials Research Society, Pittsburgh, PA) p. 120

Martin, W.H., Brocklehurst, J.E. (1964): Carbon **1,** 133

Matsubara, K., Kawamura, K., Tsuzuku, T. (1986): Jpn. J. Appl. Phys. **25,** 1016

Matsumura, K., Takahashi, A., Tsukamoto, J. (1985): Synth. Met. **11,** 9

Mayer, J.W., Eriksson, L., Davies, J.A. (1970): *Ion Implantation of Semiconductors* (Academic, New York)

Mazieres, C., Colin, G., Jegoudez, J., Setton, R. (1975): Carbon **13,** 289

McClure, J.W. (1956): Phys. Rev. **104,** 666

McClure, J.W. (1957): Phys. Rev. **108,** 612

McClure, J.W. (1960): Phys. Rev. **119,** 606

McClure, J.W. (1969): Carbon **7,** 425

McClure, J.W. (1971): In Carter, D.L., Bate, R.T. (eds.): Proceedings of the International Conference on Semimetals and Narrow Gap Semiconductors (Pergamon, New York) p. 127

McClure, J.W. (1980): Private communication

McClure, J.W., Hickman, B.B. (1982): Carbon **20,** 373

McClure, J.W., Ruvalds, J. (1964): Unpublished

McClure, J.W., Smith, L.B. (1963): In Mrozowski, S., Studebaker, M.L., Walker, Jr., P.L. (eds.): Proceedings of the 5th Conference on Carbon, University Park, PA, 1961 (Pergamon, New York) p. 3

McClure, J.W., Spry, W.J. (1968): Phys. Rev. **165,** 809

McClure, J.W., Yafet, Y. (1963): In Mrozowski, S., Studebaker, M.L., Walker, Jr., P.L. (eds.): Proceedings of the 5th Conference on Carbon, University Park, PA, 1961 (Pergamon, New York) p. 22

McKague, E.L., Bullock, R.E., Head, J.W. (1973): J. Compos. Mater. **7,** 288

McKee, D.W. (1986): Carbon **24,** 737

McKee, D.W. (1987): In Extended Abstracts of the 18th Biennial Conference on Carbon, Worcester, MA (American Carbon Society, University Park, PA) p. 448

McKee, D.W., Mimeault, V.J. (1973): In Walker, Jr., P.L., Thrower, P.A. (eds.): *Chemistry and Physics of Carbon,* Vol. 8 (Marcel Dekker, New York) p. 151

McMahon, P.E. (1973): ASTM STP 521 (American Society for Testing and Materials, Philadelphia) p. 367

McMahon, P.E. (1978): In Vinson, J.R. (ed.): *Advanced Composite Materials – En-*

*vironmental Effects*, ASTM STP 658 (American Society for Testing and Materials, Philadelphia) pp. 254–266

McNeil, L.E., Elman, B.S., Dresselhaus, M.S., Dresselhaus, G., Venkatesan, T. (1984): In Hubler, G.K., Holland, O.W., Clayton, C.R., White, C.W. (eds.): MRS Symposium on Ion Implantation and Ion Beam Processing of Materials, Boston 1983 (North-Holland, New York) p. 493

McNeil, L.E., Steinbeck, J., Salamanca-Riba, L., Dresselhaus, G. (1985): Phys. Rev. **B 31**, 2451

McNeil, L.E., Steinbeck, J., Salamanca-Riba, L., Dresselhaus, G. (1986): Carbon **24**, 73

McQuillan, B., Reynolds, G.H. (1986): In Dresselhaus, M.S., Dresselhaus, G., Solin, S.A. (eds.): Extended Abstracts of the 1986 MRS Symposium on Graphite Intercalation Compounds, Boston (Materials Research Society, Pittsburgh, PA) p. 138

Mendez, E., Misu, A., Dresselhaus, M.S. (1980): Phys. Rev. **B 21**, 827

Menjo, H., Elman, B.S., Braunstein, G., Dresselhaus, M.S. (1984): J. Chim. Phys. **81**, 835

Meschi, C., Manceau, J.P., Flandrois, S., Delhaès, P., Ansert, A., Deschamps, L. (1986): Ann. Phys. (Paris) Colloque 2, Suppl. 2, **11**, 199

Millward, G.R., Thomas, J.M. (1979): Carbon **17**, 1

Misu, A., Mendez, E., Dresselhaus, M.S. (1979): J. Phys. Soc. Jpn. **47**, 199

Moore, A.W. (1973): In Walker, Jr., P.L., Thrower, P.A. (eds.): *Chemistry and Physics of Carbon*, Vol. 11 (Marcel Dekker, New York) p. 69

Morelli, D.T., Uher, C. (1985): Phys. Rev. **B 31**, 6721

Moreton, R. (1969): Fibre Sci. Technol. **1**, 273

Moreton, R. (1976): Ph.D. Thesis, University of Surrey, Great Britain (unpublished)

Moreton, R., Watt, W. (1974): Nature (London) **247**, 360

Moreton, R., Watt, W., Johnson, W. (1967): Nature (London) **213**, 690

Morgan, W.C. (1972): Carbon **10**, 73

Mott, N.F. (1967): Adv. Phys. **16**, 49

Mrozowski, S. (1952): Phys. Rev. **85**, 609; ibid. **86**, 1056 (erratum)

Mrozowski, S. (1971): Carbon **9**, 97

Mrozowski, S. (1979): J. Low Temp. Phys. **35**, 231

Murata, M. (1982): Master's Thesis, Institute of Materials Science, University of Tsukuba

Murday, J.M., Dominguez, D.D., Moran, Jr., J.A., Lee, W.D., Eaton, R. (1984): Synth. Met. **9**, 397

Nabarro, T.K.N. (1967): *The Theory of Dislocations in Crystals* (Oxford University Press, Oxford)

Nagayoshi, H. (1977): J. Phys. Soc. Jpn. **43**, 760

Nagayoshi, H., Tsukada, M., Nakao, K., Uemura, Y. (1973): J. Phys. Soc. Jpn. **35**, 396

Nakajima, T., Kawaguchi, M., Watanabe, N. (1983): Synth. Met. **7**, 117

Nakajima, T., Watanabe, N., Kameda, I., Endo, M. (1986): Carbon **24**, 343

Natarajan, V., Woollam, J.A. (1983): In Dresselhaus, M.S., Dresselhaus, G., Fischer, J.E., Moran, M.J. (eds.): Proceedings of the Symposium on Intercalated Graphite, Vol. 20 (North-Holland, New York) p. 45

Natarajan, V., Woollam, J.A., Yavrouian, A. (1983): Synth. Met. **8**, 291

Natishan, P.N., Cahen, Jr., G.A., Stoner, G.E. (1987): Ind. Eng. Chem. Res. **26**, 125

Nemanich, R.J., Solin, S.A. (1979): Phys. Rev. **B 20**, 392

Nicholson, A.P.P., Bacon, D.J. (1975): Carbon **13**, 275

Nicklow, R., Wakabayashi, N., Smith, H.G. (1972): Phys. Rev. **B 5**, 4951

Nigrey, P.J., MacInnes, Jr., D., Nairns, D.P., MacDiarmid, A.G. (1981): J. Electrochem. Soc. **128**, 1651

Nishina, Y., Tanuma, S., Myron, H.W. (eds.) (1981): Proceedings of the Yamata 4th Conference on the Physics and Chemistry of Layered Materials, Sendai, Japan. Physica **105 B**

Nogami, T., Nawa, M., Mikawa, H. (1982): J. Chem. Soc., Chem. Commun. 1158

Null, M.R., Lozier, W.W., Moore, A.W. (1973): Carbon **11**, 81

Nye, J.F. (1960): *Physical Properties of Crystals* (Oxford University Press, Oxford) p. 134

Nysten, B., Piraux, L., Issi, J.-P. (1985a): J. Phys. **D 18**, 1307

Nysten, B., Piraux, L., Issi, J.-P. (1985b): Proceedings of the 19th International Thermal Conductivity Conference, Cookeville, TN (Plenum, New York)

Oberlin, A. (1979): Carbon **17**, 7

Oberlin, A. (1984): Carbon **22**, 521

Oberlin, A., Endo, M., Koyama, T. (1976a): Carbon **14**, 133

Oberlin, A., Endo, M., Koyama, T. (1976b): J. Cryst. Growth **32**, 335

Ohhashi, K., Amiell, J., Delhaès, P., Marêché, J.F., Guérard, D., Endo, M. (1983): In Extended Abstracts of the 16th Biennial Conference on Carbon, San Diego, CA (American Carbon Society, University Park, PA) p. 271

Ohhashi, K., Ajiro, Y., Endo, M. (1986): In Annual Meeting of the Japan Physical Society, page FB6 (in Japanese)

Ohshima, S., Dietrich, M., Linker, G. (1984): J. Appl. Phys. **44**, 1360

Olsen, L.C., Seeman, S.E., Scott, H.W. (1970): Carbon **8**, 85

Ono, S., Sugihara, K. (1966): J. Phys. Soc. Jpn. **21**, 861

Orowan, E. (1949): Rep. Prog. Phys. **12**, 185

Oshima, H., Woollam, J.A., Yavrouian, A., Dowell, M.B. (1983): Synth. Met. **5**, 113

Otani, S., Kokubo, Y., Koitobashi, T. (1970): Bull. Chem. Soc. Jpn. **43**, 3291

Owston, C.N. (1970): J. Phys. D **3**, 1615

Pacault, A., Marchand, A. (1983): Carbon **21**, 367, which contains many references to various aspects of carbon research in French laboratories

Painter, G.S., Ellis, D.E. (1970): Phys. Rev. B **1**, 4747

Palchan, I., Talinaker, M. (1987): In Dresselhaus, M.S. (ed.): *Intercalation in Layered Materials* (Plenum, New York) p. 385

Parratt, N.J. (1972): *Fibre-Reinforced Materials Technology* (Van Nostrand Reinhold, London)

Pedersen, N.E., Pedersen, J.C. (1965): In AVCO RADTN-65-67 (secret) (unpublished)

Pedersen, N.E., Pedersen, J.C., Bethe, H.A. (1969): In Proc. of Tri-Services Radar Symposium, San Diego (secret) (unpublished)

Pedersen, N.E., Waterman, P.C., Pedersen, J.C. (1985): In Kohl, R.H. (ed.): Proceedings of the 1985 CRDC Scientific Conference on Obscuration and Aerosol Research, p. 441

Pedersen, N.E., Waterman, P.C., Pedersen, J.C. (1987): Private communication

Peierls, R. (1929): Ann. Phys. (Leipzig) **3**, 1055

Pelabon, C.H. (1905): C.R. Acad. Sci. **137**, 706

Pepper, R., Nelson, D., Jarman, D., Hotham, J. (1978): Final Report to US Army Mobility Equipment Res. & Dev. Comm., Contract No. DAAK70-77-C-0155 (unpublished)

Perret, R., Ruland, W. (1969): J. Appl. Crystallogr. **2**, 209

Perret, R., Ruland, W. (1970): J. Appl. Crystallogr. **3**, 525

Perry, A.J., Phillips, K., de Lamotte, E. (1971): Fibre Sci. Technol. **3**, 317

Picraux, S.T. (1984): Phys. Today, November, p. 38

Picraux, S.T., Peercy, P.S. (1985): Sci. Am., March, p. 102

Pietronero, L., Tosatti, E. (eds.) (1981): *Physics of Intercalation Compounds*, Springer Ser. Solid-State Sci., Vol. 38 (Springer, Berlin, Heidelberg)

Pifer, J.H., Longo, R.T. (1971): Phys. Rev. B **4**, 3797

Pike, G.E., Pierson, H.O., Mullendore, A.W., Schirber, J.E. (1976): Appl. Polym. Symp. **29**, 71

Piraux, L. (1987): In Dresselhaus, M.S. (ed.): *Intercalation in Layered Materials* (Plenum, New York) p. 375

Piraux, L., Nysten, B., Haquenne, A., Issi, J.-P., Dresselhaus, M.S., Endo, M. (1984): Solid State Commun. **50**, 697

Piraux, L., Nysten, B., Issi, J.-P., Marêché, J.F., McRae, E. (1985): Solid State Commun. **55**, 517

Piraux, L., Bayot, V., Michenaud, J.-P., Issi, J.-P., Marêché, J.F., McRae, E. (1986a): Solid State Commun. **59**, 711

Piraux, L., Issi, J.-P., Salamanca-Riba, L., Dresselhaus, M.S. (1986b): Synth. Met. **16**, 93

Piraux, L., Nysten, B., Issi, J.-P., Salamanca-Riba, L., Dresselhaus, M.S. (1986c): Solid State Commun. **58**, 265

Piraux, L., Bayot, V., Gonze, X., Michenaud, J.-P., Issi, J.-P. (1987): Phys. Rev. **B 36,** 9045

Polanyi, G. (1921): Z. Phys. **7,** 323

Pompea, R.E., Bergener, D.W., Shepard, D.F., Russek, S. (1984): Opt. Eng. **23,** 149

Porod, G. (1951): Kolloid-Z. **124,** 83

Porod, G. (1952): Kolloid-Z. **125,** 51

Price, R.J. (1965): Philos. Mag. **12,** 564

Price, R.J. (1974): Carbon **12,** 159

Price, R.J., Hopkins, G.R., Engle, G.B. (1985): In Extended Abstracts of the 17th Biennial Conference on Carbon, Lexington, KY (American Carbon Society, University Park, PA) p. 348

Proctor, A.M., Sherwood, P.A. (1983): Carbon **21,** 53

Qian, X.W., Solin, S.A., Gaier, J.R. (1987): In Dresselhaus, M.S. (ed.): *Intercalation in Layered Materials* (Plenum, New York) p. 477

Rahim, I., Sugihara, K., Dresselhaus, M.S., Heremans, J. (1986): Carbon **24,** 663

Resing, H.A., Milliken, J., Dominguez, D.D., Elam, W.T. (1985): In Extended Abstracts of the 17th Biennial Conference on Carbon, Lexington, KY (American Carbon Society, University Park, PA) p. 84

Reynolds, W.N. (1971): In Gregory, J.G., Rawlins, C.J. (eds.): 3rd Conference on Industrial Carbons and Graphite, London (Society of Chemical Industries, London) p. 427

Reynolds, W.N. (1973): In Walker, Jr., P.L., Thrower, P.A. (eds.): *Chemistry and Physics of Carbon,* Vol. 11 (Marcel Dekker, New York) p. 1

Reynolds, W.N., Moreton, R. (1980): Philos. Trans. R. Soc. **A 294,** 451

Reynolds, W.N., Sharp, J.V. (1974): Carbon **12,** 103

Riggs, J.P. (1985): In *Encyclopedia of Polymer Science and Engineering,* Vol. 2 (Wiley, New York) p. 640

Riley, D.P. (1945): Proc. Phys. Soc. London **57,** 486

Robertson, J. (1986): Adv. Phys. **35,** 317

Robertson, J. (1987): Synth. Met., in press

Robertson, S.D. (1970): Carbon **8,** 365

Robertson, S.D. (1972): Carbon **10,** 221

Robinson, G. (1982): British Patent 2,095,027

Robson, D., Assabghy, F.Y.I., Ingram, D.J.E. (1971): J. Phys. D **4,** 1426

Robson, D., Assabghy, F.Y.I., Ingram, D.J.E. (1972): J. Phys. D **5,** 169

Robson, D., Assabghy, F.Y.I., Cooper, E.G., Ingram, D.J.E. (1973): J. Phys. D **6,** 1822

Roscoe, C., Thomas, J.M. (1966): Proc. R. Soc. London **A 297,** 397

Rouzaud, J.N., Oberlin, A., Beny-Bassez, C. (1983): Thin Solid Films **105,** 75

Rowe, C.R., Lowe, D.L. (1977): In Extended Abstracts of the 13th Biennial Conference on Carbon, Irvine, CA (American Carbon Society, University Park, PA) p. 170

Rüdorff, W., Schulze, E. (1954): Z. Anorg. Allg. Chem. **277,** 156

Ruland, W. (1965): Acta Crystallogr. **18,** 992

Ruland, W. (1967): J. Appl. Phys. **38,** 3585

Ruland, W. (1968): In Walker, Jr., P.L., Thrower, P.A. (eds.): *Chemistry and Physics of Carbon,* Vol. 4 (Marcel Dekker, New York) p. 1

Ruland, W. (1969a): Summary of Papers of the 9th Conference on Carbon, Boston, MA (American Carbon Society, University Park, PA) p. 72

Ruland, W. (1969b): Appl. Polym. Symp. **9,** 293

Ruland, W., Tompa, H. (1968): Acta Crystallogr. **A 24,** 93

Ruyter, I.E., Ekstrand, K. (1987): J. Phys. D **20,** 303

Saito, M.K., Kogo, Y.S. (1980): U.S. Patent 4,197,279

Sakata, H., Dresselhaus, G., Endo, M. (1987): In Extended Abstracts of the 18th Biennial Conference on Carbon, Worcester, MA (American Carbon Society, University Park, PA) p. 18

Sakata, H., Dresselhaus, G., Dresselhaus, M.S., Endo, M. (1988): J. Appl. Phys. **23,** 467

Salamanca-Riba, L. (1985): Ph.D. Thesis, Massachusetts Institute of Technology, Department of Physics (unpublished)

354

Salamanca-Riba, L., Dresselhaus, M.S. (1986): Carbon **24**, 261
Salamanca-Riba, L., Braunstein, G., Dresselhaus, M.S., Gibson, J.M., Endo, M. (1985):
  Nucl. Instrum. Methods Phys. Res. **B 7/8**, 487, IBMM Conference, July 16–20, 1984,
  Cornell University
Salamanca-Riba, L., Yeh, N.C., Dresselhaus, M.S., Endo, M., Enoki, T. (1986): J. Mater.
  Res. **1**, 177
Samuelson, L., Batra, I.P., Roetti, C. (1980): Solid State Commun. **33**, 817
Schützenberger, P., Schützenberger, L. (1890): C.R. Acad. Sci. **111**, 774
Sears, G.W. (1959): J. Appl. Phys. **30**, 358
Seldin, E.S., Nezbeda, C.W. (1970): J. Appl. Phys. **41**, 3389
Setton, R. (1988): In Selig, H., Davidov, D. (eds.): Proceedings of the 4th International
  Symposium on Graphite Intercalation Compounds, Jerusalem. Synth. Met., **23**, 511
Shacklette, L.W., Elsenbaumer, R.L., Chance, R.R., Sowa, J.M., Ivory, D.M., Miller,
  G.G., Baughman, R.H. (1982): J. Chem. Soc., Chem. Commun. 361
Sharma, M.P., Johnson, L.G., McClure, J.W. (1974): Phys. Rev. **B 9**, 2467
Sharp, J.V., Burney, S.G. (1971): In Proceedings of the International Conference on
  Carbon Fibers, Their Composites, and Applications, London. Plastics and Polymer
  Conference Supplement No. 5 (The Plastics Institute, London) p. 10
Sheaffer, P.M. (1987): In Extended Abstracts of the 18th Biennial Conference on Carbon,
  Worcester, MA (American Carbon Society, University Park, PA) p. 20
Sheehan, J.E. (1987): In *Recent Research into Carbon Composites*, Proc. of the Fourth
  Annual Materials Technology Center Conf., in press
Shindo, A. (1961a): Osaka Kogyo Gijitsu Shikuko **12**, 110, 119
Shindo, A. (1961b): Rep. Govt. Ind. Res. Inst., Osaka 317
Shindo, A. (1964): Carbon **1**, 391
Shioya, J., Matsubara, H., Murakami, S. (1986): Synth. Met. **14**, 113
Shioya, J., Yamaguchi, Y., Mizoguchi, A., Ueba, Y., Matsubara, H. (1987): In Extended
  Abstracts of the 18th Biennial Conference on Carbon, Worcester, MA (American
  Carbon Society, University Park, PA) p. 360
Shklovskii, B.I. (1978): Phys. Status Solidi B **85**, K111
Sinclair, D. (1950): J. Appl. Phys. **21**, 380
Singer, L.S. (1978): Carbon **16**, 409
Singer, L.S., Wagoner, G. (1962): J. Chem. Phys. **37**, 1812
Sittig, M. (1980): *Carbon and Graphite Fibers – Manufacture and Applications* (Noyes,
  Park Ridge, NJ)
Slonczewski, J.C., Weiss, P.R. (1958): Phys. Rev. **109**, 272
Smith, G.W. (1984): Carbon **22**, 477
Smith, R.E. (1972): J. Appl. Phys. **43**, 255
Smith, W.D., Lin, R.Y., Economy, J. (1976): Appl. Polym. Symp. **29**, 83
Solberg, W.A. (1986): Master's Thesis, Colorado State University (unpublished)
Solberg, W.A., Spain, I.L., Van Vechten, J.A., Pedersen, N.E. (1987): In Extended Ab-
  stracts of the 18th Biennial Conference on Carbon, Worcester, MA (American Carbon
  Society, University Park, PA) p. 215
Solin, S.A. (1982): Adv. Chem. Phys. **49**, 455
Sondheimer, E.H. (1952): Adv. Phys. **1**, 1
Soule, D.E. (1958): Phys. Rev. **112**, 698
Soule, D.E., McClure, J.W., Smith, L.B. (1964): Phys. Rev. **A 134**, 453
Soulen, J.R., Sthapitanonda, P., Margrave, J.L. (1955): J. Phys. Chem. **59**, 132
Spain, I.L. (1973): In Walker, Jr., P.L., Thrower, P.A. (eds.): *Chemistry and Physics of
  Carbon*, Vol. 8 (Marcel Dekker, New York) p. 1
Spain, I.L. (1981): In Walker, Jr., P.L., Thrower, P.A. (eds.): *Chemistry and Physics of
  Carbon*, Vol. 16 (Marcel Dekker, New York) p. 119
Spain, I.L. (1987): Unpublished
Spain, I.L., Ubbelohde, A.R., Young, D.A. (1964): J. Chem. Soc. **180**, 920
Spain, I.L., Ubbelohde, A.R., Young, D.A. (1967): Philos. Trans. R. Soc. London **A 262**,
  345
Spain, I.L., Volin, K.J., Goldberg, H.A., Kalnin, I. (1983a): J. Phys. Chem. Solids **44**,
  839

Spain, I.L., Volin, K.J., Goldberg, H.A., Kalnin, I. (1983b): Solid State Commun. **45,** 817

Spear, K.E. (1982): Pure Appl. Chem. **54,** 1297

Srinivasagopal, S. (1979): Ph.D. Thesis, University of Washington, Seattle. Available from University Microfilms, University of Michigan (unpublished)

Stafford, G.R. (1980): Ph.D. Thesis, University of Virginia (unpublished)

Stafford, G.R., Cahen, G.A., Stoner, G.E. (1987): Unpublished

Stamatoff, T., Goldberg, H.A., Kalnin, I.L. (1983): In Dresselhaus, M.S., Dresselhaus, G., Fischer, J.E., Moran, M.J. (eds.): Proceedings of the Symposium on Intercalated Graphite, Vol. 20 (North-Holland, New York) p. 51

Standage, A.E., Prescott, R. (1966): Nature (London) **211,** 169

Stanley, H.E. (1971): *Introduction to Phase Transitions and Critical Phenomena* (Oxford University Press, Oxford)

Stanley, H.E. (1984): In Family, F., Landau, D.P. (eds.): International Topical Conference on Kinetics of Aggregation and Gelation (North-Holland, New York)

Steinbeck, J., Dresselhaus, M.S., Dresselhaus, G., Venkatesan, T. (1986): In Dresselhaus, M.S., Dresselhaus, G., Solin, S.A. (eds.): Extended Abstracts of the 1986 MRS Symposium on Graphite Intercalation Compounds, Boston (Materials Research Society, Pittsburgh, PA) p. 129

Sugihara, K. (1970): J. Phys. Soc. Jpn. **29,** 1465

Sugihara, K. (1983a): In Dresselhaus, M.S., Dresselhaus, G., Fischer, J.E., Moran, M.J. (eds.): Proceedings of the Symposium on Intercalated Graphite, Vol. 20 (North-Holland, New York) p. 157

Sugihara, K. (1983b): Phys. Rev. **B 28,** 2157

Sugihara, K. (1984a): Phys. Rev. **B 29,** 5872

Sugihara, K. (1984b): Phys. Rev. **B 29,** 6722

Sugihara, K. (1984c): J. Phys. Soc. Jpn. **53,** 393

Sugihara, K. (1984d): In Eklund, P.C., Dresselhaus, M.S., Dresselhaus, G. (eds.): Extended Abstracts of the 1984 MRS Symposium on Graphite Intercalation Compounds, Boston (Materials Research Society, Pittsburg, PA) p. 60

Sugihara, K. (1986a): In Dresselhaus, M.S., Dresselhaus, G., Solin, S.A. (eds.): Extended Abstracts of the 1985 MRS Symposium on Graphite Intercalation Compounds, Boston (Materials Research Society, Pittsburgh, PA) p. 132

Sugihara, K. (1986b): Unpublished

Sugihara, K. (1987): Unpublished

Sugihara, K., Dresselhaus, M.S. (1986): In Dresselhaus, M.S., Dresselhaus, G., Solin, S.A. (eds.): Extended Abstracts of the 1986 MRS Symposium on Graphite Intercalation Compounds, Boston (Materials Research Society, Pittsburgh, PA) p. 135

Sugihara, K., Dresselhaus, M.S. (1987): In Extended Abstracts of the 18th Biennial Conference on Carbon, Worcester, MA (American Carbon Society, University Park, PA) p. 421

Sugihara, K., Sato, H. (1963): J. Phys. Soc. Jpn. **18,** 332

Sugihara, K., Woollam, J.A. (1978): J. Phys. Soc. Jpn. **45,** 1891

Sugihara, K., Takezawa, T., Tsuzuku, T., Hishiyama, Y., Ono, A. (1972): J. Phys. Chem. Solids **33,** 1475

Sugihara, K., Matsubara, K., Tsuzuku, T. (1984): J. Phys. Soc. Jpn. **53,** 795

Sugihara, K., Hishiyama, Y., Ono, A. (1986): Phys. Rev. **B 34,** 4298

Sun, T., McClure, J.W. (1983): In Extended Abstracts of the 16th Biennial Conference on Carbon, San Diego, CA (American Carbon Society, University Park, PA) p. 231

Taborek, P. (1987): Bull. Am. Phys. Soc. **32,** 489

Tahar, M.Z. (1985): MS Thesis, Massachusetts Institute of Technology, Department of Physics

Tahar, M.Z., Dresselhaus, M.S., Endo, M. (1986): Carbon **24,** 67

Takagi, H., Uchida, S., Kishio, K., Kitazawa, K., Fueki, K., Tanaka, S. (1987): Jpn. J. Appl. Phys. **26,** L320

Takahagai, T., Ishitani, A. (1984): Carbon **22,** 43

Takezawa, T., Tsuzuku, T., Ono, A., Hishiyama, Y. (1969): Philos. Mag. **19,** 623

Takezawa, T., Tsuzuku, T., Ono, A., Hishiyama, Y. (1971): Philos. Mag. **23,** 1241
Tamarin, P.V., Shalyt, S.S., Volga, V.I. (1969): Fiz. Tverd. Tela **11,** 1725 [English transl.: Sov. Phys. − Solid State **11,** 1399]
Tanabe, Y., Yasuda, E., Machino, H., Kimura, S. (1987): In Extended Abstracts of the Annual Meeting of the Japan Ceramic Society, Nagoya (in Japanese) p. 77
Tanaka, T., Suzuki, H. (1972): Carbon **10,** 253
Tang, M.-Y., Rice, G.G., Fellers, J.F., Lin, J.S. (1986): J. Appl. Phys. **60,** 803
Tansosen-i Konwakai 1986, Symposium on Carbon Fibers (CMC, Tokyo)
Terai, T., Takahashi, Y. (1984): Carbon **22,** 91
Tesner, P.A., Robinovich, E.Y., Rafalkes, I.S., Arefieva, E.F. (1970): Carbon **8,** 435
Thomas, J.M., Roscoe, C. (1968): In Walker, Jr., P.L. (ed.): *Chemistry and Physics of Carbon*, Vol. 3 (Marcel Dekker, New York) p. 1
Thomy, A., Duval, X. (1969): J. Chim. Phys. **66,** 1966
Thrower, P.A. (1969): In Walker, Jr., P.L. (ed.): *Chemistry and Physics of Carbon*, Vol. 5 (Marcel Dekker, New York) p. 217
Thrower, P.A., Mayer, R.M. (1978): Phys. Status Solidi **A 47,** 11
Tibbetts, G.G. (1983): Appl. Phys. Lett. **42,** 666
Tibbetts, G.G. (1984): J. Cryst. Growth **66,** 632
Tibbetts, G.G. (1985): General Motors Research Report No. GMR-5063 (unpublished)
Tibbetts, G.G. (1987): In Extended Abstracts of the 18th Biennial Conference on Carbon, Worcester, MA (American Carbon Society, University Park, PA) p. 157
Tibbetts, G.G., Beetz, Jr., C.P. (1986): General Motors Research Report No. GMR-5550 (unpublished)
Tibbetts, G.G., Beetz, Jr., C.P. (1987): J. Phys. **D 20,** 292
Tibbetts, G.G., Beetz, Jr., C.P., Olk, C.H. (1986a): General Motors Research Report No. GMR-5624 (unpublished)
Tibbetts, G.G., Endo, M., Beetz, Jr., C.P. (1986b): SAMPE J. **22−5,** 18th International SAMPE Technical Conference, Seattle, WA
Tibbetts, G.G., Devour, M.G., Rodda, E.J. (1987): Carbon **25,** 367
Tillgner, H., Ruland, W. (1987): In Extended Abstracts of the 18th Biennial Conference on Carbon, Worcester, MA (American Carbon Society, University Park, PA) p. 28
Timp, G., Dresselhaus, M.S. (1984): J. Phys. **C 17,** 2641
Touhara, H., Kadono, K., Watanabe, N., Endo, M. (1984a): J. Chim. Phys. **81,** 841
Touhara, H., Fujimoto, H., Watanabe, N., Tressaud, A. (1984b): Solid State Ionics **14,** 163
Touhara, H., Fujimoto, H., Kadono, K., Watanabe, N., Endo, M. (1987a): Electrochim. Acta **32,** 293
Touhara, H., Kadono, K., Watanabe, N., Braconnier, J.-J. (1987b): J. Electrochem. Soc. **134,** 1071
Touloukian, Y.S., Kirby, B.K., Taylor, E., Lee, T.Y.R. (1979): *Thermal Expansion*, The TPRC Data Series, Vol. 13 (Plenum, New York)
Touzain, Ph., Bagga, G.R. (1984): In Inagaki, M. (ed.): *International Symposium on Carbon Fibers* (Toyohashi Institute of Technology, Toyohashi, Japan) p. 91
Touzain, Ph., Bagga, G.R., Blazewicz, S. (1987): Synth. Met. **18,** 565
Toy, W.W., Dresselhaus, M.S., Dresselhaus, G. (1977): Phys. Rev. **B 15,** 4077
Tsukamoto, J., Matsumura, K., Takahashi, A., Sakado, K. (1986) Synth. Met. **13,** 255
Tsuzuku, T. (1961): J. Phys. Soc. Jpn. **16,** 407
Tsuzuku, T., Sugihara, K. (1975): In Walker, Jr., P.L., Thrower, P.A. (eds.): *Chemistry and Physics of Carbon*, Vol. 12 (Marcel Dekker, New York) p. 109
Tuinstra, F., Koenig, J.L. (1970): J. Chem. Phys. **53,** 1126
Ubbelohde, A.R. (1969): Carbon **7,** 523
Uchida, S., Takagi, H., Kitazawa, K., Tanaka, S. (1987): Jpn. J. Appl. Phys. **26,** L1
Uemura, Y., Inoue, M. (1958): J. Phys. Soc. Jpn. **13,** 382
Uher, C., Sander, L.M. (1983): Phys. Rev. **B 27,** 1326
van der Hoeven, B.J.C., Keesom, P.H. (1963): Phys. Rev. **130,** 1318
Van Haeringen, W., Junginger, H.G. (1969): Solid State Commun. **7,** 1723

357

Van Vechten, J.A. (1985): J. Cryst. Growth **71**, 326

Van Vechten, J.A., Solberg, W., Batson, P.E., Cuomo, J.J., Rossnagel, S.M. (1987): J. Cryst. Growth **82**, 289

Venkatesan, T., Elman, B.S., Braunstein, G., Dresselhaus, M.S., Dresselhaus, G. (1984): J. Appl. Phys. **56**, 3232

Vidno, R.P., Fischbach, D.B. (1981): In Extended Abstracts of the 15th Biennial Conference on Carbon, Philadelphia, PA (American Carbon Society, University Park, PA) p. 468

Voet, A., Morawski, J.C. (1975): In Extended Abstracts of the 12th Biennial Conference on Carbon, Pittsburgh, PA (American Carbon Society, University Park, PA) p. 87

Vogel, F.L. (1976): Carbon **14**, 175

Vogel, F.L. (ed.) (1980): Proceedings of the Second Conference on Intercalation Compounds of Graphite, Provincetown, MA 1980. Synth. Met. **2** and **3**

Vogel, F.L., Hérold, A. (ed.) (1977): Proceedings of the Conference on Intercalation Compounds of Graphite (Elsevier Sequoia, Lausanne)

Vogel, F.L., Foley, G.M.T., Zeller, C., Falardeau, E.R., Gan, J. (1977): J. Mater. Sci. Eng. **31**, 264

Volpilhac, G., Hoarau, J. (1978): Phys. Rev. **B 17**, 1445

Wada, N., Gaczi, P.J., Solin, S.A. (1980): J. Non-Cryst. Solids **35–36**, 543

Wagoner, G. (1960): Phys. Rev. **118**, 647

Wagoner, G., Smith, R.E., Bacon, R. (1987): In Extended Abstracts of the 18th Biennial Conference on Carbon, Worcester, MA (American Carbon Society, University Park, PA) p. 415

Walker, N.J. (1986): In *Carbon Fibres*, ed. The Plastics and Rubber Institute, London (Noyes, Park Ridge, NJ) p. 55

Wallace, P.R. (1947): Phys. Rev. **71**, 622

Waltersson, K. (1982): Fibre Sci. Technol. **17**, 289

Warren, B.E., Bodenstein, P. (1966): Acta Crystallogr. **20**, 602

Watt, W., Johnson, W. (1969): Appl. Polym. Symp. **9**, 215

Watt, W., Johnson, W. (1971): In Gregory, J.G., Rawlins, C.J. (eds.): 3rd Conference on Industrial Carbons and Graphite, London (Society of Chemical Industries, London) p. 417

Watt, W., Philips, L.N., Johnson, W. (1966): The Engineer **221**, 815

Watts, A.A. (ed.) (1985): ASTM Special Publication 704 (American Society for Testing and Materials, Philadelphia, PA)

Weibull, W. (1951): J. Appl. Mech. **18**, 293

White, J.L., Buechler, M. (1985): In Extended Abstracts of the 17th Biennial Conference on Carbon, Lexington, KY (American Carbon Society, University Park, PA) p. 344

Whitney, E., Kimmel, R.M. (1972): Nature (London) **237**, 93

Wicks, B.J., Coyle, R.A. (1976): J. Mater. Sci. **11**, 376

Williams, W.S., Steffens, D.A., Bacon, R. (1970): J. Appl. Phys. **41**, 4893

Woodrow, J., Mott, B.W., Haines, J.H.R. (1965): Proc. Phys. Soc. London **B 65**, 603

Woolf, L.D., Chin, J., Lin-Liu, Y.R., Ikezi, H. (1984): Phys. Rev. **B 30**, 861

Woollam, J.A. (1987): Private communication

Woollam, J.A., Chang, H., Nafis, S., Sellmyer, D.J. (1986a): Private communication

Woollam, J.A., Natarajan, V., Brandt, B. (1986b): Appl. Phys. Commun. **6**, 161

Wu, C.T.D., McCullough, R.L. (1977): In Holister, G.S. (ed.): *Developments in Composite Materials* (Applied Science, Englewood, NJ) p. 119

Wu, M.K., Ashburn, J.R., Torgn, C.J., Hor, P.H., Meng, R.L., Gao, L., Huang, Z.J., Huang, Y.Q., Chu, C.W. (1987): Phys. Rev. Lett. **58**, 908

Yasuda, E., Tanabe, Y., Machino, H., Takaku, A. (1987): In Extended Abstracts of the 18th Biennial Conference on Carbon, Worcester, MA (American Carbon Society, University Park, PA) p. 30

Yazawa, K. (1969): J. Phys. Soc. Jpn. **26**, 1407

Yetter, W.E., Beetz, Jr., C.P., Budd, G.W. (1985): In Extended Abstracts of the 17th Biennial Conference on Carbon, Lexington, KY (American Carbon Society, University Park, PA) p. 291

Yoshida, A., Hishiyama, Y., Endo, M. (1985): In Extended Abstracts of the 17th Biennial Conference on Carbon, Lexington, KY (American Carbon Society, University Park, PA) p. 297

Yoshimori, A., Kitano, Y. (1956): J. Phys. Soc. Jpn. **11**, 352

Zeitsch, K.J. (1967): In Hove, J.E., Riley, W.C. (eds.): *Modern Ceramics* (Wiley, New York) p. 314

Ziman, J.M. (1972): *Principles of the Theory of Solids*, 2nd ed. (Cambridge University Press, Cambridge)

Zimmer, J.E., White, J.L. (1982): Adv. Liq. Cryst. **5**, 157

Zunger, A. (1978): Phys. Rev. **B 17**, 626

Zureck, W., Kocik, M., Calka, W., Jakubczyk, J. (1981): Fibre Sci. Technol. **15**, 223

# Subject Index

380